T0327508

Control of Mechatronic Systems

Vice President and Editorial Director: Amanda L. Miller
Acquisitions Editors: Eric Willner and Anne Hunt
Editorial Assistant: Anne Hunt
Editorial Supervision and Production: Rosie Hayden
Production Manager: Amudhapriya Sivamurthy
Cover Manager: Hannah Lee
Marketing Manager: Sarah Brett

Control of Mechatronic Systems

Model-Driven Design and Implementation Guidelines

Patrick O.J. Kaltjob
Ecole Nationale Superieure Polytechnique
Yaounde, Cameroun

Registered Offices
John Wiley & Sons, Inc., 111 River Street, Hoboken, NJ 07030, USA
John Wiley & Sons Ltd, The Atrium, Southern Gate, Chichester, West Sussex, PO19 8SQ, UK

Editorial Office
The Atrium, Southern Gate, Chichester, West Sussex, PO19 8SQ, UK

For details of our global editorial offices, customer services, and more information about Wiley products visit us at www.wiley.com.

Wiley also publishes its books in a variety of electronic formats and by print-on-demand. Some content that appears in standard print versions of this book may not be available in other formats.

Library of Congress Cataloging-in-Publication Data

Names: Kaltjob, Patrick O. J., author.
Title: Control of mechatronic systems : model-driven design and
    implementation guidelines / Patrick O. J. Kaltjob.
Description: Hoboken, NJ : John Wiley & Sons, 2020. | Includes
    bibliographical references and index.
Identifiers: LCCN 2018051541 (print) | LCCN 2019022413 (ebook) | ISBN
    9781119505808 (hardcover)
Subjects: LCSH: Mechatronics. | Manufacturing processes.
Classification: LCC TJ163.12 .K34 2019 (print) | LCC TJ163.12 (ebook) |
    DDC 621–dc23
LC record available at https://lccn.loc.gov/2018051541
LC ebook record available at https://lccn.loc.gov/2019022413

Cover design by Wiley
Cover image: © Menno van Dijk/iStock.com

Set in 10/12pt WarnockPro by SPi Global, Chennai, India

10 9 8 7 6 5 4 3 2 1

To the Holy Trinity and Saint Mary
Special thanks to Stella, Emmanuelle, Naomi, Lukà, and David
To Aaron†, Thomas, Olive, and Anne

Contents

Preface

The control of mechatronic systems and electrical-driven processes aims to provide tools to ensure their operating performance in terms of productivity, optimization, reliability, safety, continuous operations, and even stability. This is usually achieved through hybrid control paradigms using digital or analog tools. Nowadays, digital tools are widely used to implement control systems, as they offer numerous advantages including their ability: (i) to ease control-system implementation; (ii) to design complex and built-in intelligent information processing combining multiple functions for control, fault detection and diagnostic, monitoring, and decision planning; (iii) to integrate logic and continuous control algorithms, as well as supervision programs, into hybrid control strategies; (iv) to enhance the synchronization of input and output process operations; (v) to coordinate control actions among geographically distributed systems and processes; and (vi) to achieve reliable and optimal operating conditions.

The digital control system architecture usually consists of the integration of the following functional units: a data-processing and computing unit, an electrical-driven actuating unit, a measuring and detecting unit, a data acquisition (DAQ) and transmitting unit, and a signal conditioning unit. The data-processing and computing unit can be implemented through devices such as a microcontroller (μC), a programmable logic controller (PLC) with a control function, a digital signal processing (DSP) unit, and a field-programmable gate array (FPGA).

The design of efficient control systems requires the mathematical modeling of mechatronic systems and process dynamics. This can be achieved in accordance with the operating characteristics (discrete and continuous) and objectives, as well as the technological constraints, of the related instrumentation (signal conversion, transmission, conditioning, measurement, actuation, etc.). However, in most of the current engineering literature on the design of digital control systems, the mathematical foundation of discrete-time and discrete-event systems is usually presented separately from the technological constraints of control instrumentation. For example, the operating time-delay models and signal-to-noise ratios of digital device interfaces are not usually considered. Hence, the theoretical control algorithms proposed have limited practical applicability.

Challenges in the development of a practical design approach for the control of mechatronic systems and electrical-driven processes are: (i) to size and select control instrumentation in accordance with controlled system design objectives; (ii) to develop accordingly the mathematical discrete hybrid model capturing their continuous and discrete event behavioristic characteristics; and (iii) to integrate the control systems with respect to technological constraints and operational characterization (discrete and continuous) (e.g. time delays, signal-to-noise ratios, etc.).

This book intends to revisit the design concept for the control of mechatronic systems and electrical-driven processes, along with the selection of control instrumentation. By reviewing the theory on discrete-time and discrete-event systems, as well as various elements of control instrumentation, it offers an integrated approach for: (i) the modeling and analysis of mechatronic systems dynamics and electrical-driven process operations; (ii) the selection of actuating, sensing, and conversion devices; and (iii) the design of various controllers for single- to multiple-function electrical-driven products (mechatronic systems) and processes. Furthermore, it covers some design applications from several engineering disciplines (mechanical, manufacturing, chemical, electrical, computer, biomedical) through real-life digital control system design problems (e.g. driverless vehicles, newborn incubators, elevator motion) and industrial process control case studies (e.g. power grids, wind generators, crude oil distillation, brewery bottle filling, beer fermentation).

Through this book, the reader should gain methods for: (i) model formulation, analysis, and auditing of single- to multiple-function electrical-driven products and processes; (ii) model-driven design of the software and hardware required for digital control instrumentation; (iii) sizing and selection of electrical-driven actuating systems (including electric motors), along with their commonly used electro-transmission elements and binary actuators; (iv) selection and calibration of devices for process variable measurement and computer interfaces; and (v) modeling, operating, and integrating a wide variety of sensors and actuators. Hence, the textbook is organized into eight chapters:

1) *Introduction to the Control of Mechatronic Systems.* Chapter 1 gives a brief conceptual definition and classification of mechatronic systems, electrical-driven technical processes, and control systems structure. Here, a functional decomposition of the generic control system architecture is presented, along with some examples to illustrate control instrumentation for sensing, actuating, computing, signal converting, and conditioning. Furthermore, typical functions of generic controlled systems for electromechanical products and processes are described, along with the interconnection between the control instrumentation elements. Generic requirements for control system design are outlined based on challenges to software-based (design of hybrid architecture) and hardware-based (instrumentation sizing, compliance, and selection) control system integration. This is summarized within a list of the major steps in control design projects.

2) *Physics-Based Systems and Processes: Dynamics Modeling.* Chapter 2 presents numerous examples of dynamics modeling for various electrical-driven systems and processes, including transportation systems (e.g. a sea-port gantry crane, a hybrid vehicle, a Segway, an elevator, a driverless car), production systems and processes (e.g. an energy-based wind turbine, a drilling machine, a cement-based pozzolana scratcher), chemical processes (e.g. oil distillation, a cake conveyor oven, city water treatment, fermentation, poultry scalding and defeathering), fluidic and thermal systems and processes (a mixing tank, purified water distribution, a conveyor oven, poultry scalding and defeathering thermal processes), and biomedical systems (e.g. an infant incubator, human blood glucose insulin metabolism). Systems and process behaviors can be captured through differential equations using an experimental data modeling approach and classical physical laws of conservation and continuity. The resulting models are capable of displaying multiple and nonlinear variables as well as time-variant parameter characteristics, which can further be simplified according to the system physical properties or operating boundaries. A methodology for physics-based modeling is presented through the deterministic or stochastic behavior models of commonly encountered electrical-driven systems and large-scale processes. A review on linear modeling methods such as stochastics, dynamics responses, and state space is presented in the Appendices.

3) *Discrete-Time Modeling and Conversion Methods.* Chapter 3 focuses on methods for deriving a discrete approximation of continuous systems and signals using tools such as the hold equivalent, pole-zero mapping, numerical integration, and z-transformation theorems. A technological description of computer control architecture and interface is proposed with respect to DAQ unit operations, from the bus structure to data gathering, logging, and processing with respect to signal noise reduction and approximation consideration. Critical issues related to signal conversion, such as aliasing effects, along with a methodology for the selection of a sample period, are also covered. Overall, the chapter topics include technology and methods for continuous signal digital conversion and reconstruction, such as bilinear transformation, discrete-time command sequence generation, computer control interface for data logging, conditioning, and processing, sample time selection, and computer conversion technology using various conversion techniques (successive approximation, dual-slope ADC, delta-encoded ADC, etc.), as well as processing delay effects.

4) *Discrete-Time Analysis Methods.* Chapter 4 presents methods related to discrete system dynamical analysis in the frequency and time domains. Moreover, stability definitions and tests for discrete time systems are discussed and controlled system performance assessment tools are outlined. The chapter aims to present discrete controller design specifications. Topics include frequency analysis tools such as DTFT, FFT, and DFT, discrete zero- and pole-location plots, stability tests and criteria for discrete time systems (Jury–Marden, Routh–Hurwitz), steady-state errors, performance indices (ITAE, ISE), and time and frequency properties for controller design (settling time, percentage overshoot, gain and phase margins).

5) *Continuous Digital Controller Design.* Chapter 5 presents various approaches to the design of PID controller algorithms, such as continuous time design, discrete design, direct design using roots locus, and frequency-response techniques, as well as some advanced techniques such as model predictive control. Hence, using time- or frequency-domain controller specifications, numerous examples of the design and tuning of control algorithms are described, ranging from PID family, deadbeat, feedforward, and cascade to non-interacting control algorithms. In addition to stability analysis tests, performance indices and dynamics response analysis are derived in frequency and time domains. Furthermore, the open-loop controller design for stepper motors and scalar and vector control designs for induction motors are described. Model predictive control algorithms suitable for process operations with physical, safety, and performance constraints are also presented. Comparative analyses between classical PID controllers with various state feedback topologies for DC motor speed control are performed. Overall, chapter topics include cascade control, design and tuning methods for discrete-time classical PID family controllers, and scalar and vector control. The digital state feedback controller concept is revisited for cases where it is not possible to measure all state variables. Comparatively, analyses between classical PID controllers and various state feedback topologies for DC motor speed control are presented.

6) *Boolean-Based Modeling and Logic Controller Design.* Chapter 6 presents Boolean function-based models that have been derived by using sequential or combinatorial logic-based techniques to capture the relationship between the state outputs of discrete-event system operations and the state inputs of their transition conditions. Hence, after performing process description and functional analysis, a design methodology of a logic controller for process operations (discrete event systems) is proposed. Subsequent systems behavioristic formal modeling is achieved by using techniques such as truth tables and K-maps, sequence table analysis and switching theory, state diagrams (Mealy and Moore), and even state function charts. Some illustrative examples covering key logic

controller design steps are presented, from process schematics and involved I/O equipment listings, wiring diagrams with some design strategies such as fail-safe design, and interlocks, to state transition tables, I/O Boolean functions, and timing diagrams. Examples of logic controller designs include cases of elevator vertical transportation, an automatic fruit picker, a driverless car, and biomedical systems such as robot surgery and laser-based surgery. Overall, the chapter topics cover: (i) the methodology for Boolean algebra based on the modeling of discrete event systems; and (ii) logic controller design methodology for the derivation of I/O Boolean functions based on truth tables and Karnaugh maps, switching theory or state diagrams, wiring and electrical diagrams, and P&I and PF diagrams.

7) *Hybrid Process Controller Design*. Chapter 7 presents a generic design and implementation methodology for process monitoring and control strategies (logic and continuous), with algorithms to ensure the operational safety of hybrid systems (i.e. systems integrating discrete-event and discrete-time characteristics). First, the functional and operational process requirements are outlined, in order to define hybrid control and supervision systems with respect to logic and continuous control software, data integration and process data gathering, and multi-functional process data analysis and reporting. Subsequently, a methodology is proposed for the design of monitoring and control systems. Some cases are used to illustrate the design of process monitoring and hybrid control for elevator motion, drying cement pozzolana, and a brewery bottle-washing process. Overall, chapter topics include hybrid control system design, piping and instrumentation diagrams, system operations, FAST and SADT decomposition methods, process start and stop operating mode graphical analysis, and a sequential functional chart (SFC), as well as process interlock design.

8) *Mechatronics Instrumentation: Actuators and Sensors*. Chapter 8 provides an overview of electrical-driven actuators and sensors encountered in mechatronics, including their technical specifications and performance requirements. This is covered for electromechanical actuating systems such as electric motors as well as some electrofluidics and electrothermal actuating systems. Similarly, binary actuators such as electroactive polymers, piezo-actuators, shape alloys, solenoids, and even nano devices are technically described and modeled. Additionally, a spectrum of digital and analog sensing and detecting methods are described, along with the technical characterization and physical operating principles of the instrumentation commonly encountered in mechatronic systems. Presented sensors include motion sensors (position, distance, velocity, flow, and acceleration), force sensors, pressure or torque sensors (contact-free and contact), temperature sensors and detectors, proximity sensors, light sensors and smart sensors, capacitive proximity, pressure switches and vacuum switches, RFID-based tracking devices, and electromechanical contact switches. In addition, some smart sensing instrumentation based on electrostatic, piezo-resistive, piezo-electric, and electromagnetic sensing principles are presented. Overall, chapter topics include actuating systems such as motors (AC, DC, and stepper), belts, screw-wheels, pumps, heaters, and valves, along with detection and measurement devices of process variables (force, speed, position, temperature, pressure, gas and liquid chemical content), RFID detection, sensor characteristics (resolution, accuracy, range, etc.), and nano and smart sensors.

This textbook emphasizes the modeling and analysis within real-life environments as well as the integration of control design and instrumentation components of mechatronic systems through the selection and tuning of actuating, sensing, transmitting, and computing or controlling units. Further, it looks at the matching and interconnecting of control instrumentation such as sensors, transducers, and actuators particularly the interface between connected devices and signal conversion, modification, and conditioning. As such, the reader can expect by the end of the book to have fully mastered: (i) the design requirements and design methodology for control systems; (ii) the sizing and selection of the instrumentation involved in process control, as well as microelectromechanical devices and smart sensors; (iii) the use of microprocessors for process control, as well as signal conditioning; and (iv) the sizing and selection of actuating equipment for electrical-driven systems and industrial processes. Numerous examples and case studies are used to illustrate formal modeling, hybrid controller design, and the selection of instrumentation for electrical-driven machine actuation and DAQ related to systems dynamics and process operations. Through these case studies, the reader should gain a practical understanding of topics related to the control system and instrumentation, allowing him or her to fill a control and instrument engineering position where he or she is expected: (i) to possess a good knowledge of instrumentation operating conditions and control requirements; (ii) to size and select control instrumentation; (iii) to design, develop, and implement digital controllers; (iv) to design engineering processes and electrical-driven systems; (v) to collaborate with design engineers, process engineers, and technicians in the cost- and time-based acquisition of systems and processes control equipment; and (vi) to perform technical audit to ensure instrument compliance with health-and-safety regulations.

This book was conceived to develop the reader's skills in engineering-based problem solving, engineering system design, and the critical analysis and implementation of control systems and instrumentation. It allows self-study via comprehensive and straightforward step-by-step modular procedures. In addition, examples (with their accompanying MATLAB® routines, as well as) and design- and selection- related exercises and problems are provided, along with their solutions. Furthermore, a dedicated companion website (email author at kaltjob@uwalumni.com to have access to secured website) allows the reader to download additional material for teaching, such as slide presentations on the chapter material, data files for additional laboratory sessions, example files, and innovative 2D and 3D virtual labs for physical real-life systems (i.e. model-based simulation tools that can be associated to real-life systems for in-class lab sessions).

Suggestions for teaching plans for applied control theory of mechatronic systems and electrical-driven processes would be as follows: (i) Chapter 1 through Chapter 5 (up to Section 5.3.1), for an introductory digital control-level course lasting one semester; (ii) Chapters 2, 3, and 5 (Sections 5.3 and 5.4) for advanced control students with a control theory background; (iii) Chapters 1, 3 (Sections 3.3 and 3.4), and 8 for electric-driven machine and instrumentation students with computer hardware and software programming experience; (iv) Chapters 2, 3 (Sections 3.3 and 3.4), 5 (Sections 5.2.4, 5.3, and 5.4), and 6–8 for field control and instrumentation engineers interested in the design or migration of process control of hybrid systems.

Acknowledgment

This book makes extensive use of MATLAB® routines, distributed by Mathworks, Inc. A user with a current MATLAB license can download trial products from their website. Someone without a MATLAB license can fill out a request form on the site, and a sales rep will arrange the trial for them. For additional MATLAB product information, please contact:

The MathWorks, Inc.
3 Apple Hill Drive
Natick, MA, 01760-2098 USA
Tel: 508-647-7000
Fax: 508-647-7001
E-mail: info@mathworks.com
Web: www.mathworks.com

About the Companion Website

This book is accompanied by a companion website which aims to support the teaching efforts of instructors through:

(email author at kaltjob@uwalumni.com to have FREE access to the secured website)

The website includes:
 I) Lectures material for following courses package:
    1. Digital control systems
    2. Instrumentation: sizing and selection sensors and actuators
    3. Mechatronic systems design
    4. Process automation and monitoring
    5. Advanced control systems: predictive, distributed, adaptive control strategies
    6. Electric motor/machine control: stepper, DC, AC/induction
    7. Control and instrumentation
 II) For each course listed above reading guides, other classroom resources (visual summary, course outlines/summary, animation slides);
III) For each lecture session, multiple choice questions, for each course sample exams;
IV) for each Textbook chapter, solution manual, study questions, flash cards;
 V) Solved real-life problems and projects, 2D and 3D applications for sessions of laboratory simulation.

1

Introduction to the Control of Mechatronic Systems

1.1 Introduction

The rapid expansion of automated electrically-driven systems (e.g. electromechanical machines) is related to the development of digital control strategies in order to enhance their performance and extend their functionality while significantly reducing their operating cost and complexity. However, those digital control strategies are dependent on the performance of the control instrumentation related to measurement, signal conditioning, actuating, and digital control technologies. Recent technology advancements offer a plethora of control systems instrumentation, each with design-specific requirements and compliance constraints. Hence, in addition to system modeling, the design of digital control strategies has to consider: (i) the selection of control instrumentation in accordance with performance objectives; and (ii) the integration of the control systems instrumentation and process equipment with respect to operating constraints.

Consequently, it is suitable to lay out a generic design procedure for digital control systems, especially in: (i) controlling electrically-driven systems; (ii) sizing and selecting control instrumentation related to information processing and computing, electrically-driven actuation, process sensing and data acquisition; (iii) integrating those control instrumentation with respect to controlled system performance objectives and operating constraints; and (iv) integrating multifunctional control applications.

In this chapter, the definition and classification of electrically-driven systems and technical processes are presented first. Then the functional relationship between electromechanical machine control and control within interconnected and synchronized electromechanical systems is outlined. Various components of control systems instrumentation are described along with their design requirements. Furthermore, major steps of control system migration projects are presented with some illustrative examples of industrial process control. Finally, key project management steps and the associated subsequent design documents are listed.

1.2 Description of Mechatronic Systems

Mechatronic systems are either electrically-driven products or technical processes. *Electrically-driven products* are machines transforming current, voltage, or other electrical power into mechanical, fluidic, pneumatic, hydraulic, thermal, or chemical power. Hence, those systems can be classified according to their functional objectives either as: (i) specialized machines performing specific operations; or (ii) multipurpose and adjustable machines. Control systems are a set of technologies enabling algorithmic computing or signal processing devices to use signals emitted from analog or digital detecting, sensing, and communicating devices in order

Control Of Mechatronic Systems: Model-Driven Design And Implementation Guidelines,
First Edition. Patrick O.J. Kaltjob.
© 2021 Patrick O.J. Kaltjob. Published 2021 by John Wiley & Sons Ltd.

to perform automatic operations of systems or process actuation. Such systems are expected to perform them routinely and independently of human intervention with a performance superior to manual operation.

Thus, control systems aim to provide the necessary input signals to achieve the desired patterns of variations of specific process variables. Therefore, the functions of control systems are embedded in electromechanical systems (machine or product control).

Example 1.1   Figure 1.1 shows a typical 3D printing robot for customized cooking with speed- and temperature-controlled system which could be combined with monitoring indicators for cooking time and cooking stage, as well as a control panel allowing the selection of the final mixing of the product and cooking program. This system would require:

1) the angular position control of a pressure valve delivering semi-liquefied food (paste), the *x-y* axis position control of the carriage driving the extruder head (nozzle) made of two motors with a screw mechanism, the table angular speed and the *z*-axis position control;
2) the heater temperature control (nozzle level);
3) the remote pressure and force control for the valve in charge of injecting pressured food paste feed based on environmental (e.g. space mission) and biological conditions (e.g. lower gravity forces); and
4) the logic control for the discrete selection of ingredients.

Such control design combination enhances the product or machine functionality while reducing operating and maintenance costs. This is done by integrating data processing and computing

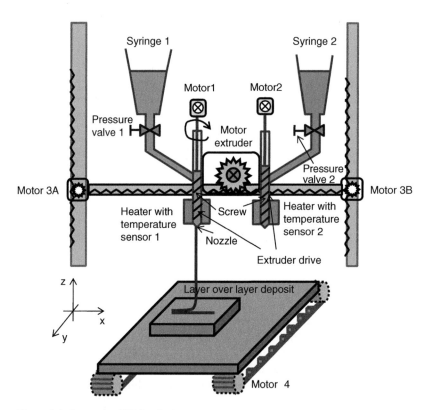

Figure 1.1 Customized 3D food printer.

operations within a field device or machine (e.g. washing machine, navigation systems etc.). Among the commonly encountered automated machines or products are those with: (i) embedded control functions; (ii) dedicated control functions; or (iii) a control function limited to a couple of sensors and actuators involved.

A technical process is the sum of all interacting machines within that process transforming and/or storing material, energy, or information. Such technical processes can be classified according to their operational objectives as follows:

1) Transportation-related processes, such as material handling processes, energy flow processes, and information transmission processes.
2) Transformation-related processes, such as chemical processes, manufacturing processes, power generation, and storage processes.

Technical processes can be characterized according to functional objectives, such as:

1) Processes characterized by a continuous flow of material or energy (e.g. cement drying process, electric power distribution, paper production). Here, the process variables are physically-related variables with a continuous range of values, such as temperatures in a heating system. The process parameters are physical properties (e.g. power transmission network impedance, liquefied gas density). Process control consists of maintaining the process state on a determined level or trajectory. In this case, process dynamics models can be obtained through differential equations.
2) Processes characterized by discrete event operations representing different process states such as device activation or deactivation during the startup or shutdown of a turbine. Here, process variables are binary signals indicating the discrete status of devices or machines involved in process operations as well as change in logic devices (e.g. activating events resulting from ON/OFF switch positioning). The process discrete event models can be obtained through Boolean functions or logic flow charts.
3) Processes characterized by identifiable objects that are transformed, transported or stored, such as silicon-based wafer production, data processing and storage operations. Here, process variables indicate the state changes of objects and can have a continuous range of values (i.e. temperature of a slab in a clogging mill, size of a part in a store) or binary variables. Those variables can also be non-physical categories (i.e. type, design, application, depot number) assigned to the objects.

Example 1.2 Figure 1.2, depicting a salt-generating solar-based thermal power plant substation, illustrates an example of non-continuous processes (discrete event- or object-related). In this power plant, solar radiation is collected by thousands of sun-tracking mirrors (heliostats), which reflect it toward a single receiver atop a centrally located tower. Solar radiation is the electromagnetic radiation emitted by the sun. As such, it is necessary to:

1) Control the collector angle and position (sun tracker) to face the sun to collect the maximum solar radiation as well as to maintain peak power despite varying climate conditions. This is done by adjusting the operating setting based on measured voltage and current outputs of the array.
2) Logically control the energy storage by switching between charging/discharging operating modes based on climatic conditions (sun availability), battery charge status, load levels, and level of energy collected through solar irradiation by mirror arrays.
3) Control the temperature of the collector used to melt a salt. The hot molten salt is stored in a storage tank to generate steam and later used to drive the turbine and attached generator.

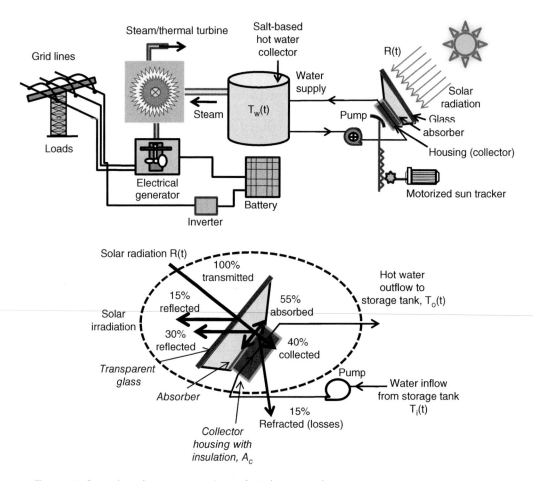

Figure 1.2 Steam-based power generation technical process schematic.

4) Control the flow of heated fluid circulating between the tank and the collector. This fluid with molten salt at a low temperature is pumped to the cold collector tower for the next thermal cycle. The operating temperature over this thermal cycle derives the quantity of energy to be extracted.

Digital control systems aim to coordinate the operations of several electrically-driven machines in order to meet specific operational objectives such as water purification, voltage control in an electrical power grid, or temperature regulation in a fermentation tank. Thus, it is usually necessary to ensure the integration of a large number of control system instruments (from data processing and computing units to measuring units). Figure 1.3 illustrates the generic components in the design of computer control systems with supervisory functions. An example of the relationship between a technical process and electromechanical system control is illustrated in Figure 1.4.

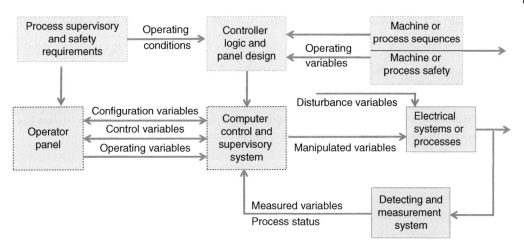

Figure 1.3 Generic controlled mechatronic systems and instrumentation block diagram. Source: Adapted from Kaltjob P.

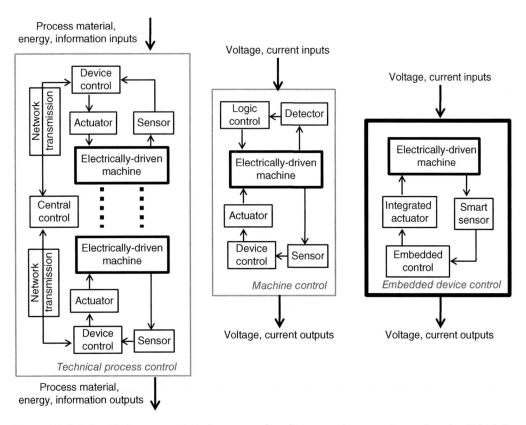

Figure 1.4 Relationship between technical process and machine control systems. Source: Based on Kaltjob P.

1.3   Generic Controlled Mechatronic System and Instrumentation Components

Through functional system decomposition, the digital control system architecture can be divided into the following functional units: data processing and computing, electrically-driven actuating, measuring and detecting, data acquisition, transmission, and signal conditioning. All these components are presented in the subsequent subsections. Figure 1.5 summarizes the connections between all the major control systems and instrumentation components.

1.3.1   The Data Processing and Computing Unit

The data processing and computing unit is used: (i) to control and regulate machine operations; (ii) to monitor machines and processes operations; and/or (iii) to coordinate operations within the same process. Data processing and computing could be performed either:

1) offline: that is, there is no direct or real-time connection between the process execution and the data processing and computing unit;
2) online for open-loop operations: that is, the protection (safety) of process operations and interlocking; or
3) online for closed-loop operations.

Commonly encountered data processing and computing devices are: digital signal processing devices, programmable logic controllers, microcontrollers, field programmable gate arrays among others, and a distributed control system (DCS) (consisting of a historian server connected to a network of field controller devices). Those devices execute program routines for:

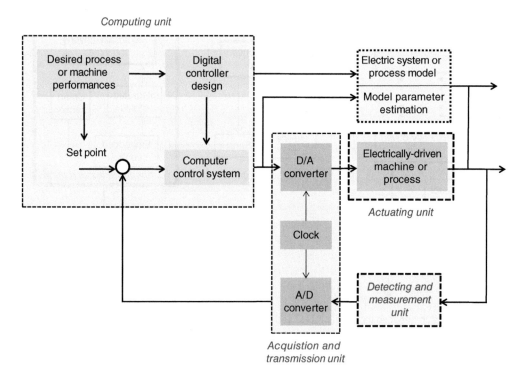

Figure 1.5 Generic control systems and instrumentation block diagram. Source: Based on Kaltjob P.

(i) the acquisition of process variables; (ii) process condition monitoring and exception handling (i.e. executing process safety operations); (iii) the control of machine operations (e.g. activation/deactivation of motor, tracking of motor speed); and (iv) the archiving and sharing of process data with other control devices through the communication network.

1.3.2 Data Acquisition and Transmission Units

Data acquisition and transmission units are used: (i) to interface with various control devices (e.g. operator panel, detecting and measuring field devices); (ii) to transport process data between network nodes; (iii) to integrate process data from different sources on a single platform; and (iv) to integrate control functions (e.g. machine control and process control). These units operate through data transfer platforms and their data distribution service protocols. They can be designed based on the Open Systems Interconnection model, which is summarized as:

1) the physical layer, being either wired or wireless connection, such as twisted-pair wiring, fiber-optic cable or radio link, and the commutation unit connecting the network to the devices (e.g. field buses for data transfer between primary controllers and field control devices);
2) the network, transmission, and transport layers performing functions such as data routing over the network, data flow control, packet segmentation and desegmentation, error control and clock synchronization. In addition, these layers provide mechanisms for packet tracking and the retransmission of failed packets; and
3) the session and presentation layers mainly used for data formatting.

1.3.3 Electrically-driven Actuating Units

Electrically-driven actuating units convert voltage or current signals from the computing unit into appropriate input forms (mechanical, electrical, thermal, fluidic etc.) for the execution of machine's and process operations. Then those converted signals produce variations in the machine's physical variables (e.g. torque, heat, or flow), or amplify the energy level of the signal, causing changes in the process operation dynamics. Some examples of actuating elements are relays, magnets, and servo motors.

1.3.4 Measuring and Detecting Units

Measuring and detecting units consist of low-power devices, such as sensors and switch-based detectors interfacing with electrically-driven machines involved in process operations. As such, they convert related physical output signals from the actuating unit into voltage or binary signals ready to be used within the data processing and computing unit. Some key functions of these devices are: (i) data acquisition related to the change of machine variables; and (ii) conversion of the machine-gathered signal into electrical or optical signals. Depending on the nature of the process signal generated, a signal conditioner can be added.

1.3.5 Signal Conditioning Units

Signal conditioning elements convert the nature of the signal generated by the sensing device into another suitable signal form (usually electrical). The signal conditioning units can also be embedded within the sensing devices. An example of such a unit is a Resistance Temperature

Detector (RTD). Here, a change in the temperature of its environment is converted into a voltage signal reflecting its resistance change through a Wheatstone bridge and the bridge is a signal conditioning module.

1.4 Functions and Examples of Controlled Mechatronic Systems and Processes

Mechatronic systems and processes have built-in intelligence through either their advanced information processing systems such as multifunctional control systems or intelligent electromechanical systems (including thermal, fluid, and mechanical processes) such as power-efficient multi-axis actuation with motion precision and detection features or miniaturized smart devices with embedded information processing capabilities. The resulting controlled mechatronic systems and processes aim to achieve various objectives: synchronize, control and sequence process operations, or detect and monitor process status.

Table 1.1 presents some typical process control objectives and their corresponding control functions along with some illustrative examples.

Example 1.3 Robot-assisted surgery is using image-guided systems to command and control operations in intravascular surgery, as depicted in Figure 1.6. Such a system has an embedded and integrated control system for its motion and direction, as well as operation monitoring, and motion synchronization between robot arms. Expected control functions include:

1) force control of a robot arm gripper;
2) synchronized angular position and velocity control of each motor-driven robot joint;
3) logic control of real-time anomalies detection (location of the abnormal cell or dysfunctional organ) and inspection using 3D imaging camera processing (color uniformity, selection based on size and shape) and laser ranging sensors;
4) path generation and motion planning (position, speed, and accelerations) for robot navigation while ensuring collision avoidance of the robot manipulator; and
5) logic control of the discrete selection of suitable cutting tools for the robot arms.

Table 1.1 Functions and implementation strategies for controlling mechatronic systems and processes.

Control system processing functions	Implementation control strategies	Examples of controlled mechatronic systems and processes
Assessing, reporting, and monitoring	Recording process variables through sensors and detectors; real-time, model-based measurement, setting parameters, and input signals.	Remote power flow measurement, configuration and voltage control (SCADA) through switchgears, transformers, and condensers in a smart power grid.
Safety compliance, detection, and diagnostics	Interlocking in case of detected failure modes, maintaining safety operations while ensuring malfunction handling.	Integrated safety and monitoring of petrochemical process variables and parameters (flowrate, temperature etc.)
Control and performance enhancement	Controlling or regulating system variables.	Position and temperature measurement as well as control of a 2D cutting machinery process.

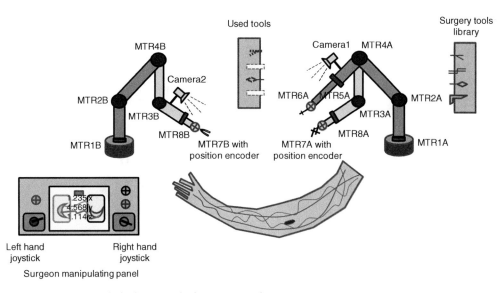

Figure 1.6 Image-guided tele-assisted robot intravascular surgery.

Example 1.4 An unmanned electric vehicle driving system is expected to have an embedded and integrated control system for speed and direction control, traffic light monitoring, and motion synchronization with other users. Consider the driverless vehicle in Figure 1.7(a): a block diagram with all relevant input and output (I/O) variables involved is depicted in Figure 1.7(b).

Example 1.5 Here, an example of control system for a crane-based vertical motion process is illustrated in Figure 1.8, while its feedback block diagram and the logic control connections are shown in Figure 1.9(a) and (b).

Example 1.6 Here, a milky beverage processing factory is illustrated in Figure 1.10. In a process of such scale, a supervisory, control, and data acquisition system (SCADA) is used to collect plant-wide data through an industrial network to archive and to ensure the execution of derived process sequences. The equivalent block diagram depicting the relevant components of such SCADA-oriented process control system is presented in Figure 1.11.

1.5 Controller Design Integration Steps and Implementation Strategies

Mechatronic systems and processes are systems embedding automatic information processing functions such as for reporting, better performance and control, and safe operation. Such functions are implemented using various control strategies through advanced control design algorithms, along with associated smart actuating or sensing devices. Because a combination of control strategies is commonly used and operated simultaneously, it is necessary to develop a design methodology to integrate control strategies with information-driven processing functions to ensure near-optimal performance under various functionalities and safe operating conditions. In addition, the control system must combine digital logic sequential control with continuous control to ensure a synergetic effect on the operation of the mechatronic system.

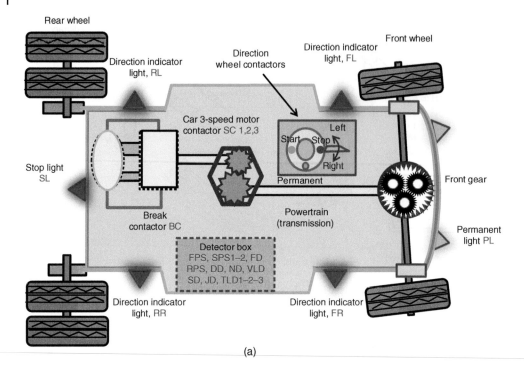

(a)

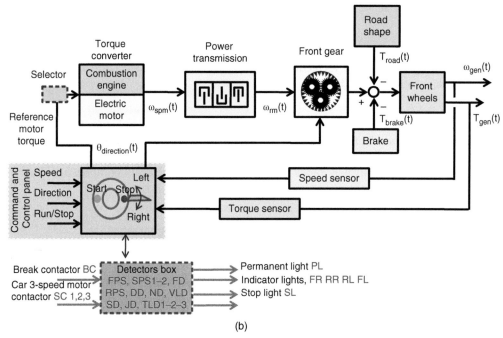

(b)

Figure 1.7 (a) Chassis of a driverless vehicle. Source: Based on Kaltjob P. (b) Hybrid control block diagram of a driverless vehicle.

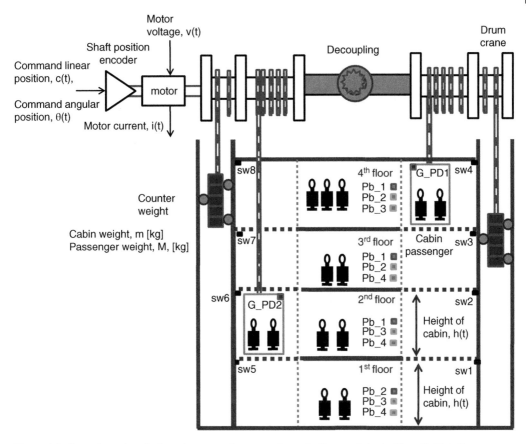

Figure 1.8 Crane-based vertical motion control system schematic. Source: Adapted from Kaltjob P.

This is usually encapsulated in an automation software solution associated with solid state computing devices (power electronics).

Furthermore, the control architecture of the mechatronic system could use either decentralized control systems, DCSs, hybrid control systems, monitoring and control systems, fault-tolerant control systems, and embedded control systems. Hence, some of them are expected to require remote and synchronization tools through data transmission and acquisition tools to ensure the coordination between operating entities as well between mechatronic system components. Recalling that any mechatronic system is by its design a combination of electrical, mechanical, and information processing technology, the control solution of a mechatronic system could combine: (i) built-in intelligence; (ii) real-time programming; and (iii) multifunctional operating characteristics. For example, this could result in higher service productivity, quality, and reliability (e.g. reduced failure rate), by embedding intelligent, self-correcting sensory feedback systems. Thus an integrated approach is suitable for the control of any multi-functional mechatronic system.

Regarding mechatronic systems and processes, three main types of integrating control design strategy can be considered: (i) spatial and structural integration; (ii) functional integration; and (iii) performance-based integration. Here, this textbook focuses on control design integration related to information processing (i.e. advanced control function along with monitoring and diagnostic features). As such, in order to achieve an efficient controlled mechatronic system, any control design project steps must encompass a coherent synergy of selection and design effort

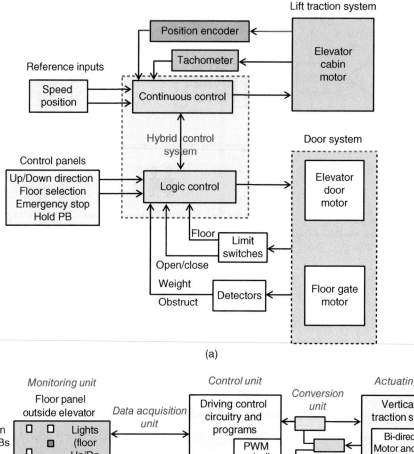

(a)

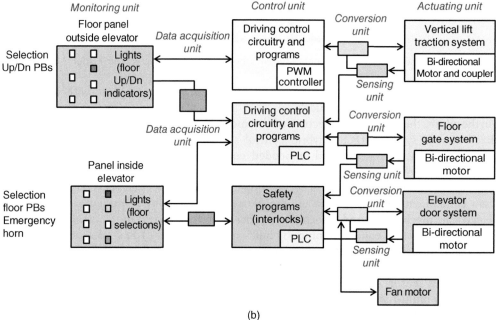

(b)

Figure 1.9 (a) Block diagram of the crane motion feedback control system. (b) Logic control connections of the crane motion feedback control system.

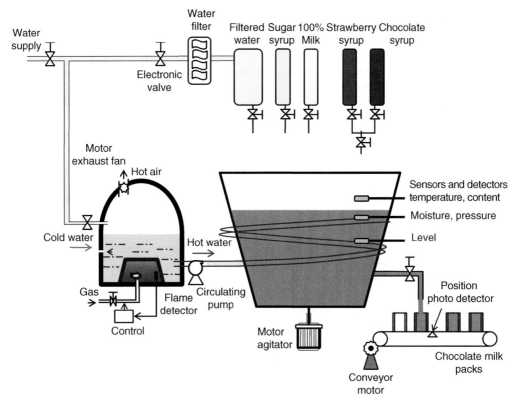

Figure 1.10 Milk-based beverage processing factory schematic. Source: Based on Kaltjob P.

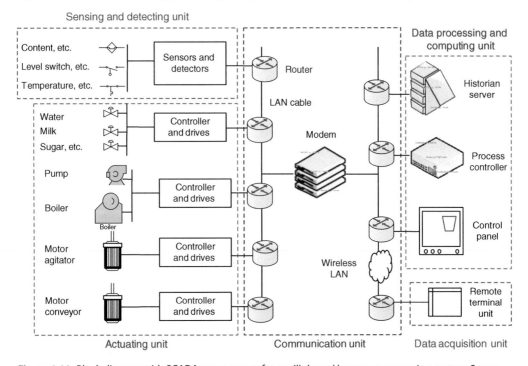

Figure 1.11 Block diagram with SCADA components for a milk-based beverage processing system. Source: Based on Kaltjob P.

between a mechatronic system and process components (actuating, sensing and data acquisition, control systems, and computing unit) under technically complex operating requirements. A project related to control system design or migration has either:

1) time-based objectives, such as improving service or productivity through an enhancement of mechatronic system functions;
2) cost-based objectives, such as reducing the cost of system operation (e.g. energy reduction) or service by improving the performance of the controlled mechatronic systems;
3) quality-based objectives, such as improving system operations efficiency or service reliability through the enhancement of control functions for mechatronic systems;
4) a combination of control system design objectives, such as delivering new services or products, replacing or extending them; or
5) monitoring and analysis objectives to be integrated into the mechanism system design.

To fulfill these objectives, the major project steps in control system design are:

1) Preliminary studies based on mechatronic system and process schematics.
2) Performance technical audit:
   a) performance assessment;
   b) performance objectives and controller specifications.
3) Functional analysis and modeling:
   a) FAST decomposition method;
   b) operating conditions and discrete time modeling;
   c) startup and shutdown sequence relationship diagram;
   d) operating description and discrete event formal modeling.
4) Hybrid controller and digital system design:
   a) continuous controller algorithm;
   b) logic controller software design;
   c) storage requirement assessment (memory size etc.), control system throughput, user interface requirements;
   d) controller hardware design (control panel design and layout, instrumentation system design and layout);
   e) logic controller circuitry design (distribution and control architecture).
5) Controller solution execution and commissioning:
   a) control feedback variable tuning;
   b) operation monitoring under failure modes;
   c) operation validation based on control system test protocol;
   d) personnel training;
   e) deployment.

Those steps could be summarized as shown in Figure 1.12. Activities related to control design or migration projects are expected to provide specific mechatronic system control documentation such as:

1) Mechatronic system and process diagrams, especially:
   a) process flow (PF) and piping and instrumentation (P&I) drawings based on the mechatronic system description, if possible;
   b) cabling and wiring diagrams (power distribution schemes, data cable routing and cable diagrams, instrument hook-up drawings);
   c) electrical schematics and connection diagrams connecting operator user interface actuators and sensors through a detailed panel layout;

d) power supply, interconnection, and distribution diagrams, and electrical device placements, such as the number of repeaters between nodes;

e) field control instrument lists with data sheet formulation for various instruments such as: flow, level, pressure, differential pressure, temperature, transmitters, converters, isolators; multipliers; control valves and so on; and

f) diagrams for safety compliance (interlock schematics).

2) Documents for mechatronic device sizing and selection, especially on:

a) actuating elements such as motor drives, control valves, orifices with estimates (calculations) of their operating conditions;

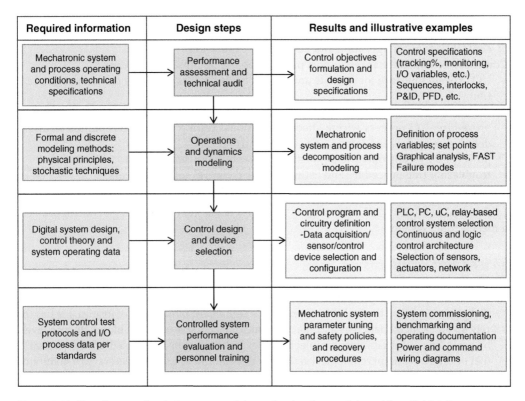

Figure 1.12 Generic controller design steps and dependencies. Source: Adapted from Kaltjob P.

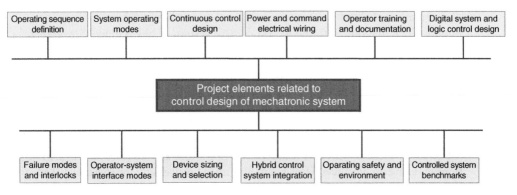

Figure 1.13 Overview of activities related to control project management. Source: Adapted from Kaltjob P.

b) sensing elements such as piezoelectric flow sensors;

c) data processing and computing elements (memory size, scan cycle etc.); and

d) power storage and utilities equipment, such as uninterruptible power source (UPS) batteries (capacity etc.).

3) Documents on sizing and selection of data network components (type, media etc.) and replacements especially related to:

a) selection of I/O modules (integrated/distributed);

b) I/O addressing;

c) data logging and process data archiving; and

d) database structure formulation.

4) Documents on hybrid controller structure, configuration, and operation:

a) continuous control and logic control design algorithm and digital system circuitry; and

b) control system configuration and instrument calibration.

5) Control system maintenance/support procedures and documents including:

a) I/O process interface for discrete and analog fields devices as well as memory organization;

b) detection and diagnosis of external field control device failure, I/O module replacement procedure; I/O process wiring check, operational testing (factory acceptance testing), definition of emergency stops, and safety control relays;

c) power supply, controller memory card replacement procedures; and

d) troubleshooting procedure and listing of hybrid control software errors.

Figure 1.12 summarizes the major steps and documents required in process control as well as those expected to be provided, while Figure 1.13 presents activities in a design control project for a mechatronic system.

Exercises and Problems

1.1 Insulin is a pancreas-generated hormone that regulates blood glucose concentration. A diabetic patient has the inability to produce a significant blood level of insulin. He has to rely on an artificial pancreas as a mechatronic device to regulate the blood glucose level by providing automatic insulin injections several times per day. Based on the patient's diet or physical and psychological activities, the amount of insulin necessary to maintain a normal blood glucose level must be determined.

a) Identify the control functions for this mechatronic system for the regulation of blood glucose level and list all input and output system variables.

b) Briefly describe the possible hybrid control strategy for this mechatronic system for the regulation of blood glucose level and list the control variables.

c) Build a block diagram with the hybrid control system of blood glucose level using a real-time blood glucose measurement device and a continuous insulin infusion pump.

d) Sketch a logic control connections of this mechatronic system instrumentation (hybrid control with control panel).

1.2 For monitoring and control objectives of the following systems (a–c):

a) electric-driven driverless vehicle with a GPS-based traffic model for trajectory guidance and a sonar-based radar tool for collision avoidance;

b) robot-aided surgery with direct current (DC) motors as actuators for 2D arm positioning, using an advanced laser beam and image processing for thermal-based cutting precision; and

c) intelligent cow traffic monitoring using a solar powered cow collar device with embedded accelerometer;

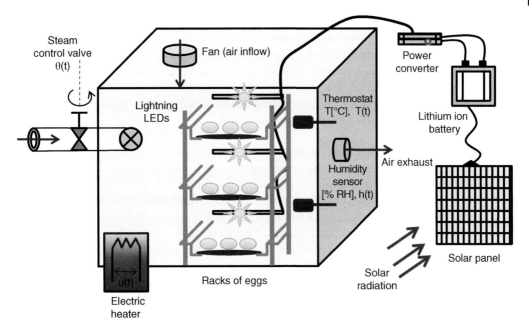

Figure 1.14 Egg incubator schematic.

perform the following technical activities:
a) Identify control project objectives and control processing functions.
b) Identify key I/O system and control variables.
c) Identify output variables and classify them as (a) measured or (b) unmeasured;
d) List some safety, environmental, and economic considerations.
e) Discuss the most suitable hybrid control strategy (logic, continuous).
f) Sketch the hybrid control block diagram.

1.3 As depicted in Figure 1.14, an incubator is used to keep eggs at a suitable warm temperature and humidity (i.e. the balance of heat over the oxygen level) to moderate conditions in order to induce hatching from embryos to mature chicks. This is done through an automatic egg-turning device by rotating the rack of eggs in order to change their position. During this incubation cycle, the embryos breathe and exchange hot air through thousands of pores in the eggshells.
a) Identify the control project objectives and possible control processing functions.
b) Identify key I/O in the incubator system and corresponding control variables.
c) Identify the output variables and classify them as (i) measured or (ii) unmeasured. Briefly describe how the unmeasured variables could be estimated.
d) List some safety considerations and interlock processes to be introduced as a result.
e) If a pump humidifier and a heater are used to supply water steam while a lamp produces radiation to heat the chamber temperature, draw the hybrid control block diagram.
f) Sketch the possible operator panel with corresponding indicator lights and logical I/O variables.

1.4 Heart surgery requires the regulation of patient's blood pressure at a desired value. This is achieved by adjusting the infusion rate of vasoactive drugs into the patient's blood vessels. In addition to the effect of vasoactive drugs, blood pressure is also affected by the level of anesthetic given to the patient.

   a) Using control strategies, sketch a detailed block diagram for a hybrid control system that enables the measurement of blood pressure and regulates the infusion rate of a vasoactive drug accordingly.

   b) From the resulting block diagram, identify devices that could be used in the measuring and detecting unit and in the actuating unit.

1.5 For the following mechatronic devices:

   a) millimeter-wave radar for the detection of a surrounding object distances in an unmanned car collision avoidance management system;

   b) blood plasma detoxification is achieved through an implanted artificial liver bioreactor. The underlying process is a filtration though a semi-permeable membrane sheet while maintaining a pressure-regulated blood circulation. A user interface is used to remotely monitor the reactor status. This is done despite human body temperature variation;

   c) an automatic electric cooking machine with speed and temperature controls combined with monitoring indicators for cooking time and cooking stage, as well as a control panel allowing the selection of spin speed and cooking program; and

   d) an automatic vacuum cleaner with a user interface for motion control. It has an antenna to communicate with the base station (remote control), four dirt sensors, each measuring the number of dirt particles, two DC motor-driven wheels, and a steering motor:

      i) List their corresponding I/O (logic and continuous) variables.

      ii) Sketch the hybrid control block diagram.

      iii) Draw the I/O variables of the control panel (with its user interface) including as much details as possible, such as the ports of I/O devices connected to external circuitry such as sensors.

1.6 A robot-based laser beam cutting system consists of two actuating devices: a joint angle motor and a fine piezoelectric transducer for head positioning. A photo-sensor measures the head position and the position error is fed to a separate controller for each actuator. Draw the block diagram for this automated cutting machine. If an optical technology is used for image processing-based feedback to achieve highest precision cutting, update the hybrid control block diagram along with an operator panel (list all I/O variables involved).

1.7 In order to design the control system for an electrically-driven dialysis blood processing system, as illustrated in Figure 1.15, a temperature and pressure are monitored from the control panel such that abnormal dialysate pressure outside the $-410-+330\,\text{mmHg}$ range (avoid exceeding the pressure in the blood compartment) with an accuracy of $\pm7\%$ triggers an alarm and flashing lights. This is also the case for membrane rupture, which could cause blood contamination. The transmembrane pressure regulates ultrafiltration (i.e. removing toxic fluid).

   a) Identify all components of the mechatronics process.

   b) List key I/O (logic and continuous) dialysis blood system variables involved.

   c) Sketch the corresponding hybrid control block diagram taking into account volumetric and flow sensor controls.

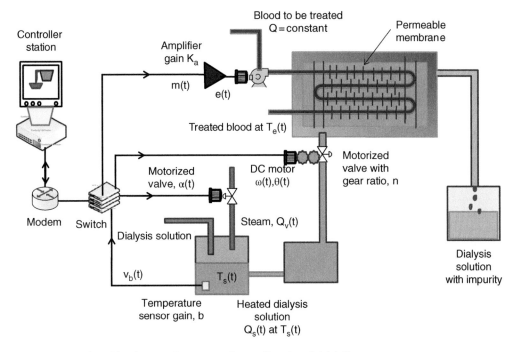

Figure 1.15 Dialysis blood processing system. Source: Based on Kaltjob P.

   d) Draw a control panel, considering variables associated with the interlocks fulfilling the safety requirements listed here.

1.8 A Global Positioning System (GPS) receiver-based autoguided missile has a precision-strike mechatronic system which is used to track a predetermined flight trajectory by regulating the angle between its axis of motion and its velocity vector. The adjustment of the missile angle ϕ is achieved through the thrust angle ß, which is the angle between the thrust direction and the axis of the missile, as illustrated in Figure 1.16.

   a) Identify the flight control processing functions for the guided missile as well as corresponding I/O variables (including GPS position, velocity, attitude, and orientation) and a gyroscope-based stability function using line-of-sight measurement.

   b) Draw a block diagram for the intelligent control of the auto positioning system for the missile angle using a gyroscope to measure its angle and a motor to adjust the thrust angle.

   c) In the case of a heat-seeking missile, a guidance control strategy is used. How would the control block diagram be modified for this?

1.9 A crude oil distillation process using a boiler-based temperature control is illustrated in Figure 1.17. The distillation process is used to separate gasoline from asphalt, controlling the reflux rate by regulating the flow of distillate composition at its top and by varying the rate of steam to the reboiler at the bottom. Among the key process variables are temperature (continuous) and flame (binary) in the combustion chamber (boiler).

   a) Identify the control and monitoring objectives of this distillation process.

   b) Identify all devices involved in the distillation column temperature and flow process control and their corresponding I/O variables.

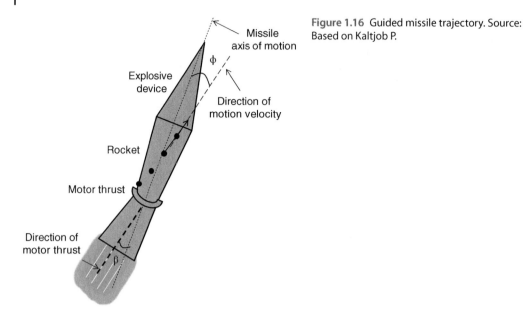

Figure 1.16 Guided missile trajectory. Source: Based on Kaltjob P.

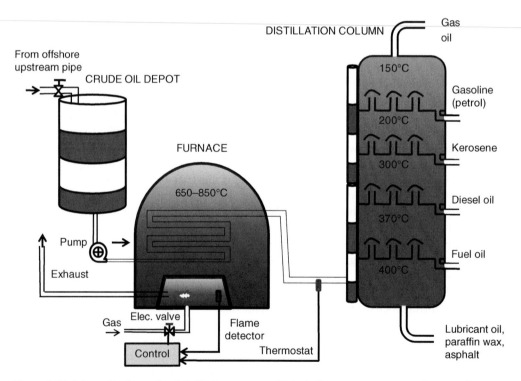

Figure 1.17 Schematic of a crude oil distillation process with its boiler temperature control system. Source: Based on Kaltjob P.

c) Sketch the block diagram for continuous pressure control, feed control, boiler temperature control, and reflux control of this distillation process.

d) List key possible I/O (logic and continuous) variables involved. Draw the corresponding hybrid control block diagrams.

Due to the multifunctional controllers required, it is desirable to have a DCS architecture with SCADA network.

e) Sketch the DCS schematic linking all required type of field control devices, data acquisition unit, and a local control unit up to the central command and control unit.

f) Sketch the schematic of the control panel and visualization system for this distillation system plant containing the execution unit of the management information system functions along all logic control connections.

1.10 Consider a control project upgrading a distributed control (no single point of failure) for an express parcel handling and dispatching center. Among the control design requirements are: (i) a parcel dimension check point; and (ii) parcel priority levels. All parcels pass through an integrated security system consisting of a series of security scanners while being directed automatically by barcode scanning points along long-distance conveyor system and a robot handling system at each end to their final transit destination via a temporary storage room.

a) Identify the control objectives for this parcel handling and dispatching center.

b) List key I/O variables (logic and continuous) for the handling and dispatching center.

c) Sketch the block diagram of this system as well as an operator panel including all I/O system and control variables.

1.11 Similar to blood glucose level or blood pressure regulation, the regulation of human body blood temperature is performed through the evaporation of water from the skin over the capillary network.

a) Identify the continuous and logic control objectives for blood temperature regulation in the human body.

b) Sketch a hybrid control block diagram for the human body blood temperature continuous control problem.

c) Integrate this system block diagram with one for human blood pressure regulation and one for human blood glucose level control.

d) From the preceding block diagram, identify devices that could be used in the measuring and detecting unit and in the actuating unit of this multi-level automated human blood regulation process.

1.12 From the cake conveyor oven depicted in Figure 1.18, it is necessary to control temperature and air moisture within the oven chamber as well as the conveyor speed in order to ensure a suitable cake baking process. An image processing camera is used for thermal sensing configuration of the chamber and an operator panel is used to define configuration settings and display operating conditions.

a) List key I/O (binary and continuous) control and system variables involved.

b) Draw the hybrid control block diagram and operator panel with I/O controls and process variables.

1.13 Among vehicle-based mechatronic systems, there are:

a) the antilock brake system (ABS) to allow vehicle wheels to smoothly stop their rotation when a mechanical brake is activated;

b) the traction control system (TCS);

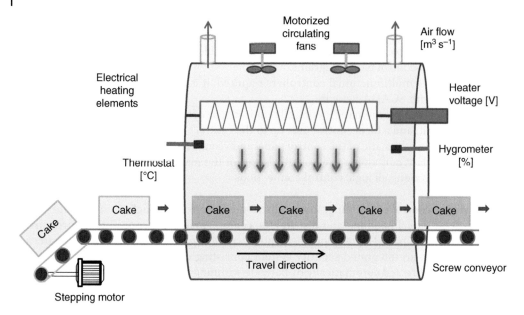

Figure 1.18 Cake conveyor oven. Source: Based on Kaltjob P.

    c) the vehicle dynamics control (VDC);
    d) the electric spark ignition for fuel-air combustion;
    e) the engine management and transmission control;
    f) the airbag activation;
    g) the air conditioning system;
    h) seat belt control;
    i) mirror control;
    j) climate and vehicle front light activation control;
    k) the parking distance control system;
    l) parallel self-parking;
    m) the averaging fuel tank level;
    n) the alcohol test and engine activation;
    o) the window lift system.
      i) Based on either operating objectives such as comfort, safety, and emission reduction, or autonomous and intelligent cruise control objectives, classify the mechatronic systems listed here.
      ii) For five mechatronic systems out of this list, define the possible I/O variables involved (hint: each variable is associated with a field control device).

1.14 Consider the horizontal motion of a helicopter hovering as depicted in Figure 1.19. The motion equipment includes an engine-driven rotor with a speed control while the pitch angle is given by the drivetrain. The wind speed can be considered a disturbance and its power is related to altitude height, also called the shear coefficient. Then, the helicopter's motion can be defined by the rotor and drivetrain. Sketch a block diagram for the speed and direction control from variations associated with the pitch rate, the pitch angle of the fuselage, and the rotor tilt angle.

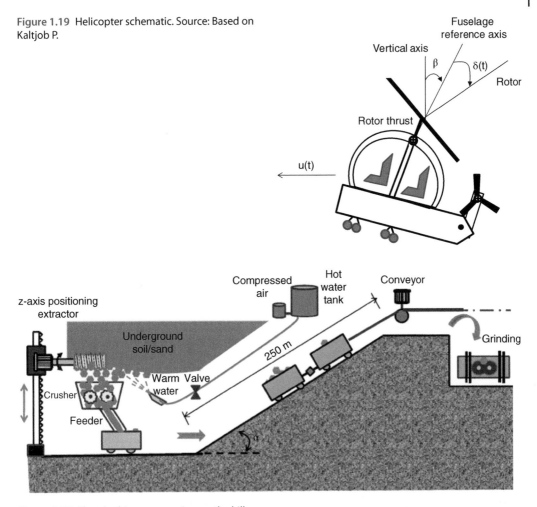

Figure 1.19 Helicopter schematic. Source: Based on Kaltjob P.

Figure 1.20 Electric-driven car moving up the hill.

1.15 Consider a gold-mining extraction operation, as depicted in Figure 1.20. It is necessary to monitor and remotely control the primary and secondary crushing motors, raw material conveyor motors, ball milling (grinding), operations related to pressurized watering, underground soil feeding in transportation trailers during extraction operations, and operations related to leading and tailings stacking.

   a) List key I/O variables involved in the monitoring and control of a gold-mining extraction process.

   b) Sketch the corresponding control system block diagram including the monitoring and control system instruments as well as I/O variables involved.

1.16 Consider robot harvesting machines that perform autonomous fruit detection in trees, picking fruit from the foliage without damage. Then the fruit is transferred directly into mobile collecting boxes as depicted in Figure 1.21.

Identify the major I/O (continuous and binary) variable systems and sketch the hybrid control block diagram with I/O variables of this harvesting robot integrating the following continuous and logic control functions:

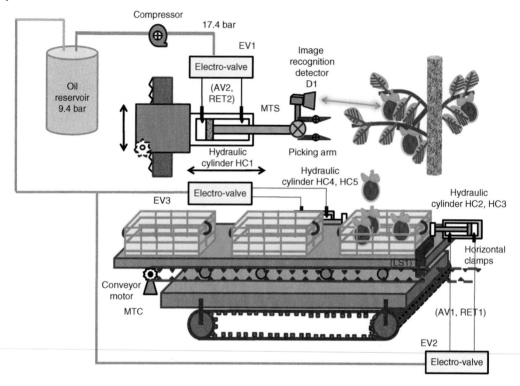

Figure 1.21 Automatic fruit harvesting robot. Source: Based on Kaltjob P.

a) Fruit detection (fruit tree location) using 3D camera image processing (color uniformity, maturity recognition using near-infrared for sugar content, selection based on size and shape) and a laser ranging sensor.

b) The path generation and motion planning (position, speed, and acceleration) for the harvesting robot's exploration using optical, magnetic, laser-guided, and GPS tools while ensuring that the robot manipulator avoids collisions.

c) The logic control for activating a solenoid in charge of a vacuum suction nozzle, especially for mature but resistant-to-vibration fruit.

d) The continuous control of the angular position of the robot picker arm fingers with interior foam sponge pads combined with a cutter for picking. The same arm logically controls the shaking magnitude and frequency to ensure reliable fruit picking.

e) The steering control of a mobile robot, with especially synchronized control of motors position, speed and acceleration in charge of the robot carrier movement (horizontal motion of the trolley), platform movement (vertical motion of the retractable self-propelled elevator system) and direction of robotic picking arm operations such as vertical inching, twisting rotation, retraction, and depositing.

f) The control of camera rotation and translation with respect to the targeted object whose features are not time-varying, while avoiding any blurring resulting from camera motion and the influence of large illumination variation in image processing using vision sensors for edge detection and square angular estimation.

1.17 In a typical nuclear power plant, as illustrated in Figure 1.22, heat is released after scission of uranium atoms within the reactor component to heat water, the steam from which generates electrical power within the generator component. Within the reactor,

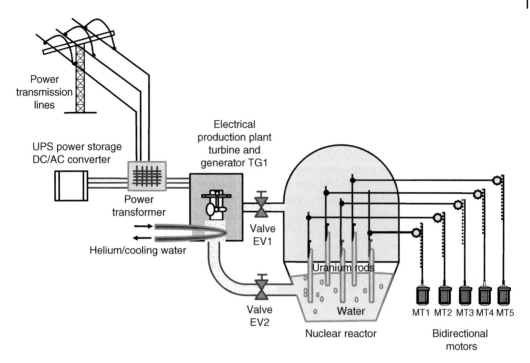

Figure 1.22 Schematic of nuclear plant.

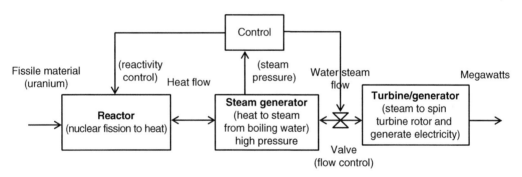

Figure 1.23 Incomplete block diagram of a nuclear plant continuous control system.

cylindrical tubes of uranium are grouped within assemblies of rods and inserted simultaneously by remotely controlled arms into the reactor core. Water flowing through the reactor core absorbs the heat energy from uranium atoms splitting through a successive radioactive fission sequence. The generated heat energy is transferred from the reactor core to the turbine component in the form of steam. This resulting high-pressure steam drives the turbine/generator unit. This highly pressurized steam causes rotation of the turbine blades and consequently rotation of the attached electric generators. For such a plant, complete the continuous control block diagram as depicted in Figure 1.23 and draw the overall hybrid control block diagram and operator panel including the following control objectives:

a) The speed and bidirectional control of the uranium rods within the reactor. The speed of the process of absorbing neutrons and the chain reaction is related to their vertical motion.

b) Control of the temperature of the reactor core due to the fact that the number of neutrons absorbed is raised, resulting in fewer neutrons being available to cause fission.

c) The control of the flowrate of circulating water within the reactor.

d) The control of the pressurized water level in the reactor where water flows in to be heated to produce steam that then flows to the turbine to generate electricity.

e) The logic control of energy operation in the reactor core by removal of the uranium rods from the water at maximum speed with respect to safety and emergency requirements. The energy generated in the water-based reactor is converted into high-pressure steam, which is used directly to turn the turbine (steam generator).

f) The logic control of the selection of a subgroup of individual rods to move upward and downward by individual attached bidirectional motors above the reactor core.

g) The logic control of energy storage by switching the charging/discharging operation based on battery charge status, load level, and the level of energy collected.

h) The control of the rate of flow of heat energy circulating between the tank and the collector.

i) The monitoring of the radiation activity within the reactor and surrounding air, ground, and liquid environment.

Bibliography

1 Åström, K.J. and Wittenmark, B. (2011). *Computer-Controlled Systems: Theory and Design*. Courier Dover Publications.

2 Bailey, D. and Wright, E. (2003). *Practical Scada for Industry*. Newnes (copyrighted Elsevier).

3 Erickson, K. and Hedrick, J. (1999). *Plant Wide Process Control*. Wiley.

4 Franklin, G.F., Workman, M.L., and Powell, D. (1997). *Digital Control of Dynamic Systems*, 3e. Boston, MA: Addison-Wesley Longman Publishing Co.

5 Goodwin, G.C., Graebe, S.F., and Salgado, M.E. (2001). *Feedback Control of Dynamic Systems*. Prentice Hall.

6 Golnaraghi, F., Kuo, B.C., and Adams, J.A. (2009). *Automatic Control*, 9e. Wiley.

7 Groover, M.P. (2007). *Automation, Production Systems and Computer-Integrated Manufacturing*, 3e. Prentice-Hall.

8 Kaltjob, P. (2018). *Mechatronic Systems and Process Automation: A Model-Driven Approach and Practical Design Guidelines*. CRC Press.

9 Kuo, B. (1995). *Digital Control Systems*. Oxford University Press.

10 Luyben, W.L. and Luyben, M.L. (1997). *Essentials of Process Control*. New York: McGraw-Hill.

11 Marlin, T.E. (2000). *Process Control: Design Processes and Control Systems for Dynamic Performance*, 2e. McGraw-Hill.

12 Ogata, K. (2004). *Modern Control Engineering*, 4e. Prentice Hall.

13 Powell, F. and Emami-Naeini, A. (2002). *Control System Design*, 4e. Prentice Hall.

14 Seborg, D., Edgar, T.F., Mellichamp, D., and Doyle, J. (2011). *Process Dynamics and Control*, 3e. Wiley.

15 Siouris, G.M. (2003). *Missile Guidance and Control Systems*. Springer.

16 Smith, C.A. and Corripio, A.B. (1997). *Principles and Practice of Automatic Process Control*, 2e. New York: Wiley.

2

Physics-Based Systems and Processes: Dynamics Modeling

2.1 Introduction

The dynamics modeling of electrically-driven systems and processes operations is the mathematical formulation capturing any spatial-temporal cause-based changes of their physical properties. Such behavior modeling aims either to analyze the physical phenomena over different operating conditions or to improve their performance using suitable and efficient automation systems. There are three approaches of behavior dynamics modeling: (i) the approach based on detailed system or process knowledge along with governing physics laws of continuity or conservation (white boxes); (ii) the approach driven by system or process data (black boxes); or (iii) the approach based on a combination of a model structure derived from the physics-based analysis with model parameters estimated over collected system or process data (gray boxes). Those resulting dynamic models can be expressed in differential or difference equation forms. Hence, depending on the system or process complexity, on the knowledge of transient and stationary characteristics, and on the empirical data available, any of those approaches can be chosen. However, those stochastic (black) or mechanistic (white, gray) approaches have led to models not being generic or having limited accuracy due to difficulties in measuring or estimating their model parameter values.

This chapter presents a generic dynamic modeling procedure illustrated through a variety of electrically-driven systems and processes related to chemical (e.g. crude oil distillation), transportation (e.g. gantry crane, pozzolana scratcher), thermal (e.g. beer fermentation, poultry scalding and defeathering), fluidic (e.g. city waste water treatment), biomedical (e.g. new infant incubator, blood glucose (BG) metabolism), and production (e.g. lathe machine, wind turbine energy generator, cake conveyor oven) applications.

2.2 Generic Dynamic Modeling Methodology

Commonly-encountered electrically-driven systems and processes transform electrical energy into mechanical, fluidic, thermal, or even electrical-related energy or work. Hence, based on the variation of their physical and dynamical properties (geometrical boundaries, operating conditions etc.…), a mathematical formulation can be derived using the principles of conservation and continuity (mass, energy, or momentum balances) or physical laws. The major steps of process dynamics modeling are:

1) *analysis of operating boundaries and conditions* in terms of defining operating objectives and framing its decomposition into subprocess operations, when possible. This consists of specifying temporal and spatial boundaries characterizing physical/chemical phenomena

Control Of Mechatronic Systems: Model-Driven Design And Implementation Guidelines,
First Edition. Patrick O.J. Kaltjob.
© 2021 Patrick O.J. Kaltjob. Published 2021 by John Wiley & Sons Ltd.

and listing all key input variables or actions that are causing change in the system, or process physical and dynamic properties corresponding under various operating conditions;

2) *development of equations for the dynamics model* using either laws of conservation and continuity from the asserted operating assumptions or experimental data and physics-based analyses (model structure and order) in order to subsequently estimate the model parameters;

3) *adjustment, refinement, and validation of the dynamics model* through sensitivity analysis and validation of assumptions using some experimental data.

Table 2.1 summarizes a step-by-step modeling procedure for electrically-driven system and process operations.

2.3 Transportation Systems and Processes

2.3.1 Sea Gantry Crane Handling Process

Gantry cranes are typical electromechanical systems used in a harbor environment for automatic container loading and unloading operations between piers and ships. During these operations, attached loads (containers) sway, which causes position and control problems. Some anti-sway devices have been developed using friction additive components to reduce dispersion but they were found to be slow to respond. In contrast, an anti-sway control strategy has been found to offer faster response during load transfer movement. However, these solutions require the dynamic modeling of the container position and real-time assessment of swaying angle using an image processing system embedded within a camera.

For its x,y,z motion, the gantry crane can be decomposed into three subsystems: (i) the gantry; (ii) the trolley attached by a cable to the gantry; and (iii) the spreader with container (load) suspended from the trolley, as illustrated in Figure 2.1(a). The container sway can be captured at both angles, $\theta(t)$ and $\beta(t)$, for the x-y and x-z planes respectively. All of these devices are driven by permanent magnet direct current (DC) motors. The handling process has two operating phases: uploading and off-loading (pick and drop), during which the gantry is either carrying the container or not. Table 2.2 summarizes the container handling process parameters and variables.

Here, the container handling process consists of a gantry moving along the rails over the pier while the trolley is moving along another set of rails attached to the gantry perpendicular to the pier. Attached to the trolley via cables, the spreader is used to hold the containers and carries out the hoisting movement by winding and unwinding the cables. The gantry crane system motion characteristics (velocity and acceleration) are summarized in Table 2.3. Depending on the obstacles, there are three possible paths (1, 2, 3) in the y-z container-spreader-trolley motion, as shown in Figure 2.1(b). The gantry crane motion is synchronized in the y-z-direction such that the container moves diagonally, meaning it is simultaneously in vertical and horizontal directions, as shown in Figure 2.1(b).

Initially, the container is hoisted, with the maximum torque, vertically up to a point where it has the maximum upward velocity. During the horizontal motion, the trolley runs at the maximum velocity with the container at a nearly constant height from the Earth's surface, depending on the swing angle $\theta(t)$ (nearly equal to zero). For simplicity only, the y-z 2D motion of the container is considered to be modeled and analyzed. The gantry motion is discarded. The choice of the shortest path is not considered. Hence the minimum travel time due to path selection and motion profile is not considered. Furthermore, it is assumed that only the translational speed of the trolley or the gantry movement would cause load swaying. All friction forces due to the

Table 2.1 Generic dynamic modeling procedure of system or process operations.

Modeling steps	Illustrative examples

Analysis of operating boundaries and conditions

1) Identify operating objectives (input forces or signals changing entity properties), define the model purpose (improve performance or understand phenomena), and list key input and output variables.

Fermentation process temperature regulation; coolant fluid flow rate and tank temperature.

2) Frame geometrical and physical operating attributes (e.g. spatial and temporal boundaries).

Fermentation tank and jacket.

3) Define physical and/or chemical phenomena characteristics evolved over process behavior and estimate some parameters and initial operating conditions.

Heat transfer (conduction and convection) as well as laminar steady cold-water flow.

$\omega(t)$ rad/sec	$T(t)$ °C
	Thermocouple 4 – 20 mA, 3 Volts)

Tachometer (0–5 V, max. 526).

4) Develop a model conceptual structure of causal relationships that mimics observed causes and effects of real operations. This should be a qualitative analysis leading to model conceptualization.

Drawing stock flow or causal diagrams (e.g. piping and instrumentation, process flow, and process schematics) with global input/output process variables and parameters (if available).

Inputs variables → | Conceptual structure and process parameters | **Outputs variables** →

5) Decompose the model conceptual structure (physical entity) into sub-models based on operating functionalities so that changes to the input variables' initial conditions can be measured or estimated by output variables.

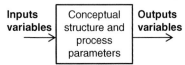

6) List and identify operational conditions for each subprocess and for the entire process. Those operating conditions should meet safety and stability requirements.

	$V(t)$ Volts	$\omega(t)$ rad/sec	$T(t)$ N.m
t = 0	1.2	0	11.3
t = ∞	13.6	32.8	131

Dynamics modeling

7) From process operating and environmental conditions, state key assumptions on model conceptual structure.

During fermentation: (1) heat losses through jacket have negligible environmental effects; (2) fluid flow properties within the jacket are nearly steady and ideal.

(continued)

Table 2.1 (Continued)

Modeling steps	Illustrative examples
8a) *Either derive*, from the analysis of the entity behavior, the model structure and order, then apply system identification methods on experimental data to estimate model parameters (gray boxes).	For example, estimation of a second-order motor with white noise Autoregressive Exogenous (ARX) model structure and its parameters (inductance and inertia) based on least squares methods, regression analysis, and curve fitting techniques.
8b) *Or apply* the principles or physical laws related to conservation and continuity (mass, energy, or momentum balances) to each subprocess (considering the constitutive relations and their initial and boundary conditions) to derive equations capturing entity properties dynamics (white boxes).	Apply Newton's and Kirchhoff's voltage laws to derive the electrically-driven motor pump dynamical model or the Reynolds transport theorem for the fluid flow dynamics in oil pipelines.
9a) Use scaling to simplify the model equations, then apply linearization model methods (if necessary). Otherwise, apply decoupling techniques.	Using operating range (e.g. $\delta = [T1,T2]$). 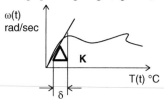 To simplify: $$\omega(t) = \alpha^3 \vec{T}(t) + \beta^2 T(t) \approx KT(t)$$
9b) Check model consistency with the real-life entity and apply dynamics model reduction techniques such as: (1) neglecting small-effect dynamics; (2) assuming time invariant process parameters; and (3) replace distribution characteristics with appropriate lumped elements.	Model order reduction or elimination; keep independent from environmental conditions; neglect uncertainty and noise in an I/O signal (avoid statistical treatment); use ordinary over partial differential equations.

Refinement and validation of the dynamics model

10) Analyze model sensitivity to slight changes in variables and parameters.	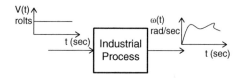
11) Refine the estimation of process parameters using statistical analysis of I/O model dynamic equations based on comparative analysis with experimental test data.	Use stochastic methods such as least squares estimation etc. and derive model parameters confidence region.
12) Evaluate uncertainties in model equations and model equation correctness.	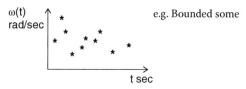 e.g. Bounded some effects of model assumptions based on amplitude ratios resulting from the I/O equation.

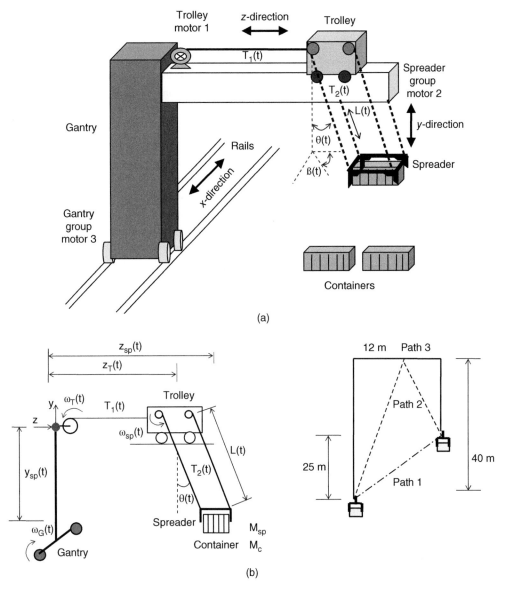

Figure 2.1 (a) Schematic of the sea port gantry crane and its components. (b) Container trolley motion and potential paths 1, 2, and 3.

contact between the guiding rails and the trolley, as well with the spreader cable, are negligible. The cables connecting the trolley and the spreader are rigid (i.e. inflexible). The trolley only moves horizontally while the spreader with container moves only in 2D (y-z-axes), such that while $\theta(t)$ is considered, $\beta(t)$ is negligible.

It is considered that the container is swinging in the opposite direction to the trolley motion. Usually, there are three fixed possible spreader cable lengths corresponding to pick position L_1, drop position L_2 and moving position L_3. When the trolley is moving toward the right direction, it is assumed that the payload moves clockwise. Because the spreader and cables have negligible

Table 2.2 Crane gantry system parameters and variables.

Parameters	Description and values	Variables	Description and values
M_c	Container mass (40 tonnes)	$\theta(t)$	Container swing angle in y-z-axis (rad)
M_t	Trolley mass (3200 kg)	$\frac{dL(t)}{dt}$	Container velocity in L-direction (m s^{-1})
M_g	Gantry mass (800 tonnes)	$z_T(t)$	Position of the trolley in z-axis (m)
M_{sp}	Spreader mass (1500 kg)	$z_{sp}(t)$	Position of the spreader in z-axis (m)
ρ	Air density at 20°C (1.2 kg m^{-3})	$\omega_G(t)$	Gantry motor angular velocity (rad s^{-1})
K	Motor torque constant	$\omega_T(t)$	Trolley motor angular velocity (rad s^{-1})
J_{con}	Container inertia (N.m.s^2 rad^{-1})	$\omega_{sp}(t)$	Spreader motor angular velocity (rad s^{-1})
L_a	Motor armature inductance (0.02 H)	$\beta(t)$	Container swing angle in x-z-axis (rad)
r	Radius of trolley pulley (0.1 m)	$T_2(t)$	Container motor generated force (N)
R_a	Motor armature resistance (27 Ω)	$T_1(t)$	Trolley motor generated force (N)
A	Container surface (m^2)	w_w	Average wind velocity (m s^{-1})
C_d	Aerodynamics coefficient (Nsm^{-1}kg^{-1})	$L(t)$	Length of spreader cable (m)
		$y_{sp}(t)$	Position of the spreader (m) in y-axis

Table 2.3 Gantry cranes speed and acceleration operating conditions.

Operating mode	Speed (m min^{-1})	Acceleration times (s)	Deceleration times (s)
Hoisting with rated load	70	2	1.5
Hoisting with 40-tonne container	100	2	1.5
Hoisting with spreader only	180	4	3
Trolley drive	250	5	5

weights compared to the container, the center of gravity of the cable-spreader-container sub-system is the center of gravity of the container.

Although the load swaying depends on the motion profile (acceleration and speed), it is assumed that the load swaying oscillates like a pendulum at a period given by:

$$T(t) = 2\pi \sqrt{L(t)/g}$$

During the load carrying phase, the velocity of the oscillation is given by:

$$\omega_n = \sqrt{1 + \frac{M_t}{M_{sp} + M_c}}$$

It is decided to model a gantry crane handling system that enables transfer of a container along the spreader-trolley direction (y-z-direction) from one location to another while minimizing the swing of the container. During the gantry crane off-loading operations (pick and drop), the trolley force $T_1(t)$ and the spreader-generated motor force $T_2(t)$ are system inputs and the spreader-container positions $y_{sp}(t)$, $z_{sp}(t)$ can be captured through variables such as: the trolley position $z_T(t)$, the cable length $L(t)$, and the swing angle positions $\theta(t)$. The following two models are presented to illustrate the complexity of system modeling.

2.3.1.1 Model 1

Along the z-direction of the trolley subsystem, applying Newton's laws yields:

$$M_t \frac{d^2 z_T(t)}{dt^2} = T_1(t) - T_2(t) \sin \theta(t)$$

Along the z-direction of the spreader-container subsystem, applying Newton's laws yields:

$$(M_{sp} + M_c) \frac{d^2 z_{sp}(t)}{dt^2} = -T_2(t) \sin \theta(t)$$

Along the y-direction of the spreader-container subsystem, applying Newton's laws yields:

$$(M_{sp} + M_c) \frac{d^2 y_{sp}(t)}{dt^2} = T_2(t)s \cos \theta(t) - (M_{sp} + M_c)g$$

From geometrical analysis, the spreader and the trolley coordinates are related as:

$$y_{sp}(t) = -L(t) \cos \theta(t)$$

$$z_{sp}(t) = z_T(t) + L(t) \sin \theta(t)$$

Taking the differentiation of these equations, the spreader cable length $L(t)$ and the swing angle $\theta(t)$ velocities yield:

$$\frac{d\theta(t)}{dt} = \frac{1}{L(t)} \left[\left(\frac{dz_{sp}(t)}{dt} - \frac{dz_T(t)}{dt} \right) \cos \theta(t) + \frac{dy_{sp}(t)}{dt} \sin \theta(t) \right]$$

$$\frac{dL(t)}{dt} = \left[\left(\frac{dz_{sp}(t)}{dt} - \frac{dz_T(t)}{dt} \right) \sin \theta(t) - \frac{dy_{sp}(t)}{dt} \cos \theta(t) \right]$$

There are five equations and five unknown variables to be determined. Substituting these equations, this system could be simplified into three equations and three unknown variables $(z_T(t), L(t),$ and $\theta(t))$ as follows:

$$(M_c + M_{sp} + M_t) \frac{d^2 z_T(t)}{dt^2} + (M_c + M_{sp}) \frac{d}{dt} \left[\cos \theta(t) + \frac{dL(t)}{dt} \sin \theta(t) \right] = T_1(t)$$

$$(M_c + M_{sp}) \left[\frac{d^2 L(t)}{dt^2} + \sin \theta(t) \frac{d^2 z_T(t)}{dt^2} - L(t) \left(\frac{d\theta(t)}{dt} \right)^2 \right] = -T_2(t) + (M_c + M_{sp})g \cos \theta(t)$$

$$\frac{d^2 \theta(t)}{dt^2} L(t) + \frac{d^2 z_T(t)}{dt^2} \cos \theta(t) + 2L(t) \frac{d^2 \theta(t)}{dt^2} = g \sin \theta(t)$$

This model with three equations and three unknowns is valid when only the trolley moves in synchronization with the spreader; its length is varying simultaneously. There is a coupling between $T_2(t)$ and $\theta(t)$.

Considering the case when the trolley moves only with the spreader, meaning it synchronizes the trolley movement with the spreader cable motion, such a system could be captured using the next model developed.

2.3.1.2 Model 2

Along the z-direction, applying Newton's laws yields:

$$M_t \frac{d^2 z_T(t)}{dt^2} = T_1(t) - T_2(t) \sin \theta(t)$$

From the spreader and container system component, this equation yields:

$$(M_{sp} + M_c)\frac{d^2 z_{sp}(t)}{dt^2} = T_2(t) \sin \theta(t)$$

$$(M_{sp} + M_c)\frac{d^2 y_{sp}(t)}{dt^2} = T_2(t)s \cos \theta(t) - (M_{sp} + M_c)g$$

The spreader and the trolley coordinates are related as:

$$y_{sp}(t) = -L(t) \cos \theta(t)$$

$$z_{sp}(t) = z_T(t) - L(t) \sin \theta(t)$$

Taking the differentiation of the previous geometric equations, the spreader cable length $L(t)$ and the swing angle $\theta(t)$ velocities yield:

$$\frac{d\theta(t)}{dt} = \frac{1}{L(t)}\left[\left(\frac{dz_T(t)}{dt} - \frac{dz_{sp}(t)}{dt}\right)\cos \theta(t) + \frac{dy_{sp}(t)}{dt}\sin \theta(t)\right]$$

$$\frac{dL(t)}{dt} = \left[\left(\frac{dz_T(t)}{dt} - \frac{dz_{sp}(t)}{dt}\right)\sin \theta(t) + \frac{dy_{sp}(t)}{dt}\cos \theta(t)\right]$$

Another approach leading to a gantry crane dynamic positioning model uses only the trolley force $T_2(t)$ as an input and captures the spreader-container position over the variables of trolley position $z_T(t)$ and swing angle $\theta(t)$, which are given by:

$$(M_c + M_{sp} + M_t)\frac{d^2 z_T(t)}{dt^2} + (M_c + M_{sp})L(t)\left[\frac{d^2\theta(t)}{dt^2}\cos \theta(t) - \left(\frac{d\theta(t)}{dt}\right)^2 \sin \theta(t)\right] = T_1(t)$$

$$\frac{d^2 z_T(t)}{dt^2}\cos \theta(t) + L(t)\frac{d^2\theta(t)}{dt^2} + g \sin \theta(t) = 0$$

Here, it is assumed that the spreader cable length $L(t)$ is constant during spreader-container motion. The previous nonlinear model can be linearized using an assumption on the stability region of $\theta(t) \approx 0$. This would result in:

$$(M_c + M_{sp} + M_t)\frac{d^2 z_T(t)}{dt^2} + (M_c + M_{sp})L(t)\frac{d^2\theta(t)}{dt^2} = T_1(t)$$

$$\frac{d^2 z_T(t)}{dt^2} + L(t)\frac{d^2\theta(t)}{dt^2} + g\theta(t) = 0$$

There are two equations and two unknown variables. In order to prevent swaying conditions, a smooth speed variation has to be considered, using a sinusoidal or parabolic approximation of the velocity profile $v_{sp}(t) = \dot{z}_{sp}(t)$, as shown in Figure 2.2. The spreader-container velocity $v_{sp}(t)$ could be given by:

$$v_{sp}(t)^2 = v_{sp\_y}(t)^2 + v_{sp\_z}(t)^2$$

where $L(t)$ is constant and

$$v_{sp\_y}(t) = -L(t)\frac{d\theta(t)}{dt}\sin \theta(t)$$

and

$$v_{sp\_z}(t) = \frac{dz_{sp}(t)}{dt} + L(t)\frac{d\theta(t)}{dt}$$

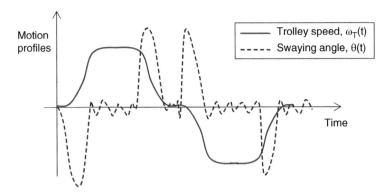

Figure 2.2 Typical variation of swaying angle and spreader speed over time.

The inertia of the spreader-container is not considered. Here, it can be noticed that a change in speed causes the variation in the load balance and swaying angle. Hence, this multiple input multiple output (MIMO) model characterizing the displacement of the container can be described using trolley and spreader cable length velocities from motor-generated force inputs $T_1(t)$ and $T_2(t)$. Usually, the Euler–Lagrange equation is used in the formulation of trolley-spreader-container motion and results in an augmented dynamic model with coupling between motor-generated forces $T_1(t)$ and $T_2(t)$. However, this is negligible in cases where the cable is not considered flexible. Furthermore, neglecting the disturbance force, the force of the wind acting on the container is given by:

$$F_w = AC_d \frac{\rho}{2} v_w^2$$

In addition to low swing angle constraints, it is suitable to achieve gantry crane loading operations with minimum energy consumption. Such constraints would require the minimum travel time. This travel time depends on the velocity and acceleration profiles as well as the path chosen between the pick and drop locations. This makes the model more complex. Also, the gantry crane is used as a load moving system, as in the lifting appliances commonly encountered in warehouses and factories. Here, the swaying could place wear on the equipment and could result in crushing of the carried load, which could also seriously harm personnel nearby. Other applications besides material transportation include a 3D printer. Here, a precise position with minimal oscillation is required to achieve high-resolution printing.

2.3.2 Vertical Elevator System

Electrical motor-driven elevators are used for vertical motion of loads. Due to variations in motion profile (velocity, acceleration), travel distance (height of lifting), load weight, cable characteristics and so on, it is suitable to develop an adequate mechanical model of elevator position. Elevator components are: (i) a passenger car (cabin), car frame, and pair of guide rails; (ii) a counterweight connected by a traction cable to balance cabin and passenger weight; and (iii) a traction system for hoisting, consisting of an electric motor, drum, and gear, as shown in Figure 2.3. Elevators use steel cable rods attached to the top of the suspended elevator car and a counterweight. These cables are passed over a drive pulley that is attached at the top of the elevator motor shaft. The elevator car and the counterweight always move in opposite directions. The suspended elevator car and the counterweight have the same weight and place an extension force on the cable. Table 2.4 summarizes the elevator system steady-state parameters and variables.

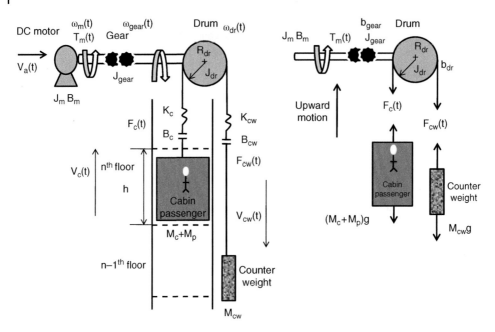

Figure 2.3 Classical elevator system schematic and its components.

Table 2.4 Some elevator system parameters and model variables.

Parameters	Description and values	Variables	Description and values
M_{cw}	Counterweight mass (1200 kg)	$V_a(t)$	Applied DC motor voltage (V)
M_c	Cabin mass (1200 kg)	$D(t)$	Passenger mass (kg)
M	Passenger mass (max. 700 kg)	$i_a(t)$	Armature current (A)
b_{dr}, b_{gear}	Gear and drum damping (N.m.s^{-1} rad^{-1})	$T_m(t)$	DC motor torque (N.m)
L_a	Motor armature inductance (H)	$\theta(t)$	DC motor position (rad)
K_m	Motor constant (1.2 V s^{-1} rad^{-1})	$\omega(t)$	DC motor velocity (rad s^{-1})
J_{gear}	Gear inertia (N.m.s^2 rad^{-1})	$\omega_{gear}(t)$	Gear velocity (rad s^{-1})
K_b	Emf motor constant (V^{-1} s^{-1} rad^{-1})	$\omega_{dr}(t)$	Drum velocity (rad s^{-1})
R_{dr}	Radius of drum (pulley) (0.3 m)	$y_{cw}(t)$	Counterweight position (m)
J_m	Motor inertia (0.053 N.m.s^2 rad^{-1})	$y_{dr}(t)$	Vertical rim position (m)
η_{gear}	Transmission ratio (50)	$y_c(t)$	Vertical cabin position (m)
L	Cable length (m)	$y_{cable}(t)$	Vertical cable position (m)
λ	Linear mass density (kg)	$F_c(t)$	Cabin applied force (N)
μ	Friction coefficient (0.11 N.m.s)	$F_{cw}(t)$	Counterweight force (N)
J_{dr}	Drum inertia (300 N.m.s^2 rad^{-1})	$T_{dr}(t)$	Drum applied force (N)
R_a	Motor armature resistance (0.61 Ω)	$f_\mu(t)$	Friction torque (N.m)
		$F_L(t)$	Passenger load force (N.m)
		$v_{cab}(t)$	Cabin linear velocity (m s^{-1})

Usually, an elevator car moves between floors by accelerating from a rest position during a first time interval, then moves at a constant speed during a second time interval until it decelerates to rest during a third time interval. During this vertical motion, the steel hoisting cables are assumed to be massless, elastic, and viscous (Calvin's design model), such that they can be approximated as a highly stiff spring combined with damping. In this study, this is all assumed to be negligible compared to the cabin and the counterweight. Furthermore, it is assumed that during each cabin journey the passenger load is constant while the cable extension and all viscous friction and contact rail friction are negligible. In addition, the passenger cabin and the counterweight are assumed to be a concentrated mass. The gear and the drum have damping due to attached bearings. Any obstacle along the vertical path that could cause cabin vibration is not considered.

It is necessary to model the elevator dynamics in order to design a refined elevator position-controlled system and to ensure the shortest travel time and the best passenger comfort, despite any disturbance such as passenger load variation and smooth motion in the presence of high acceleration and deceleration rates. As such, the elevator upward motion dynamics model, relating the input supply voltage to motor shaft speed, can be derived by applying Newton's laws to the elevator's mechanical system around the center of gravity of the drum such that:

$$J_{\text{gear}} \frac{d\omega_{\text{gear}}(t)}{dt} + J_m \frac{d\omega_m(t)}{dt} + b_m \omega_m(t) + b_{\text{gear}} \omega_{\text{gear}}(t) + T_{L \to \text{dr}}(t) = T_{m \to \text{dr}}(t)$$

Recalling that:

$$\omega_{\text{dr}}(t) = \omega_{\text{gear}}(t) = \frac{\omega_m(t)}{\eta_{\text{gear}}}$$

$$T_{m \to \text{dr}}(t) = \frac{T_m(t)}{\eta_{\text{gear}}}$$

and considering an elastic cable with a gradient of tension over the cable length, if the car is moving upward, the drum-load dynamic equation would yield:

$$J_{\text{dr}} \frac{d\omega_{\text{dr}}(t)}{dt} + b_{\text{dr}} \omega_{\text{dr}}(t) + (F_c(t) + F_{\text{cw}}(t)) R_{\text{dr}} = T_{L \to \text{dr}}(t)$$

Applying Newton's laws to the cabin-passenger load and counterweight subsystems gives:

$$M_{\text{cw}} \frac{dv_{\text{cw}}(t)}{dt} = M_{\text{cw}} g - F_{\text{cw}}(t)$$

$$(M + M_c) \frac{dv_c(t)}{dt} = F_c(t) - (M + M_c) g$$

It is assumed that:

$$v_{\text{cw}}(t) = v_c(t) = R_{\text{dr}} \omega_{\text{dr}}(t)$$

Substituting these equations into the drum-load equation yields:

$$J_{\text{dr}} \frac{d\omega_{\text{dr}}(t)}{dt} + b_{\text{dr}} \omega_{\text{dr}}(t) + \frac{d\omega_{\text{dr}}(t)}{dt} (M_{\text{cw}} + M + M_c) R_{\text{dr}}^2 + (M + M_c - M_{\text{cw}}) g R_{\text{dr}} = T_{L \to \text{dr}}(t)$$

b_{dr}, b_{gear} are damping factors due to the bearings, but are negligible. Thus, assuming that $b_{\text{gear}} = b_{\text{dr}} = 0$, the elevator upward motion dynamics model yields:

$$(\eta_{\text{gear}} J_{\text{dr}} + \eta_{\text{gear}} J_{\text{gear}} + J_m + \eta_{\text{gear}} (M_{\text{cw}} + M + M_c) R_{\text{dr}}^2) \frac{d\omega_m(t)}{dt}$$

$$+ b_m \omega_m(t) + (M + M_c - M_{\text{cw}}) g R_{\text{dr}} = \frac{T_m(t)}{\eta_{\text{gear}}}$$

at the level of the motor shaft,

$$T_m(t) = K_t i_a(t)$$

Hence,

$$i_a(t) = \frac{\eta_{\text{gear}}}{K_t}(\eta_{\text{gear}}J_{\text{dr}} + \eta_{\text{gear}}J_{\text{gear}} + J_m + \eta_{\text{gear}}(M_{\text{cw}} + M + M_c)R_{\text{dr}}^2)\frac{d\omega_m(t)}{dt}$$
$$+ b_m\omega_m(t) + (M + M_c - M_{\text{cw}})gR_{\text{dr}}$$

while from the motor's electrical component, applying Kirchoff's law yields:

$$V_a(t) = R_a i_a(t) + L\frac{di_a(t)}{dt} + K\omega_m(t)$$

Thus, during upward motion, the DC applied voltage to velocity relationship is given by:

$$V_a(t) = \frac{R_a\eta_{\text{gear}}}{K_t}\left[\begin{array}{c}(\eta_{\text{gear}}J_{\text{dr}} + \eta_{\text{gear}}J_{\text{gear}} + J_m + \eta_{\text{gear}}(M_{\text{cw}} + M + M_c)R_{\text{dr}}^2)\frac{d\omega_m(t)}{dt} \\ + b_m\omega_m(t) + (M + M_c - M_{\text{cw}})gR_{\text{dr}} \end{array} \right]$$
$$+ \frac{L_a\eta_{\text{gear}}}{K_t}\left[\begin{array}{c}(\eta_{\text{gear}}J_{\text{dr}} + \eta_{\text{gear}}J_{\text{gear}} + J_m + \eta_{\text{gear}}(M_{\text{cw}} + M + M_c)R_{\text{dr}}^2)\frac{d^2\omega_m(t)}{dt^2} \\ + b_m\frac{d\omega_m(t)}{dt} \end{array} \right]$$
$$+ K\omega_m(t)$$

During the elevator motion in the downward direction, a dynamic model would yield:

$$(\eta_{\text{gear}}J_{\text{dr}} + \eta_{\text{gear}}J_{\text{gear}} + J_m + \eta_{\text{gear}}(M_{\text{cw}} + M + M_c)R_{\text{dr}}^2)\frac{d\omega_m(t)}{dt} + b_m\omega_m(t)$$
$$+ (M_{\text{cw}} - M - M_c)gR_{\text{dr}} = \frac{T_m(t)}{\eta_{\text{gear}}}$$

It should be noted that the sign of the cabin-passenger load torque changes during the elevator upward motion. This model would not be valid in the case where the elevator moves at low velocities and for longer distances due to longitudinal oscillations. The elevator braking torque mechanism is not modeled as it should be considered to be matched against the drum inertia. The design of an elevator system that has equal cabin mass and counterweight mass allows the passenger load alone to be used to size the drive and motion profile. Furthermore, a pulse width modulation (PWM) converter used to design the variable speed drive enables the production of an input voltage toward the armature field circuit of the DC motor. The relationship of the hoisting cable configuration with longitudinal vibration is not captured in order to ease the model complexity. Also, depending on the cabin position along the hoistways, the cable weight varies and modifies the overall load in the dynamic model.

2.3.3 Hybrid Vehicle Powertrain with Parallel Configuration

The policies for reducing air pollution due to toxic emissions from internal combustion engines have led to the development of electric (EV) and hybrid electric vehicles (HEVs). HEVs use fossil-fuel and electrically-driven motor powertrains either in series, parallel, or even combined to deliver the energy required for vehicle motion. Electrochemical or electrostatic energy storage systems supply this energy to electric motors. The degree of hybridization is defined as the ratio between electric motor power and total engine power, which varies between 15 and 60%. Hence, in order to optimize energy consumption, it is suitable to develop a dynamic model of

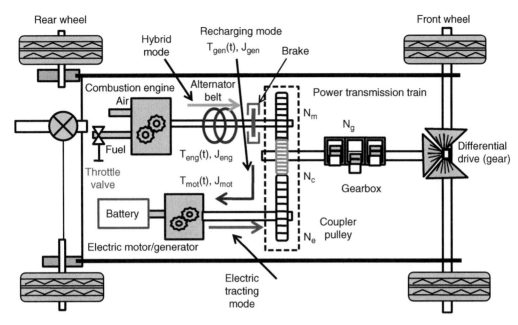

Figure 2.4 Parallel configuration of a hybrid electric vehicle.

the hybrid vehicle powertrain and the braking system in order to adjust the control strategy for efficient power to be delivered by the motor and brake activation of each wheel. In addition, compared to engines, electric motors can produce high torque at low rotational speed and have a wider operating range. This makes it possible for the powertrain to have an efficient load transfer mechanism for delivering power. Typical powertrain configurations are: front-wheel (FWD), rear-wheel (RWD), and four-wheel (4WD) drives. Consider a 2D chassis schematic for an FWD hybrid vehicle with a parallel configuration illustrating the coupling of its inside chassis components (powertrain, engine, permanent magnet DC motor, accumulators etc....), as shown in Figure 2.4. Here, the components acting on the dynamics of the electric powertrain are:

1) A gearbox, used to adapt the motor torque to the shaft velocity by the load transferring process. The load transfer is due to the alignment of the embedded gears: the sun gear, the planetary gears carrier and the ring gear. Compared to other devices such as the chain drive or belt drive, the planetary gearbox offers the best torque-to-weight ratio.
2) A differential, used to obtain a faster angular speed of the outer wheel (with a longer trajectory) over the inner wheel during the HEV cornering trajectory. It could be achieved mechanically using a gear for differential torque distribution or using electrical power differentials. The latter allows power to be applied independently to the different wheels.
3) Super capacitors, main accumulators used for electrical energy storage and are usually combined with regenerative braking. Super capacitors offer higher power than a battery due to lower heating energy dissipation from their inner resistance.
4) An alternator, used to transform the kinetic energy into electric energy by phase shifting the HEV motor current. This induced current is used to charge the accumulators, which can be batteries, capacitors or both.
5) Regenerative braking, which stores energy related to HEV braking. When the car is braking, the internal wheel is connected to the drive shaft with a clutch and decelerates due to braking momentum. During regenerative braking, the torque reverses direction and the gearbox has to be able to transfer the reversed load.

Table 2.5 HEV system parameters and model variables.

Parameters	Description and values	Variables	Description and values
N_R	Radius of the ring gear (m)	$\omega_{\text{mot}}(t)$	Motor angular velocity (rad s^{-1})
N_S	Radius of the sun gear (m)	$\omega_{\text{gen}}(t)$	Generator angular velocity (rad s^{-1})
J_{mot}	Inertia of the motor (N.m.s^2 rad^{-1})	$\omega_{\text{eng}}(t)$	Engine angular velocity (rad s^{-1})
J_{eng}	Inertia of the engine (N.m.s^2 rad^{-1})	$T_{b\,f}$	Braking torque (N.m)
J_{gen}	Inertia of the generator (N.m.s^2 rad^{-1})	$T_{\text{mot}}(t)$	Motor torque (N.m)
η_{gear}	Final gear ratio (%)	$T_{\text{gen}}(t)$	Generator torque (N.m)
r_w	Wheel radius (m)	$T_{\text{eng}}(t)$	Engine torques (N.m)
$K_{b\,f}$	Pressure torque conservation constant gain (N.m. Pa^{-1})	$T_{\text{Drive}}(t)$	Drive shaft torque (N.m)
$p_{b\,f}$	Pressure torque (Pa)	$T_{\text{Brake}}(t)$	Friction or service brake torque (N.m)
η_{trans}	Transmission axle ratio (%)	F	Interaction force between different parts of the powertrain (N)

6) HEV motor configuration, which defines the driving source and the electrical regeneration modes. HEV operating modes alternate operations between: (i) the electric mode, where its electrical motor $(T_{\text{mot}}(t), J_{\text{mot}})$ is the active driving force; (ii) the hybrid mode, where the combustion engine $(T_{\text{eng}}(t), J_{\text{eng}})$ is combined with the electric motor as the driving force; and (iii) the recharging mode, when the combustion engine is the sole active traction force while recharging the accumulator. The HEV electric generating mode (iv) uses the combination of engine and alternator $(T_{\text{gen}}(t), J_{\text{gen}})$ and is performed during either traction or braking motion. The latter is called regenerative braking. The electrical regeneration is commonly implemented using alternators so that when the HEV has reached the desired speed, the attached flywheel is disconnected from the drive shaft. Any reconnection of the flywheel to the drive shaft causes the energy stored in the wheel to be reused as a driving force into the HEV.

Hence, the modeling of the powertrain could allow development of a power-efficient HEV transport system. This would require the individual dynamics modeling of each component as presented next. Table 2.5 summarizes the steady-state HEV variables and parameters. A variable gearbox is used to adapt the motor torque characteristics to the wheel speed by a load transfer mechanism. Regarding the powertrain configuration, it is assumed that the load within the gearbox depends on the type of gearbox, the gear ratio, and the efficiency of the gearbox, and the energy losses within gearing and bearing are not considered. Only one electrical motor and combustion engine per HEV is considered, and a mechanical differential also has to be used. This mechanical differential does not add significant weight and losses to the HEV. The capacitance of the accumulator is high enough to ensure that the regenerated energy from the braking operation does not exceed the limiting voltage. Whenever the accumulators release energy, loaded energy is lost due to inner resistance. This energy is turned into heat. This is a problem since an accumulator is easily damaged if the temperature rises above 60°C.

2.3.3.1 Motor Driving and Regenerating Model

During each of the four operating modes of an HEV, the powertrain generates torque for driving motion or power regeneration purposes. This generated torque can be derived according to the operating configuration of the electric motor, the combustion engine, and even their coupling. Hence, in the case of (1) electric mode, the HEV traction torque is only provided by the electric motor while the engine is switched off (only the sun gear) such that the dynamics equation yields:

$$J_{mot} \frac{d\omega_{mot}(t)}{dt} = T_{mot}(t) - \frac{T_{D\_drive}(t) + T_{D\_brake}(t)}{\eta_{gear}} - F \times N_s$$

with F being the interaction force between different parts of the transmission train, while $F \times N$ is the reaction torque on the sun gear. Here, torques from the shaft and the brake are considered to be disturbances. In the case of (2) hybrid mode, the electric motor is combined with the engine (only the ring gear) during traction to provide an additional torque such that:

$$J_{mot} \frac{d\omega_{mot}(t)}{dt} = T_{mot}(t) - \frac{T_{D\_drive}(t) + T_{D\_brake}(t)}{\eta_{gear}} + F \times N_s$$

In the case of (3) recharging mode, the engine provides the torque for both vehicle traction and accumulator recharge (ring and sun gears), and the dynamics equation is characterized such that:

$$J_{eng} \frac{d\omega_{eng}(t)}{dt} = T_{eng}(t) - F \times (N_R + N_s)$$

In the case of (4) the regenerative braking mode, the electric motor is the drive force (only the ring gear) acting as a generator for accumulator recharge by converting the HEV kinetic energy into electrical energy such that:

$$J_{gen} \frac{d\omega_{gen}(t)}{dt} = T_{gen}(t) + F \times N_R$$

where $J_{eng}, J_{eng}, J_{mot}$ are the inertia of the engine, generator and motor, N_S, N_R are the radius of the sun and ring gears, $T_{eng}(t), T_{eng}(t), T_{mot}(t)$ are the engine, generator, and motor torques, $\omega_{eng}(t), \omega_{eng}(t), \omega_{mot}(t)$ are the engine, generator, and motor angular velocities, $T_{Drive}(t)$ is the drive shaft torque, $T_{Brake}(t)$ is the generated brake torque, r_w is the wheel radius, and η_{gear} is the final gear ratio. The HEV operating constraints are given by:

$$N_S \omega_{gen}(t) + N_R \omega_{mot}(t) = (N_S + N_R) \omega_{eng}(t)$$

and

$$\omega_{mot}(t) = \frac{\eta_{gear}}{r_w} V(t)$$

The torque converter and final differential are used to deliver a torque to the front wheel, such that η_{trans} is the axle transmission system ratio. It could be assumed that the engine speed is equal to the wheel speed scaled by the current gear (gearbox position).

2.3.3.2 Vehicle Gear Box Model
The gear box relates the throttle setting (in %) with the wheel angular velocity (rad s^{-1}). This adjustment could be done either manually or automatically. Its gear ratio, η_{gear}, is not constant but rather can vary according to:

$$\eta_{gear} = \frac{\max \cdot \omega_{motor}}{\max \cdot \omega_{wheels}}$$

2.3.3.3 Brake System Model
The pressure applied to the brake disk is converted into a braking torque with a pressure torque conservation constant gain, K_{b_f} depending highly on the speed, the temperature, and so on. Thus, each hydraulic braking system could be modeled as:

$$T_{b\_f} = p_{b\_f} K_{b\_f} \min\left(1, \frac{\omega_f}{\max \omega_f}\right) = 1.5 N \cdot m$$

Motion performance and energy efficiency of the HEV depend on the resulting control and the design parameters of accumulators, motors, gears, and differentials with an overall consideration of vehicle dynamics as well as the load due to road profile. Hence any change in these parameters could require reassessment and readjustment of the HEV motor torque and velocity. This is partially done with automatic adjustment of the gear ratio based on road profile changes when cruise control is activated on the HEV.

2.3.4 Driverless Vehicle Longitudinal Dynamics

During vehicle motion, the motor traction forces are distributed over the front and rear axles as well as left and right axles by modifying the transmitted torques at wheel level with respect to the variation of resistive forces due to acceleration or road surface. Such automatic motor torque adaptation ensures smooth vertical, lateral, and longitudinal vehicle dynamics, especially in the case of a driverless vehicle with an automatic gearbox. Hence, in order to adapt the vehicle power to changes in the force distribution (longitudinal, vertical, lateral), it is appropriate to model the coupling of the car wheel dynamics with respect to motor traction forces and resistive forces according to the road and the car motion profile (acceleration, velocity).

The vehicle dynamics can be split into longitudinal, vertical, and lateral translations and roll, pitch and yaw rotations. For simplicity, the analysis of vehicle motion is restricted to longitudinal translation. Components involved in the vehicle's longitudinal dynamics include the suspension, tire traction, engine, brakes, and steering. Considering a vehicle that moves up a hill with an angle α, as depicted in Figure 2.5. During the longitudinal acceleration or the vehicle cornering, there is a load transfer due to the weight shifting, causing unequal distribution of the tire grip (front to rear load shifting in the case of acceleration, or left to right load shifting in the case of right cornering). Therefore, the modeling of the torque applied to each wheel should be considered despite any road-wheel slipping conditions. Forces and torques involved in the planar vehicle dynamics are the related tire-road interactions, aerodynamics interaction, vehicle braking, and engine traction.

Besides the no-slip contact between the tires and the road, the road gradient is considered to be identical for all four tires. It is assumed that the tire friction is equal on each wheel and the torque is equally distributed between the wheels. This friction can be decomposed into Coulomb friction and viscous friction components. It is assumed all friction coefficients, the tire's moment of inertia and the road surface, μ_i, are dependent on the tire material. Being dry, wet, or icy, the road surface is assumed to be known in advance. The vehicle suspension and the chassis construction are not considered; rather, the model is restricted to four tires attached to a vehicle structure with a known center of gravity, to be shown in Figure 2.6. Any forces associated with vehicle dampers and springs are assumed constant and negligible. Furthermore, the steering system and a road noise model are not considered and the rolling resistance is assumed to be proportional to the mass of the vehicle. All process parameters and variables involved have been listed in Table 2.6.

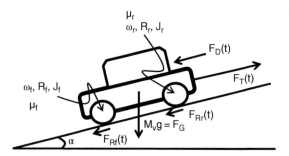

Figure 2.5 Vehicle forces schematic.

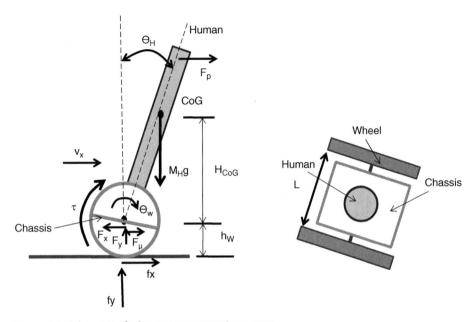

Figure 2.6 Schematic of a Segway transportation system.

Table 2.6 Some key variables and parameters for the car longitudinal dynamics model.

Parameters	Description and values	Variables	Description and values
M_v	Vehicle mass (kg)	$F_G(t)$	Gravitational forces (N)
$K_{f,r}$	Front/rear tire viscous friction coefficient (*dimensionless*)	$F_T(t)$	Traction forces (N)
μ_i	Friction coefficient between tires and road (*dimensionless*)	$F_D(V)$	Drag disturbance forces (N)
C_r	Rolling resistance coefficient (*dimensionless*)	$F_R(V)$	Frictional forces (N)
L	Wheel base (m)	$F_a(V)$	Aerodynamic resistance force (N)
R_r, R_f	Front/rear wheel radius (m)	$F_r(t)$	Rolling resistance force (N)
J_f, J_r	Front/rear tire inertia (N.m.s^2 rad^{-1})	$T_{\text{engine}\_r}(t)$	Torque provided by the engine at each wheel (N.m)
μ_f, μ_r	Front/rear tire friction coefficient (*dimensionless*)	$T_{\text{brake}\_r}(t)$	Torque applied to each tire due to the brakes (N.m)
a_1	Distance between the vehicle and center mass (m)	$T_{\text{tractive}\_r}(t)$	Reaction torque on each tire due to the tire's traction force (N.m)
a_2	Distance between the front and rear axles (m)	$T_{\text{viscous\_friction}\_r}(t)$	Rear viscous friction torque (N.m)
A	Frontal area of the vehicle (m^2)	$F_Z(t)$	Normal force (N)
ρ	Air density (kg m^{-1})	$F_{X\mu}(t)$	Friction force due to contact between tire and road surface (N)
C_d	Aerodynamic drag coefficient	$T_{\text{shaft}}(t)$	Motor shaft torque (N.m)
λ	Wheel slip ratio	$\omega_{r,f}(t)$	Rear/front tire angular velocity (rad s^{-1})
		$V_{\text{wheel}}(t)$	Tire linear velocity (m s^{-1})
		V	Vehicle linear velocity (m s^{-1})
		$\alpha(t)$	Hill angle with horizontal (rad)

During the vehicle's forward acceleration, its center of gravity generates a backward force due to its mass inertia. Hence, applying Newton's second law on all forces acting on the vehicle's longitudinal dynamics, where resistive and traction forces interact, yields:

$$M_v \frac{dv(t)}{dt} = F_T(t) - F_D(t) - F_R(t) - F_G(t)$$

Resistive forces are: (i) the drag force; (ii) tire-road friction forces; and (iii) gravitational forces. Based on fluid dynamics, the *drag force*, $F_D(V)$ increases with the square of vehicle velocity and depends on the aerodynamic resistance force $F_a(V)$ related to the wind speed and the rolling resistance forces F_r such as:

$$F_D(V) = F_a(V) + F_R(V) = \frac{1}{2}\rho A C_d V^2 + M_v g C_r V$$

Usually, from a one-wheel model, the wheel slip ratio is given by:

$$\lambda = \frac{V_{wheel} - V}{\max(V_{wheel}, V)}$$

while the wheel linear velocity is given by:

$$V_{wheel} \approx \omega_r(t)R_r$$

Tire-road friction force $F_R(V)$ affects the vehicle's throttle and brake settings system as well as the vehicle's front and rear wheel velocities. Depending on the road gradient $\alpha(t)$, the road conditions, μ, and considering $F_Z(t)$ as the normal force, the friction force due to the contact between the tire and the road surface is given by:

$$F_{x\mu}(t) = 2\mu_r F_{zr}(t) + 2\mu_f F_{zf}(t)$$

This is valid only when there is a load distribution between front and rear wheels. Hence, based on the load distribution at each point of contact and the load characteristics (vehicle geometry, road grade angle, and acceleration, deceleration, or braking mode), the friction forces yield:

$$F_{zf}(t) = F_{static\_f}(t) + F_{dynamics_f}(t) = mg\left(\frac{\alpha_2}{L}\cos\alpha(t) + \frac{H}{L}\sin\alpha(t)\right) - m\frac{dV(t)}{dt}\frac{H}{L}$$

and

$$F_{zr}(t) = F_{static\_r}(t) + F_{dynamics_r}(t) = mg\left(\frac{\alpha_2}{L}\cos\alpha(t) + \frac{H}{L}\sin\alpha(t)\right) + m\frac{dV(t)}{dt}\frac{H}{L}$$

This is valid when $\alpha(t) = \alpha = constant$. During acceleration motion, the load is carried by rear wheels, while during braking motion it is carried by the front wheel. Each wheel has a rotational speed $\omega_{r,f}(t)$ with a radius R_r, R_f and a polar moment of inertia J_f, J_r. The *gravitational force* $F_G(t)$ is given by:

$$F_G(t) = M_v g \sin\alpha(t)$$

During braking motion, the *traction force* $F_T(t)$ produced by the tire is given by:

$$F_T(\lambda) = \mu(\lambda)M_v g$$

Similarly, during driving motion, the traction force is derived from the wheel torque transmitted by the powertrain shaft torque, such that:

$$F_T(t) = \frac{T_{wheel}(t)}{R} = \frac{T_{shaft}(t)\eta_{gear}}{R}$$

Hence, during the vehicle's motion, each wheel's angular velocity can be derived by summing the applied torques, such that for the rear wheel it would yield:

$$J\omega_r = T_{\text{engine\_r}}(t) + T_{\text{tractive\_r}}(t) - T_{\text{brake\_r}}(t) - T_{viscous\,friction_r}(t)$$

The traction force is given by:

$$T_{\text{tractive\_r}}(t) = R_r F_{x\mu\_r}$$

Considering $K_{f,r}$ being the viscous friction coefficient of each wheel, the viscous friction coefficient is given by:

$$T_{viscous\,friction_r}(t) = K_{fr}\omega_r(t)$$

Vehicle cornering is characterized by lateral tire grip. This is captured by the torque applied to each tire as well as the steering angle. Furthermore, during the vehicle's motion, any bump on the road causes vertical dynamics depending on the wheel damping, the vehicle mass inertia and the suspension characteristics, which have been discarded in this study. In addition, the grip characterizing the adhesion of the wheel on the road defines the vehicle dynamics. Therefore, an accurate model of a pure tire would be nonlinear as it depends on the velocity and the slip ratio. The high variation of the slip ratio in the case of icy or wet conditions is expected to invalidate the entire dynamic vehicle traction model presented previously. This is also the case with a deteriorating suspension, a flat tire, an underperforming engine, misaligned steering, and a braking system fault.

2.3.5 Automated Segway Transportation Systems

Automated mobility devices such as the Segway have been used to transport people or objects using electrically-based guidance tools. Those devices have heavy bases to ensure low centers of gravity and to ease dynamic stability by moving between stable states. Specific to the Segway, a motorized platform or chassis is used to stabilize a passenger standing upright. Hence, by balancing forward or backward, the person dictates, respectively, the acceleration or deceleration rate, as illustrated in Figure 2.6. The Segway transportation system with a rider can be decomposed into two components: (i) chassis-human body; and (ii) wheels. Those two components rotate around the same chassis axis. Table 2.7 summarizes the steady-state parameters and variables of the Segway transportation system.

When the rider moves forward by a small angle against the vertical, this transportation device propels itself forward along the ground causing the rider to move backward to maintain his or her upright position. The rider's inclination over the chassis is used to dictate left or right motion. Hence, the synchronization of the rider balancing with the Segway motion is required, especially during the velocity change. The following assumptions are considered: (i) there is no slip between the wheel and the floor; (ii) the viscous friction of the chassis-wheel bearings is negligible; (iii) the motion in the median sagittal plane and median coronal plane is decoupled; and (iv) only the planar motion dynamics are investigated.

During the motion with a rider, the acting forces and torque in the transportation system are: motor torques, turning force, forces due to road inclination, rider active pushing force, friction forces in the wheel, the drag resistive force, the damping in the damper, and the friction force between each wheel and the motor. Using the same approach as the inverted pendulum, it is possible to derive the dynamics model of a unified passenger-Segway system as well as to ensure a stable and steady motion by avoiding a passenger falling over.

Considering that the left and right wheels have synchronized motion (meaning there is no slip) and applying the principle of moments around the chassis axis both result in the dynamic

Table 2.7 Some Segway system variables and parameters.

Parameters	Description and values	Variables	Description and values
H_{wb}	Width of wheel base (0.6 m)	$\theta_H(t)$	Rider vertical angular position (m)
H_{CoG}	Distance from rider center of gravity to wheel center of gravity (m)	$T_{rw}(t)$	Right wheel torque (N.m)
M_H	Mass of human (kg)	$x_{rw}(t), x_{lw}(t)$	Wheel linear position (left x_{lW} and right x_{rW}) (m)
M_C	Mass of chassis platform (4.9 kg)	$T_{\mu r}(t)$	Right wheels friction torque along chassis axis (N.m)
h	Thickness of wheel (0.08 m)	$T_{em}(t)$	Wheel motor shaft torque (N.m)
J_H	Inertia of rider over chassis axis (N.m.s^2 rad^{-1})	$F_p(t)$	Slip resulting force (N)
J_{rw}, J_{lw}	Left/right wheel inertia (N.m.s^2 rad^{-1})	$\theta_{rw}(t)$	Right wheels angular position (m)
M_W	Mass of wheels (2.3 kg)	$F_\mu(t)$	Slip resulting force (N)
L	Chassis platform length (0.5 m)	$F_x(t)$	Normal force (N)
R_W	Radius of wheels (0.4 m)		
H_H	Height of rider (m)		

modeling of the chassis and human body component such that:

$$(M_H H_{CoG}{}^2 + J_H)\frac{d^2\theta_H(t)}{dt^2} = M_H H_{CoG}\sin\theta_H(t) - M_H H_{CoG}\cos\theta_H(t)\frac{d^2x(t)}{dt^2}$$
$$+ (T_{wr}(t) + T_{wl}(t)) - (T_{\mu r}(t) + T_{\mu l}(t))$$

The linear to rotational motion relationship is given by:

$$x(t) = R_w\theta_{wr}(t) = R_w\theta_{wl}(t) = R_w\theta_w(t)$$

It is assumed that the mass of both wheels is such that:

$$M_{lw} = M_{rw} = M_w$$

and their inertia is given by:

$$J_{lw} = J_{rw} = \frac{1}{2}M_w R_w{}^2$$

Thus, considering a small angle $\theta_H(t)$, the principle of moments equation yields:

$$(M_H H_{CoG}{}^2 + J_H)\frac{d^2\theta_H(t)}{dt^2} = M_H g H_{CoG}\theta_H(t) - M_H H_{CoG}R_w\frac{d^2\theta_w(t)}{dt^2}$$
$$+ (T_{wr}(t) + T_{wl}(t)) - (T_{\mu r}(t) + T_{\mu l}(t))$$

Similarly, applying the principle of moments around the wheel axis with rear wheel angular velocity $\omega_r(t)$ would result in:

$$\frac{1}{2}M_{wr}R_{wr}{}^2\frac{d^2\theta_{wr}(t)}{dt^2} = T_{wr}(t) - f_{xr}(t)R_w - T_{\mu r}(t)$$

Independent of the road characteristics, the type of wheel, and the width of the wheel base, it is assumed that there is no friction at the base, therefore:

$$f_{xl}(t) = f_{xr}(t) = 0$$

Hence, similarly applying the principle of moments around the axis of the chassis, the Segway dynamics modeling is given by:

$$((M_W + M_H)R_W{}^2 + J_W)\frac{d^2\theta_W(t)}{dt^2} = M_H R_W H_{\text{CoG}} \sin\theta_H(t)\frac{d\theta_H(t)^2}{dt}$$

$$- M_H H_{\text{CoG}} R_W \frac{d^2\theta_H(t)}{dt^2} \cos\theta_H(t)$$

$$- (T_{\text{wr}}(t) + T_{wl}(t)) + (T_{\mu r}(t) + T_{\mu l}(t))$$

Assume that the combined torque is given by:

$$T_{em}(t) = T_{\text{wr}}(t) + T_{wl}(t)$$

However, considering the friction along the axis of the chassis, the viscous torque yields:

$$T_{\mu r}(t) + T_{\mu l}(t) = T_\mu(t) = K_\mu \left(\frac{d\theta_W(t)}{dt} - \frac{d\theta_H(t)}{dt} \right)$$

The rider's angular position with respect to the vertical is highly nonlinear and depends heavily on the value of his or her weight and height. As such, the shorter the rider, the lower the center of gravity. This relationship results in a steadier and more stable motion while changing the vehicle dynamics. In addition, subsequent variation of the road conditions (wet or icy) could cause one to revisit the traction, the slip force model, and the stability requirements.

2.4 Biomedical Systems and Processes

2.4.1 Infant Incubator

Despite ambient temperature and humidity variations, infant body temperature should be kept constant through the body's self-thermoregulation process, called homeostasis. This consists of regulating the blood flow into extremities (hands, head, feet, etc.) using information from the sensitive skin neuronal network and thermoreceptors connected to the hypothalamus. This energy-based physiological activity is balanced between nutrients from stored energy over feces and urine losses.

Deficiencies in this self-thermoregulation process means premature babies or newborn infants are often placed in an incubator cabinet (usually $0.5 \times 0.5 \times 1$ m^3). As illustrated in Figure 2.7, this cabinet provides the controlled environmental conditions necessary to feed the infant and also to prevent infection, hypoxia, hyperthermia (inability to maintain neonatal body temperature below 37.2°C in the presence of any microorganism infection) and hypothermia (insufficient heat metabolic production to maintain temperature above 36.5°C). Therefore, it is suitable for maintaining newborn esophageal, rectal, and abdominal temperatures around 36.5–37.2°C independent of weight, age and body water loss. This is done through the regulation of the temperature, relative humidity, and the oxygen concentration within the incubator cabinet.

In order to maintain temperatures around 36.5–37.2°C, as well as a humidity range around 15%, in the incubator system, the following equipment is assembled: (i) fans for air ventilation; (ii) a valve to control the inflow of water vapor; (iii) a radiant warmer for heat production; and (iv) insulated incubator walls to reduce energy losses. The heat is generated from the infant's body metabolism and dissipated via the skin and lungs into the surrounding surfaces within the incubator cabinet. By air motion, the heat dissipates toward the incubator walls and mattress. Also, the evaporation and latent heat from the infant's body are absorbed by the lungs and the skin before being released as moisture.

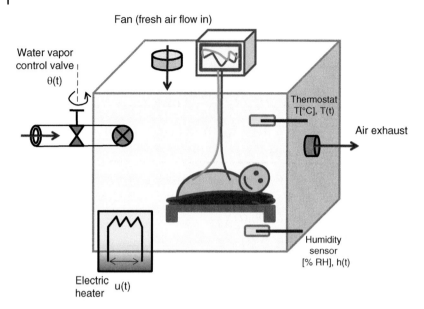

Figure 2.7 Neonatal incubator and its typical components.

For simplicity, it is assumed that: (i) the heat transfer between the infant's body and the incubator mattress is negligible; (ii) the thermal capacity of the incubator is equal to the infant's body; (iii) the surrounding air and mattress temperatures are equal to the infant's skin temperature; (iv) the air flow within the incubator is considered to be uniform; (v) despite organ activity differences, metabolic heat production over the infant's body is uniform; (vi) the air inflow rate is constant while the air temperature is time-variable; and (vii) all incubator components (walls and mattress) are homogenous throughout their material. Table 2.8 summarizes some key steady-state parameters and variables of the incubator system.

The infant's heat balance can be derived from the heat generation and dissipation within its body, the heat loss being through evaporation from the skin, the respiratory activity, and the lung-based heat loss through conduction and convection. The mass transfer and heat transfer can be derived in differential equations using the laws of conservation of heat exchange, by conduction, convection, and radiation, between the infant and the incubator's surrounding environment. Accordingly, using the principle that the balance of heat stored within the incubator is equal to the energy generated by the metabolic heat minus the energy lost through the infant's skin by radiation, convection and evaporation, a mathematical heat transfer model of the incubator wall and air can be obtained. Hence, the incubator thermoregulation model is given by:

$$C_i \frac{dT_i(t)}{dt} = Q_{mi}(t) - Q_{eb}(t) - Q_{cb}(t) - Q_{rb}(t)$$

The heat transfer per unit of time from the radiation from the infant's skin toward the incubator walls $Q_{rb}(t)$ is dependent on the infant's temperature, geometrical form, and spatial arrangement within the incubator, as well as its skin heat emissive properties. Hence, by applying the principle of conservation of energy into the three geometrical surface arrangements of the infant's body (esophageal "*es*," rectal "*re*", and abdominal "*ab*"), the total heat loss through radiation, $Q_{rb}(t)$ would yield:

$$Q_{rb}(t) = A_s[h_{es}(T_{s\_se}(t) - T_{w\_se}(t)) + h_{re}(T_{s\_re}(t) - T_{w\_re}(t)) + h_{ab}(T_{s\_ab}(t) - T_{w\_ab}(t))]$$
$$\approx h_{es}A_s(T_s(t) - T_{wa}(t))$$

Table 2.8 Some parameters and variables of infant incubator system.

Parameters	Description and values	Variables	Description and values
C_w	Heat capacity of the incubator wall (J.kg^{-1}.°C^{-1})	$T_{xa}(t)$	Temperature electrical heater (°C)
A_w	Incubator wall surface (m^2)	$T_{va}(t)$	Outside surface incubator temperature (°C)
A_s	Infant body skin surface (m^2)	$T_{wz}(t)$	Internal surface incubator temperature (°C)
C_g	Heat capacity of air (J.kg^{-1}.°C^{-1})	$T_{wa}(t)$	Incubator wall mean temperature (°C)
		$x_{hum}(t)$	Humidity outside incubator (%)
λ_w	Thermal conductivity of the incubator wall $\left(\frac{KJ}{kg}\right)$	ϕ	Air ventilation volumetric rate (m^3 s^{-1})
ρ	Density of air (m^3 s^{-1})	$T_a(t)$	Ambient temperature (°C)
K_{ee}	Evaporation coefficient (°C.mm^{-1}.Hg)	$Q_{rs}(t)$	Skin radiative heat transfer (W)
K_2	Conduction coefficient (Kg m^{-3})	$Q_{cs}(t)$	Skin convective heat transfer (W)
h_c	Respiratory rate (s^{-1})	$h_{wa}(t)$	Heat transfer coefficient (kg)
d	Incubator wall thickness (cm)	$Q_h(t)$	Generated heat flux from warmer (kg)
C_p	Specific heat of air (J.kg^{-1}.°C^{-1})	$Q_z(t)$	Heat flux during incubator ports opening (kg)
m_d	Water vapor mass flux from infant body (kg s^{-1})	$h_{wz}(t)$	Heat transfer coefficient (kg)
m_{hum}	Water vapor mass flux from radiant warmer (kg s^{-1})	$x_a(t)$	Incubator chamber humidity (vapor per kg of dry air, %)
m_z	Water vapor mass flux loss during incubator ports opening (kg s^{-1})	$T_i(t)$	Temperature inside incubator (°C)
V	Volume of incubator chamber (m^3)	$T_s(t)$	Geometrical surface mean skin temperature (°C)
C_i	Incubator thermal capacity (J°C^{-1})	$Q_{mi}(t)$	Metabolic heat generated (W)
h_w	Incubator wall convective heat transfer coefficient (W.m^{-2}.°C^{-1})	$Q_{eb}(t)$	Rate of heat lost through respiratory heat evaporation (W)
h_{ij}	Infant body (esophageal, rectal and abdominal) compartments heat transfer coefficient (W.m^{-2}.°C^{-1})	$Q_{cb}(t)$	Rate of heat loss through conduction (W)
		$Q_{rb}(t)$	Heat loss through radiation (W)

with

$$h_r = h_{se} + h_{re} + h_{ab}$$

$T_s(t)$, $T_{wa}(t)$ are, respectively, the average skin temperature (among esophageal, rectal, and abdominal temperatures) and the average wall surface temperature. Here, A_s is the total surface area of the newborn. Similarly, the respiratory heat evaporation from the infant's skin, $Q_{eb}(t)$, into the air under water vapor content is found to be proportional to the incubator rate of respiratory ventilation as well as to the metabolic heat production, $Q_{mi}(t)$, such that:

$$Q_{eb}(t) = K_{ee}h_c A_s(p_s(t) - p_a) + k_{em}Q_{mi}(t)(p_{res} - p_a)$$

p_s denotes the partial pressure of saturated water vapor at mean skin temperature and p_a the partial pressure of the water vapor in the air. Furthermore, the convective heat transfer from the infant's body toward the incubator air is given by:

$$Q_{cb}(t) = h_c A_s(T_s(t) - T_a(t))$$

This heat convective transfer consists of either natural convection, when the movement of air is caused by thermal density gradients, or forced convection due to a pressure gradient. Hence, the heat balance of the incubator wall is given by:

$$C_w \frac{dT_w(t)}{dt} = h_{wa} A_w(T_a(t) - T_{wa}(t)) + h_{wz} A_{wz}(T_z(t) - T_{wz}(t)) + Q_{rs/cb}(t)$$

Similarly, the heat balance of the air within the incubator chamber is given by:

$$\rho C_p V \frac{dT_a(t)}{dt} = h_{wa} A_w(T_{wa}(t) - T_a(t)) + \rho C_p \phi(T_z(t) - T_a(t)) + Q_{cs/cb} + Q_z(t) + Q_h(t)$$

and the mass balance of water vapor is as follows:

$$\rho V \frac{dx_a(t)}{dt} = m_d(x_z(t) - x_d(t)) + m_e + m_{hum} + m_z$$

where

$$m_d = \rho\phi$$

$$T_{wa}(t) = \frac{h_{wa} d T_a(t)/dt + 2\lambda_w T_w(t)}{h_{wa} d + 2\lambda_w}$$

$$T_{wa}(t) = \frac{h_{wa} d T_a(t) + 2\lambda_w T_w(t)}{h_{wa} d + 2\lambda_w}$$

$$T_{wz}(t) = \frac{h_{wz} d T_z(t)/dt + 2\lambda_w T_w(t)}{h_{wz} d + 2\lambda_w}$$

$$T_{wz}(t) = \frac{h_{wz} d T_z(t) + 2\lambda_w T_w(t)}{h_{wz} d + 2\lambda_w}$$

It should be noticed that this model neglects a large number of heat losses from the radiant warmer into the incubator's surrounding environment. In addition, from their design, it has been found that long hourly active radiant warmers cause infant dehydration or at least significant water losses. The mattress can be used to cool the infant's body, hence modifying the metabolic process. Such design requires the development of a new model of respiration heat losses combining the radiant warmer and mattress cooling the infant's body.

2.4.2 Blood Glucose-Insulin Metabolism

The glucose-insulin physiological metabolism regulates the blood glucose (BG) levels essentially in order to keep it at a steady-state concentration of about $70-110 \text{ mg dl}^{-1}$. The inability of the organism to regulate its BG levels due to insufficient insulin production in the pancreas or decreased insulin activity leads to serious health damage or chronic diseases such as diabetes mellitus. It can cause either hyperglycemia (when the BG level is above 270 mg dl^{-1}) due to excessive carbohydrate consumption or a low insulin level in the blood, or hypoglycemia (when the BG level is below 60 mg dl^{-1}) after too much exercise, a large insulin dosage, a low amount of carbohydrates in food, or if the person skips meals.

In the case of diabetes, the assisted regulation of BG requires modeling of the insulin-glucose metabolism. This is challenging due to parameters highly dependent on patient physiology, diet, daily physical conditions and so on. A parametric modeling approach consists of deriving the balance between the mechanisms of production and the elimination of glucose in the

Table 2.9 Some key parameters and variables of glucose and insulin metabolism.

Parameters	Description and values	Variables	Description and values
V_I	Volume of insulin per kg of patient body weight $\left(0.1421\frac{L}{kg}\right)$	$I(t)$	Plasma insulin concentration (mU ml^{-1}) (U/ml)
γ	Constant rate of insulin production by the pancreas (0.0041 mU.mg^{-1}.min^{-1})	$G(t)$	Plasma glucose concentration (mU ml^{-1})
h	Threshold value of glucose level above which pancreas cells secrete insulin (83.7 mg ml^{-1})	$X(t)$	Insulin action from remote compartment (250 mU ml^{-1})
h_n	First-order decay rate for blood insulin (0.22 mU l^{-1})	$G_1(t)$	Deviation blood glucose from basal value (mg ml^{-1}) due to insulin action
p_1	Glucose consumption rate within liver tissues (0.028735 min^{-1})	G_B	Plasma glucose basal value (200 mg dl^{-1})
p_2	Tissue glucose uptake ability at the threshold value of the glucose concentration for decrease of the insulin action (0.028344 min^{-1})	D_I	Intravenous insulin input (250 mU l^{-1})
p_3	Insulin-dependent increase in glucose uptake ability in tissue (5.03.10^{-5} min^{-2}.μU^{-1})	$D(t)$	Glucose intake (diet) in $\left(\frac{mU}{l}/min\right)$
V_G	Volume of glucose per patient body weight (0.22 l kg^{-1})	I_B	Plasma insulin concentration basal value (15 mU ml^{-1})
$m_{patient}$	Weight of a patient (102.2 kg)	$G_x(t)$	Plasma glucose with insulin injection (mg ml^{-1})
V	Volume of blood in a patient (5.5 l)	u_{1b}	Exogenous insulin infusion rate (1/min)
		$U_I(t)$	Rate of insulin absorption (mU min^{-1})

blood using variables and parameters listed in Table 2.9. An example of such an approach is the Bergman minimal model of carbohydrate metabolism. It consists of a three-compartment mathematical model, respectively: (i) the plasma insulin compartment (t) (mU ml^{-1}); (ii) the remote insulin compartment $X(t)$ (mU ml^{-1}), delaying the action of insulin on glucose; and (iii) the plasma glucose compartment $G(t)$ (mg ml^{-1}). Here, some physiological effects from stress or sickness as well as the effects of some physical activities on the metabolism of the glucose-insulin are neglected. As such any glucose addition from the liver or an intravenous insulin injection $U_1(t)$ is not considered. Furthermore, it is assumed that the injection is done from the remote insulin compartment into the circulatory system of the insulin-dependent diabetic patient.

From the plasmatic glucose compartment, the balance between the accumulation of the glucose concentration in the blood plasma and the glucose consumption by the tissue or the kidneys can be obtained from:

$$\frac{dG(t)}{dt} = \text{plasma glucose production–plasma glucose consumption}$$

which corresponds to

$$\frac{dG(t)}{dt} = D(t) - p_1 G(t) - p_2(G(t) - G_B)$$

with quantified carbohydrate input being $D(t)$ given by

$$D(t) = \frac{\text{Glucose absorption rate}}{m_{patient} V}$$

$$G(0) = G_B$$

From this equation, it should be noted that the pancreas stops producing insulin when the glucose concentration $G(t)$ is below glucose level G_B. Hence, this equation is nonlinear and insulin-glucose metabolism-dependent. Similarly, within the plasmatic insulin compartment, the balance between the accumulation of insulin from its production by the pancreas over the destruction of insulin by the insulinase enzyme can be described by:

$$\frac{dI(t)}{dt} = \text{insulin production–insulin degradation}$$

Without an insulin pump, insulin degradation is defined as being proportional to glucose variation such that:

$$\frac{dI(t)}{dt} = \gamma(G(t) - h) - n(I(t) - I_B)$$

$$I(0) = I_B$$

With an insulin pump and in the case of exogenous injection of insulin, this yields:

$$\frac{dI(t)}{dt} = \frac{U_1(t)}{V} - n(I(t) - I_B)$$

with

$$U_1(t) \approx \frac{2t^2 G_I}{t(0.05 D_I + 1.2)}$$

Hence, the insulin-glucose regulation dynamic model could be expressed as:

$$\frac{dG_X(t)}{dt} = D(t) - p_1 G(t) - p_3(G(t) - h) - p_2(h - G_B)$$

with

$$G_X(0) = G_B$$

This model is derived from a physiological understanding of the metabolism. Another approach would be to evaluate several model structures and orders, then estimate the resulting error in the process model. Here, $p_1, p_2, p_3, G_B,$ and I_B are key model parameters to be estimated based on patient data, as summarized in Table 2.10.

Here, a day's diet scenario (24 hour period) illustrated by glucose level disturbance is considered, which is distributed over three daily meals with peak values of carbohydrates (55, 80, and 65 mg dl^{-1}) corresponding to plasma glucose concentrations of 250, 310, and 280 mg dl^{-1}, respectively, for breakfast at 7.00 a.m., lunch at 12.30 p.m., and dinner at 7.00 p.m. At these meal times, it is assumed that: (i) there is an increase in glucose disturbance levels after the

Table 2.10 Typical patient information.

Patient	Gender	Age	Weight (kg)	Height (cm)	Data length (h)
1	Male	42	98	178	24

Table 2.11 Patient data.

Patient 1												
Time (hour)	3	6	7	8	9	12	13	14	16	19	20	23
Plasma glucose $G(t)$ (mg dl^{-1})	200	200	250	235	210	310	280	230	200	280	262	200
Plasma insulin $I(t)$ (mU l^{-1})	00	00	250	00	00	250	00	00	00	250	00	00
Plasma glucose with insulin injection $G_X(t)$ (mg dl^{-1})	200	200	236	216	202	291	269	218	200	267	244	200
Carbohydrates (g)			55			80				65		

Table 2.12 Model parameter estimates.

Model parameters	p_1	p_2	p_3	G_B	I_B	n	γ	h
Estimated value	0.028735	0.028344	0.00005035	200	15	0.22	0.0041	83.7

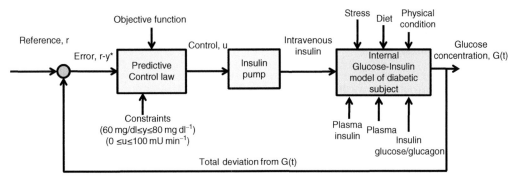

Figure 2.8 Glucose-insulin metabolism under a model prediction control paradigm.

consumption of each meal; and (ii) the BG returns to basal glucose level about 3 hours later. Equivalently, there are three regular insulin injections of 250 mU ml^{-1} at meal times, as summarized in Table 2.11.

The least-square parameter estimation method based on normalized data is presented in Appendix A. Autocorrelation of residuals techniques such as the Akaike Information Criterion (AIC) are used in the determination of the best model order. Results are listed in Table 2.12.

Patient physiological conditions such as stress, physical activity, and disease have a significant effect on glucose variation, but are rather difficult to measure and model as depicted in Figure 2.8. Thus, the integration of those physical constraints in the model would lead to a highly nonlinear and discontinuous or scenario-based stochastic model. Any such gray model is patient-dependent.

2.5 Fluidic and Thermal Systems and Processes

2.5.1 Mixing Tank

Usually, in beverage and pharmaceutical processes, it is necessary to mix two fluidic substances in a tank. This requires a precise angle and duration of electrovalve opening to ensure liquid

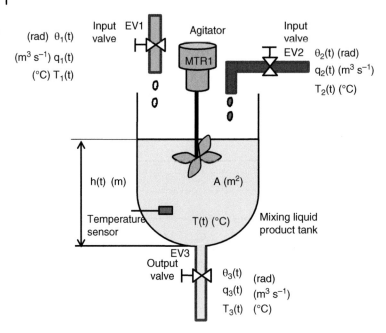

Figure 2.9 Mixing tank system.

mixture at the desired level. As illustrated in Figure 2.9, the mixing of two liquids with a slight temperature difference is achieved so that this process does not generate heat or a significant temperature variation. However, the two liquids have different densities, so the density of the mixture depends linearly on the volume fractions of both components. Such a process relates each liquid temperature, its density, and each electrovalve opening angle to a specific level of mixed liquid concentration and temperature. This can be achieved by using an MIMO dynamics model using the process parameters and variables listed in Table 2.13. During the mixing process, it is assumed that the fluid is ideally mixed as a homogeneous liquid concentration within a small fixed time interval. Also, any temperature rises due to the heat generated from the mixing process are considered negligible.

As such, it is required to investigate the behavior of the liquid concentration due to changes in throughput and mixing ratio. As the fluid temperatures are nearly equal, the mixing process quantified in terms of the liquid level in the tank $h(t)$ and the volumetric fraction of the mixture $x(t)$ can be obtained through either: (i) the total mass balance; (ii) the energy balance; or (iii) the component balance. During the mixing of liquid by the motorized agitator, there is a biochemical reaction that generates a heat constant, W_s. Hence, using the energy balance (enthalpy, H), the mixing process model yields:

$$\frac{\partial H(t)}{\partial t} = \frac{\partial H_{gen}(t)}{\partial t} + \frac{\partial H_{in}(t)}{\partial t} - \frac{\partial H_{out}(t)}{\partial t}$$

which is equivalent to:

$$C_{pm}\frac{\partial m(t)T_m(t)}{\partial t} = W_S + W_1(t)C_{p1}T_1(t) + W_2(t)C_{p2}T_2(t) - W_3(t)C_{p3}T_3(t)$$

Assume that

$$m(t) = \rho_m A h(t)$$

Table 2.13 Mixing tank process variables and parameters.

Parameters	Description and values	Variables	Description and values
C_{pi}	Heat capacity of ith liquid (kJ kg^{-1}°C)	$T_i(t)$	Liquid temperature from ith electrovalve (°C)
ρ_i	Density of ith liquid (kg m^{-3})	$T_m(t)$	Mixed liquid tank temperature (°C)
A	Conical base of the mixing tank (m^2)	$h(t)$	Height of liquid in mixing tank (m)
J	Motor inertia (N.m.s^2 rad^{-1})	$q_i(t)$	Flow rate from ith electrovalve (m^3 s^{-1})
C_d	Outflow coefficient (%)	$\omega(t)$	Rotational speed of motor agitator $\left(\dfrac{rad}{sec}\right)$
K_i	Gain from ith electrovalve (°C)	$v_3(t)$	Liquid speed at EV3 (m s^{-1})
W_s	Energy supplied to motor (kJ s^{-1})	$\theta_i(t)$	Angular position of ith electrovalve (°C)
R_a	Motor armature resistance (Ω)	$m(t)$	Mass of liquid in mixing tank (kg)
L_a	Motor armature inductance (H)	$W_i(t)$	Mass flow rate from ith electrovalve (kJ s^{-1})
V_T	Volume of tank (0.8836 m^3)	$x(t)$	Volume fraction of the mixture (%)
C_{pm}	Mixed liquid heat capacity (kJ kg^{-1}.°C)	$V(t)$	Volume of liquid in the tank (m^3)
ρ_m	Stationary density of mixed liquid (kg m^{-3})		

and recall that

$$\frac{W_i(t)}{\rho i} = q_i(t)$$

Here, the inflow temperatures are $T_1(t) = 81°C$ and $T_2(t) = 80°C$, and the steady-state inflow rates are $q_1(t) = 0.06$ m^3 s^{-1}, $q_2(t) = 0.02$ m^3 s^{-1}. Thus, the balance energy equation becomes:

$$C_{pm}\rho_m A \left(\frac{dh(t)}{dt}T(t) + \frac{dT(t)}{dt}h(t)\right)$$
$$= W_S + C_{p1}\rho_1 q_1(t)T_1(t) + C_{p2}\rho_2 q_2(t)T_2(t) - C_{p3}\rho_3 q_3(t)T_3(t)$$

which could be simplified to:

$$\frac{dh(t)}{dt}T(t) + \frac{dT(t)}{dt}h(t)$$
$$= \frac{1}{C_{pm}\rho_m A}(W_S + C_{p1}\rho_1 q_1(t)T_1(t) + C_{p2}\rho_2 q_2(t)T_2(t) - C_{p3}\rho_3 q_3(t)T_3(t))$$

Similarly, the mass balance equation is given by:

$$\frac{dm(t)}{dt} = \rho_m A h(t) = F_1(t)\rho_1 + F_2(t)\rho_2 - F_3(t)\rho_3(t)$$

while for component balance, the equation yields:

$$\frac{dV(t)}{dt} = F_1(t) + F_2(t) - F_3(t)$$

Thus, the density balance equation results in:

$$Ah_{eq}\frac{d\rho_3(t)}{dt} = F_1(t)(\rho_1 - \rho_3(t)) + F_2(t)(\rho_2 - \rho_3(t))$$

The volume fraction is given by:

$$x(t) = \frac{F_1(t)}{F_1(t) + F_2(t)}$$

Considering that each valve displays a nonlinear behavior, the relationship between the valve opening angles and flow rates is given by:

$$q_{1,2}(t) = K_{1,2}\sqrt{\theta_{1,2}(t)}$$

$$q_3(t) = C_d K_3 \sqrt{\theta_3(t)} v_3(t)$$

while the mass balance flow through the tank yields:

$$A\frac{dh(t)}{dt} + C_d K_3 \sqrt{\theta_3(t)} = K_1\sqrt{\theta_1(t)} + K_2\sqrt{\theta_2(t)}$$

In addition, the agitator energy is such that:

$$W_s = J\omega(t)^2$$

Hence, in order to derive the height of liquid in the tank $h(t)$ as a function of the angular position of each electro valve $\theta_{1,2}(t)$, the balance equation is equivalent to:

$$A\left[\frac{dh(t)}{dt}T(t) + \frac{dT(t)}{dt}h(t)\right] = \frac{C_{p1}}{C_{pm}}K_1\sqrt{\theta_1(t)}T_1(t) + \frac{C_{p2}}{C_{pm}}K_2\sqrt{\theta_2(t)}T_2(t)$$

$$-\frac{C_{p3}}{C_{pm}}C_d K_3\sqrt{\theta_3(t)}v_3 T_3(t) + \frac{J\omega(t)^2}{C_{pm}\rho_m}$$

The liquids are fully mixed in the tank after a time delay of $D = 10$ seconds and without any step change of the level of liquid in the tank. The nonlinear liquid mixing process model yields:

$$A\left[\frac{dh(t)}{dt}T(t) + \frac{dT(t)}{dt}h(t)\right] = \frac{C_{p1}}{C_{pm}}K_1\sqrt{\theta_1(t)}T_1(t - D) + \frac{C_{p2}}{C_{pm}}K_2\sqrt{\theta_2(t)}T_2(t - D)$$

$$-\frac{C_{p3}}{C_{pm}}C_d K_3\sqrt{\theta_3(t)}v_3 T_3(t) + \frac{J\omega(t)^2}{C_{pm}\rho_m}$$

$$A\left[\frac{dh(t)}{dt}T(t) + \frac{dT(t)}{dt}h(t)\right] = \frac{C_{p1}}{C_{pm}A}K_1\sqrt{\theta_1(t)}T_1(t - D) + \frac{C_{p2}}{C_{pm}}K_2\sqrt{\theta_2(t)}T_2(t - D)$$

$$-\frac{C_{p3}}{C_{pm}}C_d K_3\sqrt{\theta_3(t)}v_3 T_3(t) + \frac{J\omega(t)^2}{C_{pm}\rho_m}$$

Considering that homogeneity is achieved by the motor agitator and the outflow valve is not opened, the linearization of the balance equation around operating points, $(h_e, \theta_{1,e}, \theta_{2,e}, \theta_{3,e}, T_e)$ using the Taylor series expansion, yields:

$$h_e\delta\frac{dT(t)}{dt}T(t) - \sqrt{2g}\left(\frac{1}{2\sqrt{h_e}}\delta h(t) + h_e\right)\delta T(t)$$

$$= K_1\left(\frac{1}{2\sqrt{\theta_{1,e}}}\delta\theta_1(t) + \sqrt{\theta_{1,e}}\right)(T_1(t) - D/1)$$

$$+ K_2\left(\frac{1}{2\sqrt{\theta_{2,e}}}\delta\theta_2(t) + \sqrt{\theta_{2,e}}\right)(T_2(t) - D/1)$$

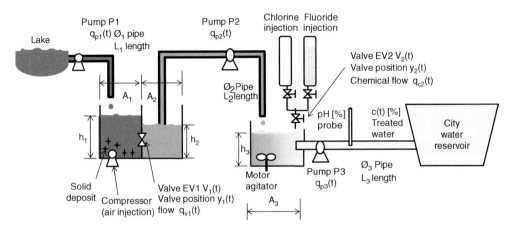

Figure 2.10 Water treatment and distribution process schematic.

2.5.2 Purified Water Distribution Process

The process of public water chemical treatment and distribution consists of a water pre-treatment stage followed by a water purification and distribution stage. First, non-purified water is uploaded from the river or lake into pre-treatment large-scale reservoirs where solid deposits are separated and removed. This is expected to take 3–7 days. Then, this pre-treated water is mixed with chemicals (such as chlorine, fluoride, or ozone oxidation) for purification. Eventually, from these treatment reservoirs, a system of pumping stations and transportation pipelines and valves convey this water to the city for public distribution, as illustrated in Figure 2.10. Such a public water distribution process must simultaneously satisfy the demand on the quantity of water estimated from its level of water at the storage station. In addition, this should meet the quality requirements of minimal water pressure at end points throughout the distribution subprocess (to prevent contamination from ground water) as well as for fire outlets. Furthermore, this process should meet water safety standards in terms of purity (such as the percentage of any residual disinfectant or cancer-causing chemicals). It is assumed that: (i) there is a physical removal of impurities through chemical treatment; and (ii) the filtering, coagulation, flocculation, and sedimentation are enough to purify the water of most microorganisms and chemical disinfectants.

Typical water treatment and distribution facilities are made of interconnected and interdependent components including: (i) reservoirs as the source of water for the process; (ii) pumping stations to raise pressure in order to circumvent drops due to elevation differences and losses due to friction; (iii) tanks for the storage of excess water within the process; and (iv) pipes and valves to regulate end water pressure and the flow rates. Hence, the modeling of the water treatment and distribution system consists of measuring the dynamics of the water volumetric flow rate, its height at the reservoirs, and the pressure along the pipes while considering constraints on water quality and quantity. Such processes can be partitioned into subprocesses (pre-treatment, inter-tank pumping, chemical treatment, pumping for city distribution). Process variables and parameters to be used are summarized in Table 2.14.

After identifying inputs and outputs of each subcomponent, each subprocess can be modeled as follows:

Reservoirs and storage tanks from the water supply subprocess are used: (i) to store a high volume of water; and (ii) to elevate water in order to keep the pressure balances over the distribution process and to ease the water flow. This model mainly captures water surface elevation.

Table 2.14 Water treatment process variables and parameters.

Parameters	Description and values	Variables	Description and values
$A_{1,2}$	Cylindrical area of tanks 1,2 (m)	$q_{p1}(t)$	Rate of liquid flow out of lake into tank 1 ($m^3\ s^{-1}$)
p_i	Valve position ($0 \le p \le 1$)	$q_{p2}(t)$	Rate of liquid flow out of tank 2 into pipe ($m^3\ s^{-1}$)
$\rho_{1,2,p}$	Fluid density (kg m^{-3})	$q_{v1}(t)$	Rate of liquid flow through valve between tanks 1 and 2 ($m^3\ s^{-1}$)
K_μ	Pipe friction coefficient (*dimensionless*)	$q_{p3}(t)$	Rate of liquid flow rate through pump out of mixing tank into pipe ($m^3\ s^{-1}$)
L	Pipe section length (m)	$v(t)$	Valve flow velocity ($m^3\ s^{-1}$)
A_p	Pipe section area (m^2)	$h_{ti}(t)$	Height of liquid in the ith tank (m)
ϕ_1	Sectional pipe 1 diameter (m)	$q_p(t)$	Pump flow rate ($m^3\ s^{-1}$)
K_v	Minor loss coefficient	$q_i(t), q_o(t)$	Water reservoir inflow outflow ($m^3\ s^{-1}$)
D	Travel time delay (s)	$h(t)$	Water reservoir head (m)
V	Volume of reservoir (m^3)	$q_{ij}(t)$	Water flow within node i to j ($m^3\ s^{-1}$)
D	Head loss steady-state value	$h_{loss}(t)$	Head loss (m)
D_{ij}	Inner diameter pipe between $i\,j$ (m)	$h_{pump}(t)$	Head developed by the pump (m)
S_T	Reservoir area (m^2)	$\omega_{motor}(t)$	Motor speed (rad s^{-1})
$h_v(t)$	Level of the reservoir valve (rad)	$\omega_{pump}(t)$	Pump shaft speed (rad s^{-1})

Using the principle of continuity, the change of water volume within the reservoir (with no leaks) is given by:

$$\rho\frac{dV(t)}{dt} = \rho_i q_i(t) - \rho_o q_o(t)$$

where ρ, ρ_i, ρ_0 are the specific density of the water in the reservoir. Assuming that $\rho_i = \rho_o = \rho$, the tank volume $V(t)$, which is dependent on the reservoir cross-sectional area A_b and the height level h_t, is given by:

$$V(t) = A_b h_t(t)$$

Substituting this into the previous equation, the height change yields:

$$\frac{dh_t(t)}{dt} = \frac{q_i(t) - q_o(t)}{A_b h(t)}$$

with $A(h(t))$ denoting the area for the water head $h(t)$ at time instant t. Here, tanks are characterized by parameters such as the capacity (diameter, height etc.) and the operational constraints defining the water level range such as the minimum and the maximum allowed water level. The flow between tanks connected by a valve depends on the level differential between them. Therefore, from Figure 2.10, it is assumed that the inter-reservoir flow is given as:

$$q_{v1}(t) = K_1 \sqrt{\left|2h_1(t) - \sum_{n=1}^{2} h_n\right| sign\left(2h_n - \sum_{n=1}^{2} h_n\right)}$$

After substituting, the dynamic equation for the reservoir is given by:

$$\frac{dh_{t1}(t)}{dt} = \frac{q_{p2}(t)}{A_b} - K_1 \sqrt{\left|2h_1(t) - \sum_{n=1}^{2} h_n\right| sign\left(2h_n - \sum_{n=1}^{2} h_n\right)}$$

Aqueducts such as canals, tunnels, or pipelines are used for water transportation between the distribution network nodes. Between any two nodes of this network, there is energy dissipation within pipes due to friction losses. Assuming that any pressure loss due to leakage is negligible, the dynamic equation describing the head-flow relationship and the friction head loss along a pipe relating node i to node j is given by:

$$\frac{dq_{ij}(t)}{dt} = \frac{gA_b}{L}(\Delta h_{ij} - h_{loss}(t))$$

$q_{ij}(t)$ is the water flow between node i and node j of the pipe with a section length L and an area A_p. Here, Δh_{ij} is the head difference between two ends of a pipe section while $h_{loss}(t)$ is the head loss. Using Bernoulli's theorem, under stationary conditions, the relationship between water velocity $v(t)$ in/out of a reservoir, gravity acceleration g, and reservoir water level $h_t(t)$ is given by:

$$v(t) = \sqrt{2gh_t(t)}$$

Assume that: (i) the main component of head loss in the pipe $h_{loss}(t)$ is related to the pipe contraction or expansion as well as the pipe bending angle; and (ii) the change in magnitude of the water velocity is negligible for long pipes (greater than 100 units in diameter). Then, the local head losses due to pipe expansions, contractions, or bends can be expressed as:

$$h_{loss}(t) = h_{ss\_loss} + \left(\frac{K_\mu}{2gA_p^2}\right)qij(t)^2$$

with h_{ss_loss}, K_μ, D_{ij} being head loss steady-state values, the friction coefficient, which depends on the physical conditions (such as inner surface roughness) and the inner diameter of the pipe section between i and j. The water is assumed to have a uniform density in the pipe (no cavitation) and the valves prevent bidirectional water flow, such that friction coefficient for a pipe surface for with a given roughness ε, shape, and flow can be calculated by:

$$\frac{1}{\sqrt{K_\mu}} = -4\log_{10}\left(\frac{\varepsilon/D_{ij}}{3.71} + \frac{2.51}{2\sqrt{2f_p N_R}}\right)$$

f_p is the friction coefficient, such that $K_\mu = 2f_p$, and N_R is the Reynolds number.

Reservoir valves are used to regulate the water flow rate or the pressure out of the reservoir. The modeling of valves depends on their size and type (i.e. opening and closing speed), so a dynamics model for the reservoir valve could be given by:

$$\frac{dh_v(t)}{dt} = -k\sqrt{h_v(t) + h_o}$$

where $h_v(t)$ is the level of the reservoir valve and k is a constant that is given by:

$$k = p_i\frac{s_{v\_max}\sqrt{2g}}{S_T}$$

s_{v_max} is the valve opening area equivalent to the area of the connected pipe and s_T is the reservoir area. As such, the head-flow dynamics model between two successive network nodes is given by:

$$q_{ij}(t) = V_{ij}G_{ij}|h_i - h_j|^{\alpha} \, sign(h_i - h_j)$$

where $0 \leq V_{ij} \leq 1$ is proportional to the opening angle of the valve, 0 when closed. Under stationary conditions, minor losses involved in the valve are given by:

$$h_m = \frac{K_v v(t)^2}{2g}$$

where $v(t)$ is the flow velocity, while minor loss coefficient K_v is estimated experimentally.

Pumps transfer the energy from the attached motor shaft into the water flow in order to overcome losses due to friction in pipes, valves, and any head difference in a distribution network. Assuming that the pump shaft speed $w_{pump}(t)$ and the motor speed $w_{motor}(t)$ are proportionally linearly dependent on the flow rate, $q_p(t)$, the total head developed by the pump $h_{pump}(t)$ yields:

$$h_{pump}(t) = a_o w_{motor}(t)^2 + b_o \frac{w_{motor}(t)}{w_{pump}(t)} q_p(t) - \frac{c_o}{w_{pump}(t)^2} q_p(t)^2$$

a_0, b_0, c_0 are pump constants that can be provided from the manufacturer's flow rate-speed characteristics. Hence, the *dynamics model pump-reservoir-valve-pipe distribution process* could be derived using the principle of mass conservation, such that the change in the water level of the reservoir is the flow entering the subprocess through, $q_{p1}(t)$ less the flow $q_{s1}(t)$ flowing out of the outlet valve or out of the interconnecting reservoir valve q_1, as illustrated in Figure 2.10. It could be obtained by:

$$\frac{dh_t(t)}{dt} = \frac{q_{p1}(t)}{A_b} - q_{s1}(t) - q_1(t)$$

For *chemical water treatment*, it could be assumed that both the water and the chemicals are fully mixed conforming to safety requirements. It is essential to maintain a specific chemical content in treated water, $c(t)$ [%] and the combined mixing chemical valve model could be given by:

$$\frac{1}{\omega_n^2} \frac{d^2 q_{c2}(t)}{dt^2} + \frac{2\zeta}{\omega_n} \frac{dq_{c2}(t)}{dt} + q_{c2}(t) = K_{cv} V_2(t)$$

The concentration of mixture (water mixed with chemical expedients), $c(t)$ is given as:

$$c(t) = \frac{q_{c2}(t)}{(q_{c2}(t) + q_{p2}(t)) - q_{p3}(t)}$$

with

$$q_{p2}(t) = a_2 h_2(t)$$

$$(q_{c2}(t) + q_{p2}(t)) - q_{p3}(t) = A_3 \frac{dh_3(t)}{dt}$$

Depending on whether the regulation is made through the chemical valve or pump P3, this equation could be subsequently simplified. Then, it is possible to integrate all equation components and to derive a suitable dynamics model of water purity.

2.5.3 Conveyor Cake Oven

Food manufacturing processes with high production rates use a conveyor oven for thermal cooking through the biochemical reaction of starchy food products such as cookies, cakes,

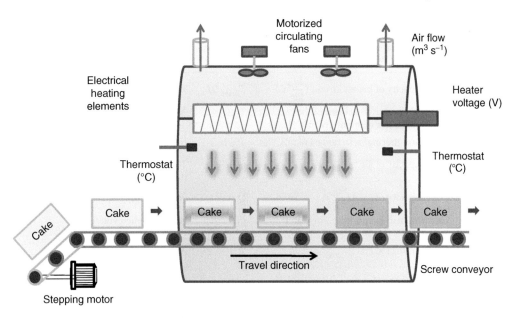

Figure 2.11 Cake conveyor-oven system.

bread, and biscuits. Such an oven can be decomposed into a three-stage baking process (preheating, cooking, cooling). Only the modeling of the cooking stage is covered in this study. In this stage, the motor-driven conveyor belt carries cakes toward an electrically heated oven. The oven temperature defines the drying time to cook cakes by convection. And depending on the cake size and production rate, the heater voltage can adjust temperatures accordingly (between 218 and 250°C). In addition, fans located at the rear of the oven move the air through the heating elements in order to increase heat homogeneity within the oven chamber, as illustrated in Figure 2.11. The changes in the baking temperature and airflow velocity cause significant variation in oven temperature. This is due to water evaporation, volume expansion, protein transformation, and even crust formation during the biochemical reaction. As such, this characterizes baking products in terms of product dimensions and shape (e.g. a volcano-like shape), moisture content, color (i.e. caramelization), and even taste, all being dependent on the oven shape and operating conditions (e.g. conveyor speed). This is a consequence of simultaneous heat and mass transfers within the product and inside the conveyor oven. Thus, in order to achieve a desired quality in the final product according to the production rate, it is suitable to model the internal temperature and the humidity of the conveyor oven.

It is assumed that: (i) the conveyor oven speed, the air temperature, and humidity are homogeneous during the baking process within the conveyor chamber; (ii) the cake exposure surface is negligible compared to the inside walls of the oven area, and both have negligible radiation properties; (iii) the air turbulence effects around the oven entrance and exit are not considered; (iv) the oven walls exposed to the outside are properly insulated; (v) the oven chamber and conveyor plate have uniform air density and humidity as well as constant air velocity; (vi) the emissivity of the oven walls and the conveyor plate is considered negligible; (vii) the temperature of the cake is uniform over its entire surface in all three dimensions and unrelated to heating time and position; (viii) there is a constant cake thermal diffusivity α with respect to time and temperature; and (ix) cakes have common shapes (cylindrical, spherical, etc.).

Consider an electrical convection tunnel oven with a working volume of 70.5 l and outer dimensions of $80 \times 65.5 \times 48 \, \text{cm}^3$. It should produce standard baked cakes with a temperature of

Table 2.15 Conveyor oven system parameters and variables.

Parameters	Description and values	Variables	Description and values
C_h	Heat capacity of heater circuit ($\frac{kJ}{kg}°C$)	$T_{heater}(t)$	Temperature electrical heater (°C)
R_{h0}	Thermal resistance to oven of the electrical heater element (W m°C^{-1})	$T_{air}(t)$	Ambient air fluid temperature (°C)
R_0	Insulation thermal resistance (W.m°C^{-1})	$T_o(t)$	Internal oven temperature (°C)
C_g	Heat capacity of air ($\frac{kJ}{kg}°C$)	$\frac{dT_{cake}(t)}{dx}$	Temperature gradient across unit thickness x
C_s	Heat capacity of cake ($\frac{kJ}{kg}°C$)	Q	Power supply to heater (4 kW)
λ	Factor of heat evaporation for water (kJ kg^{-1})	$E(t)$	Heater applied voltage
R_ω	Drying rate (1 s^{-1})	$S(t)$	Absorbed electrical heater convection (J m^{2}h^{-1})
F_s	Linear density of drying air (kg m^{-1})	$\omega(t)$	Rotational speed of drum of conveyor (m s^{-1})
F_g	Linear density of cake (kg m^{-1})	$T_{cake}(t)$	Cake surface temperature (°C)
U_V	Volumetric heat transfer coefficient ($\frac{KJ}{°C \cdot m^3 \cdot S}$)	$T_{wall}(t)$	Oven wall temperature (°C)
C_p	Oven specific heat (J.kg°C^{-1})	$q_{heater}(t)$	Quantity of heat energy generated by the electrical heater (J h^{-1})
$\dot{m}_{air}(t)$	Hot air mass flow rate (kg s^{-1})	$C_{cake}(t)$	Starchy cake moisture content (m^3 m^3)
ρ_{cake}	Cake volumetric mass (kg.m^3)	$C_o(t)$	Oven humidity or moisture content (m^3 m^3)
C_{cake}	Cake specific heat (J.kg°C^{-1})		
V_{cake}	Total volume of cake (m^3)		
$A_{conveyor}$	Conveyor inside oven surface (m^2)		

93°C at its center. The baking process combines heat and mass transfer phenomena. Heat transfer can be decomposed into: (i) radiation from the heating temperature to the air temperature inside the oven as well as from the oven walls to the air; (ii) convection inside the oven from the movement of hot air interacting with moisture on the cake surfaces; and (iii) conduction from the iron-based conveyor carrier to the cake surfaces. It should be noted that during the baking process, the cake humidity (moisture) is diffused into the oven. Similarly, water transfer results from the evaporation of the water from the cake into the air causing a water vapor–air movement in the oven.

Using parameters and variables listed in Table 2.15, it is possible to capture transient and stationary behavior of the baking and drying process based on the heater and cake temperatures as well as the air moisture within the oven chamber. Hence, from the heat transfer decomposition here, using the energy conservation principle, the thermal energy balance on the conveyor platform area is given by:

Heat energy injected + Heat energy generated = Heat energy stored − Heat energy losses

Also, the thermal energy balance of the conveyor-oven system yields:

$$\frac{dq_{heater}(t)}{dt} + \frac{dq_{generated}(t)}{dt} = \frac{dq_{stored\ cake}(t)}{dt} - \frac{dq_{wall}(t)}{dt}$$

While neglecting heat loss through the exhausted air, the quantity of heat injected by the heating element and collected by the oven is given by:

$$\frac{dq_{\text{heater}}(t)}{dt} = \dot{m}_{\text{air}}(t)C_p(T_{\text{heater}}(t) - T_{\text{air}}(t))$$

Similarly, the heat stored within the cake is given by:

$$\frac{dq_{\text{stored cake}}(t)}{dt} = \rho_{\text{cake}} V_{\text{cake}} C_{\text{cake}} \frac{dT_{\text{cake}}(t)}{dt}$$

During phase changes, the cake transforms chemical energy into thermal energy. The generated energy is given by:

$$\frac{dq_{\text{generated}}(t)}{dt} = qV_{\text{cake}}$$

with q being the volumetric density of generated energy. The hot air injected in the oven chamber heats the walls and the conveyor, thus transferring heat toward the cake either by convection, conduction, or radiation. The fans in the baking oven chamber move the air, causing forced convection. Therefore, it should be noted that natural convection is small compared to forced convection, so that the air flow characteristics display a laminar boundary layer around the cake surface. Hence, from Newton's law, the rate of the convective heat transfer is proportional to the surface area of the cake and the temperature difference between the cake surface and air, which is given by:

$$\text{Heat energy generated convection} = hA_{\text{cake}}[T_{\text{cake}}(t) - T_{\text{air}}(t)]$$

In the case of forced convection, the heat transfer coefficient through convection ($\text{W.m}^2 \,{}^\circ\text{C}^{-1}$) is given by:

$$h = 5.7 + 3.9v$$

This is valid for the air speed $v < 0.5 \text{ m s}^{-1}$ and with a distance between the heating element and the cake below of 25 cm. A_{cake} is the cake surface area in contact with the hot air within the oven chamber. The net heat transfer generated by radiation toward the cake is affected by the heater's emitting surface and its relative position. This is given by:

$$\text{Heat energy generated radiation} = \sigma \varepsilon_p A_{\text{heater}}(T_{\text{heater}}(t)^4 - T_{\text{cake}}(t)^4)$$

with σ being the Boltzmann–Stefan constant ($5.67 \ 10^8 \text{ W.m}^{-2}.\text{K}^{-4}$), ε_p is the emission coefficient of the cake surface, while A_{heater} and T_{heater} are, respectively, the cake surface area and the temperature of the combination of the heater and inner wall. The emissivity of the black object varies from 0 to 1. It is considered that, during the baking process, the boiling heat transfer causes the phase change in the cake as the water evaporates. But with a low temperature gradient, the irradiative heat transfer phenomenon is negligible. The rate of heat transferred by conduction is determined by the temperature difference between the cake and the heating medium as well as the conducting medium's resistance to heat transfer. Based on one-dimensional analysis of the cake shape and using Fourier's law of heat conduction, the heat transferred would be given by:

$$\frac{dq_{\text{generated rad}}(t)}{dt} = -kA_{\text{conveyor}} \frac{dT_{\text{cake}}(t, x)}{dx}$$

with k being the conveyor's thermal conductivity, while A_{conveyor} is the cake area in contact with the conveyor, $\frac{dT_{\text{cake}}(t)}{dx}$ is the temperature gradient across unit thickness x (negative sign because temperature decreases as thickness increases).

Similarly, the heat losses from the cake to the oven walls as well as from the oven walls to the environment are given by:

$$\frac{dq_{\text{wall}}(t)}{dt} = K_{\text{cake}}(T_{\text{cake}}(t) - T_{\text{wall}}(t)) + \frac{T_o(t) - T_{\text{wall}}(t)}{R_f}$$

with R_f being the thermal resistance of the oven walls or insulation thermal resistance, K_{oven} being the insulation coefficient, and $T_o(t)$ and $T_{\text{wall}}(t)$ being the oven wall temperatures. Usually, $K_{\text{cake}} \geq \frac{1}{R_f}$ due to the fact that the heat transfer from the water is much better than through the insulation, so it could be assumed that $T_{\text{wall}}(t) = T_{\text{cake}}(t)$. Hence the net heat transfer equation is given by:

$$\dot{m}_{\text{air}}(t)C_p(T_{\text{heater}}(t) - T_{\text{air}}(t)) + hA_{\text{cake}}(T_{\text{cake}}(t) - T_{\text{air}}(t))$$

$$+ \sigma\varepsilon_p A_{\text{heater}}(T_{\text{heater}}(t)^4 - T_{\text{cake}}(t)^4) - kA_{\text{conveyor}}\frac{dT_{\text{cake}}(t)}{dt}$$

$$= \rho_{\text{cake}}V_{\text{cake}}C_{\text{cake}}\frac{dT_{\text{cake}}(t)}{dt} + qV_{\text{cake}} - K_{\text{cake}}(T_{\text{cake}}(t) - T_{\text{wall}}(t)) - \frac{T_o(t) - T_{\text{wall}}(t)}{R_f}$$

Initially, at the beginning of the baking process, the starchy baking product is assumed to have a moisture content equivalent to the water added during the dough mixing stage. This moisture content decreases proportionally to the baking time and is called the drying rate. It is assumed to be constant. However, during this drying time period, there is water and heat diffusion from the oven chamber over the air inside. Hence the oven temperature and oven humidity or the moisture distribution model can be derived from the heat diffusion over air and water diffusion models, which yield:

$$\rho_{\text{cake}}C_{\text{cake}}\frac{\partial T_o(t)}{\partial t} = \frac{1}{nb_{\text{cake}}}\left[\frac{h_t}{A_{net}}(T_o(t) - T_{\text{cake}}(t)) + \rho_{\text{air}}\lambda_v\frac{\partial C_o(t)}{\partial t}\right]$$

and the moisture equation is given by:

$$\frac{\partial C_o(t)}{\partial t} = \frac{1}{nb_{\text{cake}}}[h_m(T_o(t) - T_{\text{cake}}(t)) + \alpha_1(\sigma\varepsilon_p(T_{\text{cake}}(t)^4 - T_o(t)^4))]$$

This one-dimensional baking process model, relating the heat and mass transfers with the cake volume expansion for various baking conditions and bakery recipes, is not covered. The coupling between the cake water and temperature is not included in these dynamic equations. More importantly, some model parameters such as heat of evaporation, convective heat and water transfer coefficients, and heat latency are difficult to estimate. However, some simplification could be made based on the baking process analysis such that convection and conduction heat transfer are negligible compared with radiation heat transfer, making the precision of the values of the absorption coefficients of the cake, oven wall surfaces, and air/steam mixtures highly important in process modeling. Furthermore, it is suitable to use correct initial and boundary oven environmental and operating conditions in order to obtain a refined model capturing the transient temperature and the water distribution in the cake through an advanced 3D cake model. This is also valid for the heat, water transfer, and flow modeling using a 3D baking chamber. Thus, the overall laws of conservation for mass, momentum, energy, and water could be applied. This model should consider operating conditions such as the airflow velocity as it affects the temperature and product texture.

2.5.4 Poultry Scalding and Defeathering Thermal Process

An automatic poultry scalding and defeathering process is illustrated in Figure 2.12. Here, supported by a motor-driven belt, the feathers are plucked from poultry by rotating plates. After

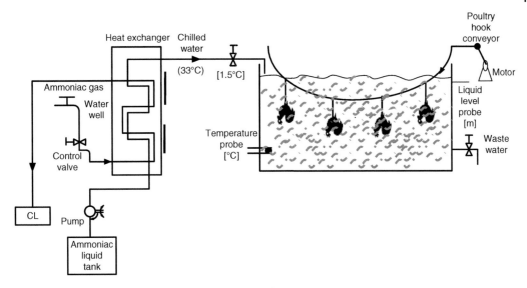

Figure 2.12 Poultry processing system with ammoniac flow.

this defeathering substation, the poultry is conveyed toward a tank filled with chilled water at a temperature around 5.5°C. At this stage, the debris on the poultry is rinsed off in running potable water, then a chemical pre-treatment is applied for meat processing before undergoing any chilling conservation. Poultry carcasses are immersed for about 3 minutes to be rinsed off in water with a chlorine content of 50–100 ppm. This is an antimicrobial processing stage critical for human pathogen control (i.e. by lowering the count of such as campylobacter, responsible for food-borne illness). A −3°C inflow of ammoniac over a heat exchanger lowers the inflow water temperature in the tank from around 6–13°C. The variation in the production volume and the poultry immersion time defines the thermal processing results. Hence, in order to meet the safety requirements compliant with product quality standards, it is necessary to predict the tank temperature despite variable production rate or poultry size. Furthermore, the tank temperature varies based on: (i) the poultry body temperature; (ii) the volume and processing rate of poultry; (iii) the heat exchanger cooling response time constant; and (iv) the heat loss through the tank walls. Process variables and parameters are listed in Table 2.16. It is assumed that: (i) the pipe-cooled water flow between the heat exchanger and the tank has negligible time delay; (ii) the heat losses along the wall and the pipe are not considered; (iii) there is a low gradient of pressure between the pipe inlet and outlet; (iv) the heat losses over the air above the tank are negligible; and (v) the heat capacity of the waste water is nearly equal to the heat capacity of the running potable water.

The occurring heat transfers are from: (i) the conduction of the ammoniac temperature over the heat exchanger toward the water; and (ii) the convection between the water temperature and the poultry temperature as well as toward the tank walls. This process can be decomposed into the following subprocess components: the tank, the processed poultry, and the heat exchanger. The expected exchanger outflow ammoniac temperature T_f is 1°C. A heat exchanger model captures the relationship between the pressures, the rates, and the temperatures of the water flows at the heat exchanger ports based on both directions of flow. The energy balance (the enthalpy) within the heat exchanger system can be given by:

$$C_w \frac{dm_w(t)}{dt} T_f(t) = \frac{dq_{\text{flow}}(t)}{dt} + \frac{dq_w(t)}{dt} + \frac{dq_a(t)}{dt}$$

Table 2.16 Some key parameters and variables of the poultry processing system.

Parameters	Description and values	Variables	Description and values
A, V	Cubic-shaped tank cross-sectional area and volume (20 m^3)	$T_h(t)$	Tank temperature (°C)
h	Height of liquid in tank (m)	$q_w(t)$	Water well valve flow rate (m^3 s^{-1})
C_w	Heat capacity of water (J kg°C^{-1})	$q_a(t)$	Ammoniac flow rate (m^3 s^{-1})
C_a	Heat capacity of input ammoniac (4.621 kJ kg^{-1})	$q_s(t)$	Wastewater valve flow rate (m^3 s^{-1})
R_p	Thermal resistance of the poultry (J m°C^{-1})	$T_a(t)$	Input tank ammoniac liquid temperature (−2°C)
H_{al}	Heat conduction coefficient (J m°C^{-1})	$T_w(t)$	Water inflow heat exchanger temperature (13°C)
C_{ww}	Processed waste water heat capacity	$\omega(t)$	Conveyor angular speed (rad s^{-1})
A_p	Skin area of poultry processed in the water tank (m^2)	$\dfrac{dm_a(t)}{dt}$	Ammonia mass flow rate (kg s^{-1})
K_{val}	Proportionality constant for the water valve (kg rad^{-1})	$\dfrac{dm_w(t)}{dt}$	Water mass flow rate (kg s^{-1})
K_{pump}	Coefficient gain for the ammoniac pump (kg m^{-3})	$T_f(t)$	Temperature of outflow ammoniac (1°C)
R_w	Tank wall thermal resistance (J m°C^{-1})	$T_{poultry}(t)$	Temperature of poultry (°C)
a	Valve discharge coefficient (m$^{2.5}$ s^{-1})	$h(t)$	Height of water in tank (m)

where the quantity of heat removed by the convection of ammoniac over the water within the exchanger is such that:

$$\frac{dq_a(t)}{dt} = C_a \frac{dm_a(t)}{dt}(T_f(t) - T_a(t))$$

and the quantity of heat removed from the water is given by:

$$\frac{dq_w(t)}{dt} = C_w \frac{dm_w(t)}{dt}(T_w(t) - T_f(t))$$

$\frac{dq_{flow}(t)}{dt}$ is the rate of the heat flow across the exchanger wall into the heat exchanger, and $\frac{dq_w(t)}{dt}$ and $\frac{dq_a(t)}{dt}$ are the changes of thermal energy in the heat exchanger due to the flow through the two ends of the heat exchanger. This heat loss over the exchanger is given by:

$$\frac{dq_{flow}(t)}{dt} = \frac{A}{R_w}(T_{ambient}(t) - T_h(t))$$

with R_w being the wall's thermal resistance given by:

$$R_w = \frac{d}{H_{al}k_f A_w}$$

and A_w, d, H_{al}, and k_f being, respectively, the area of the common wall between the inside layer, the wall thickness, the wall material thermal conductivity, and the corrugation coefficient of the wall. Thus, the energy balance of the heat exchanger component yields:

$$C_w \frac{dm_w(t)}{dt} T_f(t) = C_w \frac{dm_w(t)}{dt}(T_w(t) - T_f(t))$$

$$+ C_a \frac{dm_a(t)}{dt}(T_f(t) - T_f(t)) + \frac{A}{R_w}(T_{\text{ambient}}(t) - T_h(t))$$

Here, the mass flow rates of the water and the ammoniac toward the heat exchanger could be obtained as:

$$\frac{dm_w(t)}{dt} = K_{val}\theta(t)$$

$$\frac{dm_a(t)}{dt} = K_{pump}q(t)$$

with C_a and C_w being, respectively, the mean heat capacity of the input ammoniac and the water heat capacity while $T_a(t)$ and $T_w(t)$ are, respectively, the inflow ammoniac temperature and the water inflow heat exchanger temperature. Consider a tank with 800 ml of water per bird. From this assumption, it can be concluded that $C_w = C_{ww}$.

Hence, applying the energy balance principle to the tank component, the heat in the water tank would be equal to the heat from the circulating water minus the heat from the removed output water, the heat released through the tank wall, and the heat gained by the immersed poultry. This results in:

$$Q_T(t) = Q_{in}(t) - Q_{out}(t) - Q_{air}(t) - Q_{poultry}(t)$$

$$Q_T(t) = \frac{A_T \rho_w C_{ww}}{a^2} q_o(t)^2 \frac{dT_h(t)}{dt}$$

$$Q_{in}(t) = \rho_w C_w T_w(t)q_i(t)$$

$$Q_{wall}(t) = \frac{T_h(t) - T_a(t)}{R_w q_o(t)^2}$$

$$Q_{poultry}(t) = nR_p A_p(T_{poultry}(t) - T_h(t))$$

Thus, by substitution, it yields:

$$\frac{A_T \rho_w C_{ww}}{a^2} q_o(t)^2 \frac{dT_h(t)}{dt} = \frac{A_T \rho_w C_{ww}}{a^2} q_o(t)^2 \frac{dT_h(t)}{dt} - \rho_w C_w T_w(t)q_i(t)$$

$$- \frac{T_h(t) - T_a(t)}{R_w q_o(t)^2} - nR_p A_p(T_{poultry}(t) - T_h(t))$$

$$\frac{A_T \rho_W C_{ww}}{a^2} q_o(t)^2 \frac{dT_h(t)}{dt} = \frac{A_T \rho_W C_{ww}}{a^2} q_o(t)^2 \frac{dT_h(t)}{dt} - \rho_W C_w T_w(t)q_i(t)$$

$$- \frac{T_h(t) - T_a(t)}{R_w q_o(t)^2} - nR_p A_p(T_{poultry}(t) - T_h(t))$$

The outflow is regulated by a valve whose dynamics can be calculated as:

$$q_o(t) = a\sqrt{h(t)}$$

and

$$\frac{dh(t)}{dt} = \frac{q_i(t) - q_o(t)}{A} = \frac{K_{val}\theta(t) - a\sqrt{h(t)}}{A}$$

where R_p, A_p, C_{ww}, and n, respectively, are the thermal resistance of the poultry, the skin area of poultry processed in the water tank, the waste water heat capacity, and the number of poultry intermittently processed in batch production. Despite those assumptions, there is a coupling between the water flow rate, the ammoniac flow rate, and the poultry temperatures.

2.6 Chemical Processes

2.6.1 Crude Oil Distillation Petrochemical Process

The petrochemical process of crude oil distillation consists of separating the heated crude oil into a mixture of benzene, toluene, and xylenes. First, the crude oil is desalted to avoid corrosion of the equipment. Then, the crude oil is heated from near-ambient temperature to about 350–400°C within the boiler at the entrance of the atmospheric distillation tower. As illustrated in Figure 2.13, the distillation tower can be decomposed into the following components: (i) a distillation column with sections (oil stripping, feeding, and rectifying) and trays, respectively, for oil injection and steam separation; (ii) at the bottom, a reboiler to provide heat for oil vaporization; (iii) at the top, a condenser to cool and condense the oil vapor from the top of the column; (iv) a reflux drum to hold the condensed vapor so that liquid reflux can be recycled back from the top of the column; and (v) a condenser. In the tower, the crude oil is continuously processed into boiling range fractions of vaporized oil through several trays spaced at regular intervals. These distillation trays contain perforations and bubble caps. Hence, at each tray, fractions of rising vaporized oil can be collected from the sides of the distillation column at a specific temperature corresponding to the cutpoint for a desired product (i.e. kerosene, light gas oil, heavy gas oil, etc.). At the distillation tower overhead, the remaining vaporized crude oil is cooled down and condensed into a liquid gas (i.e. LPG). The unvaporized oil or asphalt residue drops off toward the bottom of the tower to be reheated or drawn off (i.e. asphalt). After leaving the tower, all of the fractions are routed toward oil-refining downstream processing units. The residue at the bottom is fed through the vacuum unit.

It is assumed that: (i) the separation processes are isothermal in all phases at 100°C; (ii) there is a constant pressure throughout each tray that varies linearly with the height of the column; (iii) the overhead vapor is totally condensed; the vapor holdup on each tray is negligible compared to the liquid holdup in each tray, the condenser, and the reboiler, which are all constant; (iv) there is a complete mixing at every tray; (v) the feed can be either a liquid, a vapor, or a mixture; (vi) there is a vapor-liquid phase equilibrium at the crude oil surface; and (vii) the molar flow rates of the vapor and the liquid through the stripping and rectifying sections are constant.

Consider a five-stage distillation tower of 1.95 m in diameter, 5 m in height, with trays spaced 60 cm apart. Here, a reboiler provides the heat for oil vaporization from the bottom of the column while a condenser is used to cool and to condense the vapor from the top of the column. The reflux drum is holding the condensed vapor in order to move the liquid reflux back from the top of the column. It should be recalled that the crude oil distillation process is used for purification of final products. The levels of purification of the final output products are given by the degree of purity for the distillate x_D, higher than or equal to 98%, and the degree of impurity at the bottom, x_B less than or equal to 2%. The concentration of the final product outputs x_B and x_D should be measured to predict and to maintain the product quality based on the feed flow F, the feed concentration c_F, the liquid flow rate L, and the vapor flow rate V.

At each stage, after combining the reaction and the fractionation of the pressured crude oil, some fractions are condensed into derivatives of p-xylenes while the remaining steam of the crude oil is moved to the next stage. The resulting condensed oil product is stored to be used as

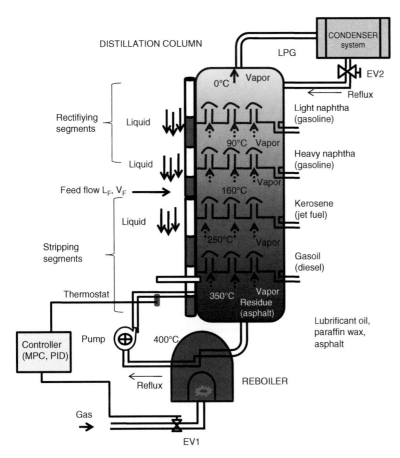

Figure 2.13 Distillation tower schematic.

commercial products such as fuel, diesel, kerosene, or gas. The residual crude oil at the bottom of each stage is used for reboiling. Hence, the pressure and temperature of the crude oil define the proportions of the separated p-xylene products such as ethylene, propylene, and butylenes. In order to maintain the final product quality and proportions of p-xylene products over all derivatives of crude oil, the dynamics model of the temperature and internal flow rate of the distillation process should be developed.

Here, the process components considered are the reboiler, a five-stage distillation tower, and a condenser over which the changes in input and output flow rates, temperatures, and pressures are quantified. Here, among the process parameters and variables listed in Table 2.17, for each stage of the feed flow there are furnace coil flow, distillation column pressure, temperature, and flow rate. Also, preheating temperature should be below 400°C to avoid carbon deposition in the pipeline. The liquid flow rate L and the vapor flow rate V are process inputs, as well as disturbance inputs such as the feed flow F and the feed concentration c_F, while the product outputs are the concentrations x_B and x_D.

As shown in Figure 2.14, the distillation process is designed with $N = 5$ stages and operates at a constant pressure. The feed can be considered as a mixture of ligas and naphtha. Here, the inputs are condenser work, reboiler work, reflux mass flow rate, and distillate mass flow rate. Hence, applying the vapor-liquid equilibrium principle for the five distillation stages, as shown in Table 2.18, so that $x_i(t)$ and $y_i(t)$ are, respectively, the liquid concentration and the vapor

Table 2.17 Some key variables and parameters of the distillation process.

Parameters	Description and values	Variables	Description and values
P	Distillation tower inside (Pa)	$L_S(t)$	Stripping liquid flow rate (kg h^{-1})
z_F	Molar fraction of light element in the feed (%)	$V_S(t)$	Stripping vapor flow rate kg h^{-1}
q_F	Constant of (liquid/vapor) feed (%)	$L_R(t)$	Rectification liquid flow rate (kg h^{-1})
λ	Heat evaporation of oil (kJ kg^{-1})	$V_R(t)$	Rectification vapor flow rate (kg h^{-1})
x_D	Molar fraction of distillate (%)	$F(t)$	Oil upstream feed molar flow rate (kg h^{-1})
x_B	Molar fraction of light element in the bottom (%)	$D(t)$	Condensate flow rate (kg h^{-1})
M_B	Quantity of crude oil in reboiler (kmole)	$B(t)$	Distillate bottom flow rate (kg h^{-1})
M_D	Quantity of crude oil in condenser (kmole)	$Q_B(t)$	Heat provided in the reboiler (kJ h^{-1})
M_T	Quantity of crude oil in tower stages (kg)	$L_D(t)$	Reflux flow rate (kg h^{-1})
c_B	Heat capacity of the bottom (kJ kg.°C^{-1})	t_{in}, t_{out}	In/out temperature of reboiler (°C)
M_n	nth stage molar fraction retained (%)	M_D	Condenser duty (W)
x_n	nth stage molar fraction of distillate within liquid (%)	M_R	Reboiler duty (W)
y_n	nth stage molar fraction of distillate within vapor (%)	$x_D(t)$	Feed molar flow rate (kg h^{-1})
T_B	Outlet temperature (°C)	$x_n(t)$	Liquid oil concentration at n^{th} stage (kg l^{-1})
T_S, T_E	Input and output residue reboiler temperatures (°C)	$y_n(t)$	Vapor oil concentration at n^{th} stage (kg l^{-1})
λ	Heat of vaporization (kJ kg^{-1})	$V_f(t)$	Distillate flow rate (kg h^{-1})
ρ_L, ρ_G	Density of liquid and vapor phases (kg m^{-3})		
C	Correction factor depending flow rates of two-phase flows		

concentration on the ith stage, while α is the relative volatility. This yields the following generic equation:

$$\frac{dM_i x_i(t)}{dt} = L_{i+1}(t)x_{i+1}(t) - L_i(t)x_i(t) + V_{i-1}y_{i-1}(t) - V_i y_i(t)$$

So, at the top and the bottom of the stages, this yields:

$$M\frac{dx_i(t)}{dt} = \frac{dM_i x_i(t)}{dt} - x_i(t)\frac{dM_i(t)}{dt}$$

while at other stage including (feed stage):

$$M\frac{dx_i(t)}{dt} = \frac{dM_i x_i(t)}{dt}$$

Thus, at the condenser (top stage, $n = 1$):

$$M_D\frac{dx_n(t)}{dt} = (V + V_F)y_{n+1} - Lx_n - Dx_n$$

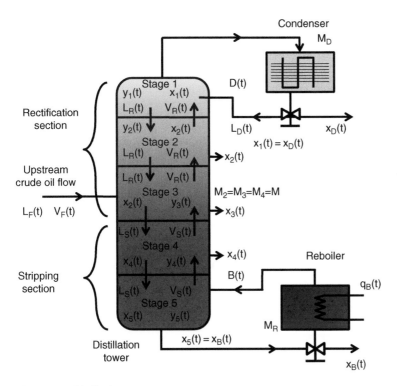

Figure 2.14 Distillation tower.

Table 2.18 Average crude oil fractioning temperature.

Fraction	Average boiling temperature (°C)	Final product
Gas oil	190–350	Diesel
Kerosene	180–260	Diesel, jet fuel
Heavy naphtha	90–195	Gasoline
Light naphtha	45–90	Gasoline
LPG	−30 to 0	Propane fuel
Residue	+450	Coke, asphalt

which corresponds to:

$$\frac{dx_1(t)}{dt} = \frac{V_R}{M_D}(y_2(t) - x_1(t))$$

Similarly, at the feed stages ($m = 2.3.4$), i.e. all stages below the condenser top stage and above the reboiler bottom stage, and assuming that $M_1 = M_2 = M_3$, the generic dynamics equation is given by:

$$M_D\frac{dx_m(t)}{dt} = V(y_{m-1} - y_m) + L(x_{m+1} - x_m) + L_F(x_F - x_m)$$

which is equivalent to:

$$M\frac{dx_2(t)}{dt} = L_R(t)x_2(t) + V_R(t)y_3(t) + L_R x_1(t) - V_R(t)y_3(t)$$

$$M\frac{dx_3(t)}{dt} = -L_S(t)x_3(t) + V_5(t)y_4(t) - V_R(t)y_3(t) + L_R(t)x_2(t) + Fz_F(t)$$

$$M\frac{dx_4(t)}{dt} = L_R(t)x_3(t) - L_5(t)x_4(t) + V_5(t)y_5(t) - V_5(t)y_4(t)$$

At the reboiler bottom stage ($k = 5$), the generic equation is given by:

$$M_B\frac{dx_k(t)}{dt} = (L + L_F)x_2(t) - Vy_k(t) - Bx_k(t)$$

which is equivalent to:

$$M_B\frac{dx_5(t)}{dt} = -L_s(t)x_4(t) - V_s(t)y_5(t) - B(t)x_5(t)$$

Thus, the vapor boilup $V_S(t)$ kmol h^{-1} generated by the heat input to the reboiler can be derived such that:

$$V_s(t) = \frac{Q_B(t) - C_B B(t_S - t_E)}{\lambda}$$

$$V_R(t) = F(1 - q_F(t)) + V_s(t)$$

$$L_R(t) = L_D(t)$$

$$y_i(t) = \frac{\alpha x_i(t)}{1 + (\alpha - 1)x_i(t)}$$

$$L_S(t) = Fq_F(t) + L_R(t)$$

$$F(t) = B(t) + D(t)$$

$$z_F(t) = x_B B(t) + x_D D(t)$$

Hence,

$$D(t) = F(t)\left(\frac{z_F(t) - x_B(t)}{x_D(t) - x_B(t)}\right)$$

Also, the vapor velocity phase arises in the column and is given at each stage by:

$$\omega_v(t) = C\sqrt{\frac{\rho_L - \rho_G}{\rho_G}}$$

The reflux drum is estimated by:

$$M_D = \frac{5(L_f + V_f)}{60}$$

where L_f is the reflux flow rate (kg h^{-1}); V_f is the distillate flow rate (kg h^{-1}). At the nth stage, the rate of the material accumulation is equal to the quantities entered and generated, less the quantities leaving and consumed within. Hence, the constant relative volatility throughout the column and the vapor-liquid equilibrium relation can be expressed by:

$$y_n(t) = \frac{\alpha x_n(t)}{1 + (\alpha - 1)x'_n(t)}$$

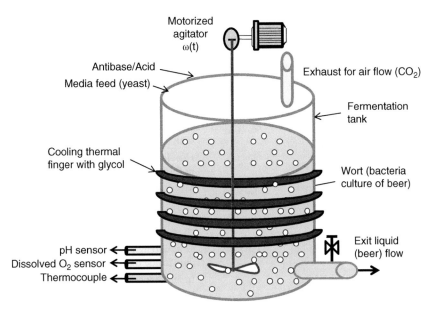

Figure 2.15 Fermentation tank schematic.

The distillation system dynamics are described by nonlinear differential equations expressing the vapor-liquid equilibrium relationship between $y_n(t)$ and $x_n(t)$. A typical distillation tower is expected to operate at nominal steady-state conditions so that the feed stream contains 30% benzene, 40% toluene and 30% xylenes and the feed flow rate is 500 kmol h^{-1}, while the distillate contains 95% benzene and 5% toluene in the feed. Therefore, in this model it can be assumed that the variables deviate only slightly from normal operating conditions. Then, it can be linearized using a Taylor series and the operating conditions. The resulting model can also be reduced. Furthermore, in this model any changes in the feed stream, the feed flow rates, and the feed compositions are not considered. As a consequence, any pressure fluctuations would cause the modeling to be more challenging and even almost impossible.

2.6.2 Lager Beer Fermentation Tank

In addition to the brewery industries, fermentation processes are used in bakery, dairy, and even pharmaceutical industrial applications. Beer fermentation is a three-stage chemical process where the sugar, in the presence of microorganisms, breaks down into alcohol and carbon dioxide. First, the yeast slurry is added to the fermentation tank. At this stage, sterile air or oxygen is injected into the bottom of the tank to support the living yeast and to start the consumption of all sugars by that yeast. This chemical reaction generates alcohol and heat partially cooled by a lower temperature liquid circulating within a finger around the fermentation tank as well as by exhausting carbon dioxide into the air, as illustrated in Figure 2.15. Process variables are summarized in Table 2.19. The fermentation process stage takes between 3 and 5 days. Lager beers are cooled at around 1°C and aged for 7–24 days. Table 2.20 summarizes the remaining fermentation stages.

The wort temperature has to be predictable and maintained within a range in order to maximize yeast activity and to obtain the growth of a mixed culture of lactic acid bacteria ensuring a consistent end-product taste. For example, a tank temperature above 14°C generates

Table 2.19 Some key parameters and variables of the fermentation process.

Parameters	Description and values	Variables	Description and values
V_t	Fermentation tank volume ($100\ \mathrm{m}^3$)	$Q_{\text{generated}}(t)$	Heat generated from culture growth with fermentation tank (kcal $\mathrm{h}^{-1}\ \mathrm{l}^{-1}\ \mathrm{s}^{-1}$)
$M = BD/26$	Mass of malt used (14 423 kg)	$Q_{\text{evacuated}}(t)$	Heat evacuated from the fermentation tank (kcal $\mathrm{h}^{-1}\ \mathrm{l}^{-1}\ \mathrm{s}^{-1}$)
D	Alcohol content (5%)	$Q_{CO_2}(t)$	Heat losses through CO_2 gasses produced (kcal $\mathrm{h}^{-1}\ \mathrm{l}^{-1}\ \mathrm{s}^{-1}$)
B	Quantity of beer desired (75 000 l)	$\omega_m(t)$	Motor speed (rad s^{-1})
$E = B * 1.25$	Quantity of water (93 750 l)	$Q_{RET}(t)$	Heat produced through quantity chemical reaction (kcal $\mathrm{h}^{-1}\ \mathrm{l}^{-1}\ \mathrm{s}^{-1}$)
$H = 0.1B/20$	Quantity of hops (375 kg)	$T_o(t)$	Output temperature of the cooling fluid inside thermal finger (°C)
$L = 14.5B/20$	Quantity of yeast (55 kg)	$T_i(t)$	Input temperature of the cooling fluid inside thermal finger (°C)
$M_t = L + H + M$	Total mass within the tank (93 kg)	$T_a(t)$	Ambient air temperature (°C)
d	Humidity of air within entrance fermentation tank in (%)	$q_{\text{finger}}(t)$	Heat transfer rate from wort to finger (kcal $\mathrm{h}^{-1}\ \mathrm{l}^{-1}$)
k_f	Heat capacity through convection (kJ °C kg^{-1})	$q_{tank}(t)$	Heat evacuation rate from the fermentation tank to outside ambient (kcal $\mathrm{h}^{-1}\ \mathrm{l}^{-1}$)
α_m	Velocity-heat gain	$q_{\text{motor}}(t)$	Heat generated rate through the motor agitation of fluid (kcal $\mathrm{h}^{-1}\ \mathrm{l}^{-1}$)
C_M	Wort and tank heat capacity (kcal $\mathrm{h}^{-1}\ \mathrm{l}^{-1}\ °\mathrm{C}^{-1}$)	$V_F(t)$	Air volume per volume of fermentation tank (%)
M_F	Cooling fluid mass flow rate (kg s^{-1})		

Table 2.20 Fermentation temperature profile.

Temperature, $T_o(t)$ (°C)	Time duration (d)	Stages
14	3–5	Adding yeast to sugar to produce alcohol
1	2	Cooling beer process
14	7–24	Aging beer process

unwanted by-products and affects the final product taste and quality. This is done by removing the generated heat by circulating glycol solution in jackets around the tank. The quality of the beer depends on the temperature, pressure, and pH control, as well as the regulated air inflow while mixing the solution with pitched yeast through a motorized ON/OFF agitator. Hence, a dynamic thermal model of the tank is suitable to achieve desired temperatures during the fermentation process.

By applying the energy balance (the enthalpy) within the fermentation tank and the thermal finger system, the difference between energy generated and energy evacuated would be given by:

$$Q_{generated}(t) - Q_{evacuated}(t) = C_M \frac{dT_o(t)}{dt}$$

Recalling that during beer fermentation, there is an increase of the beverage temperature within the tank that can be quantified by the heat produced through the chemical reaction Q_{RET} for *n*-alkanes so that:

$$y_1[C_u O_v H_w N_t] + y_2 O_2 + y_3 NH_4^+ \rightarrow z_1 C_X H_Y O_Z N_q + z_2 CO_2 + z_3 H_2 O - Q_{RET}$$

Furthermore, the motor agitation of the fluid transforms its kinetic energy into heat for fermentation proportional to the motor speed $\omega_m(t)$. Hence, the overall heat generated is given by:

$$Q_{generated}(t) = Q_{RET}(t) + \frac{dq_{motor}(t)}{dt} = Q_{RET}(t) + \alpha_m \omega_m(t)$$

Part of the thermal energy generated is transferred by convection of heat from the dissolved oxygen within the air. This increases the moisture level within the tank. The accumulation of dissolved carbon dioxide (CO_2) within the tank can be expressed based on the variation of air volume and moisture level such that:

$$\frac{dQ_{CO_2}(t)}{dt} = V_F \left(\frac{100 - d}{100} \right)$$

A remaining part of the energy is transferred by conduction over the thermal finger and the tank wall. The overall quantity of heat evacuated over time is given by:

$$Q_{evacuated}(t) = \frac{dQ_{CO_2}(t)}{dt} + \frac{dq_{finger}(t)}{dt} + \frac{dq_{tank}(t)}{dt}$$

$$= V_F \left(\frac{100 - d}{100} \right) + M_F(T_i(t) - T_o(t)) + k_f(T_o(t) - T_a(t))$$

with *d* being the humidity of air within the entrance of the fermentation tank as a percentage. Thus, substituting the overall heat balance equation yields:

$$C_M \frac{dT_o(t)}{dt} + (k_f - M_F)T_o(t) = \alpha_m \omega_m(t) + Q_{RET}(t) - V_F \left(\frac{100 - d}{100} \right) + M_F T_i(t)$$

The quantities of yeast, sterile air, and dissolved carbon dioxide (CO_2) content used during fermentation define the quantity of heat generated, $Q_{RET}(t)$. Hence, the difficulty of quantifying such variables limits the predictability of the beverage temperature when using this modeling approach. An alternative using the gray-box modeling approach could be useful to refine its real-time estimate under specific operating conditions. Also, there is a correlation between pH levels and the speed of sugar conversion into alcohol. But the coupling between these variables requires further knowledge of reaction kinetics, thermodynamics, transport, and physical properties in order to refine the fermentation process model.

2.7 Production Systems and Processes

2.7.1 Single Axis Drilling System

Drilling process operations consist of making round holes of various sizes and depths with a specified accuracy. Hence, this process proceeds with drilling holes within specific tolerances

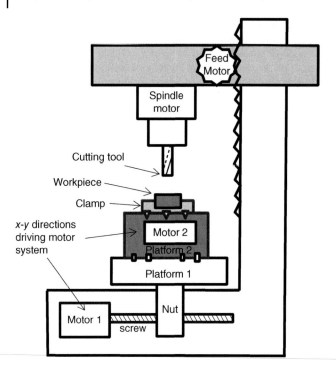

Figure 2.16 Drilling machine system.

(not oversize or off-center) by avoiding the deflection of the thrust force or limiting the vibration of the drilling structure (cutting force vibration). Indeed, some consistent vibrations can even cause tool breakages. Consider a drilling machine tool with its (x-y) platform on which the workpiece is clamped and positioned by motors 1 and 2, as depicted in Figure 2.16. The feed motor ensures a feed speed of the tool toward a clamped workpiece while the spindle motor defines the cutting feed rate f through a cutting torque $F_c(t)$ and a thrust force $F_T(t)$ in the z-direction. Applying an excessive thrust force can cause drill bucking while high cutting torque may cause hole edge chipping and heating. It is assumed that: (i) the cutting fluid, cutting speed, and helix angle effects on the cutting process are negligible; (ii) the effect of the cutting speed on the cutting force can be ignored; (iii) the friction acting on the motor shaft is viscous; and (iv) nonlinear friction at near-zero velocities is not considered. A drilling operation producing a hole with specific depth and size is related to the cutting feed rate and the spindle speed of the drilling machine tool, as well as the cutting torque and the thrust force. The drilling process variables and parameters are listed in Table 2.21.

Consider an AISI 1045 steel roll-shaped workpiece, in which it is necessary to cut a 3 mm diameter hole with a depth of 5.5 mm using a twist drill tool with a hardness H_B of 180. This is expected to be achieved on a drilling machine with nominal operating conditions such as a cutting feed rate f of 0.004 rev s^{-1} and a spindle cutting speed N of 1350 rpm.

From the mechanistic model, the acting forces during the cutting operations are usually decomposed into tangential $F_{tangent}(t)$, feed $F_{feed}(t)$, and radial $F_{radial}(t)$ forces. It is assumed that the cutting force $F_c(t)$ acts on the spindle shaft and is only related to the tangential force, $F_{tangent}$, as $F_c(t) \approx F_{tangent}(t)$. From the assumption on the friction properties of the drilling process, only stationary behavior is considered. Hence, at constant nominal speed (during

Table 2.21 Some key variables and parameters of the drilling machine.

Parameters	Description and values	Variables	Description and values
$J_{1,2}$	Inertia for motors 1, 2 (N.m.s^2 rad^{-1})	$T_{m,1,2}(t)$	Torque for motor 1, 2 (N.m)
K_i	Torque constant motors 1, 2, s (V.s rad^{-1})	$T_{m,s}(t)$	Spindle motor torque (N.m)
η_{pulley}	Motor 2 pulley transmission ratio	$T_c(t)$	Cutting torque (N.m)
η_{screw}	Motor 1 screw transmission ratio	$\theta_c(t)$	Cutting tool angular position (rad)
$L_{a,i}$	Armature inductance motors 1, 2 (H)	$\omega_i(t)$	Velocity of motors 1, 2, 3 (rad s^{-1})
$R_{a,i}$	Armature resistance motors 1, 2, s (Ω)	F_c	Cutting force (N.mm^{-1}.rev)
b_i	Damping ratio motors 1,2 (N.m.s rad^{-1})	$T_T(t)$	Thrust force (N.m)
J_M	Feed and spindle motors and gearbox inertia (N.m.s^2 rad^{-1})	$T_{\text{Load},1,2}(t)$	Load torque for i^{th} motor (N.m)
B_M	Feed and spindle motor damping ratio (N.m.s rad^{-1})	N	Spindle cutting speed (rad s^{-1})
K_M	Motor torque constant (V.s rad^{-1})	$T_{m,f}(t)$	Feed motor torque (N.m)
V_f	Vertical feed velocity (m s^{-1})	$F_{\text{radial}}(t)$	Radial force from the tool (N)
n	Spindle speed (rad s^{-1})	$F_{\text{tangent}}(t)$	Tangent force from the tool (N)
f	Cutting feed rate (rad s^{-1})	$F_{\text{feed}}(t)$	Feed force from the tool (N)

steady-state operations), the cutting force can be expressed as:

$$F_C = K_S d^\beta (V_f)^\gamma f^\alpha$$

such that the drill diameter is d, the feed rate is f, and the vertical feed velocity is V_f. Hence, the corresponding cutting torque $T_c(t)$ is acting on the spindle shaft such that it is given by:

$$T_C(t) = r_{\text{tool}} F_c = r_{\text{tool}} K_S d^\beta (V_f)^\gamma f^\alpha$$

with r_{tool} being the tool radius. The thrust force $F_T(t)$ acting along the z-direction can be expressed as:

$$F_T(t) = k_1 F_{\text{feed}} + k_2 F_{\text{radial}} = K_p D^a f^b$$

Variables such as the depth of cut d, the cutting speed V_f or V, the feed rate f, have nonlinear relationships with the cutting force. In order to derive the model parameters such as the cutting constants α, β, γ, the force coefficient K_s, the thrust force coefficient K_p, and parameters a, b, the least squares curve fitting technique can be used with a frequency response function (FRF) plot of the cutting test data. These data are dependent on the workpiece material's hardness, the cutting tool material, its geometry, and the cutting conditions (coolant used, cutting temperature, etc.). Each parameter should be estimated for specific operating conditions (temperature, lubrication, workpiece material, etc.). Also, the deformation is assumed to be perfectly plastic. Considering a friction force $F_\mu(t)$ and all generated resistive forces during the drilling operation, the cutting $T_c(t)$ and friction $T_\mu(t)$ torques should be at least equal to the spindle motor torque, $T_{m,s}(t)$, for the cutting operations to take place, only if:

$$T_\mu(t) + T_c(t) = T_{m,s}(t)$$

If $T_{m,s}(t)$ is less than $T_\mu(t) + T_c(t)$, the spindle motor will stop and eventually the tool will break. It should be recalled that

$$\frac{f}{V_f} = \frac{60}{n}$$

with n being the spindle speed. In this system model, the impact dynamics, the tool workpiece friction, and the tool wear are neglected. Hence, the equation describing the motion driving system of the secured tool in the drilling machine is given by:

$$J_M \frac{d^2\theta_c(t)}{dt^2} + B_M \frac{d\theta_c(t)}{dt} = T_{\text{Load}}(t) = T_{m,s}(t) - T_c(t)$$

Recalling and manipulating the classical electric component of the DC spindle motor, this yields

$$T_{m,i}(t) = K_{t_i} i_i(t); E_i(t) = Ke_i \omega_i(t); V_i(t) = R_{a,i} i_i(t) + L_a \frac{di_i(t)}{dt} + E_i(t)$$

Thus,

$$\frac{L_{a,s} J_M}{K_{t,s}} \frac{d^3\theta_c(t)}{dt^3} + \left(\frac{L_{a,s} J_M}{K_{t,s}} + \frac{L_s B_M}{K_{t,s}}\right)\frac{d^2\theta_c(t)}{dt^2} + \left(\frac{R_{a,s} B_M}{K_{t,s}} + K_{e_s}\right)\frac{d\theta_c(t)}{dt}$$

$$= V_s(t) - \frac{R_{a,s}}{K_{t,s}} T_{\text{Load},s}(t)$$

At low velocities, the dynamics equations relating the spindle motor torque and forces acting during the cutting are highly dependent on nonlinear viscous friction. This is due to the dependence of the viscous friction coefficient on periodic shaft velocity. Furthermore, the gray-box modeling of the transient characteristics during the cutting process is not considered due to the challenging large numbers of parameters and variables involved, especially those related to the operating conditions.

2.7.2 Cement-Based Pozzolana Portal Scraper

In a cement milling station, a portal scraper with a capacity of 200 ton h^{-1} is used to convey cement raw materials such as gypsum or wet pozzolana from storage to the grinding or drying stations. This portal scraper operates in a storage warehouse of raw material deposits. As illustrated in Figure 2.17, this portal scraper uses six motors to ensure the transversal positioning process above the deposit and pozzolana removal by the scraping process. Once the transversal positioning of the scraper above the raw material deposit is completed by MTR5 and MTR6, the scraping can start using two motorized arms (MTR1, MTR4), as illustrated in Figure 2.17(a). The primary (MTR4) and secondary (MTR1) arms of the scraper move from the right to the left of the deposit, then the conveyor carries the cement product toward other subprocesses. The operation of this handling system requires a synchronized positioning of both arms to avoid a transversal deflection or even breakage. In addition, the precise positioning for both scraper arms just above the deposit aims to maximize the removal rate and to minimize the required torque of the scraping arm. Hence, it is suitable for modeling the resistive torques effect and the synchronized positioning of the scraper. Table 2.22 lists some key process variables and parameters involved.

Here, the scraping motor torque $T_{m1}(t)$ of the secondary arm would operate if and only if it is equal to or higher than the resistive torques (especially the viscous friction $T_{v1}(t)$ and the static friction $T_{\mu 1}(t)$). These resistive torques depend on the humidity rate of the pozzolana deposit as well as the level of contact given by the pozzolana deposit angle $\theta_d(t) \approx \theta_3(t)$ and the

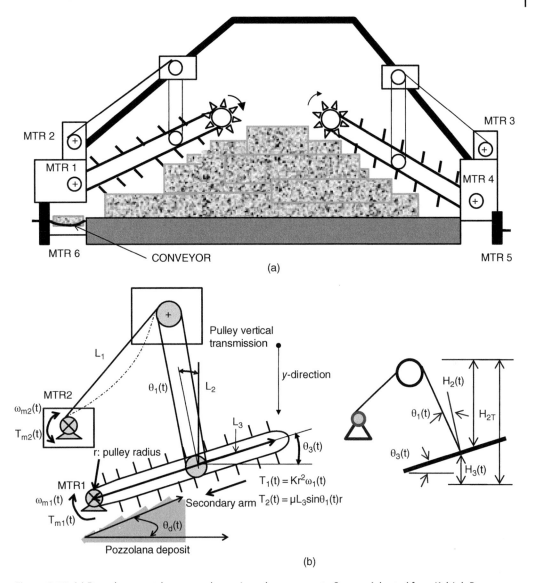

Figure 2.17 (a) Pozzolana portal scraper schematic and components. Source: Adapted from Kaltjob P. (b) Secondary arm of the scraper.

primary scraper arm angle $\theta_1(t)$. Similarly, the secondary arm motor torque $T_{m2}(t)$ is based on the scraper weight and the primary scraper arm angle $\theta_1(t)$. It is considered that the distributed viscous forces (dynamic friction and static friction) depend on the weight of the secondary arm and the contact between the deposit and the scraper μ_2, as well as the humidity, H_h. Thus, applying Newton's law on the secondary arm, as illustrated in from Figure 2.17(b), the dynamics equation of the MTR1 scraping process is given by:

$$(J_1 + \eta_{\text{pulley}}{}^2 J_{\text{pulley}})\frac{d\omega_{m1}(t)}{dt} + b_1\omega_{m1}(t) = T_{m1}(t) - T_{\mu1}(t) - T_{v1}(t)$$

Assume that

$$T_{\mu1}(t) = \mu_1 M_{AS}g\cos\theta_3(t)L_3\sin\theta_1(t) - K(\theta_3(t) - \theta_d(t))^\alpha H_h{}^\beta$$

Table 2.22 Some keys variables and parameters for the translation and scraping process.

Parameters	Description and values	Variables	Description and values
M_{AS}	Mass of secondary arm (kg)	$\theta_{m,i}(t)$	Angular position of the ith motor ($i = 1, 2, 3, 4, 5, 6$) (rad)
$L_{a,i}$	Armature inductance for i^{th} motor ($i = 1, 2, 3, 4$) (H)	$\omega_{m,i}(t)$	Angular velocity of the ith motor ($i = 1, 2, 3, 4, 5, 6$) (rad s^{-1})
$R_{a,i}$	Armature resistance for i^{th} motor ($i = 1, 2, 3, 4$) (Ω)	$V_{a,i}(t)$	Input voltage of the ith motor ($i = 1, 2$) (V)
b_i	Damping ratio for i^{th} motor ($i = 1, 2, 3 ,4$) (N.m.s rad^{-1})	$T_{m,i}(t)$	Motor torque of the ith motor ($i = 1,2,3,4$) (N.m)
μ	Static friction coefficient (N.m s^{-1})	$i_i(t)$	Current in the ith motor ($i = 1, 2, 3, 4, 5, 6$) (A)
r_i	Radius of pulley over i^{th} motor ($i = 1, 2$) (m)	$E_i(t)$	Back emf voltage of the ith motor ($i = 1, 2, 3, 4, 5, 6$) (V)
K	Dynamic friction coefficient	$T_i(t)$	Rope tension ($i = 1, 2$) (N)
J_i	Inertia for i^{th} motor ($i = 1, 2, 3, 4$) (N.m.s^2 rad^{-1})	$T_{v,1}(t)$	Viscous friction torque for MTR1 (N.m)
K_i	Torque constant for i^{th} motor ($i = 1, 2, 3, 4$) (V.s rad^{-1})	H_{2T}	Total height (m)
η_{pulley_i}	i^{th} motor pulley transmission ratio (%)	$T_{\mu,1}(t)$	Static friction torque for MTR1 (N.m)
L_3	Length of the secondary arm (m)	H_h	Humidity level of the deposit (%)
J_{pulley_i}	i^{th} motor inertia of the pulley (N.m.s^2 rad^{-1})	$H_3(t)$	Height of the primary arm (m)
L_2	Rope length of secondary arm (m)	$\theta_d(t)$	Pozzolana deposit angle (rad)
		$\theta_2(t)$	Angle of the rod over the vertical (rad)
		$\theta_3(t)$	Angle of the scraper arm over the horizontal (rad)

and

$$T_{v1}(t) = Kr^2 \omega_{m1}(t)$$

Recall that the shaft torque of motor 2 $T_{m,1}(t)$ is given by:

$$V_{a,1}(t) - E_1(t) = R_{a,1} i_1(t) + L_{a,1} \frac{di_1(t)}{dt} = \frac{1}{K_{i,1}} \left[R_{a,1} T_{m,1}(t) + L_{a,1} \frac{dT_{m,1(t)}}{dt} \right]$$

It is assumed that the center of gravity of the secondary arm is located at its middle so that during the upward motion of the secondary arm, MTR2 positioning could be given by:

$$\begin{cases} H_2(t) = L_2 \cos \theta_1(t) \\ H_3(t) = \frac{L_3}{2} \cos \theta_3(t) \end{cases}$$

while the total height is given by:

$$H_2(t) + H_3(t) = L_2 \cos \theta_1(t) + \frac{L_3}{2} \cos \theta_3(t) = H_{2T}$$

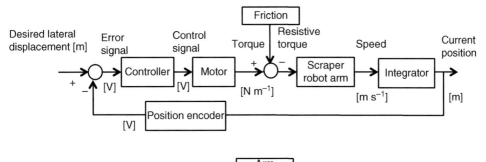

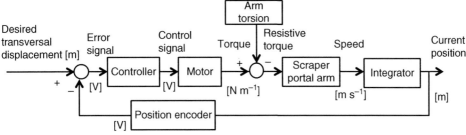

Figure 2.18 Subprocess function block diagrams for the pozzolana portal scraper..

Hence, considering the cable tension force $T_1(t)$ as the load on the secondary arm such that:

$$T_{\text{Load}}(t) = T_1(t)\eta_{\text{pulley}}$$

Then, applying the principle of moments around $\theta_1(t)$, the dynamics equation yields:

$$T_1(t)L_2 \cos\theta_1(t) - M_{AS}g\frac{L_3}{2}\cos\theta_3(t) = M_{AS}L_3{}^2\frac{d^2\theta_1(t)}{dt^2}$$

Hence, the dynamics equations for the secondary arm yields:

$$J_2\frac{d\omega_{m2}(t)}{dt} + b_2\omega_{m2}(t) = T_{m2}(t) - T_{\text{Load}}(t)$$

Figure 2.18 presents the block diagrams of the components involved in the transversal motion and scraper processes. Motors MTR5 and MTR6 are used for transversal motion. These motors need to have synchronized speeds $\omega_{m,5}(t)$, $\omega_{m,6}(t)$ to avoid a transversal deflection.

Similarly, the modeling of the scraper positioning for the primary arm can be done by specifying the MTR3 and MTR4 motors' kinematics. Hence, it could be possible to apply synchronized commands on both motorized arms. Eventually, this portal scraper modeling would lead to a nonlinear MIMO model with a control scheme, as presented in Figure 2.18.

2.7.3 Variable Pitch Wind Turbine Generator System

A renewable energy system such as a wind turbine generator system converts naturally generated mechanical energy into an electric energy source. This is done by using a drivetrain system to convert the time-varying wind speed into the corresponding speed of rotating blades through a transmission train gearbox. These rotating blades are attached to a generator system in order to produce the induced field current necessary to generate electric power. Hence, in order to maintain a higher $\frac{electric\ power}{wind\ power}$ performance ratio as well as to ensure voltage stability and uninterrupted operation of the wind turbine, it is desirable to adjust and to maintain the rotor speed with respect to the wind speed fluctuation. This could be easily achieved with a wind turbine

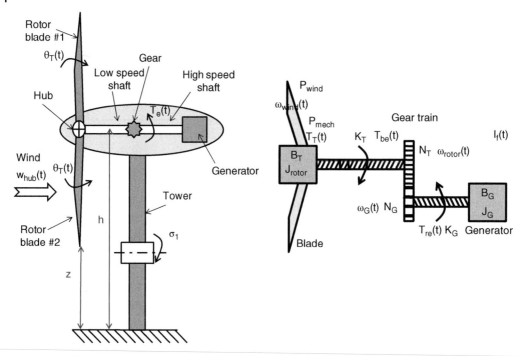

Figure 2.19 Wind turbine generator system.

dynamic model determining the relationship between the wind speed and the generated rotor speed. The three-bladed wind turbine system is depicted in Figure 2.19. It is assumed that the tower is flexible, while there is no blade flexibility. Hence, the variation of the blade pitch angle is negligible. Also, the pitch angle variation is negligible. In addition, the geographical configuration is such that the wind direction or the nearest wind turbine effect does not have a prevailing effect on the aerodynamic interactions. The turbine is assumed to run in a normal operating mode (after an initial period), without facing high wind speeds.

From the configuration of the wind turbine, the relationship between the wind speed variations and the height can be expressed as:

$$w_{\text{wind}}(t, z) = w_{\text{hub}}(t)\left(1 + \frac{z}{h}\right)^m$$

where z is the height difference between the hub and the ground, $w_{\text{hub}}(t)$ is the wind speed at the hub height h and the coefficient m is the power wind-shear coefficient. Based on the rotor angular speed $\omega_{rotor}(t)$, the air density ρ, the rotor radius $R = h - z$, and the linear wind speed $w_{\text{wind}}(t)$, the aerodynamic wind transformation coefficient (also the called tip speed ratio) is given by:

$$\alpha(t) = \frac{\omega_{rotor}(t)R}{w_{\text{wind}}(t, z)}$$

The mass air flow rate, $\dot{m}_{\text{wind}}(t)$, can be derived as:

$$\dot{m}_{\text{wind}}(t) = \rho \pi R^2 w_{\text{wind}}(t)$$

Hence, the wind power is given by:

$$P_{\text{wind}}(t) = \frac{1}{2}\dot{m}_{\text{wind}}(t)w_{\text{wind}}(t)^2 = \frac{1}{2}\rho \pi R^2 w_{\text{wind}}(t)^3$$

Table 2.23 Some key parameters and variables of the wind turbine generator system.

Parameters	Description and values	Variables	Description and values
R	Rotor radius (m)	$V_f(t)$	Generator voltage (V)
ρ	Air density (kg m^3)	$w_{wind}(t,z)$	Linear wind speed (m s^{-1})
J_T	Turbine blade inertia (N.m.s^2 rad^{-1})	$\alpha(t)$	Tip speed ratio
J_G	Turbine inertia (N.m.s^2 rad^{-1})	$\omega_{rotor}(t)$	Turbine angular speed (rad s^{-1})
K_T, K_G	Turbine and generator stiffness coefficients (N.m rad^{-1})	$\dot{m}_{wind}(t)$	Mass air flow rate (kg s^{-1})
B_T, B_G	Turbine and generator damping coefficients (N.m.s rad^{-1})	$\omega_G(t)$	Generator rotor speed (rad s^{-1})
N_T, N_G	Turbine and generator gearbox numbers of teeth	$P_{wind}(t)$	Wind power (kW)
C_p	Portion of wind power captured by the blade	$P_{mech}(t)$	Generated mechanical power (kW)
A	Rotor blades swept area (m^2)	$T_T(t)$	Turbine torque (N.m)
z	Hub height (m)	$T_G(t)$	Generator torque (N.m)
h	Height difference between the hub and the ground (m)	$P_{elec}(t)$	Generator power output (kW)
m	Power wind-shear coefficient	$\theta_T(t), \theta_G(t)$	Turbine and generator angular position (rad)
		$w_{hub}(t)$	Wind speed at the hub height (m s^{-1})

Consider that the portion of wind power captured by the blade, C_p, is known. Thus, applying the energy conservation principle to the wind-blade system, so that the transformation of the wind motion into blade rotation corresponds to the transfer of a portion of the wind power into mechanical power, yields:

$$P_{mech}(t) = \frac{1}{2}\rho\pi C_p R^2 w_{wind}(t)^3$$

At the steady state, the aerodynamic torque or the turbine torque during the blade rotation is given by:

$$T_T(t) = \frac{P_{mech}(t)}{\omega_{rotor}(t)}$$

Hence, the turbine torque transmitted by the wind to the rotor blade regardless of its mass is given by:

$$T_T(t) = \frac{1}{2}\rho\pi \frac{C_p}{\alpha^3} R^5 \omega_{rotor}(t)^2$$

At the steady state, the rotor speed can be expressed as (Table 2.23):

$$\omega_{rotor}(t) = 2\pi \frac{N_T(t)}{60} n$$

From the gearbox ratio configuration, the configuration of the turbine rotor torque over the generator torque relationship results in:

$$T_T(t) = \frac{N_T}{N_G} T_G(t) = \frac{\omega_{rotor}(t)}{\omega_G(t)} T_G(t)$$

Around the rotational axis of the blade and the generator shaft, their dynamics equation yields:

$$J_T \frac{d\omega_{rotor}(t)}{dt}(t) + B_T\omega_{rotor}(t) + K_T\theta_{rotor}(t) = T_T(t) - T_{be}(t)$$

$$J_G \frac{d\omega_G(t)}{dt} + B_G\omega_G(t) + K_G\theta_G(t) = T_{re}(t) - T_G(t)$$

Recall that the generated torque is related to the generator power as:

$$T_G(t) = \frac{N_G}{N_T} \frac{P_G(t)}{\omega_G(t)}$$

After substitution and simplification, the equation relating the blade's rotational speed to the power generated is given by:

$$\left(J_T + \left(\frac{N_G}{N_T}\right)^2 J_G\right) \frac{d\omega_G(t)}{dt} + \left(B_T + \left(\frac{N_G}{N_T}\right)^2 B_G\right)\omega_G(t) + \left(K_T + \left(\frac{N_G}{N_T}\right)^2 K_G\right)\theta_G(t)$$

$$= \frac{P_{mech}(t)}{\omega_{rotor}(t)} - \left(\frac{N_G}{N_T}\right)^2 \frac{P_{elec}(t)}{\omega_G(t)}$$

where the torque equations referring to the generator component are:

$$J_T \frac{d^2\theta_T(t)}{dt^2} + B_T\left(\frac{d\theta_T(t) - d\theta_G(t)}{dt}\right) + K_T(\theta_T(t) - \theta_G(t)) = T_T(t) - T_G(t)$$

$$J_G \frac{d^2\theta_G(t)}{dt^2} + B_G\left(\frac{d\theta_G(t) - d\theta_T(t)}{dt}\right) + K_G(\theta_G(t) - \theta_T(t)) = T_G(t) - T_T(t)$$

The electric power generated is transformed into a field current through the following equation:

$$P_{elec}(t) = K_T\omega_G(t)KI_f(t)$$

In this case the exciter or induced field voltage is given by:

$$V_f(t) = R_f I_f(t) + L_f \frac{dI_f(t)}{dt}$$

Exercises and Problems

2.1 Consider an elevator system that has an inclined tracking rail with viscous friction, $F_v(t)$, and a counterweight, as illustrated in Figure 2.20. It is necessary to develop the elevator dynamics model relating the permanent-magnet DC motor applied voltage and the cabin vertical position. Using the system schematic in Figure 2.20 and Table 2.24 of parameters and variables, answer the following questions. State any assumptions made.

   a) By applying the appropriate physical laws of conservation and continuity, what is the elevator system dynamics model?
   b) How can this model be refined or reduced?
   c) What is the effect of stiction phenomena on this dynamics model?

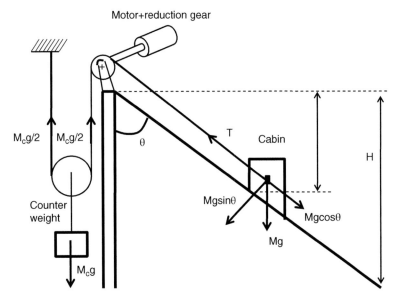

Figure 2.20 Inclined elevator schematic and its components.

Table 2.24 Variables and parameters of the elevator driven by a permanent magnet DC motor.

Parameters	Description and values	Variables	Description and values
M_{cw}	Mass of counterweight (kg)	$V_a(t)$	Applied DC input voltage (V)
M_c	Mass of cabin (kg)	$D(t)$	Mass of passenger (kg)
M	Mass of passenger of (kg)	$i_a(t)$	Armature current (A)
h	Inter-floor height (4 m)	$T_m(t)$	Motor torque (N.m)
L_a	Motor armature inductance (H)	$\omega(t)$	Motor velocity (rad s^{-1})
		$\theta(t)$	Motor position (rad)
K_m	Motor armature constant (V s^{-1} rad^{-1})	$\omega_{gr}(t)$	Gear velocity (rad s^{-1})
J_{gear}	Gear inertia (N.m.s^2 rad^{-1})	$\omega_{dr}(t)$	Drum velocity (rad s^{-1})
K_b	Emf Motor constant (V s^{-1} rad^{-1})	$F_{cw}(t)$	Counterweight force (N)
r	Radius of drum (pulley) (m)	$F_\mu(t)$	Friction torque (N.m)
J_m	Motor inertia (N.m.s^2 rad^{-1})	$T_{dr}(t)$	Drum applied force (N)
η_{gear}	Transmission ratio	$y_c(t)$	Vertical cabin position (m)
L	Cable length (m)	$y_{cable}(t)$	Vertical cable position (m)
λ	Linear mass density	$F_c(t)$	Cabin applied force (N)
μ	Friction coefficient		
J_{drum}	Drum inertia (N.m.s^2 rad^{-1})		
R_a	Motor armature resistance (Ω)		

2.2 Infant incubator

Water vapor and fresh air are injected into a heated chamber to reproduce an appropriate and humidity temperature for a premature infant, as illustrated in Figure 2.7. Consider that the physiological activity of a premature infant generates humidity and temperature within the chamber is negligible. For safety reasons, a control formulation that decouples chamber temperature from humidity is required.

a) Write a set of differential equations relating temperature and humidity variation based on the water vapor valve position $\theta(t)$ and the heater input voltage $u(t)$ as well as the heat generated by the infant. State any assumptions made. (Hint: assume that the chamber temperature is uniformly distributed.)

b) Decouple the incubator's temperature and humidity output variables. Please detail your reasoning.

c) Solve the dynamic (differential or difference) equations for the temperature and the humidity.

d) Sketch the response of the incubator's temperature. State an expression for the steady state in terms of system parameters.

2.3 Consider a heating station (boiler) illustrated in Figure 2.21 as part of the crude oil upstream and distillation processes shown in Figure 2.22.

It should be recalled that crude oil from the upstream process is expected to be preheated to a temperature range of 390–400°C within the gas-powered furnace prior to any distillation process. Hence, depending on the crude oil quality or purity and the level of obstruction in the pipeline, maintaining the steady outflow temperature could be challenging, forcing some plants to perform regular and costly maintenance operations. Thus, it is desired to develop a temperature model of this crude oil preheating process.

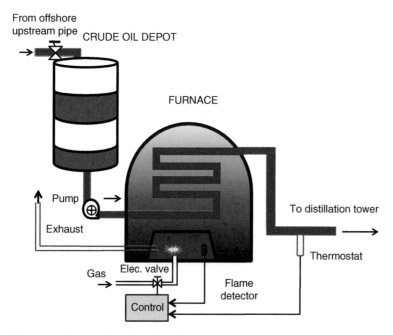

Figure 2.21 Furnace-based crude oil heating system.

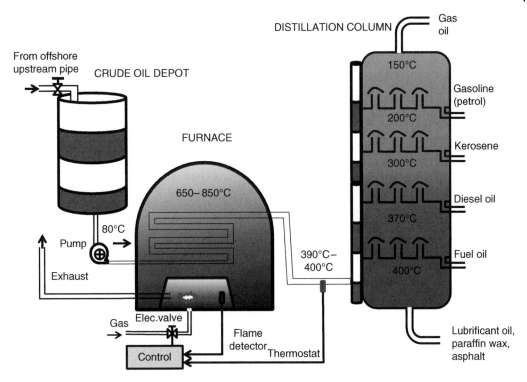

Figure 2.22 Crude oil preheating and distillation.

Answer the following questions:

a) What is the differential equation relating the furnace chamber temperature $T_f(t)$ to the upstream crude oil flow rate $F_i(t)$ and the flow rate of the gas supply to the furnace $q(t)$? State any assumptions made.

b) Similarly, derive the relation between the furnace chamber temperature, $T_f(t)$, and the temperature of the crude oil exiting the furnace, $T(t)$.

c) What are the possible process disturbances?

d) State any assumptions about heat losses within the furnace. Applying the energy conservation principle within the furnace, derive a refined model determining the variation of the temperature of the crude oil exiting the furnace, $T(t)$, from the variations in the furnace temperature and the flow rate and pressure of the crude oil. The model should consider the heat capacity due to the crude oil's chemical composition, the furnace's heat capacity, dust in the serpentine pipe, and temperature losses from poor insulation. It is assumed that the mass balance of crude oil is given by:

$$\frac{\rho dV(t)}{dt} = \rho F_o(t) - \rho F_i(t) = 0$$

e) What is the key variable affecting the level of heat of the crude oil leaving the furnace? How can its time constant be derived?

f) What would be the difference in the dynamical model if this process was a batch process instead of a continuous process (Table 2.25)?

2.4 Among medical surgery applications, laser-based thermotherapy uses generated heat to destroy a lesion in a patient with tumor cells. This is achieved through the absorption of the generated laser light by tissue and its transformation into heat. In order to

Table 2.25 Some key variables and parameters of the crude oil preheating process.

Parameters	Description and values	Variables	Description and values
R_0	Insulation thermal resistance (kg)	$T(t)$	Crude oil temperature exiting furnace (°C)
C_g	Heat capacity of air (kJ kg^{-1}°C^{-1})	$T_u(t)$	Crude oil upstream temperature (°C)
C_s	Heat capacity of crude oil (kJ kg^{-1}°C^{-1})	$T_f(t)$	Furnace temperature (°C)
λ	Heat evaporation of oil (kJ kg^{-1})	$q(t)$	Flow rate of gas supply to furnace (m s^{-3})
V	Crude oil volume in the furnace (m³)	$F_i(t)$	Upstream crude oil flow rate (m s^{-3})
T_{ref}	Desired furnace exit temperature 390–400°C	$\omega(t)$	Speed of upstream pump (rad s^{-1})
K_1	Convection coefficient	$F_o(t)$	Downstream crude oil flow rate (m s^{-3})
K_2	Conduction coefficient (kg m^{-3})		
ρ	Crude oil volumetric mass (kg m^{-3})		
D	Diameter of serpentine pipe (m)		
L	Length of serpentine pipe (m)		

deliver efficient local treatment, as well as to prevent thermally induced damage on a healthy organ or vessel of the patient, it is necessary to understand this thermo-biological phenomenon. Using a super-luminescent diode (SLD) light source, a laser wavelength between 450 nm and about 1600 nm is directed in a series of pulses toward the human tumor cell. It absorbs this transferred energy until it completely melts after the breaking of its chemical bonds.

Among the factors defining the laser-based cancer treatment process, there are: (i) the optical properties of laser light generation, propagation, and absorption and transmission by tissues; (ii) the thermal properties of the energy transferred over tissue by heat radiation and conduction; and (iii) the physiological properties of tissue destruction (melting and vaporization). This treatment process should also ensure healthy vessel or organ damage prevention, as illustrated in Figure 2.23. Thus, for safe laser surgery, it is required to describe the heat transfer relationship linking the energy-based laser-light absorption by the patient's tumor tissue with the subsequent surrounding temperature distribution.

a) Perform a process decomposition of the laser-based treatment and list some key input/output process variables in a table, as well as some key parameters related to optical signal diffusion and temperature distribution.
b) Using the optic light diffusion law, derive the tissue absorptivity model to quantify the percentage of light absorbed $I_{abs}(t)$, scattered, reflected, and refracted over the laser light transmitted through two layers of skin tissue.
c) Assuming the focus diameter d, the laser's penetration depth h and scanning speed, the latent heat (for fusion and evaporation), the power, and the laser-focused spot diameter are known and constant, and considering that the laser-tissue interaction when the laser light energy $Q_I(t)$ absorbed by tissue is converted into heat, $Q(t)$, the model for heat generation by light absorption is given by:

$$Q_I(t) = I_{abs}e^{-\mu_a b}\beta$$

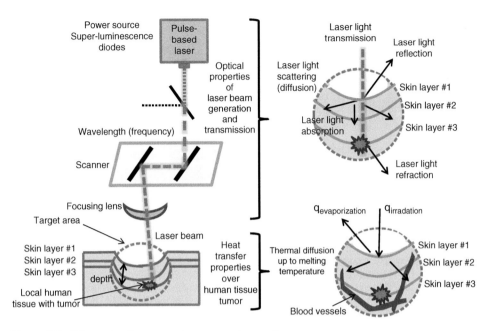

Figure 2.23 Schematic of laser-based surgery on human tissue.

where β is the given scattering coefficient, b is the distance between the laser beam and the tissue end point, and μ_a is the transmitted laser photon absorption or dissipation coefficient, that is, the ratio of the number of photons absorbed in the given grid cell divided by the total number of photons transmitted times the volume represented by the grid cell.

Also, consider the following assumptions: (i) the tissue is heated to melting temperature; (ii) all layers of skin tissue are opaque and have constant properties (including thermal conductivity and absorptivity); (iii) change of tissue phase from solid to vapor occurs at a single evaporation temperature; (iv) tissue evaporation does not interfere with the transmitted laser signal; (v) the spot size of the laser signal is constant; (vi) any power losses from the beam reflection by mirrors used to change the laser beam direction are negligible; and (vii) the metabolic heat generation per unit volume is negligible. By applying the 2D Pennes's bioheat equation given by:

$$\rho C_p \frac{\partial T(t)}{\partial t} + K\left(\frac{\partial^2 T(t)}{\partial x^2} + \frac{\partial^2 T(t)}{\partial y^2}\right) = Q_{\text{conv}}(t) + Q_l(t) + Q_{\text{blood}}(t)$$

with K being the thermal conductivity, $Q_{\text{conv}}(t)$ being the convection between the tissue surface and surrounding air, $Q_l(t)$ the heat energy transmitted by laser generation, and $Q_{\text{blood}}(t)$ is the metabolic blood heat absorption and perfusion, update the list of parameters and variables accordingly and derive the solution for the temperature $T(t)$ surrounding the tissue under treatment. Please state two additional assumptions.

2.5 A lathe is a turning machine tool that is used to cut a cylindrical workpiece. First, the workpiece is clamped by a tailstock. Then, through the feed-based horizontal penetration of the cutting tool into the rotating workpiece at spindle speed $\omega(t)$ over its longitudinal axis, a part of the workpiece material is removed, as illustrated in Figure 2.24. This turning

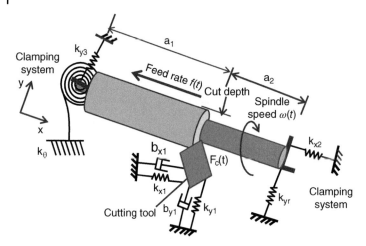

Figure 2.24 Equivalent workpiece cutting process with the clamping system.

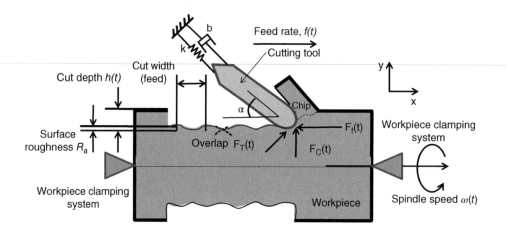

Figure 2.25 Cross-sectional view of the cutting process with a machine tool.

process is performed for several successive revolutions causing cutting over the work-piece surface. Thus, the surface roughness is dependent on the feed rate, the cut depth, the cutting speed, the angle of approach, and the hardness of the workpiece material. It is also related to chatter phenomenon from tool oscillations.

During this process, the forces acting on the workpiece are the radial thrust force $F_r(t)$, the feed force $F_f(t)$, and the cutting force $F_c(t)$ provided by the cutting tool. In order to model this cutting tool structure, a lumped mass-spring-damper system with two degrees of freedom is considered, as shown in Figures 2.24 and 2.25. Consider the following assumptions: (i) the cutting tool edge is sharp; and (ii) the deformation of the workpiece due to the tool structure is negligible.

a) Upon decomposition of the lathe machining process into: (i) the feed drive system, and (ii) the cutting process, list all involved key variables and parameters in a table.

b) Based on experimental cutting process data analysis on a 6061-aluminum alloy work-piece, the required quality of the surface, R_a, can be expressed as:

$$R_a = 0.23e^{0.24F}$$

It is required to derive the model relating the surface roughness R_a with the cutting force $F_c(t)$, the thrust force $F_t(t)$, the cut depth, the cutting tool material, the function feed rate f, and the speed $\omega(t)$. Assuming that the combined resistive forces of tool penetration are:

$$F_x = k_x h(t)$$

$$F_y = k_y h(t)$$

where k_y, k_x are the cutting stiffness in the x- and y-directions, while the cut depth yields:

$$h(t) = h_0 + (1 - \xi)h_c(t)$$

with h_0, ξ, and $h_c(t)$ being, respectively, the nominal cut depth, the overlap constant, and the cutting tool lateral displacement in the x-axis direction. Hence, using the schematics depicted in Figure 2.24, considering all forces resisting the tool penetration, and assuming that the machine tool structure is only flexible in the y-direction, derive the 2D dynamics vibration model of the lathe machining process on the workpiece. (Hint: the workpiece clamping system at both ends is represented by translational and rotational springs.)

c) Deduce the equation defining the surface finishing model of the 6061-aluminum alloy workpiece over the resistive forces and operating variables (feed rate, speed).

2.6 Solar radiation systems have been used to deliver hot water or air for the central heating of living installations (e.g. in apartment buildings). Here, the energy from the solar radiation is converted into heat through solar collectors heating the fluid (usually water or air). This is subsequently transferred into a storage tank. Such systems use pumps or gravity to move the fluid between the collector and the storage tank. The plate collector consists of an absorber made of a metal sheet with high thermal conductivity (i.e. aluminum or copper) covered by glass to increase light intensity and reflexivity, as illustrated in Figure 2.26.

In this exercise, this system is used to create suitable climate characteristics (humidity and temperature) for a barn with double rooms, from a time-dependent energy source (i.e. the sun). Consider 50 × 30 × 20 m barn spaces that have a solar water heating system with a pump regulating the hot water flow rate to the radiators in the double-room barn as illustrated in Figure 2.27.

a) Recall that the transferred mass of water circulating from the solar collector to the heated barn is the energy carrier. Perform a functional decomposition of the solar-based heating system for the barn unit into process components.

b) Based on this decomposition, fill out all the relevant process variables and parameters in Table 2.26.

c) From the following assumptions, state, when valid, their implications:
   i) All heat transport phenomena are taken in the 1D flow direction, except for the heat carried.
   ii) Process component physical properties are dependent on temperature.
   iii) The energy is only transferred in the flow direction by mass transfer.

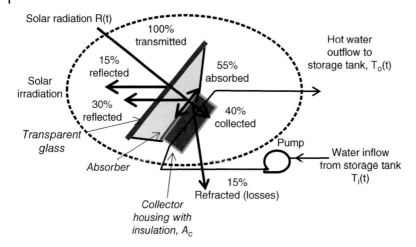

Figure 2.26 Detail of the plate collector of solar heating system.

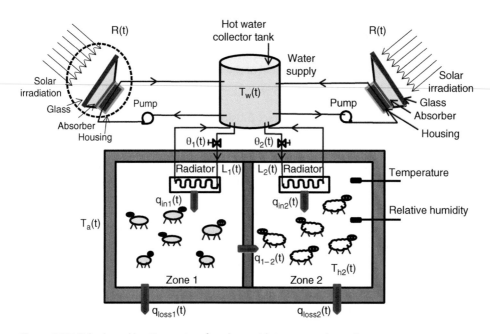

Figure 2.27 Solar-based heating system for a barn with two rooms (zones).

    iv) The properties of glass and insulation are independent of temperature.
    v) Solar radiation and ambient conditions are time-dependent.
    vi) Dust and dirt on the collector are negligible.
    vii) The mass flow rate in the collector tube is uniform.
    viii) Heat is lost to the atmosphere either by convection or by radiation from the heating infrastructure.

  d) Based on the energy balance principle, develop a 1D heat transfer mathematical model for each component of the solar-based heating system. It is assumed that: (i) the heat capacity of the air gap between the cover and the absorber plate is negligible; (ii) the longitudinal heat conduction in the absorber plate and pipe wall is

Table 2.26 Solar heating process variables and parameters.

Symbol	Process parameters	Symbol	Process variables
–	Collector area (m²)	$q_{in1,2}(t)$	Heat transferred into barn (mJ h⁻¹)
C_w	Collector heat loss constant (W m⁻² °C⁻¹)	–	Absorbed solar irradiation per unit collector area (kJ m² h⁻¹)
–	Water heat capacity (kJ kg⁻¹ °C⁻¹)	–	Ambient temperature (°C)
–	Thermal resistance of barn walls (W m⁻¹ °C⁻¹)	–	Barn temperature (°C)
–	Specific heat capacity of air heat capacity of (kJ kg⁻¹ °C⁻¹)	$T_w(t)$	Temperature of water in tank (°C)
		–	Temperature of water exiting the collector (°C)
		$T_i(t)$	Temperature of water returning to the collector (°C)

also negligible; and (iii) windows, walls, doors, and ceilings of the double-room barn have the same heat transfer coefficient.

e) Develop a linear model of the heating system describing the dynamics relationship between process output variables, such as the water tank temperature, and input variables, such as the average animal temperature, irrelevant the barn's interior and exterior temperatures.

f) Analyze the heating system's response to: (i) negative energy collected; and (ii) large changes in ambient temperature and solar radiation.

g) For safety reasons, if the pump is deactivated, derive the corresponding linear model of the water tank temperature found in (e).

2.7 Pharmaceutical, food, or even cosmetic industries use extraction approaches to separate specific chemical entities from their natural sources, such as plants. Among these, a maceration technique is used for the production of non-synthetic medicinal drugs. Here, medicinal plants are soaked within a tank in a specific solvent such as ethanol for a considerable time. The efficiency of the extraction process is improved by continuously circulating the solvent through a maceration (extractor) vessel, as illustrated in Figure 2.28. In this process, the solvent is pumped from its bottom tank vessel toward the top of the maceration vessel, where it is sprayed over the surface of the macerated material. Such solvent circulation results in a uniform distribution of the drug extract concentration (equilibrium concentration) in a shorter amount of time. Depending on the solvent used (i.e. water, alcohol, chloroform, hexane etc.), as well as the medicinal plant involved, the equilibrium drug concentration varies. This steady-state concentration defines the extracted product quality. Furthermore, after each solvent circulating cycle, the solvent viscosity is increased while the flow profile is expected to be more turbulent. Hence, there is a need to adjust the pump speed accordingly to maintain the same process time constant.

For the drug extraction schematic depicted in Figure 2.28, it is required to develop a model relating the flow rate to the pump speed with respect to the change in solvent viscosity.

a) Using the double tank system depicted in Figure 2.28, perform a functional decomposition of the extraction process into subcomponents and list all relevant process variables and parameters involved in a table.

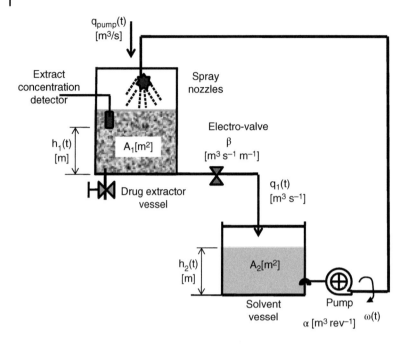

Figure 2.28 Drug extractor double-tank process.

b) By applying laws of continuity to fluid flow and considering the nonlinear valve angular model $\beta(t)$ so that the rate of flow through the valve is given by:

$$q(t) = \beta(t)\sqrt{h(t)}$$

derive the differential flow dynamics equations relating liquid level in each tank to the pump flow rate $q_{pump}(t)$.

c) Assuming that the viscosity, $\delta(t)$, doubles every five solvent circulating cycles, causing a non-laminar flow profile, update these dynamics equations.

d) If the extract concentration, $\phi(t)$, is expected to double after each three solvent circulating cycles, estimate the relationship between the initial vessel levels and the time to reach the drug extraction equilibrium concentration ($\phi(t) > 56\%$). (Hint: find the number of cycles required.)

e) Derive a method to estimate the number of cycles required to reach the steady-state extract concentration.

2.8 Motorized gyroscopic stabilizing an aircraft autopilot system

Autopilot systems are used in aircraft to ensure appropriate flight navigation during poor weather or rough air conditions, such as regions of turbulent winds. This requires a stabilizing system to automatically maintain a desired compass heading and altitude of the aircraft. Such a flight stabilizing system uses gyroscopic stabilizers associated with each axis (yaw, pitch, and roll) to detect the aircraft's change in navigation level. According to any angular deviation, appropriate torque-based compensation is provided through electro-hydraulic actuators. Consider a passenger plane using a DC motor located under the passenger cabin floor to generate the necessary stabilizing torque. It enables plane

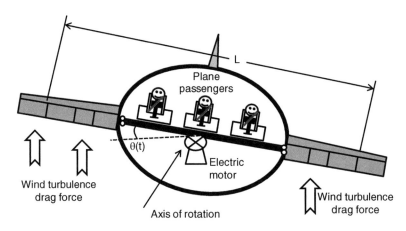

Figure 2.29 Gyroscopic aircraft stabilizing system.

stabilization once the gyroscope has detected angular (roll direction only) variation, as illustrated in Figure 2.29.

a) For a fixed number of passengers with a given total weight M_p and a platform inertia J_p, derive the linear platform dynamics relating its angular position $\theta(t)$ to the motor-generated torque $T_m(t)$. (Please the state all necessary assumptions with respect to your model, so that the wind force is acting at the center of each aircraft wing.)

b) Assuming that an abrupt wind is applied on the left wing (only in the roll direction) of the plane in autopilot mode for a short duration, derive the expected response of the system. (Hint: impulse as input signal.)

c) Describe how the system dynamics model could be modified if the wind affects its motion both in the yaw and roll directions while in the automatic pilot mode.

2.9 Gamma radiation-based food sterilization using a robot-based package-handling system: food and surgical devices are among those products that are legally required to be cleaned using a sterilization process to effectively kill almost all living microorganisms (i.e. bacteria, fungi, etc.). Typical sterilization techniques include those using pressurized vapor, ethylene oxide, gamma radiation, gas plasma, or peracetic acid. In the case of gamma radiation, a dose of gamma rays (between 1 and 30 MeV) is directed toward a product for a time duration depending on its thickness and volume, as well as the type of microorganism to be killed. Gamma radiation causes the breakage of molecular bonds in all material that is exposed. This requires handling with extreme care to avoid being harmful to humans. As such, a human-free operating system is essential for such a sterilization process.

Consider an automated sterilization process using a planar robot arm and a conveyor, as depicted in Figure 2.30. The conveyor moves the food product boxes to be irradiated to the pickup (upload) location at a predetermined speed. Then, the robot arm picks and drops them into the irradiation area, which is immersed in an underground water chamber. The gamma rays pass through the encapsulated food product and the sterilization treatment takes place for a certain time before the robot raises it back to the other side of the conveyor, as illustrated in Figure 2.30. Synchronization of the pick and drop operation of the robot arm should be considered, by computing the motion of two (x-y-axis) motorized robot arms with the displacement of the conveyor carrying the product boxes to be processed.

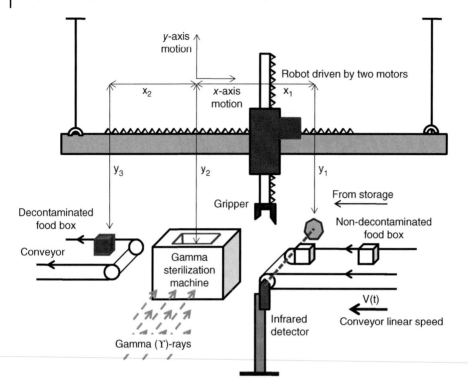

Figure 2.30 Robot handling system for gamma radiation-based food sterilization process.

Table 2.27 Typical robot handling system parameters and variables.

Symbol	Process parameters	Symbol	Process variables
m	Mass of product box (kg)	$T_{mx}(t)$	x-axis robot arm torque (N.m)
K_i	Torque constant for x-axis motor (*dimensionless*)	$T_{my}(t)$	y-axis robot arm torque (N.m)
n_{screw}	y-axis motor screw transmission ratio	$T_{\text{Load}}(t)$	Load torque (N.m)
M	Mass of robot arm (kg)	$\omega_x(t)$	x-axis angular velocity (rad s^{-1})
Δt	Sterilization processing time (s)		

a) Perform a functional decomposition of the conveyor-robot handling system depicted in Figure 2.30 into various subprocesses and list all relevant assumptions with respect to the motion profile of the food product boxes.

b) Based on this decomposition, complete Table 2.27 with all relevant process variables and parameters.

c) In order to synchronize the conveyor speed, derive a dynamic model for the 2D axis positioning of this robot-conveyor handling system that enables transfer of a food product box from a fixed arrival conveyor location $P_1(x_1, y_1)$ to a determined irradiation location $P_2(0, y_2)$. Then, it moves to the departure conveyor position $P_3(x_3, y_3)$.

2.10 Electric vehicles with rechargeable batteries have limited traveling distances and require frequent power recharging. In an attempt to improve this, it has been considered to

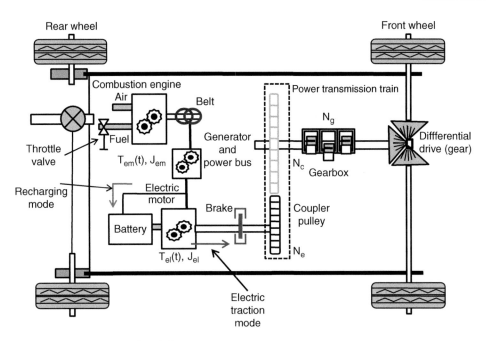

Figure 2.31 Series hybrid diesel-electric powertrain.

equip hybrid vehicles with an internal combustion engine coupled to an electrical motor/generator with a recharging function. This hybrid vehicle with a series configuration is such that the engine is capable of supplying the energy for recharging the battery based on the power requirements.

Consider a hybrid vehicle that has two operating modes: (i) an electric mode, when the electric motor generates the traction force; and (ii) a recharging mode, when the combustion engine delivers the traction force while charging the battery, as depicted in Figure 2.31. Hence, in order to reduce the fuel consumption and toxic emissions, the hybrid vehicle must optimize the battery recharging cycles over its driving range. Based on the vehicle's traction torque requirements and the battery's recharging power requirements, an optimized strategy would consist of splitting the required traction force between the engine and the electric motor.

a) This hybrid system can be decomposed into: (1) energy-related components, such as (i) energy storage (i.e. fuel storage tank and ultra-capacitors), and (ii) energy conversion (i.e. power amplifier, clutch/torque converter, combustion engine, generator, electric motor, transmission, and drive); and (2) dynamics processing components including motor traction, aerodynamic drag reduction, and rolling resistance tires. Torque couplers are used as a link between these components. The degree of hybridization is the ratio (between 15 and 60%) between the electric motor power and the engine power. Derive the block diagrams of each component with the corresponding variables and parameters involved.

b) Among the following assumptions, discuss their implications:
   i) the power losses in the final transmission are negligible with respect to other sources of power losses;
   ii) all vehicle powertrain components connecting the motors to wheels are rigid;
   iii) temperature variation affects the rate of reaction and the ionization of the electrolyte within the ultra-capacitors.

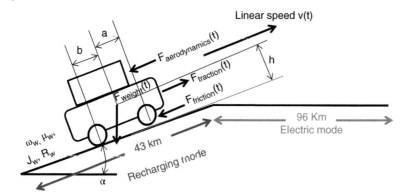

Figure 2.32 Vehicle longitudinal motion dynamics.

c) Based on the schematics of the powertrain in Figure 2.31 and the vehicle in motion in Figure 2.32, consider that the net vehicle dynamics could be summarized as:

$$m\frac{dv_{\text{vch}}(t)}{dt} = F_{\text{traction}}(t) + F_{\text{resistive}}(t)$$

$$= F_{el}(t) + F_{em}(t) + F_{\text{friction}}(t) + F_{\text{weight}}(t) + F_{\text{aerodynamics}}(t)$$

and consider that the NiCd batteries' rate of discharge state change is given by:

$$\frac{SOC(t)}{dt} = \frac{-V_{oc} + \sqrt{V_{oc}^2 - 4P_{\text{batt}}R_{\text{batt}}}}{2C_{\text{batt}}R_{\text{batt}}}$$

where V_{oc}, R_{batt}, and C_{batt} are, respectively, the battery open circuit voltage, its internal resistance, and the capacity of the battery. Hence,

$$P_{\text{batt}} = P_{\text{mot}} + P_{gen} + P_{\text{motor-loss}} + P_{\text{gen\_loss}}$$

Thus, derive the vehicle dynamics equations of all related system components, especially the battery discharging model, the torque generation model, the vehicle dynamics model, and the energy consumption model.

Develop the dynamics equation relating the discharging of the battery to the traction torque produced by the electric motor during electric mode.

2.11 Indoor farming methods aim to increase agricultural food production using reduced amounts of water and space compared to traditional methods. They also limit negative environmental impact through reduced soil degradation. However, this indoor cultivation for artificial plant growth, independent from the weather and soil conditions (e.g. winter, desert etc.), requires: (i) lighting exposure through Light Emitting Diodes (LEDs) to replace sunlight for photosynthesis; (ii) suitable temperature, humidity, and watering levels; and (iii) specific nutrients. These installations are very costly, especially due to the electric power required for the lighting. It is possible to use solar cells to supply such an energy load.

Consider a multilevel hydroponics installation (with three tiers of shelves), capable of growing indoor fruit and vegetables (e.g. lettuce, strawberries, tomatoes, eggplants, cabbage, sweet potatoes, or red and green chili peppers). It uses an LED light source, as illustrated in Figure 2.33. Here, the warm air and water vapor are injected into the farming installation, where lettuces are grown on soils containing nutrient solutions (fertilizer, salts, pH adjustments) and enriched minerals.

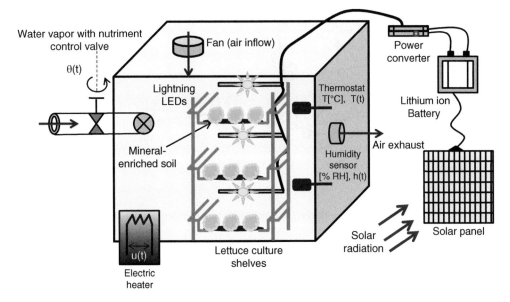

Figure 2.33 Indoor lettuce farming.

a) Considering indoor farming process variables such as the relative humidity, temperature, lighting quantity and quality (e.g. wavelength, intensity, fluence rate, photoperiod), as well as plant growth level (leaves), perform the functional decomposition of the farming process into subcomponents and list all relevant process parameters and variables.

b) Discuss the implications of the following assumptions:
   i) plant transpiration through proper air ventilation has a negligible heat transfer;
   ii) the interior temperature is normally distributed;
   iii) the farming configuration uses a single type of building material for insulation.

c) Develop the dynamics model capturing the relationship between the water vapor valve, the electric heater voltage, and the temperature and humidity inside an indoor farm with a multilevel installation. What will happen if the electric heater operates intermittently (discontinuously)?

d) Assuming that: (i) blue and red lights have been used preferentially for their optimal effect on the illumination absorption of plant morphology and photosynthesis; and (ii) an intermittent light operating mode improves plant photosynthesis (i.e. when lettuce leaves are exposed to light, they transform the water and carbon dioxide into glucose, water, and oxygen); recalling that each 3-day, the distance between the LED and plant need to be readjusted such that lumen per meter square of pleant leaf remained constant, develop the model of LED lighthing using the model for light absorption inside the leaf given by:

$$\frac{Lumens}{m^2} = \frac{I(t)}{\left(h(t) - 0.025\frac{a(t)}{h_o}\right)^2} = constant$$

where h is the distance between the lighting device and the leaf surface, h_o is the leaf thickness, I_o is the incident irradiance (blue and red), and a is the leaf surface area.

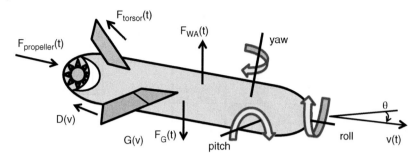

Figure 2.34 Electrically-driven UV.

e) Present a methodology to derive the relationship between lettuce leaf growth and light.

f) Derive the methodology to size power related to the battery based on the lighting requirements.

2.12 An underwater vehicle's (UV) navigation dynamics over roll (forward motion), pitch (up/down motion), and yaw (left/right motion) are highly dependent on the pressure, depth, tide, weather (fluid density and speed), propeller diameter, torque, and volumetric flow rate. Consider a simplified UV driven by an electric propeller on the tail and three motor-driven wing tips, as illustrated in Figure 2.34. As such, the synchronization of all the UV motors' speeds and directions is required.

a) Perform the functional decomposition of the 3D UV navigation system into subcomponents.

b) List all relevant UV navigation system parameters and variables when it is only moving due to its propeller, which is capable of generating the required hovering force.

c) Discuss the implications of the following assumptions:
   i) the center of mass coincides with the center of buoyancy;
   ii) the mass distribution is homogeneous;
   iii) third- and higher-order hydrodynamic drag terms are neglected;
   iv) yaw, pitch, and roll motions can be neglected;
   v) weight force and the upward Archimedes force are not applied at exactly the same point.

d) Recall from fluid dynamics, an immersed object displaces a certain volume of surrounding fluid, called the added mass $M_{added_{mass}}$. Consider $u(t)$ as the linear velocity, m_{uv} as the mass of the UV, cross-flow drag, and propeller thrust force $F_{propeller}(t)$ ρ is the water density and K_f is the axial flow form factor, which has to be identified. By applying Newton's first principle to the center of gravity of the UV depicted in Figure 2.34 for one degree of freedom, derive the dynamics equation relating its speed and all forces involved.

Table 2.28 Typical operating conditions of a solar-based heating process.

Time (hour)	Chamber 1 temperature $T_c(t)$ (°C)	Chamber 1 humidity %	Ambient/ chamber 2 temperature $T_a(t)$ (°C)	Ambient/ chamber 2 humidity $H_a(t)$ (%)
6 a.m.	0.0	−6	1	0
7 a.m.	0.0	−7	1	0
8 a.m.	0.7	−5	1	0
9 a.m.	1.8	−2	1	0.39
10 a.m.	2.2	1	0	0.63
11 a.m.	2.4	3	0	0.77
12 a.m.	3.1	3	0	0.86
1 p.m.	3.0	4	0	0.52
2 p.m.	2.5	5	0	0.31
3 p.m.	2.0	5	0	0.19
4 p.m.	1.3	4	1	0.11
5 p.m.	0.3	0	1	0.07
6 p.m.	0.0	−2	1	0.04

2.13 Consider the climate change modeling of a barn with two rooms, developed as depicted in Figure 2.27. The pig barn is divided into distinct climatic zones with specific humidity and temperature characteristics.

a) Discuss the implications of the following assumptions:
   i) zone climates affect each other by the presence of internal air flow;
   ii) there is no significant climate gradient within the barn;
   iii) the pressure dynamics are an order of magnitude greater than the climate dynamics.

b) Consider temperature changes within the pig barn building for a 1-day period summarized in Table 2.28. The stochastic model structure and order are given by:

$$C_1 \frac{dT_1(t)}{dt} = \frac{T_1(t) - T_2(t)}{R_1} + \frac{T_a(t) - T_1(t)}{R_o} + \phi + A\Phi$$

where Φ is the solar radiation, ϕ the heating system, C_1, the heat capacity, $T_1(t)$ the temperature in the first zone of the barn and $T_2(t)$ the temperature in the second zone of the barn, R_1 is the resistance to heat transfer to the second zone, $T_a(t)$ is the ambient temperature, and A is the area to the outside. R_o is the resistance to heat transfer between inside and outside. Based on process data and this model structure, apply the least-square estimation method to derive the model.

2.14 A hot rolling mill process consists of reducing the thickness of a metal sheet by passing it several times between two parallel rolls that compress it, as illustrated in Figure 2.35. Here, the screws attached to the roll and driven by two DC motors are used for large

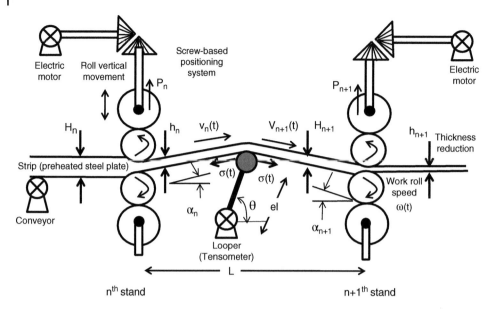

Figure 2.35 Roll mill two-stand schematic.

position changes between passes. The metal plate thickness depends on the roll position, the rolling force, the plate thickness, and its temperature.

Considering that plate roll eccentricity has to be avoided:

a) Discuss the implications of the assumption that the temperature variations are due to inhomogeneous heating by reheating furnaces and inhomogeneous cooling by the roller tables during plate transportation.

b) Elaborate a thickness and tension modeling strategy by capturing the relationship between tension and speed (Hint: the rolling force equation is a function of the entry thickness, exit thickness, and tension at the front of the stand.)

c) Using the rolling mill illustrated in Figure 2.35, apply Newton's law to derive the dynamics equations relating the metal plate thickness reduction to the roller position, rolling force, pressure, conveyor speed, and plate temperature after each pass.

Bibliography

1 Baik, O.D., Grabowski, S., Trigui, M. et al. (1999). Heat transfer coefficients on cakes baked in a tunnel type industrial oven. *Journal Food Science* 64 (4): 688–694.

2 Bergman, R.N., Ider, Y.Z., Bowden, C.R., and Cobelli, C. (1979). Quantitative estimation of insulin sensitivity. *American Journal Physiology-Endocrinology and Metabolism* 236: E667–E677.

3 Burton, T., Sharpe, D., Jenkins, N., and Bossanyi, E. (2001). *Wind Energy Handbook*, vol. 1. Chichester, UK: Wiley.

4 Carsonc, E. and Cobelli, C. (2001). *Modeling Methodology For Physiology and Medicine*. San Diego, CA: Academic Press.

5 Cinar, A. (2003). *Batch Fermentation: Modeling, Monitoring, and Control*. CRC Press.

6 Omstead, D.R. (1989). *Computer Control of Fermentation Processes*. CRC Press.
7 Friis-Jensen E. *Modeling and Simulation of Glucose-Insulin Metabolism*. Ph.D. Thesis, Technical University of Denmark, Kongens Lyngby, Denmark, May, 2007.
8 Fox, J.A. (1977). *Hydraulic Analysis of Unsteady Flow in Pipe Networks*. London, UK: The MacMillan Press Ltd.
9 Furnass, R.J. and Ulsoy, A.G. Dynamics modeling of thrust force and torque for drilling. In: *Proceedings of the American Control Conference*, 384–390. Chicago, IL.
10 Gay, S.E., Emadi, A., Ehsani, M., and Gao, Y. (2004). *Modern Electric, Hybrid Electric, and Fuel Cell Vehicles: Fundamentals, Theory, and Design*. CRC Press.
11 Groover, M.P. (2007). *Fundamental of Modern Manufacturing*. Wiley.
12 Lyon, A. and Püschner, P. (2010). *ThermoMonitoring: A Step Forward in Neonatal Intensive Care*. Germany: Drägerwerk AG & Co.
13 Marcotte, M. (2007). Heat transfer in food processing. In: *WIT Transactions on State of the Art in Science and Engineering*, vol. 13 (ed. S. Yanniotis and B. Sundén), 27. WIT Press https://doi.org/10.2495/978-1-85312-932-2/08.
14 Perez, J.M.R., Golombek, S.G., Fajardo, C., and Sola, A. (2013). A laminar flow unit for the care of critically ill newborn infants. *Medical Devices: Evidence and Research* 6: 163–167.
15 Boothnyd, G. and Knight, W.A. (1989). *Fundamentals of Metal Cutting and Machine Tools*. New York: Marcel Dekker.
16 Kalpakjian, S. and Schmid, R.S. (2001). *Manufacturing Engineering and Technology*. Prentice Hall.
17 Olsen, B., Shaw, S.W., and Stepan, G. (2003). Nonlinear dynamics of vehicles. *Vehicle System Dynamics* 40 (6): 377–399.
18 Rath G., *Model based Thickness Control of The Cold Strip Rolling Process*, D.S. thesis, University of Leoben, Austria, 2000.
19 Roffel, B. and Betlem, B. (2006). *Process Dynamics and Control Modeling for Control and Prediction*. Wiley.
20 Simon-Santos A. *Contribution a la Conception des Sous-Marins Autonomes : Architecture des Actionneurs, Architecture des Capteurs d'Altitude, et Commandes Référencées Capteurs*, Ph.D. thesis, Ecole Nationale Supérieure des Mines de Paris, 1995.
21 Shaw, M.C. (1984). *Metal Cutting Principles*. Oxford University Press.
22 Sorensen J.T., *A Physiologic Model of Glucose Metabolism in Man and Its Use to Design and Access Improved Insulin Therapies for Diabetes*, Ph.D. thesis, MIT, Cambridge, MA, 1985.
23 Wellstead, P.E. (1979). *Introduction to Physical System Modeling*. Academic Press.

3

Discrete-Time Modeling and Conversion Methods

3.1 Introduction

Typical electrically-driven system and process signals (e.g. volts, pressure etc.) require a digital approximation in time and quantization in magnitude before being processed in computer operations such as signal sampling, control program execution, and converted signal (data) storage. This is achieved by using signal conversion instrumentation that enables transducing and scaling of analog into digitally-processed signals or vice versa. These devices allow any converted discrete signal to be a good approximation of the equivalent analog signal processing. In addition, signal conversion has to be carried out in real time. Subsequently, the characterization or modeling of discrete-time process signals can be derived by using difference equation, z-transform, discrete Fourier transform, or signal converting devices (e.g. transformations such as impulse-invariant or step-invariant, hold elements such as zero-order, first-order, etc.).

In this chapter, preliminary discrete approximation tools of continuous systems are reviewed. Here, modeling techniques to derive discrete approximations of continuous systems, digital signal processing, and approximation of an analog signal to derive a discrete dynamics model are discussed. A technological-based description of computer control architecture and interface is given with respect to data acquisition (DAQ) unit operations (e.g. successive approximation, dual-slope analog-to-digital converters (ADC), Delta-encoded ADC conversion techniques etc.) from bus structure to data gathering, logging, and processing with respect to signal noise reduction and approximation considerations. The limitations on the sampling period with respect to the quality of discrete model approximations and its effect on controller computing delays, along with methodology for selection of the sample period, are also covered. Overall, topics include continuous signal digital conversion and reconstruction technology and discrete-time command sequence generation, computer control interface, and sample time selection.

3.2 Digital Signal Processing Preliminaries

3.2.1 Digital Signal Characterization

Typical digital signal processing and process signals manipulation are illustrated in Figure 3.1. Here, a periodic detection of a continuous signal $x(t)$, by switching the opening and closing of a device at a frequency time $1/T$ for a very small time duration p, could lead to a sampled chain of impulses $x_p^*(n)$ with a magnitude nearly equal to the sampled value of $x(t)$ at the sampling instants. Such signals are called discrete-time signals or digital signals that could be real or complex, and a time-varying chain of numbers $X(n)$, usually at equally spaced intervals, so

Control Of Mechatronic Systems: Model-Driven Design And Implementation Guidelines,
First Edition. Patrick O.J. Kaltjob.
© 2021 Patrick O.J. Kaltjob. Published 2021 by John Wiley & Sons Ltd.

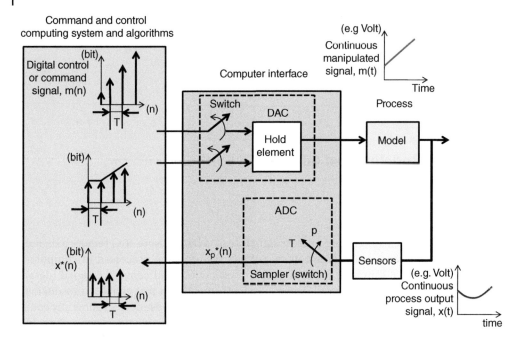

Figure 3.1 Generic digital processing of continuous process signals.

that $t = nT$ leads to $x(t) \approx x(nT)$. Hence, these signals can be derived either through continuous signal sampling and quantizing at equal time intervals, or by collecting discrete values of variables at equal time intervals over a period of time.

For the purpose of model analysis, it is useful to have a mathematical description of these digital signals. These signals can be mathematically captured either using: (i) their discrete-time value signal characteristics such as their sample amplitudes and their corresponding time values; or (ii) their discrete-time sinusoidal signal characteristics such as their frequency, amplitude and phase values. The sequence of sample signals is characterized by a temporal or spatial domain representation in the first case, whereas in the second case it is done through a frequency domain representation. Signals are converted from the temporal or space domain to the frequency domain, usually through the Fourier transform. The Fourier transform converts the signal information to a magnitude and phase component of each frequency. Often, the Fourier transform is converted to the power spectrum, which is the magnitude of each frequency component squared.

For the temporal and spatial domain representation, Figure 3.2(a) illustrates the signals unit sample train (sequence) from an ideal sampler such that:

$$\delta(n) = \begin{cases} 0 & for \quad n \neq 0 \\ 1 & for \quad n = 0 \end{cases} \tag{3.1}$$

Each unit sample signal $\delta(n)$ has unity value at $n = 0$ and zero everywhere else. Hence, it is used to describe an output sampler signal sequence $x^*(n)$, which is a train of unit samples that have nonzero values only between the time delay T at $0, 1T, 2T, 3T, \ldots, nT$ and the corresponding amplitudes $A(0), A(1T), A(2T), A(3T), \ldots, A(nT)$, as shown in Figure 3.2(b).

Then, from the graphical representation of the discrete-time signal in Figure 3.2(b) and assuming after each sampling the value of $x^*(n)$ remains constant until the following next sample, a mathematical representation for the sequence $x^*(n)$ yields:

$$x^*(n) = \ldots + A(-2T)\delta(n+2) + A(-T)\delta(n+1) + A(0)\delta(n) + A(T)\delta(n-1) + \ldots \tag{3.2}$$

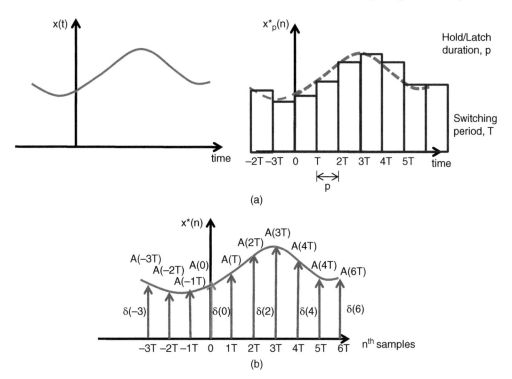

Figure 3.2 (a) Continuous function $x(t)$ and output of a generic sampler $x_p^*(n)$. (b) Equivalent discrete-time signal $x^*(n)$.

which could be equivalent in the generic form of:

$$x^*(n) = \sum_{-\infty}^{+\infty} A(kT)\delta(n - k) \tag{3.3a}$$

while from the frequency representation, the sampler sequence $x^*(n)$ would yield:

$$x^*(n) = r^n e^{j\theta n} = r^n(\cos\theta_n + j\sin\theta_n) \tag{3.3b}$$

Defining for any n being an integer $n = t/T$, and any T sample period, and using a backward shift z^{-k} in Equation (3.1), this yields:

$$\delta(n - k) = z^{-k}\delta(n) \tag{3.4}$$

with k being an integer. Therefore, a function $x(t)$ can be described by summing the translated and scaled unit sequence of samples in the form of:

$$x^*(n) = \sum_{-\infty}^{+\infty} x(kT)\delta(n - k) = \sum_{-\infty}^{+\infty} x(kT)z^{-k}\delta(n) \tag{3.5}$$

with $x(kT)$ being the value of function $x(t)$ at $t = kT$.

Example 3.1   The sample sequence of the ramp input in Figure 3.3 is given by:

$$x^*(n) = \ldots + 0\delta(n+1) + 0\delta(n) + A(T)\delta(n-1) + A(2T)\delta(n-2) + A(3T)\delta(n-3) + \ldots$$

with a slope of A

$$x^*(n) = \ldots + 0\delta(n+1) + 0\delta(n) + AT\delta(n-1) + 2AT\delta(n-2) + 3AT\delta(n-3) + \ldots$$

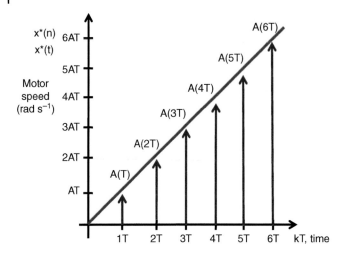

Figure 3.3 Discrete-time ramp-like signal.

which could be simplified as:

$$x^*(n) = AT(\ldots + 0\delta(n+1) + 0\delta(n) + \delta(n-1) + 2\delta(n-2) + 3\delta(n-3) + \ldots)$$

equivalent to:

$$x^*(n) = AT(\ldots + 0z + 0z^0 + 1z^{-1} + 2z^{-2} + 3z^{-3} + \ldots)\delta(n)$$

and from the infinite series, the function can be simplified to

$$x^*(n) = \frac{ATz^{-1}}{(1 - z^{-1})^2}\delta(n)$$

Hence, any time-varying signal sampled at equally spaced time intervals can be represented by sequence process data information called a sample sequence or time series. Indeed, time-varying process inputs can be described as time series, which could be simplified. A more detailed table of equivalence between the differential equation and difference equation is shown in Appendix C, while some examples are depicted in Table 3.1.

Table 3.1 Examples of signals equivalence between continuous and discrete domain.

Continuous-time domain	Laplace domain	Discrete-time domain
h(t)	H(s)	H(z)
$u(t)$	$\dfrac{1}{s}$	$\dfrac{z}{(z-1)}$
t	$\dfrac{1}{s^2}$	$\dfrac{Tz^{-1}}{(1-z^{-1})^2}$
$\dfrac{t^2}{2}$	$\dfrac{1}{s^3}$	$\dfrac{T^2 z(z+1)}{2(z-1)^3}$
e^{-at}	$\dfrac{1}{s+a}$	$\dfrac{z}{z - e^{-aT}}$

3.2.2 Difference Equation: Discrete-Time Signal Characterization Using Approximation Methods

It is possible to characterize a discrete process dynamical by a model of the relationship of any process input signal sequence to an output signal sequence, called the difference equation. There are several algorithms for obtaining difference equations from differential equations. These are also known as numerical methods for solving differential equations, and a first-order differential equation is used in the form of:

$$\frac{dx(t)}{dt} = f(x(t), t) \tag{3.6}$$

This could be extended for the higher derivatives. As depicted in Figure 3.4, by using numerical integration, there are three common ways to approximate the area:

- forward rectangle (approximately by looking forward from)
- backward rectangle (approximately by looking backward from)
- trapezoid (approximately by average)

Recall that the signals are considered discrete in time for discrete values $t = kT = t_k$ with $k = 0, 1, 2, \ldots$ Hence, the discrete-time first derivative is approximated. Then, in order to derive the discrete-time model equivalent of the system differential equation, numerical integration techniques could be used as follows.

3.2.2.1 Numerical Approximation Using Forward Difference

By definition, the first derivative for a continuous function $x(t)$ at a point t is:

$$\frac{dx(t)}{dt} = \lim_{\Delta t \to 0} \frac{x(t + \Delta t) - x(t)}{\Delta t} \tag{3.7}$$

If the time is discretized and the time increment equals the sampling period such that $\Delta t = T$ and it is small enough, the derivative can be replaced by:

$$\frac{dx(t)}{dt} \approx \frac{x(t_k + \Delta t) - x(t_k)}{\Delta t_k} = \frac{x(kT + T) - x(kT)}{T} = \frac{x_{k+1} - x_k}{T} \tag{3.8}$$

This recursive iterative algorithm depends on the sampling period. The higher the sampling period, the wider the approximation error using this method will be. This causes the discrete

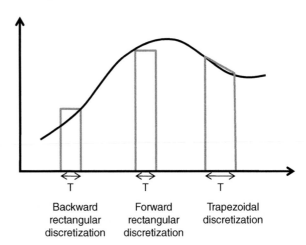

Figure 3.4 Discrete integration using backward, forward, and bilinear transformation.

equivalent signal to be less accurate. However, the selection of a proper sampling or step size is bounded by the speed of the digital processing unit. The remarks concerning the selection of sampling period are also valid in this case. This method approximates the derivative by a forward difference. It can be applied to determine the next discrete-time value (e.g. solving a model off-line) but it cannot be used to estimate the derivative's value at a moment when only previous samples are available (e.g. in the case of on-line real-time applications such as control or identification). In these cases, the backward difference method is used.

3.2.2.2 Numerical Equivalence Using Backward Difference
The derivative can also be approximated with the expression:

$$\frac{dx(t)}{dt} \approx \frac{x(t_k) - x(t_k - \Delta t)-}{\Delta t_k} = \frac{x(kT) - x(kT - T)}{T} = \frac{x_k - x_{k-1}}{T} \tag{3.9}$$

This recursive equation can be easily implemented on a computer as it is based on previous process data. The method can be easily extended such that, for a second-order differential, it results in:

$$\frac{d^2x(t)}{dt^2} = \lim_{\Delta t \to 0} \frac{\frac{dx(t)}{dt} - \frac{dx(t-\Delta t)}{dt}-}{\Delta t} \cong \frac{\frac{\Delta x(t_k)}{\Delta t_k} - \frac{\Delta x(t_k - \Delta t_k)}{\Delta t_k}}{\Delta t_k} = \frac{\frac{x_k - x_{k-1}}{T} - \frac{x_{k-1} - x_{k-2}}{T}}{T}$$

$$= \frac{x_k - 2x_{k-1} + x_{k-2}}{T^2} \tag{3.10}$$

that is, a second-order (two steps behind) expression. Using the backward shift z^{-1}, the equivalence yields:

$$z^{-1} = e^{-sT_s} \approx 1 - sT_s \Rightarrow s = \frac{1 - z^{-1}}{T_s} \tag{3.11}$$

Thus, this discretization maps the left-half of the s-plane to a circle of radius of 0.5 on the right half of the z-plane.

3.2.2.3 Numerical Equivalence Using Bilinear Transform
The function $x(t)$ is given by integrating it between two successive sampling instants, which results in:

$$\int_{t_k}^{t_{k+1}} \dot{x}(t)dt = \frac{\dot{x}(k) + \dot{x}(k+1)}{2} \approx \frac{x(k+1) - x(k)}{T} \tag{3.12}$$

Thus, the trapezoidal approach gives a refined signal approximation. If the first-order Padé approximation of z^{-1} is used instead of a first-order series approximation as mentioned above, then the approximation yields:

$$z^{-1} = e^{-sT} \approx \frac{1 - \frac{T}{2}s}{1 + \frac{T}{2}s} \Rightarrow s = \frac{2(1 - z^{-1})}{T(1 + z^{-1})} \tag{3.13}$$

By translating into a frequency domain, some warping in frequency occurs when the sample frequency is low with respect to system dynamics. Hence, each discrete-time left-shift by n corresponds to the z^n multiplying factor in the pulse transform or the z-transform and each d^n/dt^n in continuous-time domain corresponds to the $a_n s_n$ multiplying factor in the Laplace transform. It should be noticed that there are higher distortions with the forward and backward rule. The trapezoid rule offers a better approximation, even if it requires a prewarping, to maintain the power at some specified frequency. Table 3.2 summarizes some typical approximation methods.

Table 3.2 Table of equivalence based on numerical approximation techniques.

Approximation method	Differential equation to be derived	z- to s-domain approximation	s- to z-domain approximation	Equivalent function in the z-domain H(z)
Forward difference	$\dot{x}(k) \approx \dfrac{x(k+1) - x(k)}{T}$	$z = 1 + Ts$	$s = \dfrac{z-1}{T}$	$H_f(z) = \dfrac{a}{\frac{z-1}{T} + a}$
Backward difference	$\dot{x}(k+1) \approx \dfrac{x(k+1) - x(k)}{T}$	$z = \dfrac{1}{1 - Ts}$	$sT = \dfrac{z-1}{z}$	$H_b(z) = \dfrac{a}{\frac{z-1}{Tz} + a}$
Trapezoid rule (bilinear transform)	$\dfrac{\dot{x}(k) + \dot{x}(k+1)}{2} \approx \dfrac{x(k+1) - x(k)}{T}$	$z = \dfrac{1 + \left(\frac{T}{2}\right)s}{1 - \left(\frac{T}{2}\right)s}$	$\dfrac{s + sz}{2} = \dfrac{z-1}{T}$	$H_t(z) = \dfrac{a}{\frac{2}{T}\frac{z-1}{z+1} + a}$

Example 3.2 *Discretization of a Nonlinear Model*

Consider a process model given by:

$$\dot{x}(t) = -\sqrt{2x(t)} + Au(t)$$

Using the backward difference, this leads to a nonlinear expression given by:

$$x(k+1) \approx x(k) + T(-\sqrt{2x(k)} + Au(k))$$

While using the forward difference, it would yield:

$$x(k) \approx x(k-1) + T(-\sqrt{2x(k-1)} + Au(k-1))$$

Such equations can be solved iteratively.

Example 3.3 *Comparison of Discrete Equivalences*

Consider a low-pass filter with a cutoff frequency of 3 rad s^{-1} being equivalent to:

$$H(s) = \frac{3}{s+3}$$

The magnitude and angle of $H(s)$ at the cutoff frequency at $s = j\omega = j3$ such that:

$$|H(j\omega)|\big|_{j\omega=j3} = \left|\frac{3}{j3+3}\right| = \frac{\sqrt{3^2}}{\sqrt{3^2+3^2}} = 0.7071 \leftrightarrow 20log_{10}0.707 = -3.01 \text{ dB}$$

$$\angle H(j\omega) = \angle H(j3) = \angle\frac{3}{j3+3} = \arg(3) - \arg(3i+3) = -\frac{\pi}{4} = -45^{\circ}$$

This angle lies between $\pm\pi$(real axis).

```
angle((3)/(3+3i));
abs((3)/(3+3i));
```

Choosing a sampling frequency of $f_s = 100$ Hz, thus $T = 0.01$ s.
Forward rectangular rule $s = \frac{z-1}{T} = 100(z-1)$
Therefore,

$$H_F(z) = \frac{3}{100z - 100 + 3} = \frac{3}{100z - 97}$$

This simulation has no zeros, a pole at $z = 0.97$, and a DC gain of 0.309. To find the magnitude and angle at 3 rad s^{-1}, let

$$z = e^{j\omega T} = e^{j3(0.01)} = e^{j(0.03)} = \cos(0.03) + j\sin(0.03) = 0.999 + j0.03$$

And by substitution it yields:

$$H_F(e^{j(0.03)}) = \frac{3}{2.9 + j3}$$

$$|H_F(z)| = 0.7190 = -2.8654\,dB$$

$$\angle H_F(z) = -45.9710°$$

Backward rectangular rule $\left(s = \dfrac{z-1}{Tz} = 100\left(\dfrac{z-1}{z}\right)\right)$

Therefore,

$$H_B(z) = \frac{3z}{103z + 100}$$

This simulation has a zero at $z = 0$ and a pole at $z = 1.03$. Using

$$z = e^{j\omega T} = e^{j3(0.01)} = 0.999 + j0.03$$

and

$$H_B(e^{j3(0.01)}) = \frac{2.997 + j0.09}{2.897 + j3.09}$$

$$|H_B(e^{j3(0.01)})| = 0.7079 = -3.006\,dB$$

$$\angle H_B(e^{j3(0.01)}) = -45.13°$$

Trapezoidal rule $\left(s = \dfrac{2}{T}\dfrac{z-1}{z+1} = 200\left(\dfrac{z-1}{z+1}\right)\right)$

Therefore,

$$H_T(z) = \frac{3(z+1)}{203z - 197}$$

This simulation has a zero at -1 and a pole at 0.9704. Using $z = e^{j\omega T}$, the transfer function yields:

$$H_T(e^{j3(0.01)}) = \frac{5.997 + j0.09}{5.797 + j6.09}$$

$$|H_T(e^{j3(0.01)})| = 0.7133 = -2.9346\,dB$$

$$\angle H_T(e^{j3(0.01)}) = -45.552°$$

At a high sampling frequency, the trapezoidal and backward are the preferred rules to approximate the system frequency response. However, there is a large variety of differential equations as well as several other numerical methods to solve them, such as the predictor-corrector method and the Runge–Kutta family of algorithms. The predictor-corrector algorithm is based on the Euler method but has a greater accuracy. It combines the Euler "prediction" method with the trapezoidal "correction" method of integration.

Table 3.3 DC motor speed values for different times.

t	$\omega(t) = 4(1 - e^{-t})$
0	0
1	2.5285
2	3.4587
3	3.8009
4	3.9267

Example 3.4 Consider the *first order model* of the DC motor given by:

$$\frac{Rb + K_{em}^{2}}{K_{em}}\Omega(t) + \frac{RJ}{K_{em}}\frac{d\Omega(t)}{dt} = V(t)$$

The corresponding differential equation and solution are, respectively:

$$\omega(t) + \tau\frac{d\omega(t)}{dt} = KV(t)$$

with $\tau = 1$ min and $K = 0.4$ rev min^{-1} V^{-1}, $v(t) = 10$ V for $t \geq 0$, and the initial condition is $\omega(0) = 0$.

$$\omega(t) = 4(1 - e^{-t})$$

Applying numerical values for 1-minute increments, it yields the results summarized in Table 3.3.

For $t = T$, the solution is given by

$$\omega(T) = Ce^{-T/t} + Kv_0$$

with C being determined by the initial condition as:

$$C = \omega_0 - Kv_0$$

Thus,

$$\omega(T) = \omega_0 e^{-T/\tau} + Kv_0(1 - e^{-T/\tau})$$

The first-order difference equation is then given by (Figure 3.5):

$$\omega(n) - e^{-T/\tau}\omega(n - 1) = K(1 - e^{-T/\tau})v(n - 1)$$

If the sample period is chosen as $T = 1$ min, the constants in the difference equation are:

$$e^{-T/\tau} = e^{-1} = 0.368$$

$$K(1 - e^{-T/\tau}) = 0.4(0.632) = 0.2528$$

Thus, the difference equation is given as:

$$\omega(n) = 0.368\omega(n - 1) + 0.2528v(n - 1)$$

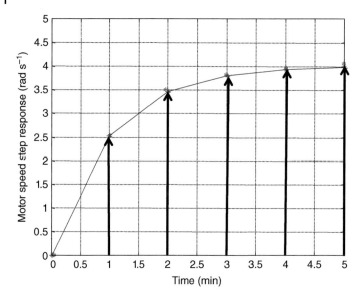

Figure 3.5 Continuous-time and discrete-time equivalent step response.

with an initial condition $\omega(0) = 0$ and with $v(n) = 10$ V for $n = 0, 1, 2, 3$, by using the previous difference equation, the following discrete values can be derived:

$$\omega(0) = 0$$

$$\omega(1) = 0(0.368) + 0.2528(10) = 2.528$$

$$\omega(2) = 2.528(0.368) + 0.2528(10) = 3.458$$

$$\omega(3) = 3.458(0.368) + 0.2528(10) = 3.8006$$

$$\omega(4) = 3.8006(0.368) + 0.2528(10) = 3.9266$$

This recursive method, consisting of computing the sampled output from the previous sampled input, produces exactly the same result as the solution of the differential equation. But this method, in contrast to the differential equation, does not hold information about the response of the process at times other than the sample times $t=0, 1, 2,\ldots$min.

```
clear all;clf; w0 = 0; K = 0.4;  v = 10;
for i = 1:1:6;
w(1)=w0; w_cont(1)=w0;
w(i+1) = exp(-1)*w(i)+K*v*(1-exp(-1));
w_cont(i+1) = ((w0-K*v)*exp(-i))-(w0-K*v);
end;
t=[0:1:6];figure(1);plot(t,w,'*red',t,w_cont,'-blue');xlabel('Time (min)');
ylabel('Motor speed step response (rad/sec)');grid;axis([0 5 0 5]);
```

Example 3.5 As illustrated in Figure 3.6, in the case of a varying input between the sample periods and with $v(n) = \{10, 5, 2\}$, V for $n = 0, 1, 2$ and for $n \geq 3$ $v(n) = 0$.

Then, using the recursive method, this yields:

$$\omega(0) = 0$$

$$\omega(1) = 0(0.368) + 1.264(10) = 1.264$$

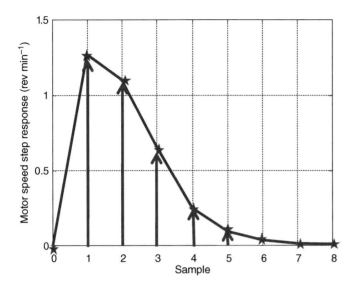

Figure 3.6 Continuous-time and discrete-time equivalent step response for varying inputs.

$$\omega(2) = 1.264(0.368) + 1.264(5) = 1.097$$

$$\omega(3) = 1.097(0.368) + 1.264(2) = 0.656$$

$$\omega(4) = 0.656(0.368) + 1.264(0) = 0.242$$

$$\omega(5) = 0.242(0.368) + 0 = 0.089$$

$$\omega(6) = 0.089(0.368) + 0 = 0.033$$

The assumption of constant input between sample periods is required. Some solutions from the process difference equation model are listed in Table C2 in Appendix C.

```
clear all;clf;w0=0;K=0.2;v1=10;v2=5;v3=2;v=0;u(1)=0;
for i = 1:1:4;
w(1)=w0;
w_cont(1)=0; w_cont(2)=0; w_cont(3)=0; w_cont(4)=0;w_cont(5)=0;
w(2)  = exp(-1)*w(1)+K*v1*(1-exp(-1));
w(3)  = exp(-1)*w(2)+K*v2*(1-exp(-1));
w(4)  = exp(-1)*w(3)+K*v3*(1-exp(-1));
w(5)  = exp(-1)*w(4)+K*v*(1-exp(-1));
w(i+5)  = exp(-1)*w(i+4)+K*v*(1-exp(-1));
w_cont(i+5)  = ((w(1)-K*v)*exp(-i))-(w(1)-K*v);
end;
t=[0:1:8];plot(t,w,'*red');hold on;plot(t,w,'blue');xlabel('Sample');
ylabel('Motor speed step response (rev/min)'); grid; axis([0 8 0 1.5]);
```

Example 3.6   Consider the *dead time process* inherent in most of the processes with a fluid transportation lag or dead time, *D*. Here, the output is identical to the input, except that the output lags behind the input by some samples period *d*.

$$T_{out} = T_{in-d}$$

where $d = D/T$ must be an integer; if not it should be rounded up to the highest integer.

A Discrete-Time Approximation Method using Zero-Pole Matching Equivalence

Another discrete approximation method consists of matching poles in the Laplace transform of a continuous-time signal with those in z of signal samples according to $z = e^{sT}$ equivalence. The idea of zero-pole matching is to use this mapping to determine the location of zeros as well. The matching equivalence procedure is outlined next.

1) All poles of $H(s)$ are mapped to poles of $H(z)$ using $z = e^{sT}$. For example, $s - a = 0$ maps to $z = e^{-aT}$;

2) Similarly, map finite zeros (if any) of $H(s)$ to zeros of $H(z)$ using $z = e^{sT}$. For example, $s - b = 0$ maps to $z = e^{-bT}$;

3) Map zeros at infinity (recall that $H(s)$ always has an equal number of poles and zeros, with perhaps some zeros at ∞) to $z = -1$. Basically, zero at $j\omega = \infty$ maps to $z = e^{-jp} = -1$ (representing the highest frequency);

4) Let m and n be the degrees of numerator and denominator of a continuous-time transfer function, then,

$$H(s) = \frac{(s+d_1)(s+d_2)\dots(s+d_m)}{(s+a_1)(s+a_2)\dots(s+a_n)} \tag{3.14}$$

If $m < n$, then the system will have $n - m$ zeros at infinity. Each continuous-time zero at $s = 1$ is mapped to a discrete-time zero at $z = j1$. The rationale is that it is preferable that the highest continuous-time frequency of $s = j1$ corresponds to the highest discrete-time frequency of $2z = e^{j\frac{T}{4}}$ for $T = -1$.

5) Set the DC gain of $H(z)$ to match the gain of $H(s)$ at the critical frequency (typically $s = 0$). Here, $H(s)$ at $s = 0$ is equivalent to $H(z)$ at $z = 1$ (i.e. $H(z)|_{z=1} = H(s)|_{s=0}$). Thus, this consists of adjusting the gain of the discrete equivalent to equal continuous-time and discrete-time systems gains at a pre-specified frequency, in the form of the following equality:

$$H_{zp}(e^{j\omega_o T}) = H(j\omega_o) \tag{3.15}$$

Often in practice, this requires DC gains to be equal so that it yields:

$$H_{zp}(1) = H(0) \tag{3.16}$$

Therefore, from a generic continuous system transfer function factored in the form:

$$G(s) = \frac{k_c(s+d_1)(s+d_2)\dots(s+d_i)}{(s+a_1)(s+a_2)\dots(s+a_n)} \tag{3.17}$$

where the poles of the process transfer function are $-a_1, -a_2, \dots, -a_n$ and the zeros are $-d_1, -d_2, \dots, -d_j$. Some of the poles may be complex conjugates, as well as some of the zeros. Using the relationship $z = e^{sT}$, the discrete transfer function becomes:

$$G(z) = \frac{k'_c(1 - e^{-d_1 T}z^{-1})(1 - e^{-d_2 T}z^{-1})\dots(1 - e^{-d_i T}z^{-1})}{(1 - e^{-a_1 T}z^{-1})(1 - e^{-a_2 T}z^{-1})\dots(1 - e^{-a_n T}z^{-1})} \tag{3.18}$$

The gain k'_c is usually set so that the magnitude of the discrete controller frequency response is the same as the continuous controller at some critical frequency. If this frequency is $\omega = 0$, then it can be derived as:

$$k'_c = \frac{k_c(1 - e^{-a_1 T}z^{-1})(1 - e^{-a_2 T}z^{-1})\dots(1 - e^{-a_i T}z^{-1})}{(1 - e^{-d_1 T}z^{-1})(1 - e^{-d_2 T}z^{-1})\dots(1 - e^{-d_n T}z^{-1})} \tag{3.19}$$

Results from the pole-zero mapping simulation are less accurate than the trapezoidal and prewarped rules. The mapping between the s- and the z-planes breaks down (actually it becomes non-unique) as $s \to \frac{j\pi}{T}$, or $s \to \frac{j\omega}{2}$, half the sampling frequency. From resulting Bode plots, the forward and backward methods are the worst, one being too high in magnitude and the other too low. The trapezoidal, prewarped, and pole-zero mapping simulations are similar. Note that all methods behave very poorly as the frequency approaches one-half the sampling frequency or $\omega = 2\pi f \approx 63$ rad s^{-1}.

```
SYSD = C2D(SYSC,Ts,METHOD) converts the continuous-time LTI
model SYSC to a discrete-time model SYSD with sample time Ts.
The string METHOD selects the discretization method among the following:
'zoh' Zero-order hold on the inputs
'foh' Linear interpolation of inputs (triangle approx.)
'imp' Impulse-invariant discretization
'tustin' Bilinear (Tustin) approximation
'matched' Matched pole-zero method (for single input single output (SISO) sys-
tems only).
'prewarp' Tustin approximation with frequency prewarping.
The critical frequency Wc (in rad s⁻¹) from
SYSD = C2D(SYSC,Ts,'prewarp',Wc)
The default is 'zoh' when METHOD is omitted.
```

3.2.3   Z-Transform and Inverse Z-Transform: Theorems and Properties

Consider a discrete-time signal $x(t)$ sampled every T seconds with amplitudes of $X(0), X(1T)$, and so on.

$$x(t) = X(0)\delta(t) + X(T)\delta(t - T) + X(2T)\delta(t - 2T) + X(3T)\delta(t - 3T) + \ldots \tag{3.20}$$

Recalling that the Laplace transform of a unit impulse is given by:

$$L\{\delta(t)\} = \int_0^\infty \delta(t)e^{-st}dt = \lim_{\Delta t \to 0}\left(\frac{1 - e^{-s\Delta t}}{s\Delta t}\right) = \lim_{\Delta t \to 0}\left(\frac{se^{-s\Delta t}}{s}\right) = 1 \tag{3.21}$$

It could be derived that

$$L\{\delta(t - kT)\} = e^{-sKT} \tag{3.22}$$

The z-transformation of complex variables can be defined such that:

$$z = e^{sT} = e^{(\sigma+j\omega)T} = e^{\sigma T}\cos\omega T + je^{\sigma T}\sin\omega T \tag{3.23}$$

Hence, considering $n = kT$ and the function $x(t)$ being a train of impulses equally spaced by $\Delta t(\Delta t \to 0)$ with constant amplitudes are $X(T), X(2T)\ldots$, its Laplace transformation would yield:

$$L\{x(t)\} = X(s) = X(0) + X(1T)e^{-sT} + \ldots + X(2T)e^{-s2T} = \sum_{k=0}^{\infty}X(kT)e^{-skT}$$

$$= \sum_{k=0}^{\infty}X(kT)z^{-n} = X(z) \tag{3.24}$$

This is equivalent to the conversion from the Laplace transform to the z-transform of the continuous-time signal $x(t)$ sampled with an interval of T such that:

$$X(z) = X(0) + X(1T)z^{-1} + X(2T)z^{-2} + X(3T)z^{-3} + \ldots = \sum_{n=0}^{\infty}X(n)z^{-n} \tag{3.25}$$

Using only variable z, the z-transform of a sampled sequence $x(n)$ is given by:

$$X(z) = x(0) + x(1T)z^{-1} + x(2T)z^{-2} + x(3T)z^{-3} + \ldots = \sum_{n=0}^{\infty} x(n)z^{-n} \tag{3.26}$$

For a SISO discrete-time system, where it is considered a discrete signal sequence expressed by $x(kT)$ and a sampling period T as a unity, the following theorems can be stated.

Theorem on addition and subtraction:

$$Z[x_1(kT) \pm x_2(kT)] = X_1(z) \pm X_2(z) \tag{3.27}$$

Theorem on multiplication by a constant α:

$$Z[\alpha x(kT)] = \alpha Z[x(kT)] = \alpha X(z) \tag{3.28}$$

Theorem on real translation (time delay and time advance), with n being a positive integer:

$$Z[x(kT - nT)] = z^{-n}X(z) \tag{3.29}$$

and

$$Z[x(kT + nT)] = z^{+n}Y(z) \tag{3.30}$$

Theorem on complex translation with α being a constant:

$$Z[e^{\pm\alpha kT}x(kT)] = X(ze^{\pm\alpha T}) \tag{3.31}$$

Theorem on initial value theorem, if the limit exists:

$$\lim_{k\to 0} x(kT) = \lim_{z\to\infty} X(z) \tag{3.32}$$

Theorem on final value theorem, for a function having no pole on and outside the unit circle in the z-plane:

$$\lim_{k\to\infty} x(kT) = \lim_{z\to 1}(1 - z^{-1})X(z) \tag{3.33}$$

Theorem on real convolution:

$$X_1(z)X_2(z) = Z[x_1(kT)x_2(NT - kT)] = Z\left[\sum_{k=0}^{N} x_2(kT)x_1(NT - kT)\right]$$

$$= Z\left[\sum_{k=0}^{N} x_1(kT)^* x_2(kT)\right] \tag{3.34}$$

For the inverse z-transformation or transformation back to the time domain, the power series expansion can be used. The objective is to find the time domain function:

$$x^*(t) = Z^{-1}\{X(z)\} \tag{3.35}$$

For $N = 0, 1, 2, \ldots$ if the function $G(z)$ has the form

$$X(z) = \frac{N(z)}{D(z)} = \frac{c_0 + c_1 z^{-1} + c_2 z^{-2} + \ldots}{b_0 + b_1 z^{-1} + b_2 z^{-2} + \ldots} \tag{3.36}$$

then a long Euclidian division can be used to divide the numerator $N(z)$ by the denominator $D(z)$ to obtain the power series form of the sampled signal given by:

$$X(z) = X(0) + X(1)z^{-1} + X(2)z^{-2} + \ldots \tag{3.37}$$

Then, the transfer function is factored into lower order components using partial fraction expansion. Table C2 in Appendix C can be used to obtain the inverse transformations. If an item in the tables matches the z-transform, the corresponding function $x(nT)$ is the required inverse transformation. It is important to note that the inverse transformation results in $x^*(t)$ rather than $x(t)$ and that $x(nT)$ is not defined between sample instants.

Example 3.7 Consider a function given by the relationship between the output y and the input u such that:

$$y(k) = -a_1 y(k-1) - a_0 y(k-2) + b_1 u(k-1) + b_0 u(k-2)$$

Taking the z-transform of both sides of the difference equation:

$$Z\{y(k)\} = Z\{-a_1 y(k-1) - a_0 y(k-2) + b_1 u(k-1) + b_0 u(k-2)\}$$

or

$$Z\{y(k)\} = -Z\{a_1 y(k-1)\} - Z\{a_0 y(k-2)\} + Z\{b_1 u(k-1)\} + Z\{b_0 u(k-2)\}$$

or

$$y(z) = -a_1 z^{-1} y(z) - a_0 z^{-2} y(z) + b_1 z^{-1} u(z) + b_0 z^{-2} u(z)$$

or

$$[1 + a_1 z^{-1} + a_0 z^{-2}] y(z) = [b_1 z^{-1} + b_0 z^{-2}] u(z)$$

or

$$\frac{y(z)}{u(z)} = \frac{b_1 z^{-1} + b_0 z^{-2}}{1 + a_1 z^{-1} + a_0 z^{-2}} = \frac{b_1 z + b_0}{z^2 + a_1 z + a_0}$$

3.2.4 Procedure for Discrete-Time Approximation of the Continuous Process Model

Defining the impulse-invariant z-transform being the Laplace transform of sampled impulse process response after the variable change using $z = e^{sT}$. Using the fact that an impulse signal over the latch-based digital-to-analog convertor (DAC) results in a step signal for the hold time period, it is possible to derive the corresponding discrete equivalent signal, as presented in Table 3.4.

Using a table of equivalence of z-transforms, it is possible to derive the discrete-time model directly from the model in the Laplace domain obtained in step (ii) or from step (iv) from the generic procedure before or from that step.

3.2.4.1 Z-Transfer Functions and Block Diagram Manipulation
Models in the form of difference equations can be transformed directly into z-transfer functions. A block diagram can be manipulated in the discrete-time domain as in the continuous-time domain. However, when there is a mixture of system models in both domains, so that when there is a sampler (with sampling period T) between the transfer function of the cascaded continuous elements, the following procedure can be used:

1) consider the outputs of samplers as inputs to the system along with all inputs of the system;
2) treat all other inputs of the system as outputs;
3) write, respectively, the inputs and the outputs relationship of the system using the gain formula;
4) take the pulsed transform or the z-transform of the equations obtained in step 3, and manipulate these equations to get the pulse-transfer function or the z-transfer function.

Table 3.4 General steps for developing discrete-time model of linear time invariant process.

Steps	Illustrative examples
1) Develop a physical process dynamics model, if possible using differential equations	Use stochastic approach or apply physics principles or laws for each process component. e.g. valve flow rate $\dfrac{d\theta(t)}{dt} = \dfrac{q(t)}{A}$.
2) From an ordinary differential equation (ODE) description, derive its corresponding Laplace transformation	Select manipulated input and process output variables from physical model: $L(ODE) = H(s) \Rightarrow f(input = M(s))$ including the effects of initial conditions e.g. $L\left\{\dfrac{d\theta(t)}{dt} = \dfrac{q(t)}{A}\right\} \Rightarrow s\varphi(s) - \varphi(0) = \dfrac{Q(s)}{A}$
3) Derive the model of manipulated step input and substitute into process model in Laplace domain	$L\{M_{step}input\} \Rightarrow M_{step}(s) = M_{step}(t = 0)\dfrac{1}{s}$ e.g. $Q_{step}(s) = \dfrac{q_{desired}(kT)}{s}$ then for $\dfrac{d\theta(t)}{dt} - \dfrac{q(t)}{A} \Rightarrow \varphi(s) = \dfrac{\varphi(0)}{s} + \dfrac{q_{desired}(0)}{As}\dfrac{1}{s}$
4) Find its corresponding continuous-time process response solution	$L^{-1}\{\varphi(s)\} \Rightarrow \theta(t)$ being the response to a step input given initial conditions. e.g. $\theta(t) = \theta(t = 0) + \dfrac{q_{desired}(t = 0)}{A}t$
5) Set and find the sampled process response	Sample at time $t = T$, derive the discrete-time equation for $\theta(T)$ using $M_{step}(t = 0)\dfrac{1}{s}$ and initial conditions. e.g. $\theta(T) = \theta(t = 0) + \dfrac{q_{desired}(t = 0)}{A}T$
6) Find the difference equation models in the discrete (sampled) time domain	Generalize this equation here to an arbitrary time kT, using $n = kT$ e.g. $\theta(n+1) = \theta(n) + \dfrac{q_{desired}(n)}{A}T \Rightarrow \theta(n) =$ $\theta(n-1) + \dfrac{q_{desired}(n-1)}{A}T$
7) Using backshift properties, find the equivalent model z-transform	Setting $Z[x(kT - nT)] = z^{-n}Y(z)$ to transform the difference equations using backshift properties e.g. $x(z) = z^{-1}x(z) + \dfrac{q_{desired}(z)}{A}T$
8) Find the discrete-time transfer function for the desired sample outputs given the (latched) manipulated step inputs	Convert the z-transform of difference equations for the desired rational fraction format, e.g. $\dfrac{x(z)}{q_{desired}(z)} = \dfrac{z^{-1}}{1 - z^{-1}}\dfrac{T}{A}$
9) Convert it into rational fraction using long division to check correctness by comparison with z-transform table of equivalence	Invert the transfer functions to obtain independent difference equations. See Appendix C.
10) Identify the independent or cross coupled difference equations as needed and repeat from step (iii);	e.g. If independent equations exist then, $\dfrac{G(z)}{H(z)} = \dfrac{G(z)}{T(z)}\dfrac{T(z)}{H(z)} \rightarrow \dfrac{G(z)}{T(z)} = \dfrac{G(z)}{H(z)}\left(\dfrac{T(z)}{H(z)}\right)^{-1}$

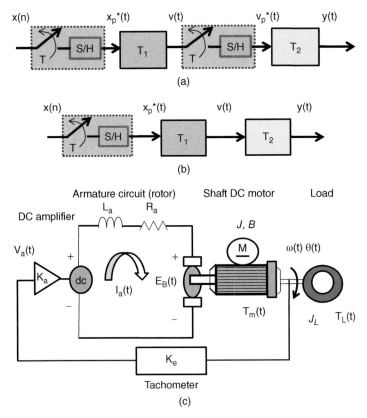

Figure 3.7 (a) System with two samplers. (b) System with a cascaded continuous process with a single sampler. (c) Schematic of an armature-controlled DC motor.

Some properties of cascaded (series) interconnected continuous-time systems. As illustrated in Figure 3.7(a), when the two systems are separated by a sampler, the equivalent pulse transform of the two systems T_1, T_2 would be equal to the product of the pulse transform of the two systems. As such, taking the successive pulse transform of both systems would result in:

$$y(t) = T_1^* T_2^* x(n)$$

As illustrated in Figure 3.7(b), when the two systems T_1, T_2 are not separated, they have to be treated as one system when taking the pulse transform. Taking the pulse transform results in:

$$y(t) = [T_1 T_2]^* x(n)$$

where in the Laplace domain, it is given by:

$$[T_1 T_2]^* = \frac{1}{T} \sum_{n=-\infty}^{\infty} T_1(s + jn\omega_s) T_2(s + jn\omega_s)$$

Example 3.8 A process consists of two integrations in series such as:

$$\left. \begin{array}{l} \dfrac{dy(t)}{dt} = K_1 m(t) \\[2mm] \dfrac{dc(t)}{dt} = K_2 y(t) \end{array} \right\} \Rightarrow \dfrac{d^2 c(t)}{dt^2} = K_1 K_2 m(t)$$

The equivalent difference equation using backshift approximation is given by:

$$\frac{y(n) - y(n-1)}{T} = K_1 m(n-1)$$

$$\frac{y(z)}{m(z)} = \frac{K_1 T z^{-1}}{1 - z^{-1}}$$

and

$$\frac{c(z)}{y(z)} = \frac{K_2 T z^{-1}}{1 - z^{-1}}$$

but

$$\frac{c(z)}{m(z)} \neq \frac{y(z)c(z)}{m(z)y(z)} = \frac{K_1 K_2 T^2 z^{-2}}{(1 - z^{-1})^2}$$

Because the variable $m(t)$ is not constant between two consecutive samples as there is no sampler between $y(t)$ and $c(t)$. The correct procedure is to use the combined integration equations such that:

$$\frac{d^2 c(t)}{dt^2} = K_1 K_2 m(t)$$

therefore,

$$\frac{c(z)}{m(z)} = \frac{\frac{K_1 K_2 T^2}{2} z^{-1}(1 + z^{-1})}{(1 - z^{-1})^2}$$

Generally, the continuous-time system model in series must be combined into one continuous-time model before the discrete transfer function or the difference equation is obtained.

Example 3.9   As depicted in Figure 3.7(c), consider a DC motor with a permanent magnet configuration having a linear torque-speed operating range, as illustrated in Figure 3.7(c). Table 3.5 summarizes the process variables and parameters involved.

The dynamics model of the DC motor is given by:

$$\omega(t) + \frac{R_a J_{eq} + L_a B}{R_a B + K_b K_m} \frac{d\omega(t)}{dt} + \frac{L_a J_{eq}}{R_a B + K_b K_m} \frac{d^2\omega(t)}{dt^2} = \frac{K_m}{R_a B + K_b K_m} V_a(t)$$

which could be rewritten as:

$$\frac{d\theta(t)}{dt} + \frac{R_a J_{eq} + L_a B}{R_a B + K_b K_m} \frac{d^2\theta(t)}{dt^2} + \frac{L_a J_{eq}}{R_a B + K_b K_m} \frac{d^3\theta(t)}{dt^3} = \frac{K_m}{R_a B + K_b K_m} V_a(t)$$

Table 3.5 Some key variables and parameters values of DC motor.

Variables	Description	Parameters	Description and values
$v(t)$	Amplifier input voltage (V)	B	Motor damping (3 N m.s^{-1} rad^{-1})
$e(t)$	Amplifier output voltage (V)	R_a	Armature resistance (2 Ω)
$i(t)$	Motor current (A)	L_a	Armature inductance (3.10^{-2} H)
$T(t)$	Motor torque (N m)	K_b	Back emf constant (12 10^{-2} V s^{-1} rad^{-1})
$\omega(t)$	Motor velocity (rad s^{-1})	K_m	Motor constant (4.2 10^{-2} N m A^{-1})
$\theta(t)$	Motor position (rad)	J_{eq}	Load-motor inertia (40 10^{-2} N m A^{-1})

By using values from Table 3.5, the third-order transfer function can be decomposed into a first-order and second-order system, or several first-order systems, it could be rewritten as:

$$\frac{d\theta(t)}{dt} + 0.0042\frac{d^2\theta(t)}{dt^2} + 1.8 \times 10^{-6}\frac{d^3\theta(t)}{dt^3} = 0.009V_a(t)$$

Considering that the position is sampled at the rate $T = 0.02$ s, it is suitable to use the encoder-based absolute position measurement rather than the derived average velocity, such that taking the Laplace transform yields:

$$s\theta(s) + \frac{R_aJ_{eq} + L_aB}{R_aB + K_bK_m}s^2\theta(s) + \frac{L_aJ_{eq}}{R_aB + K_bK_m}s^3\theta(s) = \frac{K_m}{R_aB + K_bK_m}V_a(s)$$

which is equivalent to

$$s\theta(s) + 0.0042s^2\theta(s) + 1.8 \times 10^{-6}s^3\theta(s) = 0.009V_a(s)$$

Then, the partial fraction expansion would yield:

$$\frac{\theta(s)}{V_a(s)} = \frac{0.009}{s + 0.0042s^2 + 1.8 \times 10^{-6}s^3}$$

Then, by decomposing the transfer function from before into:

$$\frac{\theta(s)}{V_a(s)} = \frac{\theta_1(s)}{V_a(s)} + \frac{\theta_2(s)}{V_a(s)} + \frac{\theta_3(s)}{V_a(s)}$$

This can be rewritten as:

$$\frac{\theta(s)}{V_a(s)} = \frac{0.009}{s} + \frac{0.0013}{0.00047s + 1} - \frac{0.0103}{0.0038s + 1}$$

Then, this would correspond to:

$$0.0038\frac{d\theta_1(t)}{dt} + \theta_1(t) = -0.0103V_a(t)$$

$$0.00047\frac{d\theta_2(t)}{dt} + \theta_2(t) = 0.0103V_a(t)$$

$$\frac{d\theta_3(t)}{dt} = 0.009V_a(t)$$

Using the equivalence table with $T = 0.02$ s, each of these equations would yield:

$$\frac{\theta_1(z)}{V_a(z)} = \frac{-0.01025z^{-1}}{1 - 0.005179z^{-1}}$$

$$\frac{\theta_2(z)}{V_a(z)} = \frac{0.0013z^{-1}}{1 - 3.307 \times 10^{-19}z^{-1}}$$

$$\frac{\theta_3(z)}{V_a(z)} = \frac{0.00018z^{-1}}{1 - z^{-1}}$$

Corresponding to

$$\theta(z) = \theta_1(z) + \theta_2(z) + \theta_3(z)$$

This results in an equivalent difference equation given by:

$$\frac{\theta_3(z)}{V_a(z)} = \frac{0.007001z^{-1} + 0.017284z^{-2} + 0.002475z^{-3}}{1 - 1.7820z^{-1} + 0.9060z^{-2} - 0.1240z^{-3}}$$

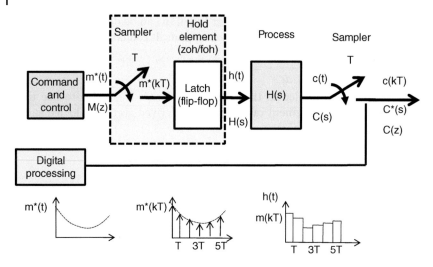

Figure 3.8 Process control with sampler and hold circuits.

3.2.5 Conversion and Reconstruction of the Continuous Signal: Sampling and Hold Device

3.2.5.1 Sampler and Hold-Based Process Model

It is possible to convert and reconstruct the process signal using signal conversion devices. This is done by inserting a sample and hold device (latch) between a digital processing (computing unit) and an existing continuous process, and a sampler between the process signal output and the digital processing device, as illustrated in Figure 3.8. Here, an output signal, $h(t)$ can be derived by a signal from a D/A conversion through a sampler and hold circuit $m(kT)$. This would result in a discrete approximation of the continuous-time from a continuous-time command input $m^*(kT)$. Similarly, it is possible to use signal sampling from an ADC to obtain the equivalent digital signal $c(kT)$. There is a variety of signal conversion devices classified into latches, such as zero-order or first-order hold. The circuit design of these conversion devices is covered in Chapter 6 as well as the signal resolution and the sampling rate selection issues.

Zero-Order Hold (ZOH) The ZOH is a commonly encountered DAC. It receives a binary word equivalent electrical signal $m(t)$ and converts it into an output voltage or current signals, $h(t)$.

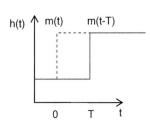

Viewing it as a filter, a typical input and output (I/O) relationship can be defined by setting $h(t) = m(t) - m(t - T)$, as illustrated in Figure 3.9. Taking the Laplace transform of the ZOH that holds the magnitude of the signal carried for the entire sampling period kT, this would yield:

$$H(s) = L[h(t)] = L[m(t) - m(t - T)]$$

$$= L[u(t) - u(t - T)] = \frac{1 - e^{-Ts}}{s} \tag{3.38a}$$

Figure 3.9 Time delay effect on signal processing by zero-order hold elements.

Similar results can be obtained by decomposing the signal into step signals, $u(t - nT)$ and $u(t - (n + 1)T)$ with an amplitude defined by $m(nT)$, as illustrated in Figure 3.10.

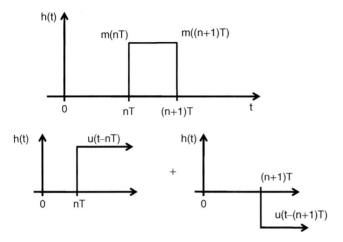

Figure 3.10 Step signal conversion of sampling of manipulation input.

The time series of this signal is given by:

$$h(t) = m(0)(u(t) - u(t - T)) + m(T)(u(t - T) - u(t - 2T))$$
$$+ m(2T)(u(t - 2T) - u(t - 3T)) + \ldots \tag{3.38b}$$

$$h(t) = \sum_{n=0}^{\infty} m(nT)(u(t - nT) - u(t - (n+1)T)) \tag{3.38c}$$

Then, the Laplace transformation of this signal yields:

$$H(s) = \sum_{n=0}^{\infty} m(nT) \left(\frac{e^{-snT}}{s} - \frac{e^{-s(n+1)T}}{s} \right) = \frac{1 - e^{-sT}}{s} \sum_{n=0}^{\infty} m(nT)e^{-snT} = \frac{1 - e^{-sT}}{s}M^*(s)$$
$$\tag{3.38d}$$

Transfer function with the ZOH.

When the ZOH, captured by its model $G_{ZOH}(s)$, is connected in cascade with a linear time invariant process characterized by a transfer function $G_p(s)$, the z-transform of the convolution of these transfer functions can be written as:

$$G_p(z) = Z[G_{ZOH}(s)G_p(s)] = Z \left(\frac{1 - e^{-TS}}{s} G_p(s) \right) \tag{3.39}$$

Thus, using the time delay property (backshift) of the z-transformation from the ZOH can be simplified in the generic form of:

$$G_p(z) = (1 - z^{-1})Z \left(\frac{G_p(s)}{s} \right) \tag{3.40}$$

Figure 3.11 presents a typical block diagram of the process with a sampler and hold equivalent model.

Example 3.10   *Third-Order Process with ZOH*
Consider the case of a DC motor/amplifier system that is manipulated by a voltage signal $v(t)$ from a DAC. Take the DAC to be modeled as being a ZOH to transform a signal $m^*(t)$, from

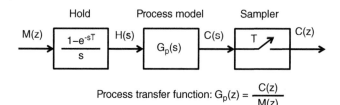

$$\text{Process transfer function: } G_p(z) = \frac{C(z)}{M(z)}$$

Figure 3.11 Generic block diagram of process with sampler and hold equivalent.

collecting the following signals: the motor velocity $\omega(t)$, its position $\theta(t)$, and the voltage input $V_a(t)$, so the model is given by:

$$\frac{1}{\omega_n^2}\frac{d^2\omega(t)}{dt^2} + \frac{2\xi}{\omega_n}\frac{d\omega(t)}{dt} + \omega(t) = k_m V_a(t)$$

$$\frac{d\theta(t)}{dt} = \omega(t)$$

The corresponding Laplace transformation is:

$$\Omega(s) = \frac{k_m}{\left(\frac{s^2}{\omega_n^2} + \frac{2\xi}{\omega_n} + 1\right)} V_a(s)$$

$$\theta(s) = \frac{1}{s}\Omega(s)$$

Using Equation (3.39), the discrete-time motor position transfer function using the zero-order hold consists of:

$$\theta(z) = Z\left(G_{ZOH}(s)\frac{1}{s}\frac{k_m}{\left(\frac{s^2}{\omega_n^2} + \frac{2\xi}{\omega_n} + 1\right)}\right) M(z) = (1 - z^{-1})Z\left(\frac{1}{s}\frac{1}{s}\frac{k_m}{\left(\frac{s^2}{\omega_n^2} + \frac{2\xi}{\omega_n} + 1\right)}\right) M(z)$$

while the discrete-time motor velocity transfer function results in:

$$\Omega(z) = (1 - z^{-1})Z\left(\frac{1}{s}\frac{k_m}{\left(\frac{s^2}{\omega_n^2} + \frac{2\xi}{\omega_n} + 1\right)}\right) M(z)$$

It should be noticed that:

$$\theta(z) \neq Z\left(\frac{1}{s}\right)\Omega(z)$$

This is because there is no sampler between the motor velocity and the motor position, as illustrated in the Figure 3.12.

Triangle Hold Equivalent

A first-order hold extrapolates between two samples using straight lines, as shown in Figure 3.13. Similar to the latch zero-order equivalent models, the first-order equivalent $H_{tri}(z)$ of a continuous-time transfer function $H(s)$ would result in:

$$H_{tri}(z) = \frac{(1 - z^{-1})^2}{Tz} Z\left(\frac{H(s)}{s^2}\right) \tag{3.41}$$

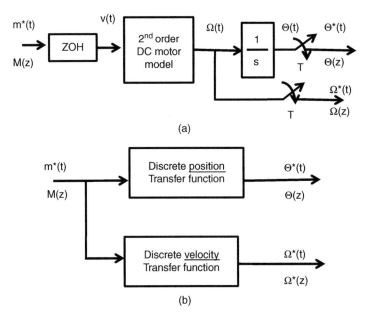

(a)

(b)

Figure 3.12 (a) and (b) Decomposition of the discrete third-order process model into two second-order processes.

Figure 3.13 Time-delay effect of triangle (first-order) hold element.

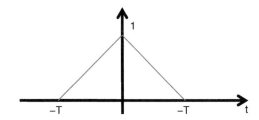

Time Delay Associated with Hold Equipment

In the case of ZOH, the average delay associated with the hold can be estimated as equivalent to a continuous signal delayed by $T/2$, as illustrated in Figure 3.14.

3.2.5.2 Construction Methods of a Continuous Signal from a Data Sequence

The continuous signals can be reconstructed from the sampled data using a suitable numerical interpolation method or DAC. In addition to covering a DAC relationship with the sampling theorem in this chapter, numerical interpolation methods are also presented. It should be noted that this is commonly used to approximate the digital command signal through an approximated curve from the data sequence using piecewise line segments. Among some of the interpolation methods used in constructing command profile curves, there are:

1) linear interpolation between set points to connect two data points for a constant rate of change in the case of either a single command variable or a synchronization between multiple variables;
2) cubic polynomial representation and generation for complex paths and for multiple axes;
3) advanced methods using spline functions.

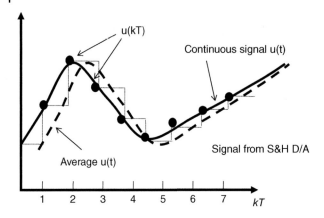

Figure 3.14 Time-delay of effect on continuous signal discrete approximation.

Other interpolation methods, such as the circular arc, the incremental actuator for open-loop control of a stepping motor, quadratic functions, and non-uniform rational B-splines (NURBSs), are not covered in the textbook. For some practical considerations, the construction of digital command signals is usually required to meet constraints on one or more of the following performance measures: (i) residual vibration amplitude; (ii) transient characteristics such as the rise time and settling time; (iii) robustness to modeling uncertainties; and (iv) peak actuator effort. There are also constraints based on the computation time or precision of the geometric representation.

Signal Reconstruction Using Linear Interpolation Considering a set of sampled data between t_O and t_k given by $[u(t_0), u(t_1), \ldots, u(t_k)]$, a linear interpolation can be used to generate a straight-line relationship between consecutive points at time interval T given by $t = t_0 + nT$, for $n = 0, 1, 2, \ldots N$ and

$$N = \frac{t_k - t_0}{T} \tag{3.42}$$

The linear interpolation represents an estimation of intermediate points between these sampled data. This linear interpolation for any $t_{i-1} \le t \le t_i$ results in:

$$u(t) = u(t_i) + (t - t_i)\frac{u(t_i - t_{i-1})}{t_i - t_{i-1}} \tag{3.43}$$

A linear interpolation recursive algorithm can be developed using Equation (3.41). This should result in the discontinuous curve shown in Figure 3.15. It is expected to have some deviation errors over the portion of ending segments due to constraint considerations such as the velocity to position profiles.

Signal Construction Using Cubic (Third-Order) Polynomials Another approximation method involves interpolating some sampled data at discrete-time $t = t_1, t_2, \ldots, t_k$ using some third-order polynomials. Consider a command signal $c(t)$ that can be captured by:

$$c(t) = r\theta(t) = b_1 t^3 + b_2 t^2 + b_3 t + b_4 \tag{3.44a}$$

Moreover, b_1, b_2, b_3, b_4 are the ith polynomial coefficient that is valid for $t_{i-1} \le t \le t_i$. The derivatives of $c(t)$ along the ith segment are:

$$\begin{cases} r\omega(t) = \dot{c}(t) = 3b_1 t^2 + 2b_2 t + b_3 \\ r\alpha(t) = \ddot{c}(t) = 6b_1 t + 2b_2 \end{cases} \tag{3.44b}$$

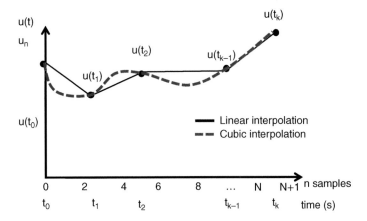

Figure 3.15 Typical generated curves using linear and cubic interpolations.

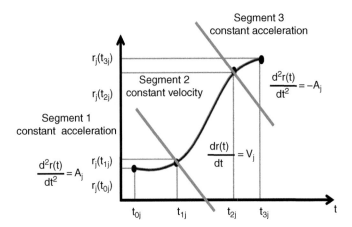

Figure 3.16 Signal reconstruction using three-spline functions for synchronized motion profile.

The coefficients b_1, b_2, b_3, b_4 can be derived using the boundary conditions and constraints at the end points of the ith segment. In the case of a linear motor position with motor radius r, such constraints would be to have the motor position, velocity, and acceleration at the end matching point in order to avoid motion discontinuity. This should be done at both segment ends and should result in six boundary conditions. Given $c(t)$ and $\dot{c}(t)$ at time t_{i-1} and t_i, it is possible to derive the coefficients.

Signal Construction Using Three Spline Functions Another command signal construction method could also consist of restructuring the three represented profile segments as illustrated in Figure 3.16. Here, there are constraints such as all velocity and acceleration segments should ensure synchronized motion, as shown in Figure 3.16. For the first segment, there is acceleration at its maximum rate until the maximum velocity is reached, then at the second segment motion is maintained at the same maximum velocity, while in the third segment it decelerates to a stop at the maximum rate.

The three segments can be defined using Δr, V_j, A_i such as:

$$\Delta t_{1ij} = t_{1j} - t_{0j} = \frac{V_j}{A_i} \tag{3.45}$$

$$\Delta t_{2ij} = t_{2j} - t_{1j} = \frac{|\Delta r|}{V_j} - \Delta t_{1ij} \tag{3.46}$$

$$\Delta t_{3ij} = t_{3j} - t_{2j} = \Delta t_{1ij} \tag{3.47}$$

With respect to the boundary conditions, the command profile for each segment can be derived such that:

$$\begin{cases} at\ t = t_{oj} \quad thus \quad\quad r(t_{oj}) = r_{oj} \quad\quad\quad\quad\quad\quad\quad\quad and \quad \dfrac{dr(t_{oj})}{dt} = 0 \\[2mm] at\ t - t_{1j} \quad thus \quad r(t_{1j}) = r_{0j} + \dfrac{1}{2}A_j\,\Delta t_{1ij}{}^2\,sign(\Delta r) \quad and \quad \dfrac{dr(t_{3j})}{dt} = V_j sign(\Delta r) \\[2mm] at\ t = t_{2j} \quad thus \quad r(t_{2j}) = r_{3j} - \dfrac{1}{2}A_j\,\Delta t_{3ij}{}^2\,sign(\Delta r) \quad and \quad \dfrac{dr(t_{2j})}{dt} = V_j sign(\Delta r) \\[2mm] at\ t = t_{3j} \quad thus \quad\quad r(t_{3j}) = r_{3j} \quad\quad\quad\quad\quad\quad\quad\quad and \quad \dfrac{dr(t_{3j})}{dt} = 0 \end{cases} \tag{3.48}$$

Example 3.11 In the case of a DC motor shaft providing an up and down elevator cabin displacement, the model kinematics are also expected to be compliant with profile constraints. Here, the elevator motion between two or more successive floors has the following constraints: (i) smooth and comfortable elevator ride; (ii) travel time as short as possible; and (iii) if possible, good positioning accuracy. Usually, such a command input of motion profile is represented by discrete values over time. Then, using an interpolation or curve fitting methods, it is possible to derive the functions corresponding to the best curve fitting these sequences of data points, as illustrated in Figure 3.17. A typical up and down position profile can be decomposed into motion profile segments. As an illustrative example, the motion profile in Figure 3.18 can be depicted in seven segments within three phases: (i) acceleration phase from rest to maximum velocity consisting of ramping up, holding, and ramping back; (ii) constant speed phase; and (iii) acceleration phase from rest to maximum velocity consisting of ramping up and holding and ramping back.

Here, it is considered that the position motion profile is a cubic polynomial, in order to avoid step changes in velocity and acceleration. A smooth ride can be achieved through parabolic

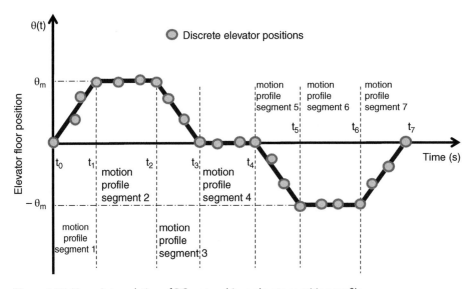

Figure 3.17 Linear interpolation of DC motor-driven elevator position profile.

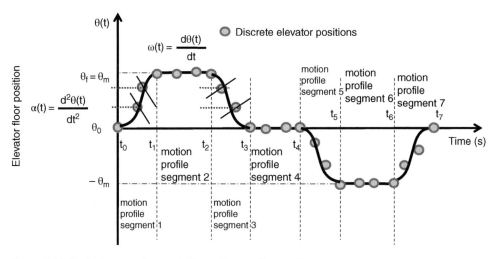

Figure 3.18 Equivalent angular parabolic position motion profile.

acceleration and deceleration profiles with three spline functions representing the motion segments 1a, 1b, and 1c, as illustrated in Figure 3.18. The motion can be characterized by an angular position with a trajectory initial time $t_0 = 0$, and a final time t_f, such that the initial angular position $\theta(t_0) = \theta_0$ and the final angular velocity of the trajectory $\theta(t_f) = \theta_f$. Similarly, the initial angular velocity $\omega(t_0) = \omega_0$ and the final angular position of the trajectory $\omega(t_f) = \omega_f$. Hence, by using a third-order cubic polynomial approximation method that could satisfy the motion constraints, the angular position and velocity are given by:

$$\begin{cases} \theta(t) = a_0 + a_1 t + a_2 t^2 + a_3 t^3 \\ \omega(t) = a_1 + 2a_2 t + 3a_3 t^2 \\ \alpha(t) = 2a_2 + 6a_3 t \end{cases} \tag{3.49}$$

The coefficients are given as:

$$\begin{cases} a_0 = \theta_0 \\ a_1 = 0 \\ a_2 = \dfrac{3}{t_f^{\,2}}(\theta_f - \theta_0) \\ a_3 = \dfrac{2}{t_f^{\,3}}(\theta_0 - \theta_f) \end{cases} \tag{3.50}$$

Recalling that the relationship between the wheel linear and angular rotation of radius R from the drum attached to the motor shaft is given by:

$$x(t) = R\theta(t) \text{ for } t \in [t_0, t_f] \tag{3.51}$$

$$x(t) = R\left[\theta_0 \; 0 \; \tfrac{3}{t_f^{\,2}}(\theta_f - \theta_0) \; \tfrac{2}{t_f^{\,3}}(\theta_0 - \theta_f)\right] = R\left(\theta_0 + \frac{3}{t_f^{\,2}}(\theta_f - \theta_0)t^2 + \frac{2}{t_f^{\,3}}(\theta_0 - \theta_f)t^3\right) \tag{3.52}$$

while velocity yields:

$$v(t) = R\omega(t) = R\left(\frac{6}{t_f^{\,2}}(\theta_f - \theta_0)t + \frac{6}{t_f^{\,3}}(\theta_0 - \theta_f)t^2\right) \tag{3.53}$$

and the acceleration yields:

$$\alpha(t) = R\left(\frac{6}{t_f^{\,2}}(\theta_f - \theta_0) + \frac{12}{t_f^{\,3}}(\theta_0 - \theta_f)t\right) \tag{3.54}$$

Thus, for the dynamics model of seven-segment up and down elevator motion, the kinematics boundary time conditions t_0, t_1, t_2, t_3 can be defined as:

$$\begin{cases} t_1 = \dfrac{\alpha_m}{j_m} \\[2mm] t_2 = \dfrac{\omega_m}{\alpha_m} \\[2mm] t_3 = \dfrac{\alpha_m}{j_m} + \dfrac{\omega_m}{\alpha_m} \end{cases} \tag{3.55}$$

Hence,

$$t_2 - t_1 = \frac{\omega_m}{\alpha_m} - \frac{\alpha_m}{j_m} \tag{3.56}$$

Thus, the position profile at each time (t_0, t_1, t_2, t_3) is defined as

$$\begin{cases} \theta(t) = \theta_1(t) = \omega(0)t + \sum \omega_{\text{mean}}t = \omega(0)t + \dfrac{1}{6}j_m t^3 \ldots \forall t \in [t_0, t_1[\\[3mm] \theta(t) = \theta_2(t) = \omega(0)t + \sum \omega_{\text{mean}}t + \sum \dfrac{1}{2}\alpha_{\text{mean}}t^2 \ldots \forall t \in [t_1, t_2[\\[3mm] \theta(t) = \theta_3(t) = \omega(0)t + \sum \omega_{\text{mean}}t + \sum \dfrac{1}{2}\alpha_{\text{mean}}t^2 \ldots \forall t \in [t_2, t_3[\end{cases} \tag{3.57}$$

This is equivalent to:

$$\begin{cases} \theta(t) = \theta_1(t) = \omega(0)t + \sum \omega_{\text{mean}}t = \omega(0)t + \dfrac{1}{6}j_m t^3 \ldots \forall t \in [t_0, t_1[\\[3mm] \theta(t) = \theta_2(t) = \omega(0)t + \dfrac{1}{6}j_m t^3 + \omega_1(t)(t - t_1) + \dfrac{1}{2}\alpha_1(t)(t - t_1)^2 \ldots \forall t \in [t_1, t_2[\\[3mm] \theta(t) = \theta_3(t) = \theta_2(t) + \omega_2(t)(t - t_2) = \dfrac{1}{2}\alpha_m(t - t_2)^2 + \dfrac{1}{6}j_m(t - t_2)^3 \ldots \forall t \in [t_2, t_3[\end{cases}$$

The velocity motion profile is given by:

$$\begin{cases} \omega(t) = \omega_1(t) = \omega(0) + \sum \alpha_{\text{mean}}t = \omega(0) + \dfrac{1}{2}\alpha(t)t = \omega(0) + \dfrac{1}{2}j_m t^2 \ldots \forall t \in [t_0, t_1[\\[3mm] \omega(t) = \omega_2(t) = \omega(0) + \sum \alpha_{\text{mean}}t = \omega(0) + \dfrac{1}{2}\alpha(t)(t - t_1) \ldots \forall t \in [t_1, t_2] \\[3mm] \omega(t) = \omega_3(t) = \omega(0) + \dfrac{1}{2}\alpha(t)t + \alpha(t - t_1) - \dfrac{1}{2}\alpha(t)(t - t_2) \ldots \forall t \in]t_2, t_3] \end{cases} \tag{3.58}$$

This is equivalent to:

$$\begin{cases} \omega(t) = \omega_1(t) = \omega(0) + \sum \alpha_{\text{mean}}t = \omega(0) + \dfrac{1}{2}\alpha(t)t = \omega(0) + \dfrac{1}{2}j_m t^2 \ldots \forall t \in [t_0, t_1[\\[3mm] \omega(t) = \omega_2(t) = \omega(0) + \dfrac{1}{2}j_m t^2 + \alpha_m(t - t_1) \ldots \forall t \in [t_1, t_2] \\[3mm] \omega(t) = \omega_3(t) = \omega(0) + \dfrac{1}{2}j_m t^2 + \alpha_m(t - t_1) - \alpha_m(t - t_1) - \dfrac{1}{2}j_m(t - t_1)^2 \ldots \forall t \in]t_2, t_3] \end{cases} \tag{3.59}$$

Angular velocity (rad.s^{-2})

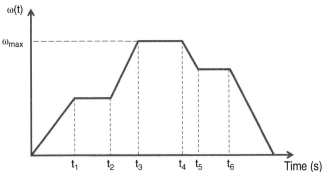

Figure 3.19 Generic motion profile.

The acceleration motion profile is given by:

$$
\begin{cases}
\alpha(t) = \alpha_1(t) = j_m t \ldots \forall t \in [t_0, t_1[\\
\alpha(t) = \alpha_2(t) = \alpha_m \ldots \forall t \in [t_1, t_2[\\
\alpha(t) = \alpha_3(t) = \alpha_m - j_m(t - t_2) \ldots \forall t \in [t_2, t_3]
\end{cases}
\tag{3.60}
$$

The jerk motion profile is given by:

$$
\begin{cases}
j(t) = j_m \ldots \forall t \in [t_0, t_1[\\
j(t) = 0 \ldots \forall t \in [t_1, t_2[\\
j(t) = -j_m \ldots \forall t \in [t_2, t_3]
\end{cases}
\tag{3.61}
$$

with operating conditions such that: $a_{mean} = \frac{1}{2}a(t)$ for a ramp-like accelerating and decelerating phase, $a_{mean} = a_{max}$ during the constant speed phase; $\omega(0) = 0$ is the starting and final position zero velocity; and $\omega_{mean} = \omega_{max}$. It should be recalled that segments (t_4, t_5, t_6, t_7) have the same equations with a negative sign. The rated velocity is given by:

$$
\omega_{nom} = 2\pi N_M
\tag{3.62}
$$

with N_M being the rated speed on the nameplate of the motor. For some applications, the motor speed is constant while for others it varies by level. Hence, regardless of the applications, the velocity profile is very useful to derive the required motor torque, as the load torque is associated to dynamic torque. It is possible to distinguish several types of motion profile generics, as illustrated in Figure 3.19.

Example 3.12 *Process Command Data Sequence Interpolation*

Figure 3.20 illustrates a process command input sequence using three types of signal approximation:

1) the ramp command trajectory where there is a simple constant velocity ramp input up to the desired position;
2) the trapezoidal command trajectory where there is a short initial constant positive acceleration period n_1 to n_2, leading to a constant velocity period n_2 to n_3, (ramp) which ends with a constant negative acceleration period n_3 to n_4, up to the desired speed;
3) the harmonic (sinusoidal velocity) command trajectory where there is a harmonic (offset cosine wave) velocity command. The continuous-time equivalent would be:

$$
\omega^*(t) = \frac{\omega_{peak}}{2}(1 - \cos\omega_0(t))
$$

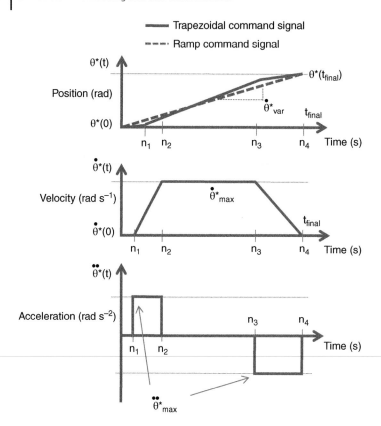

Figure 3.20 Command input trajectories (position, velocity and acceleration).

meaning that there are offsets of the cosine where ω_o is the frequency of the waveform. However, the parametric form uses T for the sampling period and $n_{\text{final}} = n_4$ for the final time step. In addition, it is assumed that when the final time steady state position has been reached, it is maintained. For the ramp trajectory approximation, with t_{final} known, the number of samples n_{final} to reach the final value is given by:

$$n_4 = n_{\text{final}} = roundup \left\{ \frac{t_{\text{final}}}{T} + 1 \right\}$$

Therefore, from an impulse input $\delta(z)$, the ramp trajectory is given by:

$$\frac{\theta^*(z)}{\delta(z)} = \dot{\theta}_{\text{var}} T \frac{z^{-1}}{(1 - z^{-1})} - \dot{\theta}_{\text{var}} T \frac{z^{-1}}{(1 - z^{-1})^2} z^{-n_{\text{final}}}$$

with $\dot{\theta}_{\text{var}} = \frac{\theta_{\text{final}} - \theta_{\text{initial}}}{Tn_{\text{final}}}$. It should be noticed that only three parameters are used to model the ramp trajectory $(T, n_{\text{final}}, \Delta\theta^* = \theta_{\text{final}} - \theta_{\text{initial}})$. For the trapezoidal trajectory approximation, the corresponding digital command signal can be given by:

$$\frac{\theta^*(z)}{\delta(z)} = \frac{1}{2} \ddot{\theta} T^2 \frac{(1 + z^{-1})z^{-1}}{(1 - z^{-1})^{-3}} (1 - z^{-n_2} - z^{-n_3} + z^{-n_4})$$

Knowing that the velocity profile can be segmented into three, as shown in Figure 3.20, for time interval $[Tn_2 - Tn_1]$ and $[Tn_4 - Tn_3]$, there is a change in velocity causing acceleration

Table 3.6 Table of parameters used to build various process command input sequences.

Types of command input sequence	Parameters required
Step	1 parameter (y_{final})
Ramp	2 parameters (y_{final}, $\dot{y}_{max}$)
Parabolic	3 parameters (y_{final}, $\dot{y}_{max}$)
Harmonic	3 parameters (y_{final}, $\dot{y}_{max}$, signal frequency)

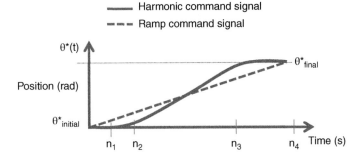

Figure 3.21 Equivalent harmonic command trajectory.

and deceleration. The maximum velocity and acceleration are given by ($\dot{\theta}_{max}$, $\ddot{\theta}_{max}$):

$$\dot{\theta}_{max} = \frac{\Delta\theta^*}{2(n_3 + n_4 - n_2)T}$$

and

$$\ddot{\theta}_{max} = \frac{\dot{\theta}_{max}}{(n_2)T}$$

For the sinusoidal/harmonic trajectory approximation, the position profile is given by:

$$\frac{\theta^*(z)}{\delta(z)} = K(\omega_0 T \frac{z^{-1}}{(1 - z^{-1})^2} - \frac{\sin(\omega_0 T)z^{-1}}{(1 - 2\cos(\omega_0 T)z^{-1} + z^{-2})}(1 - z^{-n_{final}})$$

Here, only the parameters ($n_{final} = n_4$, T, $\Delta\theta^* = \theta_{final} - \theta_{initial}$) are required to model the trajectory, where $K = \frac{\omega_{max}}{2\omega_0}$ such that $\omega_{max} = \frac{2\Delta\theta^*}{n_4 T}$ and $\omega_0 = \frac{2\pi}{n_4 T}$. For the step trajectory approximation, the position profile is given by:

$$\theta^*(z) = \theta_{final}\frac{1}{(1 - z^{-1})}$$

Table 3.6 summarizes the type of command input and corresponding parameters required (see Figure 3.21).

3.3 Signal Conditioning

Active transducers convert non-electrical energy into an electrical output signal without requiring external excitation to operate. This is the case, for a thermocouple in temperature measurement. In contrast, passive transducers change the electrical property value of a device such as resistance, inductance, or capacitance according to changes in the physical quantity being measured. For example, with strain gauges, the resistivity varies proportionally to the applied stress, and with Linear Variable Differential Transformers (LVDTs), the inductance changes proportionally to the resulting displacement. Hence, the high-voltage signals of many transducer outputs require a signal conditioning unit to be properly read by other control devices. Thus, the signal conditioning consists of processing a transducer output by demodulating, amplifying, filtering and linearizing, range quantizing, and isolating from the signal from the noise before transmitting it to the comparator in a form compatible with the reference system input. The functions performed by signal conditioners can be described by:

1) signal scaling (amplification or attenuation) in order to use a reasonable portion of the analog-to digital (A/D) range and to avoid the A/D saturation effect;
2) signal filtering (noise elimination or isolation) in order to ensure a better restitution of the initial signal received from analog filters or digital filters;
3) phase shifting or reference shifting in order to ensure proper wave shaping;
4) signal buffering in order to capture a useful output signal content within an analog input board;
5) bridge balancing in order to calibrate output signals or to detect signal traffic overload;
6) signal demodulation through a mathematical manipulation of the signal such as differentiation, division, integration, multiplication, root finding, squaring, subtraction, or summation;
7) signal linearizing used in the case of a nonlinear relationship to the physical quantity measured.

In summary, signal conditioning is achieved by the variation of input signal amplitude and frequency. Such a signal conditioner output could be directly used by the ADC unit. Table 3.7 summarizes the functions required from typical signal conditioners for some measurement systems.

Typical transducers require external voltage or current excitation from the signal-conditioning unit such as strain gauges, thermistors, and resistance temperature detectors (RTDs). Connection between the analog measurement device and the A/D device could be single-ended (to avoid offset), or differential (to measure signals with large voltages range). In order to minimize the effects on the control system instrumentation and the process performance, the selection of a signal conditioner should consider the sample rate through:

Table 3.7 Types of transducers and associated signal-conditioning functions required.

Temperature	Amplification
	Cold junction
	Filtering
Strain gauge	Voltage excitation
	Filtering
	Linearization
	Bridge configuration
	Excitation
High-voltage	Isolation
Absorption transducer	Offset removal
	Amplification

(i) sensor response speed and range; (ii) conversion time; and (iii) frequency of sample signal itself.

Example 3.13 A 1 mV °C^{-1} resolution of a thermometer output signal requires a minimum of 2.5 °C before a 12-bit A/D device indicates a change in digital value. Hence, by amplifying the output signal of a thermometer by ×1000, each bit on the A/D device is equivalent to 0.001 mV, which corresponds to 0.0025°C, thus avoiding A/D device saturation. If the temperature variations are measured with ×1000 amplification, 22°C is measured as 22 V, which is much greater than the A/D input range. Therefore, it is possible to adjust the output of the amplifier with an offset of −22 V so that the output signal would be 0 V.

3.4 Signal Conversion Technology

Typical signal conversion includes: DC to AC, AC to DC, frequency to voltage, voltage to frequency, digital-to-analog (D/A), and analog-to-digital (A/D). In this section, only the last two conversion types are covered.

3.4.1 Digital-to-Analog Conversion

A DAC consists of a resistive ladder network connecting switches within an analog circuit and an operational amplifier receiving the equivalent set of input digital values to produce a corresponding analog voltage signal. Based on the binary-driven switch positions (logic 1/closed or logic 0/open), the current flows across specific resistors within this ladder network. The operational amplifier is used to sum all voltage potentials across the set of those specific resistors and to generate an output voltage. For example, using operational amplifiers and a D-flip-flop, a set of switch positions equivalent to a binary word is converted into an output voltage signal. Figure 3.22 depicts an 8-bit unsigned binary word DAC. Consider an 8-bit binary word given by $b_0, b_1, b_2, b_3, b_4, b_5, b_6, b_7$, such that $b_i\{1, 0\}$ is sent out of the computing

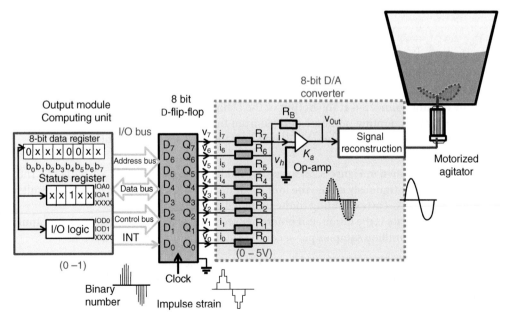

Figure 3.22 Example of an 8-bit resistive ladder DAC.

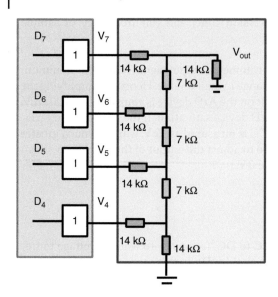

Figure 3.23 Resistive-based ladder of a DAC.

unit toward the D flip-flop through its I/O address buses. Then, the D flip-flop I/O data bus collects this binary number. Depending on the binary word, the states of the switches are closed or open, allowing the current to pass through their connected resistors. This generates voltages $v_0, v_1, v_2, v_3, v_4, v_5, v_6, v_7$ of the D flip-flops. These voltage outputs of D flip-flop are fed into the summing operational amplifier. The resistor values on the operational amplifier are chosen such that its output voltage is given by:

$$V_{\text{out}} = v_h \left(\frac{R_B}{R/2^{k-1}} b_{k-1} + \frac{R_B}{R/2^{k-2}} b_{k-2} + \dots + \frac{R_B}{R/2} b_1 + \frac{R_B}{R} b_0 \right) \tag{3.63}$$

with $v_h = v_0 = v_1 = v_2 = v_3 = v_4 = v_5 = v_6 = v_7$ being the logic-1 output voltage of the D flip-flops. Equation (3.63) can be rewritten as:

$$V_{\text{out}} = v_h \left(\frac{R_B}{R} N \right) \tag{3.64}$$

with the binary number output to DAC given from $N = b_{k-1}2^{k-1} + b_{k-2}2^{k-2} + \dots + b_1 2^1 + b_0 2^0$. In addition, the gain of the DAC $K_a = \frac{R_B}{R}$ with $R = R_0 = R_1 = R_2 = R_3 = R_4 = R_5 = R_6 = R_7$ being any resistor connected at the output of the D flip-flop. The voltage levels produced by DAC are given by the counting numbers derived from the number of binary bits generated. In this case, b_0 is the most significant bit (MSB) and b_7 is the least significant bit (LSB).

Example 3.14   As shown in Figure 3.23, a resistive ladder network is used as a 4-bit DAC. The binary input is sent continuously across the four separate data lines v_4 through v_7, which results in an output of the DAC corresponding to the intended discrete voltage value.

   The DAC sampling rate is the number of points per second that a DAC can generate as output analog values (up to a million samples per second). Among characteristics of DAC are:

1) the *range*, which is the voltage interval of values that the DAC can output, given by $[V_{\text{max}}, V_{\text{max}}]$. The maximum output voltage, V_{max} is produced when all bits are 1 $N = 2^k - 1$ and is given by:

$$V_{\text{max}} = V_h \left(\frac{R_f}{R} (2^k - 1) \right) \tag{3.65}$$

The minimum output voltage, V_{min}, is achieved when all bits are zeros ($N = 0$) and it results in $V_{min} = 0$. When a signed analog output signal is obtained from the DAC, the number of points in the area derived for k bits is estimated by inverting the sign bit and adding a biased voltage of $-V_h(R_f/R)2^{k-1}$ to the output v_0. Here, the interval is given:

$$V_{max} = V_h\left(\frac{R_f}{R}(2^{k-1} - 1)\right) \tag{3.66}$$

and

$$V_{min} = -V_h\left(\frac{R_f}{R}(2^{k-1})\right) \tag{3.67}$$

2) The *accuracy* and *linearity* of the DAC are proportional to the precision of the input resistors on the operational amplifier. Furthermore, stabilizing the temperature of the resistor ladder network circuit can also improve the accuracy.

3) The *resolution* of a DAC, V_q, is the number of k digital bits (number of discrete values) corresponding to the range of analog values. The minimum nonzero analog voltage produced is obtained when $N = 1$. As such, the resolution as the number of discrete values or "levels" is written as a power of 2 given by:

$$resolutions/combinations = (2^k) \tag{3.68}$$

Then, the resolution, which depends on the resistor network ladder configuration, is given by:

$$V_q = V_h\frac{R_f}{R} \tag{3.69}$$

When the resolution and the range are known, the minimum number of bits required in the DAC is given by:

$$k = \frac{\log\left(\frac{V_{max} - V_{min}}{V_q} + 1\right)}{\log 2} \tag{3.70}$$

with k being an integer or results from the equation rounded up to the next highest integer. For example, an ADC with a resolution of 8 bits can encode an analog input in $2^8 = 256$ different levels, representing the ranges from 0 to 255 (i.e. unsigned integer) or from -128 to 127 (i.e. signed integer), depending on the application. The resolution can also be defined electrically and expressed in volts. In the case of the encoder pulse signal counting, it is given by:

$$\overline{\omega}_{res} = \frac{\theta_{res}}{T} \tag{3.71}$$

while in the case of the clock pulse signal counting, it is given by:

$$\omega_{ccp} = \frac{\theta_{res}}{N_x T_{CLK}} \tag{3.72}$$

with N_x and T_{CLK} being, respectively, the count number and the clock sampling time. Then, the resolution yields:

$$\overline{\omega}_{res} = \omega_{ccp\ pos+} - \omega_{ccp\ pos-} = \frac{\theta_{res}}{N_x T_x} - \frac{\theta_{res}}{N_{x\mp1} T_x} \tag{3.73}$$

Consider a sampling rate T. The average velocity sensed (or weighted moving average) is given by:

$$\overline{\omega} = \frac{\theta(k) - \theta(k - 1)}{T} \tag{3.74}$$

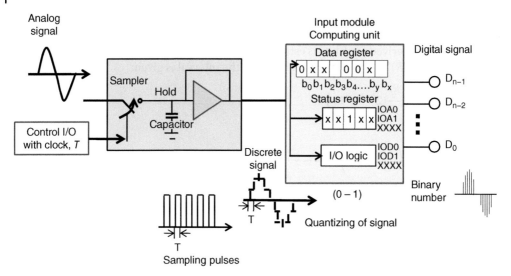

Figure 3.24 Analog-to-digital conversion technique.

Thus, by increasing the sampling period, $(T, \overline{\omega})$ appears to be closer to ω with degraded resolution. This may cause *quantization*. However, using the clock pulse counting, the resolution yields:

$$\overline{\omega}_{res} = \omega_{pulse} = last \frac{\theta_{pulse}}{T} \tag{3.75}$$

4) The *settling time (slew time)* is the conversion time delay of the input DAC binary word within the expected accuracy voltage level.
5) The *offset error* is the difference between the expected DAC output analog signal and the real value when the digital value of 0 is applied. The *gain error* is the linear deviation from the ideal transfer line of a DAC, which corresponds to a change of slope from the ideal converter.

3.4.2 Analog-to-Digital Conversion

The ADC is a device that samples an analog input voltage and encodes that voltage as a binary number (1 or 0). An ADC is implemented either using an ADC software conversion or through an ADC hardware conversion process. In the case of hardware conversion, the voltage level input signal is the reference voltage. Hence, the two-step converter consists of a sampling and holding (S/H) unit along with a quantizing and encoding (Q/E) unit, as illustrated in Figure 3.24. The quantization is the partitioning of the reference voltage signal range into a number of discrete quanta, followed by the matching of the input signal with the correct quantum. The encoding assigns a unique digital code to each quantum and allocates this digital code to the input signal.

The ADC allows the computer to sample proportionally the analog feedback voltage from transducers. The resulting digital voltage is achieved through a sample and hold circuit. Sample and hold circuitry is used by the flip-flop memory. This can be achieved by using a comparator between the input voltage V_i and a variable digital voltage V_0. The time from V_0 to V_i is the conversion time. The main ADC characteristics are the speed and the resolution of the conversions.

The ADC *resolution* is the smallest detectable change in an analog voltage signal resulting in a variation in digital output, given by:

$$\Delta V = \frac{V_r}{2^N} = \frac{V_{RH} - V_{RL}}{2^n} \tag{3.76}$$

with V_r being the reference voltage range, N the number of bits in digital output, 2^N the number of states, n the number of digits, V_{RH} the higher value of input voltage, and V_{RL}, the lower value of input range. The resolution represents the quantization of the error inherent in the conversion of the signal to a digital form. For example, consider a 6-bit ADC corresponding to $2^6 = 64$ different values so that in a 1 V scale voltmeter with 0.000001 V steps, it could measure from -0.999999 to 0.999999 V. The resolution of an ADC is degraded by an increase in temperatures and resistances within its components. A voltage divider or amplifier is used to fit the input signal without changing the resolution. *The input range* of an ADC is the interval of voltage values that are properly converted. The ratio of the highest over the lowest voltage values converted by an ADC is called the *dynamic range* and is expressed in decibels (dB) by:

$$\text{dynamic range} = 20 \log \left(\frac{S}{N} \right) \tag{3.77}$$

with S being the highest voltage level and N the noise level. *The rate of accuracy* of an ADC is the smallest resolution of analog voltage (LSB). Recall that the ADC consists of a successive sampling and holding process of a chain of digital values (bits). This is achieved by a switch and a capacitor. The minimum sample rate is obtained when the capacitor is discharged (no value held), while the maximum sample rate is derived from the speed at which the capacitor is charged. The time required to commute the sample switch is called the aperture time or delay t_a. Any deviation between the analog voltage value and the current voltage value input while the switch is open causes an error in accuracy. The maximum input frequency f_{MAXLSB} is given by:

$$f_{MAXLSB} = \frac{2}{2\pi t_a 2^n} \tag{3.78}$$

The accuracy of an ADC can be improved either by increasing the resolution or by increasing the sampling rate that causes an increase in the maximum frequency. A pulse signal from a clock is used to synchronize the timing of the ADC conversion process. Commonly encountered ADCs are ramp, successive approximation, dual-slope, and parallel or "flash."

A *digital ramp ADC* produces a saw-tooth signal that ramps up and goes to zero. A timer starts counting when ramp starts. When the ramp voltage reaches the input value, a comparator fires and this transition of the comparator stops the binary counter and holds the digital value corresponding to the analog voltage.

A *successive approximation ADC* uses a counter circuit known as a successive approximation register (SAR), which is an internal high-speed n-bit DAC iterating a series of levels and comparing them to the input voltage, as depicted in Figure 3.25. Here, instead of counting up in a binary sequence, this register counts all values of bits starting with the MSB and finishing at the LSB. The successive approximation ADC is faster than the digital ramp ADC due to its digital logic to converge on the value closest to the input voltage. Such an n-bit ADC requires 2^n cycles to perform a conversion.

This conversion process consists of:

1) initializing the MSB as 1;
2) converting the digital value to an analog value using DAC;

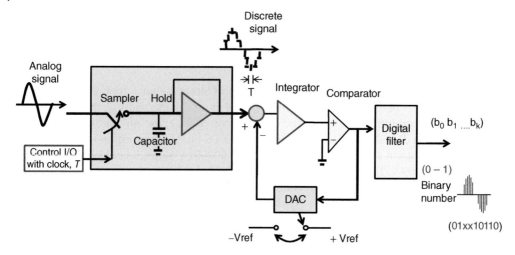

Figure 3.25 Successive approximate A/D conversion circuitry.

3) comparing the recorded value to the analog input value; and
4) checking if $V_{in} > V_{DAC}$:
   a) if yes, set bit 1
   b) if no, bit is 0 and test the next bit.

Example 3.15 It is desired to convert a 0–12 V range command input signal into an 8-bit ADC or $2^8 = 256$ binary counts (or values) from 0 to 255 with a count increment of 1/256. The 0000 0000 binary number, for 0 V corresponds to 0 values, and similarly, 6 V corresponds to 128 count values, and the 1111 1111 binary number for 11.95 V corresponds to 255 count values. A 10-bit ADC would offer a greater resolution as there are $2^{10} = 1024$ binary counts, which is equivalent to a count increment of 1/1024. Hence, the case of a 0–5 V range command input signal with a 10-bit ADC corresponds to a resolution of 4.88 mV per LSB, while for 8-bit ADC it would be 19.5 mV.

Example 3.16 Consider an 8-bit resolution A/D device used in the measurement of a 0–10 V signal. The input signal can be measured in steps of $10/256 = 39$ mV (i.e. ADC can only detect changes greater than 0.039 V). Thus, a 10 V analog input corresponds to the digital number 255 and a 0 V analog input corresponds to 0. Any input signal increase or decrease causes a change by 1 from the previous number (e.g. 9.961 V is digitally represented by 254). A 12-bit A/D device is more sensitive to variations in the input voltage since its minimum resolution is given by $10\,V/4096 = 0.00244$ V. If the input signal exceeds 10 V, the A/D produces a flat line corresponding to the saturation from the analog DAQ device.

Another ADC device is the *dual-slope ADC* (DS-ADC), which consists of an amplifier for integration, a signal comparator, a counter, a clock, some controlled switches, and a capacitor, as illustrated in Figure 3.26.

The ADC derives the voltage by measuring time and using digital logic to compute the input voltage. Initially, switch S_1 is set to ground, S_2 is closed, the counter is set at 0 and the capacitor is discharged. Then the conversion starts, S_2 is open, and S_1 is set such that the input to the integrator is V_{in}. DS-ADC integrates an unknown input voltage V_{in} during a time delay T_{d1}, then "de-integrates" it using a known reference voltage V_{ref}. At the end of the time delay T_{d1}, the input is applied across the capacitor and it starts to charge for a period of T_{d1}. After T_{d1}, the capacitor

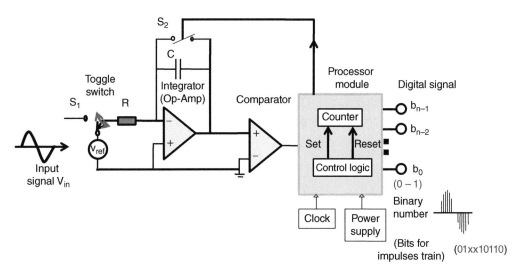

Figure 3.26 Dual-slope A/D conversion circuitry.

commutes to a negative reference voltage and starts to discharge at a rate proportional to the reference. The timer is started when S_1 is set such that the counter begins to count clock pulses for a period of T_{d1}, after which the counter resets to zero. A comparator is used to determine when the output voltage of the integrator crosses zero.

An advantage of a dual-slope ADC over the single-slope ADC is that the conversion result is insensitive to errors in the component values, as any error from a component value during the integrate cycle is cancelled out during the de-integration phase. This method is insensitive to noise.

In the case of a *flash ADC* converter with n bits of resolution, there are $(2^n - 1)$ comparators, (2^n) resistors, and a control logic unit, as depicted in Figure 3.27. The 2^n resistors divide the reference voltage into (2^n) equal intervals (i.e. one for each comparator), which correspond to the input range, while the $(2^N - 1)$ comparators determine in which of these (2^N) voltage intervals the input voltage V_{in} lies. Then, V_{in} is compared to each voltage level simultaneously so that the comparator would set output either to 1 for all voltages below the input voltage or to 0 otherwise. The resulting chain of digital values is fed into an encoder with the combinational logic to translate comparator outputs into the corresponding n-bit values. The flash ADC conversion time is bounded by the settling time of the comparators and the propagation time of the combinational logic.

Pipeline ADC or *multistage* ADCs are similar to the flash converter, with a lower number of comparisons. It uses a coarse conversion so that the difference in the input signal is determined with a DAC and the results are combined in a last step. This type of ADC is fast, has a high resolution, and only requires a small die size.

A *delta-encoded ADC's* main components are resistors, capacitors, comparators, control logic, and a DAC, as shown in Figure 3.28. During its operation, an input signal is oversampled and goes to an integrator. The integrated output signal and the DAC are channeled to a comparator where they are compared to the ground. The comparator controls the counter. The circuit uses negative feedback from the comparator to iteratively adjust the counter (delta encoder) until the DAC's output is close enough to the input signal. This delta modulator adjustment of the output signal ensures that the average error at the quantizer output is zero. After iteration, it results in a digital signal output in which a sequence of binary numbers is

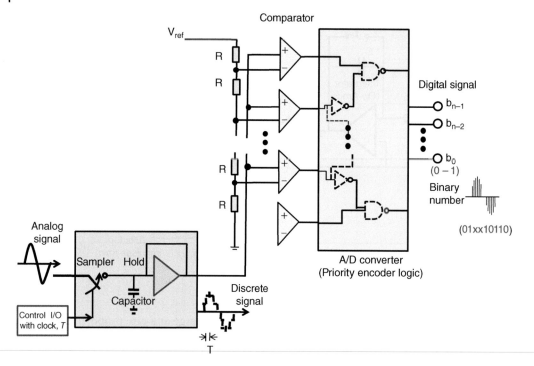

Figure 3.27 Flash A/D conversion circuitry.

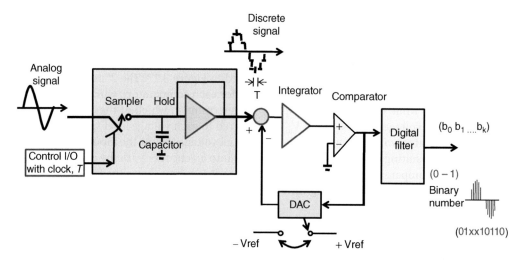

Figure 3.28 Delta-encoded ADC block diagram.

proportional to V_{in}. A delta-encoded ADC has an up-down counter that feeds a DAC. This device has a high resolution and very wide ranges, and does not require precision components. Nevertheless, the conversion time is dependent on the input signal level. Furthermore, it requires a high sampling rate. Some ADCs successfully combine the delta and successive approximation approaches in the case of a signal with high frequencies and low magnitude.

A *sigma-delta* (SD) ADC is a high-sampling-rate 1-bit ADC combining a low-pass filter (to cut or shape higher frequency electronic noise) with a DAC in a feedback loop. The resulting converted 1-bit sequence is digitally filtered through a low-pass filter to improve resolution and reduce the data flow rate.

Example 3.17 The n-bit AD-converter receives from a speed sensor an analog voltage signal of $y_a(t)$ value in the range $[Y\text{min}, Y\text{max}]$, to convert it into a digital signal of y_d number of bits. Generally, $y_a(t)$ would be converted to a digital signal in the form of a set of bits given by:

$$y_d \approx b_{n-1}b_{n-2} \dots b_2 b_1 b_0$$

where each bit b_i has a value of 0 or 1. These bits are interpreted as coefficients in a number with base 2 given by:

$$y_d \approx b_{n-1}2^{n-1} + b_{n-2}2^{n-2} + \dots + b_2 2^2 + b_1 2^1 + b_0 2^0$$

with b_0 being the LSB, while b_{n-1} is the MSB.

Example 3.18 Consider a 12-bit ADC used to convert a signal in the range $[0\text{ V}, 10\text{ V}]$. The digital signal is given by:

$$y_d = Y_{\min} = 0 \cdot 2^{11} + 0 \cdot 2^{10} + \dots + 0 \cdot 2^1 + 0 \cdot 2^0 = 000000000000_2 = 0_2 = 0(\textit{decimal})$$

$$y_d = Y_{\max} = 1 \cdot 2^{11} + 1 \cdot 2^{10} + \dots + 1 \cdot 2^1 + 1 \cdot 2^0 = 111111111111_2 = 1000000000000_2 - 1$$

$$Y_{\max} = 2^{12} - 1 = 4095(\textit{decimal})$$

The resolution is the interval represented by the LSB, which is:

$$y_{d\_res} = \frac{Y_{\max} - Y_{\min}}{Number\ of\ intervals} = \frac{Y_{\max} - Y_{\min}}{2^n - 1} = \frac{10V - 0V}{4095} = 2.44mV$$

3.5 Data Logging and Processing

3.5.1 Computer Bus Structure and Applications

Beside DAC and signal conditioning, a knowledge of computing unit architecture and interfacing to process actuating and sensing devices is highly valuable for process control system design. Indeed, the computing unit either sends instructions encapsulated in voltage signals to the process actuating devices, or receives binary encoded signals from sensing devices. Such instructions or encoded signals can handle a set of bits corresponding to I/O field device connections or addresses. Such an exchange of instructions and process data is done between the computing unit components. The three main components of the computing unit architecture are the processor, memory, and bus structure, as depicted in Figure 3.29. While the processor component is responsible for the execution of program instructions using data stored in the memory component, similarly, the bus structure ensures the communication links for data exchanged between the processor, the memory, and the input/output (I/O) devices. This is achieved through the storage or the execution of the interrelated input/output logic level of data or program. More specifically, there are synchronous buses using a clock-based protocol to activate and deactivate exchanges between asynchronous buses using a handshake-based protocol. Furthermore, during such a process, data and instructions are transferred between the computing unit components and the field devices through the bus structure. Communication lines and function-specific buses include: address, data, and control. The address line specifies

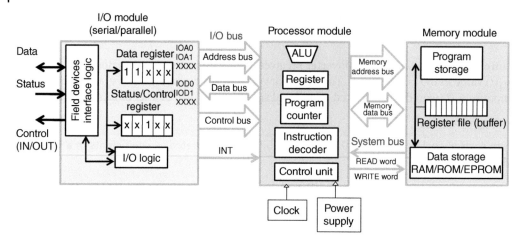

Figure 3.29 Computer data and instruction addressing structure.

Table 3.8 Logic level signals of I/O data bus.

IN	OUT	Function
0	0	No I/O operation is to be performed
0	1	I/O address and data valid for output operation
1	0	I/O address is valid for input operation
1	1	Not used

which I/O (with a unique address) device communicates with the computing unit at a given time. It consists of a number of separate signals, each specifying one bit in the address of the I/O device to be accessed by the processor. Data I/O devices are transferred to/from the computing unit using a data line based on IN and OUT signals representing write or read requests on the control line. As presented in Table 3.8, IN and OUT signals are used to indicate when the addresses and data specified by the I/O address bus and the I/O data bus are valid. The I/O data bus is bidirectional, depending on whether it is an input or output operation performed by the processor. The number of lines in the I/O data bus is usually the same as the computing unit word size. (e.g. 8, 16, 24, 32, 64 or more bits). It should be noticed that the bus creates a bottleneck, limiting the I/O data throughput.

The I/O device interfacing mechanism can be captured through the memory register or the buffer mapping activities, which include synchronizing signals from system clocks, interrupt lines, and status lines. For example, in order to alert the processor when an I/O operation is complete, interrupt circuitry and programming are used. As such, READ and WRITE instructions from and to I/O devices are executed using techniques for I/O service operations including programmed I/O (polling), interrupt-driven I/O, or direct memory access. With a programmed I/O technique, a program routine requests an a I/O service operation by checking the status of the I/O devices periodically. In the case of an interrupt-driven technique, an interrupt communicates with the processor through a dedicated control bus based on an interrupt service routine program for enabling or disabling the active signal from or to field devices. This allows the processor to interrupt normal program execution momentarily to process the I/O service operation

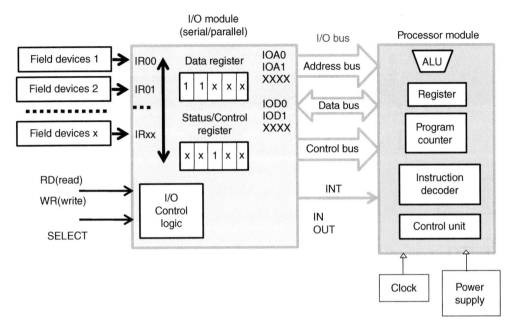

Figure 3.30 Input/output field devices interfacing with computing unit.

rather than waiting for an I/O operation to complete, as is done with polling of the I/O port. The direct memory access technique uses a specialized processor to ensure execution of I/O service operations. Figure 3.30 illustrates a generic I/O interfaces for an x-bit computing unit.

Example 3.19 Consider push button (PB) and limit switch (SW) input devices that are used to activate a light emitting diode (LED) by checking their status (open or closed) through the computing unit, as illustrated in Figure 3.31. From the execution of the READ instruction, the LED can be activated or deactivated when the ID_0 and ID_1 bits are loaded over the data bus (bus driver) onto the register in the processor. Then, after the status is checked, the computing unit interface generates a SELECT signal to transfer ID_0 and ID_1 bit data toward the I/O data bus and the address of the LED interface on the I/O address bus. The processor can set the OUT signal to logic 1 or 0, indicating a change of LED status through variation of both signal (OUT

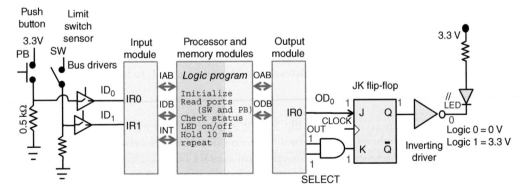

Figure 3.31 Computing unit interface for the LED activation.

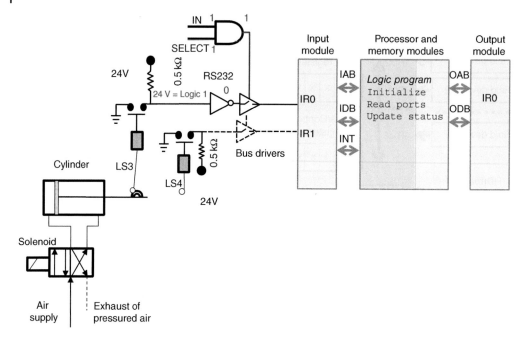

Figure 3.32 Pneumatically-driven process with limit switche I/O interface circuitry.

and SELECT signals). This causes the JK flip-flop to sample and hold the level (0 or 1) of the OD_0 at the output Q connected to the RS232 driver. If the logic level at the output of the driver connected to the LED is at 1 (3.3 V), the current through the LED is zero, as there is no voltage differential across it, and the LED is OFF. Otherwise, the LED is ON with the power source and the resistance defining the LED brightness.

Considering another example of a pneumatically-driven cylinder whose travel is bound by two limit switches LS3 and LS4, as depicted in Figure 3.32. Solenoid activation drives the pneumatic piston rod of the cylinder. Any contact between either switch closes the corresponding switch contact. An open switch would correspond to the lowest bit 0 sent to the processor, and a closed switch to bit 1. Here, the bus driver (RS232) could connect the corresponding all-bit data from all limit switches to the I/O interface over the data line only when both signals (SELECT and IN) are at logic 1. Then, it is immediately disconnected from the I/O data line on the resetting of the IN signal to logic 0 by the processor; subsequently the loaded LS3 and LS4 data are checked and their statuses are updated in the register of the processor. In the case where a limit switch is open, the inverting driver receives 24 V (logic 1), which is inverted into logic 0. Then, the corresponding input bit data is sent over the data bus in the register of the processor. In a case where any switch is closed, the bus driver would be set to logic 1, to be sent to the lowest bit of the processor.

Example 3.20 Using an asynchronous flip-flop, it is possible to store and hold the output value for a time interval when the input conditions vary (limit switches change). This is called debouncing circuitry in reference to the mechanical contact bounce effect when the switch rapidly changes position (i.e. repeatedly open and closed). It is useful to avoid a reactivation output value despite input change, as in the case illustrated in Figure 3.33. Here, a toxic liquid has to interlock its input control valve to avoid harming personnel or destroying equipment in

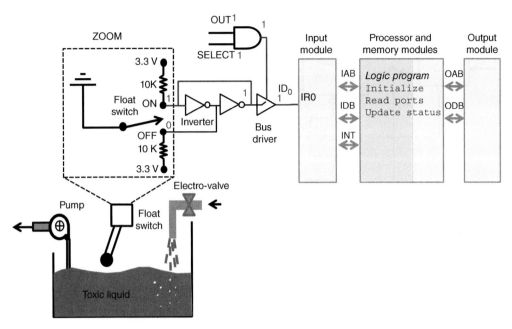

Figure 3.33 Double switch position interface with debouncing.

a facility. A safe design uses a float switch with two positions: ON and OFF. When the switch is in the OFF position, the logic level 0 is set to the register of the processor. Otherwise, it is logic level 1, corresponding to the switch in the ON position. This is done using two successive inverting drivers to eliminate contact bounce effects. When it is in the ON or OFF position, the output of the inverter in series flip-flop is logic 1 or 0, respectively. For a time interval, any subsequent variation of switching contacts would not affect this logic output. Such debouncing circuitry could be implemented using a D flip-flop. This design is useful for a system that has an input with more than two positions.

3.6 Computer Interface and Data Sampling Issues

Within digital control architecture, the sample period is the rate at which process data are sampled and sent to the computing unit. A *clock* connected to the DACs and ADCs supplies a pulse train every T seconds to synchronize the D/A and A/D closing and opening bus operations. Figure 3.34 illustrates a typical DAQ and control realization process. Here it can be noted that computer process control systems use multiple entities and their resulting operations are time- or event-triggered. These operations are also synchronized by the computer clock to achieve parallel or simultaneous tasks, especially for entities such as:

1) the *sensor node* that is time-triggered. A zero-order hold circuit usually samples an analog sensor;
2) the *link to the controller*, which is time triggered or event-triggered, and can be a dedicated link or a bus. The link is often modeled to be the source of delays and jitter created by a contention in the communications or by transient faults, for example. This is a kind of disturbance model;
3) the *controller node*, which could be time-triggered or event-triggered. The control law is computed once every control period, during which, in many cases, the controller node often

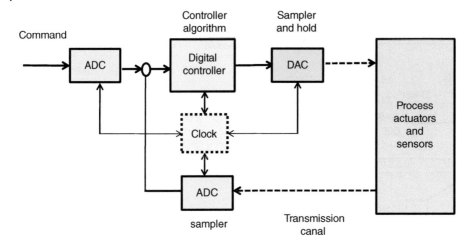

Figure 3.34 Generic digital data acquisition and control processing system.

runs other tasks. Hence it can be modeled as a source of delay and jitter (i.e. a disturbance model);

4) the *computing unit*;

5) the *link to the actuator* is similar to the link between the sensor and the controller. The difference is that the information of time delay and jitter lies in the future from a perspective of control law calculation;

6) the *actuator node* can be time- or event-triggered. It is rare that the model of a distributed system adds delays and jitter in the actuator node, but delays and jitter are often inherited from a preceding task.

Consider $u(k)$, being the signal being sampled and held in order to produce a continuous signal $u_{as}(t)$. The latter should be held constant during a time interval kT to $(k+1)T$. Figure 3.35 illustrates the components of data processing time delays. However, to achieve efficient computer-controlled regulation, this periodic operation of sampling and holding $u_{as}(t)$ must consider some time constraints, such as:

1) The *control execution period or scan time*, which is the time taken by the controller to compute data (read, calculations, save, and output). While the read, save, and output times are constants, the calculation time depends on the length of the executed program and the execution speed of the computer. Therefore, the control period cannot be considered constant.

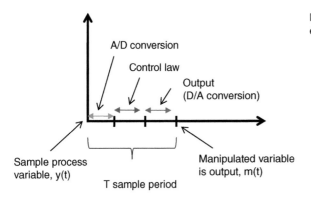

Figure 3.35 Timing structure for the execution of computer control algorithm.

The scan time is equal to:

$$T_{scan} = T_{rd} + T_{ex} + T_{sw} \tag{3.79}$$

with T_{rd} being the read time or the time taken by the computer to access/read the data coming from the process; T_{ex} being the time of memory data program execution computation depending on the length of the executed program and the CPU speed being given by:

$$T_{ex} \approx \frac{\text{Number of instructions}}{\text{Frequency of CPU}} \tag{3.80}$$

and T_{sw} being the time for saving, adding, writing, and transmitting data, which is equivalent to the time taken to save data and transmit them to the process;

2) the *control transmission delay*, T_t, is the propagation time and the time taken by signals to move along transmission canals (wireless, cables etc.), including the transmission return sensor time;
3) *transient errors* are failures that occur in the overall system. There are static failures (e.g. a hardware failure that renders a utilization factor greater than one) and dynamic failures;
4) the *jitter* being time-related to abrupt, spurious variations in the duration of any specified related interval that could arise, such as due to clock drift, branching the entire four timing problems listed. It can affect the control and process behavior;
5) the *holding element* introduces time delays such as T_a, which is the acquisition time (duration from the sampling instant of the command signal to the time of signal output to the S/H), and T_p, which is the aperture time (time between the start of the hold command and the time the sampler switch is opened);
6) the *feedback signal sampling element* or *sensor interface (A/D)*, which is based on the latch-based conversion technique, as illustrated by Figure 3.36. Furthermore, the sensor resolution is based on physics and is expected to be infinite. Examples of the technical encoding principle are depicted in Figures 3.37 and 3.38.

The clock pulse counting interface uses less lag than the encoder pulse counting interface that has a $\frac{1-z^{-1}}{T}$ delay. A signal conversion circuit such as a D flip-flop (latch) causes a time shift of mT in signal, as illustrated in Figure 3.39.

Therefore, the procedure to compute the $\varphi(z)$ latch output from the modeling procedure is as follows:

1) Using the physical model of the latch, consider the unit pulse response as a step input.
2) Put the process model in an ordinary differential equation (ODE) form.
3) Find the equivalent Laplace transform; that is, $Ł$ (ODE) = Φ (s).

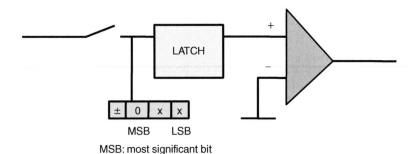

Figure 3.36 Latch-based analog-to-digital signal encoding technique.

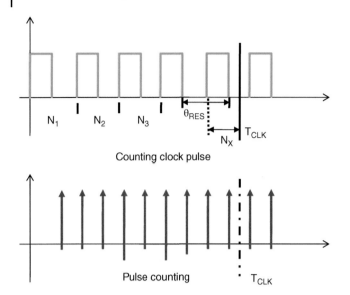

Figure 3.37 Pulse counting principle of thermocouple generated signals.

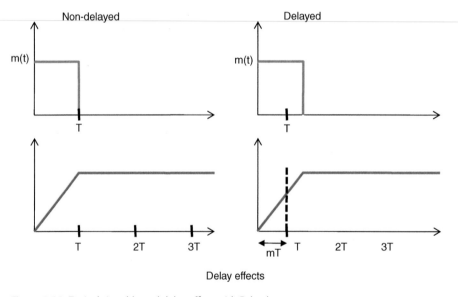

Figure 3.38 Typical signal-based delay effect with D-latch.

4) Then, perform a direct unit step response calculation of the process output response using T (without delay) and mT (when considering delays) (hint: take the impulse itself and use the Taylor expansion).

Example 3.21   Consider an integration type process model given by $G(s) = \frac{1}{s+a}$; the signal with converter time delay is illustrated in Figure 3.40, while Table 3.9 summarizes various possible sampled responses in the z-transform.

Figure 3.39 Typical input impulse signal holding.

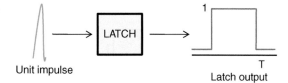

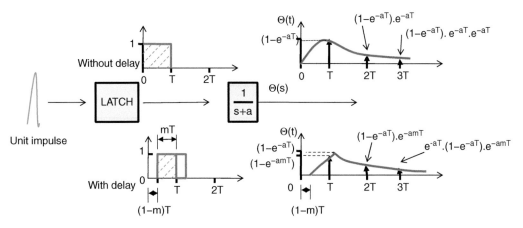

Figure 3.40 Time delay in discrete response of first-order process.

Table 3.9 Sampled response in *z*-transform for a first-order process.

Without delay	With delay
$G(z) = \dfrac{(1 - e^{-aT})z^{-1}}{1 - e^{-aT}z^{-1}}$	$G(z) = (1 - e^{-amT})z^{-1} + \dfrac{(1 - e^{-aT})e^{-amT}}{1 - e^{-aT}z^{-1}}z^{-2}$
(Consider as a combination of a step and a delay ramp)	$G(z) = \dfrac{(1 - e^{-amT})z^{-1}(1 - e^{-aT}z^{-1}) + (1 - e^{-aT})e^{-amT}z^{-2}}{1 - e^{-aT}z^{-1}}$

Example 3.22 *Multiple Sampling Rates*

Consider a DC motor block diagram that illustrates two cascaded state feedbacks, as depicted in Figure 3.41. Here, there are two sampling rates due to many cascade feedbacks. The sampling time of the current $i(t)$ is 10 times faster than the sampling time of the position loop. Therefore, either as a deadbeat process or as a first-order process model, the current inner loop has enough time to converge and settle between two successive samples of the motor position. Thus, it does not affect either the outer position loop or the associated generated motor torque step change, as illustrated in Figure 3.41.

However, if the current inner loop is reduced to be only three times faster than the out-loop DC motor position, information on the DC motor dynamics is required, as illustrated in Figures 3.42 and 3.43. This causes a loss of information, as illustrated in red. To circumvent it, it is possible to use an observer to decouple the velocity. However, the modeling of the quantization of the effect on the sampling period is developed later.

In summary, the discrete-time delay in sampling, T_{delay}, is the time taken by the sampling and hold operation. This time delay is the dominant magnitude in the sampling time selection, which means that the sampling rate has a secondary effect. This delay, T_{delay}, and the transmission and

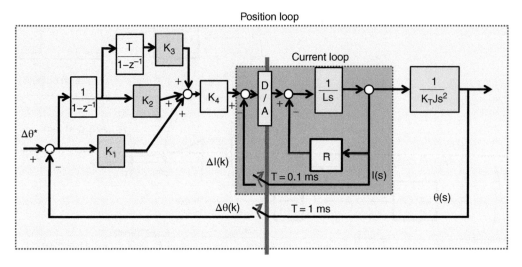

Figure 3.41 DC motor block diagram with current and position cascade feedback.

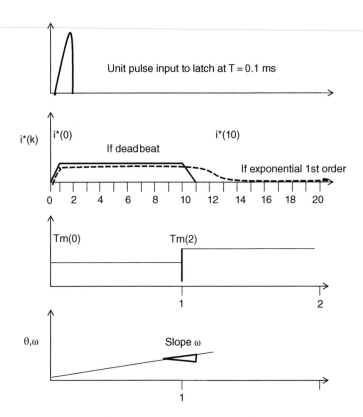

Figure 3.42 Typical DC motor current, position, and torque responses to unit pulse impulse.

Figure 3.43 Typical response of the current loop for a slower sampling period of inner current loop.

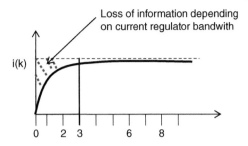

holder signal processing times are given by:

$$T_{\text{delay}} = T_a + T_p + T_{\text{scan}} + T_t \tag{3.81}$$

It should be noticed that the clock synchronizes all measurements and controls the signal. Hence, the control delay is not constant. For example, when data are transmitted to the computer and the computer has not completed the processing of the data received before, a problem will appear. Therefore, the transmission time delay must be enough for the computer to perform calculations before receiving the next data. Hence, the selection of the sampling period consists of considering all the existing, delays, even in data transmission, so that the control performance will not be affected. The selected delay represents the minimum value of the sampling period.

3.6.1 Signal Conversion Time Delay Effects

In addition to the hard constraints related to the process and the control system physical structure, the sample period should also fulfill the process performance requirements. Therefore, it should be noted that the sample period chosen for a computer control system can have a significant impact on its performance, especially on:

1) *The disturbance rejection time issue.* The magnitude of the process response is dependent on the sample period. For a short sample period, disturbances are detected quickly and compensated before they can significantly affect the process output. On the other hand, with a long sample period, the disturbance affects the process for a relatively long time before a compensation is applied, making the magnitude deviations of the process output likely to be unacceptably large. Therefore, it is suitable to consider which disturbances can occur in a process and to select a sample period that is short enough. Hence, disturbances can be rejected by the controller before they adversely affect the system performance. The maximum sampling period capable to ensure the rejection of disturbance depends on the types of process (boiler, motor, cooler etc.) and the kind of process behavior at stake. Furthermore, the knowledge on disturbance periodical characteristics and tolerance limits in output changes the process control strategy;

2) *The process dynamics stability.* A system is dynamically stable if all its roots have a negative real part in the s-domain or if they belong to the unity circle in the z-domain. This is because any time delay introduces a phase shift and even causes process instability. The roots of the characteristic equation determine the system stability, and the value of these roots is a function of the sample period T in most cases. If T is made too large, the magnitude of one or more of the roots can be less than 1 and the resulting system is unstable. However, a stability analysis can establish a limit on the sample period during stationary operating conditions without necessarily finding the shortest sample period required for the desirable transient results;

3) *The speed of process response.* The sample period chosen for a discrete control system must be equal to or shorter than the input response time of the system. Before the settling time (time taken by any system to reach its final value) T_S, the system experiences its transient behavior, and after T_S, the system is in its steady state. The computer control must be able to follow the transient behavior, so it is necessary to set a sampling time lower than the settling time T_S. The transient behavior of a given system is characterized by its rise time T_r and it must satisfy $T_s \leq T_r$ in order to ensure that the system evolves. In short, several factors usually compete when selecting a sampling rate, such as:

   a) sample as fast as possible, to maximize accuracy or to provide adequate response time;
   b) sample as slow as possible, to preserve the processing time while avoiding the noise from dominating the input signal;
   c) sample at a rate that is a multiple of the control algorithm frequency to minimize the jitter.

Based on the typical conflicting process requirements listed previously, the sample period should be derived with care when the discrete controller is designed, to ensure that a favorable system performance is achieved. It is usually preferable to make the sample period as short as possible in order to achieve rapid response and minimize the adverse effects of the sample period on stability. On the other hand, calculations must be performed by the control computer every sample period. The shorter the sampled period, the greater the computational load placed on the control computer.

4) *Aliasing effects.* When a continuous-time signal is sampled, some information can be lost, as the generated discrete-time signal is defined only for the sampling instants. For a sinusoid with a period of T, as illustrated by Figure 3.44, if the sampling period $T_{S1,2}$ is greater than the half-period, meaning $T/2$, it is possible to completely miss one lobe of the sinusoid. Even more, if the sampling period is twice per sampling period, the same values are obtained in time for each period and cannot be distinguished. With a sampling period $T_{S1,2} < T/2$, each lobe can be sampled at least once and the oscillation will be detected. However, if $t \to \infty$, all amplitude values of the sinusoid will be detected.

Before sampling or digitalizing a signal, it is suggested to pass it through a low-pass filter as a continuous-time signal; or else, the process data might be corrupted by high-frequency signals appearing as low frequency.

3.6.1.1 Nyquist Sampling Theorem and Shannon's Interpolation Formula

A signal $x(t)$ can be reconstructed from samples $x(n) = x(nT)$ when the sampling frequency, f_S, is greater than twice the maximum frequency of the signal being sampled, $f_{\text{sampled frequency}}$,

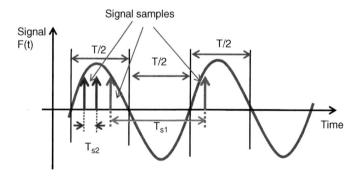

Figure 3.44 Sampling period effect on sinusoid signal.

equivalently, when the Nyquist frequency (half the sample rate) exceeds the highest frequency of the signal being sampled.

Oversampling Case A case is considered where a sampling frequency, f_s, is required to be more than twice the desired system bandwidth so that a digital filter can be used in exchange for a weaker analog anti-aliasing filter. For $f_s > f_{\text{sampled frequency}}$, the signal $x(n)$ can be reconstructed via the Shannon interpolation formula given by:

$$x(t) = \sum_{n=-\infty}^{\infty} x(n) \frac{\sin[2\pi f_s(t - nT)]}{2\pi f_s(t - nT)} \tag{3.82}$$

Undersampling (Below the Nyquist Rate) A signal of frequency $f_{\text{sampled frequency}} \pm nf_s$ (n being an integer) is generated. This undesired frequency interaction and translation due to sampling is called aliasing. If aliasing occurs, no signal processing operation downstream of the sampling process can recover the original continuous-time signal. Here, sampled high-frequency signals can appear to be lower-frequency signals because of aliasing and consequently become indistinguishable in the discrete-time domain. The apparent lower frequencies, f_a, due to aliasing are:

$$f_a = |f_{\text{sampled signal}} - nf_s| \tag{3.83}$$

Practically, it is suitable to use a high sampling frequency with an analog filter matching this frequency. Then, after the ADC, the signal is down-sampled to the frequency used by the controller. Recalling that the pre-filter introduces a phase shift and ω_b, the desired bandwidth of the closed-loop system to avoid aliasing effects, is set as $5\omega_b \leq 2\pi T \leq 100\omega_b$, the (non-ideal) anti-aliasing filter has a cutoff frequency ω_f in between ω_b and ω_c. The approximate relation is given by:

$$t_s\omega_c = \frac{5}{\xi} \tag{3.84}$$

where t_s is the settling time (1%) and ξ is the damping factor, which could be rewritten as $\frac{t_s}{100} \leq T \leq \frac{t_s}{5}$. A related appropriate choice is also to let $T = \frac{t_r}{10}$, where t_r is the rise time of the open-loop system. Here, sample rates should be about 20 times the bandwidth or faster in order to ensure that the digital controller matches the performance of the continuous controller.

3.6.2 Estimation of the Minimum Sampling Rate to Be Selected

The signal reconstruction requires the time step T of the discrete-time signal to be relatively small in order to behave approximately as the original continuous-time signal. For the design of process control system, the sampling period must be compatible with rate specifications of the process I/O devices, interfaces used, and with design specifications. Thus, an option could be to select the sampling period to be long enough so that sufficient time is available for the execution of the controller algorithm and I/O operations during each discrete-time step. In addition, it is suitable to select the sample period to be as short as possible in order to achieve rapid response and minimize the adverse effects of the sample period on stability. Therefore, depending on the applications of the control system and instrumentation for either: (i) digital filter design; (ii) digital controller design; or (iii) process simulation, the choice of the sampling period is expected to achieve tradeoffs between conflicting with sample period objectives. Here, some guidelines (rules of thumb) for sampling period selection are discussed.

With digital filters for signal processing, the objective is to make the difference between the original and the reconstructed signal as small as possible, usually using DSPs (digital signal processors). Thus, sampling must fulfill the requirement of:

$$\frac{1}{T} \geq 5f_H \tag{3.86}$$

Here, f_H is the highest frequency for the discrete-time filter to have almost the same characteristics as the original continuous-time filter. For example, in the case where the low-pass filter f_H is five times the bandwidth, a filter up to 25 times the bandwidth would have the characteristics similar to that of the original continuous-time filter.

With a digital controller and holding element, the objective is to make the closed-loop system behave according to performance expectations, usually implemented using digital controllers (µController, programmable logic controller etc.). Thus, the selection of a sample time T usually depends on:

1) the execution time of the control algorithm;
2) the interface speed of the A/D and DACs;
3) the desired process dynamics;
4) the desired closed-loop dynamics (system responses, command trajectory, disturbance input);
5) the stability (all systems can become unstable);
6) the smallest sample time that is suitable for the disturbance rejection;
7) the avoidance of aliasing effects (Nyquist frequency larger than the closed-loop bandwidth);

Therefore, based on the process closed-loop bandwidth, ω_b, the system response rise time, T_r, the frequency response of the continuous-time system, ω_c, and the computing time of the digital controller, T_{scan}, the following rules of thumb for sampling time, T, selection are presented:

Rule 1a. The DAC holds the computed controller signal during a time step, which should be:

$$T \geq 0.2T_{\text{res}} \tag{3.87}$$

where the response time, T_{res} is 63% of rise time T_r, which is derived from the set point step response of the system. For a system having a dominating time constant, τ, the response time is approximately equal to this time constant ($T_{\text{res}} = \tau$). If the bandwidth of the process closed-loop control system is ω_b (rad s^{-1}), the response time of the control system can be estimated by:

$$T_{\text{res}} \approx \frac{1}{\omega_b} \tag{3.88}$$

Rule 1b. The sampling time T for a discrete-time controller could be derived from the desired speed of the closed-loop system, such that the number of samples N taken by the closed-loop system per rise time T_r is 4–10:

$$N = \frac{T_r}{T} \approx 4 - 10 \tag{3.89}$$

Rule 1c. The sampling period, T, is based on the rise time, T_r, of the system response to a step such that the input to the discrete controller should be sampled approximately six times while the sampling interval can be adjusted to two to three times per rise time. It could be estimated by:

$$\frac{T}{2} \leq \frac{T_r}{10} \tag{3.90}$$

This corresponds to a sampling frequency that is 10–20 times the system's closed-loop bandwidth.

$$T \approx \frac{(10 \, to \, 20)}{\omega_b} \tag{3.91}$$

Rule 2. Considering the closed-loop bandwidth f_b(Hz) of the feedback system: this is related to the speed at which the feedback system can track the command input as well as the amount of attenuation the feedback system must provide in the face of plant disturbances. Such a sampling period should be chosen in the range of:

$$\frac{1}{30 f_b} < T < \frac{1}{5 f_b} \tag{3.92}$$

Rule 3. Another criterion for the selection of the sampling period is based on the rise time T_r of the feedback system, to provide smoothness in its time response. For a *first-order linear system* of the form $H(s) = \frac{1}{\tau s + 1}$, the rise time is given by:

$$T_r = 2.2\tau \tag{3.93}$$

Then, the sample period should be bound by:

$$0.095 T_r < T < 0.57 T_r \tag{3.94}$$

For a *second-order linear system* of the form $H(s) = \dfrac{\omega_n^{\,2}}{s^2 + 2\xi\omega_n s + \omega_n^{\,2}}$, the rise time of the canonical second-order system is given by:

$$T_r = \frac{\pi - \sin^{-1}(\sqrt{1 - \xi^2})}{\omega_n \sqrt{1 - \xi^2}} \tag{3.95}$$

For a damping ratio $\zeta = 0.707$, the rise time yields:

$$T_r = \frac{3.33}{\omega_n} \tag{3.96}$$

Then the sampling time, T, should be bounded by:

$$0.06 T_r < T < 0.4 T_r \tag{3.97}$$

Rule 4. A time delay of up to a full sample period is suitable between two digital controller computed values. Such time delay should be kept at about 10% of the rise time corresponding to the sampling period to satisfy:

$$T < \frac{0.05}{f_b} \tag{3.98}$$

Rule 5. Based on the frequency response of the continuous-time system, the sampling rate, T_{ω_0} should be selected such that:

$$0.15 < f_{\omega_0} < 0.5 \tag{3.99}$$

where f_{ω_0} is the gain crossover frequency of the continuous-time system in radians per second.

Rule 6. The sampling period must result in a Nyquist frequency (half the sampling rate) greater than the highest input signal. With ω_b being the desired bandwidth of the closed-loop system, in order to avoid some aliasing effects, the sampling period should be bound as:

$$5\omega_b < 2\pi T < 100\omega_b \tag{3.100}$$

When the settling time (1%) t_s is known, it could be rewritten as:

$$\frac{t_s}{100} \le T \le \frac{t_s}{5} \tag{3.101}$$

A relatively good choice would be:

$$T = \frac{t_r}{10} \tag{3.102}$$

Rule 7. The sample period is constrained by the performance of the computing unit such that:

$$T \ge T_{\text{scan}} \tag{3.103}$$

In addition, this sampling period must be compatible with the rate of data updates through I/O interfacing devices (such as DAC or ADCs). Here, the selection of the sample period is achieved in the knowledge that the computing unit must perform calculations at every sample period and that the shorter the sampled period, the greater the computational load placed on the computing unit. Additional free time should be available to allow future enhancements to the controller algorithm.

With process simulators, the sample period should be

$$T \le \frac{0.1}{|\lambda|_{\text{MAX}}} \tag{3.104}$$

where $|\lambda|_{\text{MAX}}$ is the largest of the absolute values of the eigenvalues in the system model. Consider a matrix A to be the state-space model given by: $\frac{dx}{dt} = Ax + Bu$. For the transfer function model, the poles instead of the eigenvalues (the poles and the eigenvalues are equal for most systems that do not have pole-zero cancelations) are used. If the model is nonlinear, it must be linearized before calculating eigenvalues or poles. If possible, it is suggested to use a simulation of the system to test the effect of the value of T before implementation.

3.6.2.1 Remarks on Sample Periods

High-sampling rates are suitable to control temperatures for heating systems (e.g. boilers). This could take many seconds as storage tank temperature change is slow.

Small sampling rates are suitable for motion control applications. This could be less than a microsecond.

The sampling rate period could introduce some numerical errors in digital controller design. A solution could be to have the sample period large enough in order to avoid numerical error issues.

Example 3.23 Consider the load torque disturbance $D(t)$ and an input voltage $v(t)$. The motor shaft position has been approximately modeled using:

$$\frac{d\theta(t)}{dt} = K_m v(t) + K_t D(t)$$

The corresponding process difference equation is:

$$\theta_n - \theta_{n-1} = K_m T v_{n-1} + K_t T d_{n-1}$$

Any change in motor shaft position, $\Delta\theta$, from an applied disturbance of the load torque of magnitude D during that sample period, T, would be given by:

$$\Delta\theta = K_t T D$$

Considering that for a load torque of 90 N.cm, the deviation in shaft position due to these torques is less than 0.01 revolutions. Hence, with $K_t = 0.3$ rpm/N.cm, the sample period, T_{dist} with respect to the disturbance rejection capability of the system would be bounded by:

$$T_{dist} \leq \frac{0.01 rev \left(\frac{60 \sec}{\min}\right)}{0.3\frac{rpm}{N.cm}(90 \text{ N.cm})} = 0.022 \sec$$

The sample period for a discrete controller must be chosen such that it is equal or shorter than the system response time to inputs. For a first-order system, time constant is equal to 0.03 s. From the rule of thumb, the sample period, T_{tri} should be $T < 0.03$ s to ensure a good tracking of input signals. With respect to system time response, the sampling period should be:

$$T_{tri} = 0.03 \text{ s}$$

With respect to the stability criteria, where an output voltage of 5 V is sent by the controller toward a motor causing its rotation at 950 rpm with no load torque being applied, the motor gain would be given by:

$$K_m = \frac{950 \text{ rpm}}{5 \text{ V}} = 190 \text{ rpm V}^{-1}$$

Then, the closed-loop transfer function yields:

$$\frac{\theta(z)}{r(z)} = \frac{TK_m z^{-1}}{1 - (1 - K_m T)z^{-1}}$$

and the stability condition is satisfied if

$$0 < K_m T < 2$$

The sample rate, T_{sa} with respect to the stability criteria would be

$$T_{sa} < \frac{2 \times 60}{190} = 0.63 \text{ s}$$

Consider that the delay associated with the control algorithms, execution and I/O interfacing T_a is estimated to be:

$$T_a = 3 \text{ μs}$$

From the analyses of the three sample periods (T_{dist}, T_{tri}, T_{sa}), the sample period must satisfy $T_a < T < \min(0.022; 0.05; 0.63)$ equivalent to 3 μs $< T < 0.022$ s. It is possible to select $T = 1$ ms.

Exercises and Problems

3.1 a) Derive the number of bits a DAC, requiring a voltage in the ±8 V range and a resolution of ±0.05 V, which can be delivered for process control.
  b) Consider a process signal output range of ±8 V and a resolution 0.05% of this range. Derive the number of bits of A/D required.
  c) Derive the number of bits a DAC can provide for the velocity command of a bidirectional motor between 2.5 m s^{-1} and 0.1 m s^{-1}, with a possible step change of 0.05 m s^{-1}, (Hint: non-zero velocity is 0.1 m s^{-1}).
  d) Derive the number of bits a DAC can deliver for the temperature command profile of a heater system, such that the maximum temperature is 280°C and the minimum non-zero temperature is 45°C with an incremental value of 0.2°C.

3.2   a) Consider a motor shaft operating in the velocity range of $0.90 - 2.30$ m s^{-1}. Determine the number of bits provided by an ADC with a 150 Hz sampling frequency to ensure a resolution of 0.02 m s^{-1}.

    b) Consider an 8-bit ADC with $V_{ref} = 5$ V. For the input signals of 1.5 and 5 V, derive the resulting digital signal (code) in the case of bidirectional and unidirectional operations, respectively.

    c) From the list of physical phenomena that follows, list two (contact and contactless) sensors used to measure them.

      i) Flow rate of liquids in pipelines
     ii) Speed of an elevator motor
    iii) Pressure of an underwater vehicle
    iv) Fermentation tank temperature.

3.3   Considering a temperature sensor of a liquid in a tank delivering at 150°C to a 9.5 V signal output with a maximum tank temperature of 210°C:

    a) derive the ADC voltage increment required to achieve a resolution of 0.5°C;

    b) derive the number of bits required as an ADC for $0 - 180$°C temperature measurement;

    c) select the gain of an amplifier located between the temperature sensor and an 8-bit ADC for $0 - 150$°C temperature measurement.

3.4   Design and sketch the circuitry of a 10-bit binary signed DAC capable of outputting a ± 10 V signal.

3.5   Consider an 8-bit ADC with $V_{ref} = 5$ V and a differential input channel. The positive input voltage V+ and the negative voltage V− are, respectively, in the range of [0.99, 1.03] and [1.02, 1.025] V. Answer the following questions:

    a) What is the input range of the resulting differential input?

    b) What percentage of the full input range does that cover?

    c) How large is the quantization error as a percentage of the differential input range? If the ADC offers the gains $\{0.1, 0.2, 2, 5, 20, 50, 100, 200\}$, which gain would you select?

    d) How does your selected gain affect the quantization error (again as a percentage of the differential input range)?

3.6   Follow these instructions to design an anti-aliasing filter for an unknown waveform amplitude:

    a) identify the anti-aliasing frequency at normal operation, a motor produced a 65 Hz signal, and the A/D sampling rate is 3 Hz;

    b) estimate the apparent frequency;

    c) based on the Shannon theorem, determine at what sampling frequency the aliasing effect does not occur.

3.7   Consider a motor that rotates at a nominal speed of 1500 rpm, with each revolution of the motor shaft involving 55 pulsation signals (commutation) toward the velocity sensor. What will the maximum allowable sampling period be for the ADC to avoid an aliasing effect?

3.8   Consider a 100 Hz ADC with a resolution on the angular position sensor of 0.005 rev. Derive the resolution of the average velocity. If the A/D sampling frequency is doubled, what will the new velocity resolution be?

3.9 a) What value is ¨1101¨ in the decimal number system? What would be the minimum resolution expected from this DAC if it received this binary number for a [1.4, 3.2] V operating range?
b) What functions are is provided by a DAC?
c) List some of the advantages and disadvantages of the resistive ladder network.
d) Consider an 8-bit "resistive ladder" DAC. Draw its complete schematic and write a program that will step through 256 different analog voltages.

3.10 A motor generates an output voltage signal, as shown in Figure 3.45.
a) If the signal is captured by an ADC with a 150 ms sampling period, design an anti-aliasing filter for this converter.
b) In the case where a 50 Hz signal is sampled at 1.5 Hz, what is the lowest apparent frequency in the sampled data that arises from aliasing? What range of sample frequencies ensures that aliasing does not occur when a 50 Hz is being sampled?

3.11 A ball is attracted magnetically, as illustrated in Figure 3.46.
Find the linear discrete model that relates distance $h(t)$ between the ball and the solenoid (in meters) to induced current $i(t)$. State all assumptions. Similarly, derive the discrete model relating the voltage $V(t)$ (in volts) and the current $i(t)$ (in amperes). Identify any significant limitations in these models.

3.12 a) Following the general discrete procedure presented in Table 3.4, obtain the discrete model for a process defined by:

$$\tau_1 \frac{d^2 y(t)}{dt^2} + \tau_2 \frac{dy(t)}{dt} = Kx(t)$$

Check the results with the equivalent z-transform in Appendix C and conclude.

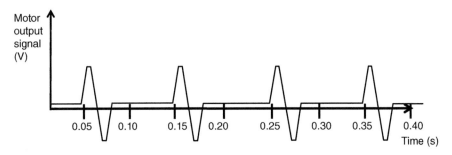

Figure 3.45 Motor voltage output signal.

Figure 3.46 Magnetically suspended ball.

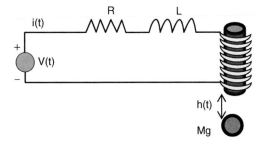

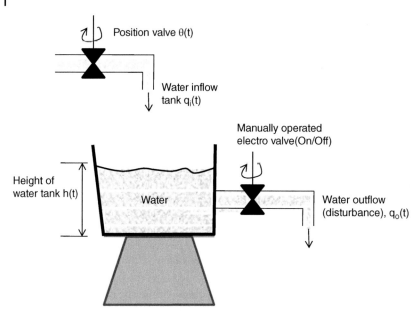

Figure 3.47 Water tank with inlet valve computer controlled.

b) The water tank illustrated in Figure 3.47 has a manually operated outlet chute and a computer-controlled inlet valve. The diameter of the tank is 24 m. The valve allows water to flow into the tank $q_i(t)$ at a rate of 2.8 m^3 s^{-1} when it is fully opened. The flow out of the tank $q_o(t)$ is to be modeled as $q_o(t) = p(t)\sqrt{2g}\sqrt{h(t)}$. For a sample period of 45 s, derive the difference equation describing the height of water in the tank $h(t)$ as a function of the percentage of valve opening $p(t) = a = constant$ and the disturbance flow $q_o(t)$.

3.13 Reconsider the tank illustrated in Figure 3.47 and the process model relating the valve position $\theta(t)$ with the height level of water in the tank $h(t)$.
   a) Derive the process difference equation.
   b) Plot the process response $h(n)$ to a valve position step change of of 0.8 m amplitude. All initial conditions are assumed to be zero. Assume the sample period is 0.1 min.
   c) Derive and plot the continuous solution of the process model from the differential equation for the same input. Compare the results.
   d) With a sample period of 0.5 min, compute the first six process response $h(n)$ values from the discrete step input depicted in Figure 3.48. Plot the results.

3.14 The spacecraft shown in Figure 3.49 has a mass of 3500 kg, while the rocket has an initial mass of 9000 kg. The rocket carries a propulsion engine that generates a thrust of 750 000 N. With a sample period of 5 ms, develop a discrete time dynamic model for the spacecraft position (x-axis) (in kilometers). Assume that when the engine is set to ON (input variable $f(t)$), the rocket mass consumption rate is 150 kg s^{-1}.

3.15 For each of the continuous functions shown in Figure 3.50, develop their equivalent discrete-time command input for two sampling periods of 0.01 and 0.02 s.

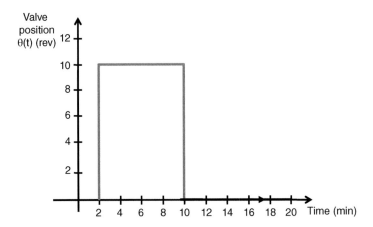

Figure 3.48 Step input to tank filling system.

Figure 3.49 Spacecraft motion with an angle α.

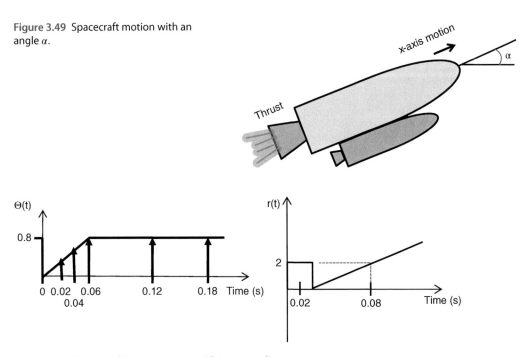

Figure 3.50 Command input sequence with intersampling.

3.16 Using the input signal sequence in Figure 3.51,
  a) derive the equivalent discrete command input obtained by sampling with a time period of 1 s;
  b) for a sample rate of 1 Hz, derive the corresponding $r(n)$ for the periodic command input sequences shown in Figure 3.52. (Hint: the term $1/(1 - z^k)$ generates a sample sequence that has a non-zero term every k samples.)

3.17 Consider two linear motion rods, each attached to a 12-steps-per-revolution stepper motor in a 2D printer, as depicted in Figure 3.53.

Figure 3.51 Command input sequence.

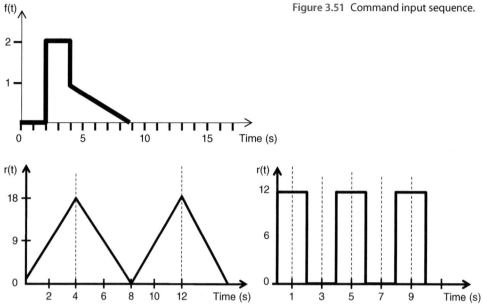

Figure 3.52 Periodic command inputs.

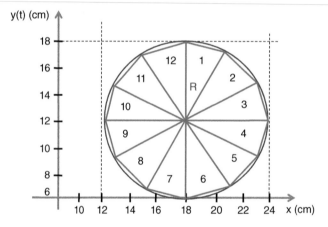

Figure 3.53 Circle curve point-to-point motion.

The motors' operating characteristics are: motor torque rating of 0.9 N m; screw drive pitch of $3.2\,\mathrm{rev\,cm^{-1}}$; and inertia of 0.03 N.m.s^2. Generate the synchronized x and y command input model corresponding to an open control algorithm for the 2D printer positioning system.

3.18 Develop a fully parametric, point-to-point position ($\theta_{\mathrm{initial}}$ to θ_{final}) discrete command input signal using a trapezoidal velocity trajectory with three speed profiles (constant acceleration segment, constant velocity segment, and constant deceleration segment). Consider the upper limits $\pm\,\dot{\omega}_{\mathrm{max}}$ and ω_{max} are known. This means that the parametric position command discrete equation would be:

$$\theta^*(z) = f(\theta^*_{\mathrm{initial}}, \theta^*_{\mathrm{initial}}, T, \dot{\omega}_{\mathrm{max}}, \omega_{\mathrm{max}})$$

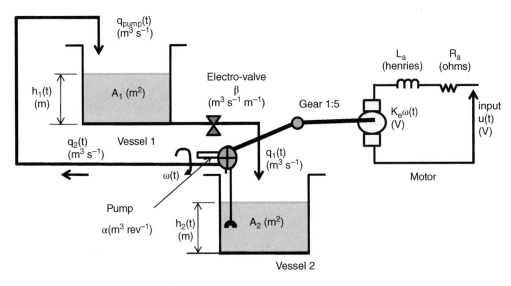

Figure 3.54 Two-vessel system with pump.

NB: The total time to reach the final desired position is not specified but rather is a function of the desired distance to travel, the maximum acceleration, and maximum velocity. In addition, the initial and final velocities are zero. Finally, the final position value θ^*_{final} should be held indefinitely.

3.19 For a given sample period of 25 ms, develop a discrete-time model for the two-tank system shown in Figure 3.54.

3.20 a) Find the z-transform for the following process models using, respectively: (i) a ZOH; and (ii) a first-order hold. Plot their resulting unit step responses for $T = 0.5$ s.

$$G_p(s) = \frac{12.5}{s}$$

$$G_p(s) = \frac{1.5}{0.25s + 1}$$

$$G_p(s) = \frac{0.15}{s^2 + 3.5s + 4.5}$$

b) Perform a comparative analysis of the results obtained.

3.21 It is desired to accelerate a motor from 0 up to 300 rpm in 50 ms, to operate it at that speed for 5 s, and then to decelerate it to a stop in 50 ms. The shaft position profile, $\theta(t)$ is to be represented by the following three cubic polynomials. Determine the spline coefficients.

$$\begin{cases} \theta(t) = a_3 t^3 + a_2 t^2 + a_1 t + a_0 \dots \forall t \in [0, t_1] \\ \theta(t) = b_3 t^3 + b_2 t^2 + b_1 t + b_0 \dots \forall t \in]t_1, t_2] \\ \theta(t) = c_3 t^3 + c_2 t^2 + c_1 t + c_0 \dots \forall t \in]t_2, t_3] \end{cases}$$

3.22   a) A system response is given by $y(kT) = kT$ for $k \geq 0$. Find $y(z)$ for this response among one of following the solutions:

i) $y(z) = \dfrac{Tz}{(z-1)^2}$

ii) $y(z) = \dfrac{z}{T^2(z-1)^2}$

iii) $y(z) = \dfrac{T^2z}{(z-1)}$

iv) $y(z) = \dfrac{Tz}{(z-1)^2}$

b) Find the response at the first four sample instants $(T(0), T(1), T(2), T(3))$ of:

$$T(z) = \frac{z^3 + 1.2z^2 + 4}{z^3 + 3.2z^2 + 0.45z}$$

(Hint: Long division)

c) A 50 Hz signal is sampled at 2.5 Hz. Derive the lowest apparent frequency in the sampled data that arises from aliasing. What is the range of the sample frequencies to ensure that aliasing does not occur when a 50 Hz signal is being sampled?

Bibliography

1 Ackermann, J. (1985). *Sampled Data Control Systems: Analysis and Synthesis, Robust System Design*. Springer-Verlag.

2 Åström, K.J. and Wittenmark, B. (2011). *Computer-Controlled Systems: Theory and Design*. Courier Dover.

3 Bollinger, J.D. and Duffie, N.A. (1989). *Computer Control Machines and Processes*. Addison-Wesley.

4 Brockwell, P.J. and Davis, R.A. (2009). *Time Series: Theory and Methods*, 2e. Springer.

5 Fadali, S. and Visioli, A. (2012). *Digital Control Engineering*, 2e. Academic Press.

6 Isermann, R. (1989). *Digital Control Systems*. Berlin: Springer-Verlag.

7 Jury, E. and Tsypkin, Y.Z. (1971). On the theory of discrete systems. *Automatica*, Elsevier 7 (1): 89–107.

8 Powell, G.F.J.D. and Workman, M. (2006). *Digital Control of Dynamic Systems*, 3e. Ellis-Kagle Press.

9 Golnaraghi, F., Kuo, B.C., and Adams, J.A. (2009). *Automatic Control*, 9e. Wiley.

10 Lorenz D.R. (1999). *Advances in Electric Drive Control, IEEE International Conference Electric Machines and Drives*, pp. 9–16.

11 Patterson, D.A. and Hennessy, J.L. (2013). *Computer Organization and Design MIPS Edition: The Hardware/Software Interface*. Morgan Kaufmann.

12 Paraskevopoulos, P.N. (1996). *Digital Control Systems*. Prentice Hall.

13 Polderman, J.W. and Willems, J.C. (2013). *Introduction to the Mathematical Theory of Systems and Control*. Springer.

14 Santina, M., Stubberud, A., and Hostetter, G. (1994). *Digital Control System Design*. Oxford University Press.

15 Skogestad, S. and Postlethwaite, I. (2005). *Multivariable Feedback Control: Analysis and Design*, 2e. Wiley.

4

Discrete-Time Analysis Methods

4.1 Introduction

Based on the discrete-time process models, it is possible to perform discrete-time analysis using graphical tools such as discrete zero and pole location plot or discrete frequency response (Fourier transforms). In addition, stability definitions and tests for discrete-time systems are covered based on the specific parameter system model. Hence, the design specifications for discrete controller algorithms are deduced from derived discrete-time system characteristics (steady and transient) in time-domain properties (e.g. rise time, settling time, overshoot, etc.) as well as in frequency domains (phase margins [PM], and gain margins [GM]). This chapter aims to define discrete controller design specifications by revisiting the analysis of discrete system dynamics using graphical tools such as root locus and frequency response (discrete-time Fourier transform [DTFT]; fast Fourier transform [FFT]; and discrete Fourier transform [DFT]) as well as stability tests and criteria for discrete-time systems (the Jury–Marden test and Routh–Hurwitz criterion). Accordingly, time and frequency properties for controller design (settling time, percent of overshoot, gain, and PMs, etc.) are defined. Then, some indices for the dynamic performance evaluation and benchmarking of the mechatronic system and process are discussed.

4.2 Analysis Tools of Discrete-Time Systems and Processes

4.2.1 Discrete Pole and Zero Location

The root locus is the determination of the closed-loop pole locations under varying gain K, or more precisely, it represents the values of the poles of the transfer function $G_p(z)$ for which the condition $1 + KG_p(z) = 0$ is satisfied as K varies from 0 to infinity. This means that the root locus includes all the points in the z-plane that correspond to a phase of $KG_p(z)$ being equal to $180°$. This may be written as two algebraic constraints so that the magnitude condition is given by:

$$|KG_p(z)| = 1 \tag{4.1}$$

and the angle condition is given by:

$$\angle G_p(z) = 1\angle 180° \text{ or } (1 + 2m)180° \text{ for } m = 0, \pm1, \pm2, \ldots \tag{4.2}$$

These two conditions allow graphical solution of the roots (poles) from the characteristics equation based on the controller gain and controller pole/zero elements. This is different from the eigenvalue migration graphical representation. Hence, the various locations of the roots

Control Of Mechatronic Systems: Model-Driven Design And Implementation Guidelines,
First Edition. Patrick O.J. Kaltjob.
© 2021 Patrick O.J. Kaltjob. Published 2021 by John Wiley & Sons Ltd.

(the contour) of the characteristics equation of the closed-loop system after varying system parameters, such as gain K from zero to infinity, give an idea of the choice of open-loop pole and zero to achieve a suitable system dynamic response.

Consider in the z-domain the generalized form of the discrete closed-loop transfer function, which can be written as:

$$G_p(z) = \frac{c_i'z^i + \cdots c_2'z^2 + c_1'z + c_0'}{b_n'z^n + \cdots b_2'z^2 + b_1'z + b_0'} \tag{4.3}$$

The corresponding characteristics equation is obtained from the denominator of the transfer function and can be expressed as:

$$1 + KG_p(z) = 1 + \frac{K(z + a_1')(z + a_2')\cdots(z + a_n')}{(z + p_1'')(z + p_2'')\cdots(z + p_p'')} = 0 \tag{4.4}$$

This would be equivalent to

$$(z + b_1')(z + b_2')\cdots(z + b_m') = 0 \tag{4.5}$$

Hence, to locate any point on the root locus, the angle condition could be used such that:

$$(\angle(z + a_1') + \angle(z + a_2')\cdots + \angle(z + a_n')) - (\angle(z + p_1'') + \angle(z + p_2'')\cdots + \angle(z + p_p''))$$
$$= \pm 180(1 + 2m) \tag{4.6}$$

Recalling that for any roots $z = c + jd$, the angle would be $\angle(z + a_2) = \arctan\frac{d}{c - a_2}$. The gain K corresponding to any resulting point z_p on the root locus could be derived using the magnitude condition such that:

$$\frac{K|z_p + a_1'||z_p + a_2'|\ldots|z_p + a_n'|}{|z_p + p_1''||z_p + p_2''|\ldots|z_p + p_p''|} = 1 \tag{4.7}$$

For a second-order system model, the roots are:

$$\begin{cases} z_1 = e^{-a_1'T} \\ z_2 = e^{-a_2'T} \end{cases} \tag{4.8}$$

If a root is real and less than zero, there is no equivalent root in the *discrete* domain. If the roots are complex-conjugate pairs, these equations become:

$$\begin{cases} z_1 = e^{-\omega_n(\xi + j\sqrt{1-\xi^2})T} \\ z_2 = e^{-\omega_n(\xi - j\sqrt{1-\xi^2})T} \end{cases} \tag{4.9}$$

Then, the location of these roots in the z-plane for varying values of ζ and ω_n could be sketched in a z-plane (real and imaginary axes). The procedure for sketching the various root locations is presented in Table 4.1.

Example 4.1 *Applying the Magnitude and Angle Condition for Root Locus Sketching*
Consider the open-loop transfer function given by:

$$H(z) = \frac{z}{z^2 + \frac{z}{2} + \frac{5}{16}} = \frac{z}{(z - p_1)(z - p_2)}$$

with $p_1 = \frac{1 + 2j}{4}$ and $p_1 = \frac{1 - 2j}{4}$

Figure 4.1 illustrates the resulting pole zero, departure angle, magnitude, and phase for a test point, as well as the resulting roots locus sketching.

Table 4.1 Generic rules for manually sketching the root locus in the z-plane.

Rules and definitions

Rule 1: Locate the open-loop zeros and poles of the system model given by $G_p(z)$ in the graphic representation over the z-plane, such that **X** would denote a pole and **O** would denote a zero.

Rule 2: The number of branches is equal to the order of characteristic equation (i.e. $1 + KG_p(z) = 0$) or the number of poles. All n branches of the locus start at the poles of open-loop $G_p(z)$. This corresponds to when the gain K is equal at zero. It ends at either the zeros of open-loop $G_p(z)$ or at infinity when the system parameter gain reaches infinity. Furthermore, the root loci on the real axis are on the left of an odd number of (real axis) poles and zeros. Also, in the case of complex conjugates poles, the root loci are symmetrical with respect to the real axis.

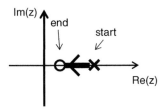

Rule 3: Asymptotes of the loci are dependent on the poles and zeros of $G_p(z)$, and are given when roots go to infinity. If asymptotes exist, they are characterized by their intersection point with the real axis, also called the centroid, which is given by:

$$\sigma = \frac{\sum p_i - \sum z_i}{no.of\ poles - no.of\ zeros}$$

When the root loci move to infinity toward zero, the root loci are asymptotic lines radiating out from the centroid centered at σ with angles ϕ_i given by:

$$\phi_i = \frac{180(1 + 2m)}{no.of\ poles - no.of\ zeros}$$

with $i = 1, 2, \ldots, no\ of\ poles, no\ of\ zeros$, and m being any integer. If there are fewer zeros than poles, then the loci end at infinity as K increases to infinity.

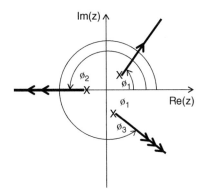

Rule 4: The departure (arrival) angles of each branch of the locus from any complex pole (toward any complex zero) is computed by the application of the angle condition that is given by:

$$\sum_i \psi_i - \sum_i \phi_i = (1 + 2m)180° \text{ with } m = 0, \pm1, \pm2, \ldots$$

Recall that the angle of departure of the root locus from the pole $\phi_{l,\,dep}$ is the difference between the net angle from all other poles and zeros and the angle condition of $180(1 + 2m)°$. In the case of angle of departure, the test point is the complex pole, while in the case of angle of arrival, the test is the complex zero. Consider the angle to a test point from zero index i being ψ_i, and the angle to same test point from any pole index i being ϕ_i:

$$\phi_{l,dep} = \sum_i \psi_i - \sum_i \phi_i \pm 180(1 + 2m)$$

(continued)

Table 4.1 (Continued)

Rules and definitions

Similarly, the angle of arrival toward a complex zero is given by:

$$\psi_{l,\text{arl}} = \sum_i \phi_i - \sum_i \psi_i \pm 180(1 + 2m)$$

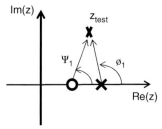

Rule 5: The root locus crosses the real axis through the breakaway or break-in points. If the characteristics equation is given by $1 + KG_p(z) = 0$, then the break-in and breakaway points could be derived from

$$K = -\frac{1}{G_p(z)}$$

This equation can be used to search for the maximum value (breakaway point) and the minimum value (arrival or break-in point) by:

$$\frac{dK}{dt} = \frac{d}{dz}\left(-\frac{1}{G_p(z)}\right) = 0$$

Rule 6: The locus crosses the imaginary jw-axis at points found by applying Routh's criterion for a value $z_c = ja$. This is done by replacing z_c in the characteristics equation and by setting the resulting real and imaginary component of the equation to be equal to zero. Then derive the imaginary axis value α and the value of the corresponding gain K.

Rule 7: The root locus plot correctness is such that any closed-loop pole corresponds to a point on the root locus.

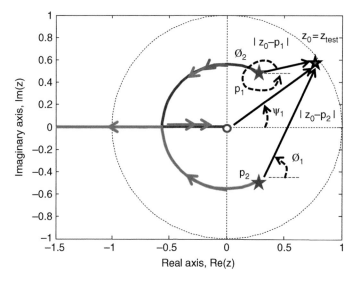

Figure 4.1 Root locus sketch for desired pole phase and magnitude estimation with $T = 0.1$ s.

Example 4.2 Consider a system model given by:

$$G(s) = \frac{1}{s^2}$$

such that the corresponding discrete-time transfer function is:

$$KG(z) = \frac{T^2}{2} \frac{K(z+1)}{(z-1)^2}$$

If $T = \sqrt{2}$, the characteristics equation is given by:

$$1 + KG(z) = 1 + \frac{K(z+1)}{(z-1)^2} = 0$$

Consider $z_c = \sigma$; solving for K yields:

$$K = -\frac{(\sigma-1)^2}{(\sigma+1)} = F(\sigma)$$

From its derivatives,

$$\frac{dF(\sigma)}{d\sigma} = 0, \text{so } \sigma = 1 \text{ and } \sigma = -3$$

Using the magnitude condition and phase condition, the root locus is illustrated in Figure 4.2. Consider an open-loop step response of the discrete transfer function given by:

$$H(z) = \frac{1}{(1 - e^{-a_1 T}z^{-1})(1 - e^{-a_2 T}z^{-1})\dots(1 - e^{-a_n T}z^{-1})}$$

Thus, the closed-loop step response of the discrete transfer function $H(z)$ is given by:

$$H_{\text{closed loop}}(z) = \frac{H(z)}{1 + H(z)} = \frac{\frac{1}{(1-e^{-a_1 T}z^{-1})(1-e^{-a_2 T}z^{-1})\dots(1-e^{-a_n T}z^{-1})}}{1 + \frac{1}{(1-e^{-a_1 T}z^{-1})(1-e^{-a_2 T}z^{-1})\dots(1-e^{-a_n T}z^{-1})}}$$

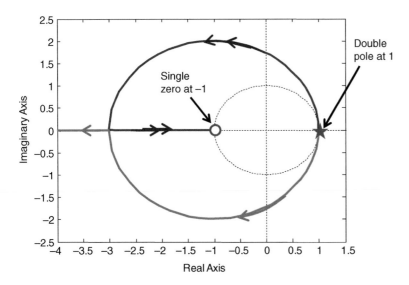

Figure 4.2 Root locus sketch for double poles.

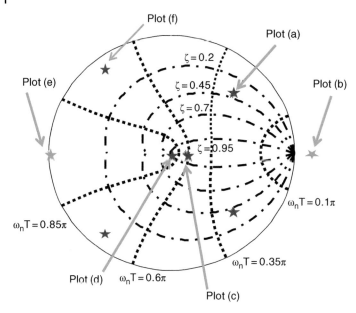

Figure 4.3 Root locations and associated impulse response in the z-plane for various ξ and ω_n.

such as the poles $-a_1, -a_2, -a_3, \ldots, -a_n$, which are illustrated in the z-plane plots next described. The basic relationship between the characteristics equation roots and the transient response of discrete-time system are shown in Figure 4.3 where:

1) In the case where the *roots are on the positive, real axis in the z-plane* and inside the unit circle of the z-plane, the impulse response would result in a rise of responses that decay exponentially with an increase of kT, as illustrated in Figure 4.4(c). It should be noted that the closer the roots are to the unit circle, the slower the decay response. When the root is at $z = 1$, the response has a constant amplitude. Roots outside would display unstable system responses and the responses increase with kT, as shown in Figure 4.4(b).

2) In the case of *complex conjugate with the roots* being inside the unit circle of the z-plane, this will lead to oscillatory responses that decay with an increase in kT, as shown in Figure 4.4(a) and (f). When the roots are closer to the unit circle, the response will decay more slowly. As the roots move toward the second and the third quadrants, the oscillating frequency of the response increases. Usually, the zeros of the discrete-time transfer function represent the derivatives of the input, thus changing the transient characteristics, such as speeding up the response and increasing overshoot. It should be noted that a zero at $z = 0$ corresponds to a unit advance, just as a pole at $z = 0$ is a unit delay.

3) In the case where the *roots are on the negative, real axis in the z-plane,* this corresponds to the boundaries of the periodic strips in the s-plane, as illustrated in Figure 4.4(d) and (e). For example, consider:

$$s = -\xi\omega_n \pm j\omega_n T\sqrt{1-\xi^2} = -\sigma_1 \pm j\omega_s$$

The complex-conjugate points are on the boundaries of the primary strip in the s-plane. The corresponding z-plane points are:

$$z = e^{-\sigma_1 T}e^{\pm j\omega_s} = e^{-\sigma_1 T}\angle j \pm \omega_s$$

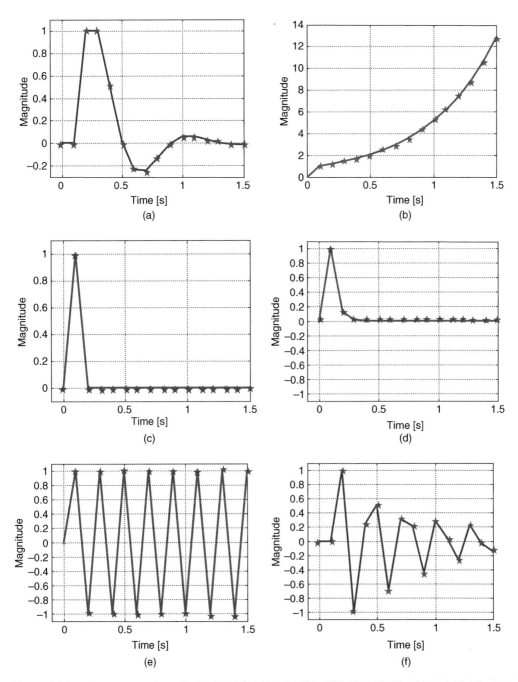

Figure 4.4 Impulse responses for various values of various ξ and ω_n. Plot (a) stable. Plot (b) unstable. Plot (c) stable. Plot (d) stable. Plot (e) limited stability. Plot (f) stable.

which are on the negative real axis of the z-plane. Here, ξ represents the damping ratio and ω_n represents natural frequency. Thus, the angle of the pole is given by:

$$\omega_s = \omega_n T \sqrt{1 - \zeta^2} = \frac{\sqrt{1-\zeta^2}}{\zeta} |\ln(|z|)|$$

which are logarithmic spirals for ξ at constant value starting at $\omega_s = 0$ and $|z| = 1$, as shown in Figure 4.3. From response plots in Figure 4.4, it can be seen that rate of decay of the exponential response is determined by the radius r of the pole location. If $r > 1$, the sequence of the output increases. The larger r causes a faster impulse response. If $r = 1$, the sequence of the output does not vary. If $r < 1$, the sequence decays such that the closer r is to zero, the faster the decay response. A typical roots locus on the z-plane with its unit circle is shown in Figure 4.3.

```
clear all;clf;
numz=[1];denz=conv([1 0.5000-0.71i],[1 0.5000+0.71i]);
sys_D=tf(numz,denz,0.1); % closed-loop transfer function
Ts=0.1;T=0:Ts:1.5;[y,t]=impulse(sys_D,T);figure(1);plot(t,y,'*r');
hold on;plot(t,y,'blue');grid;xlabel('Time [sec]');ylabel('Magnitude');
axis([-0.1 1.5 -1.1 1.1]);
numz1=[1];denz1=[1 -1/8];sys_D1=tf(numz1,denz1,0.1);
[y1,t1]=impulse(sys_D1,T);
figure(2);plot(t1,y1,'*b');hold on;plot(t1,y1,'red');grid;
xlabel('Time [sec]');ylabel('Magnitude');axis([-0.1 1.5 -1.1 1.1]);

numz2=[1];denz2=[1 1];sys_D2=tf(numz2,denz2,0.1);[y2,t2]=impulse(sys_D2,T);
figure(3);plot(t2,y2,'*b');hold on;plot(t2,y2,'red');
xlabel('Time [sec]');ylabel('Magnitude');grid;axis([-0.1 1.5 -1.1 1.1]);

numz4=[1];denz4=[1 -1.2];sys_D4=tf(numz4,denz4,0.1);
[y4,t4]=impulse(sys_D4,T);figure(4);grid;plot(t4,y4,'*b');
hold on;plot(t4,y4,'red');xlabel('Time [sec]');ylabel('Magnitude');grid;

numz5=[1];denz5=[1 0];sys_D5=tf(numz5,denz5,0.1);[y5,t5]=impulse(sys_D5,T);
figure(5);plot(t5,y5,'*b');hold on;plot(t5,y5,'red');xlabel('Time [sec]');
ylabel('Magnitude');grid;axis([-0.1 1.5 -0.1 1.1]);
numz6=[1];denz6=conv([1 -0.5000-0.4940i],[1 -0.5000+0.4940i]);
sys_D6=tf(numz6,denz6,0.1);[y6,t6]=impulse(sys_D6,T);
figure(6);grid;plot(t6,y6,'*r');hold on;plot(t6,y6,'blue');
xlabel('Time [sec]');ylabel('Magnitude');grid;axis([-0.1 1.5 -0.3 1.1]);
hold off;k = 0;
%to draw root location of closed-loop transfer function (k=0)
r = rlocus(sys_D,k);r1 = rlocus(sys_D1,k);r2 = rlocus(sys_D2,k);
r4 = rlocus(sys_D4,k);r5 = rlocus(sys_D5,k);r6 = rlocus(sys_D6,k);
figure(8);plot(real(r),imag(r),'*r');hold on;plot(real(r1),imag(r1),'*m');
hold on;plot(real(r2),imag(r2),'*g');hold on;plot(real(r4),imag(r4),'*c');
hold on;plot(real(r5),imag(r5),'*b');hold on;plot(real(r6),imag(r6),'*r');
hold on;zgrid(0.2:0.25:1,0.1*pi:0.25*pi:pi);axis([-1.3 1.3 -1.1 1.1])
```

4.2.2 Discrete Frequency Analysis Tools: Fourier Series and Transform (DFT, DTFT, and FFT)

The shape of any time-domain waveform is not relevant, but rather the information on the frequency, the phase, and the amplitude of its component sinusoids. The DFT is used to extract this information. As such, the frequency response of a system can be described as a sum of weighted complex exponentials. For a linear time-invariant system, it is the system steady-state response to varying frequency parameters of sinusoidal input signal $x(t) = X\sin(\omega t)$. Hence the resulting system response $y(t) = Y\sin(\omega t + \phi)$ is described by $Y = X|G(j\omega)|$ and $\phi = \angle G(j\omega)$. Consequently, they are given as $|G(j\omega)| = \frac{|Y(j\omega)|}{|X(j\omega)|}$ and $\angle G(j\omega)\angle\frac{Y(j\omega)}{X(j\omega)}$. Usually, the logarithmic or Bode, polar, or Nyquist plots can be used to illustrate system response magnitude and phase change over frequency variation. These weights are the impulse response samples. This DFT series is a special case of a family of representations of signals that are collectively called *Fourier transforms*. Among these transforms, there is the *DTFT*, used even when the discrete-time signals are not periodic. Consider a discrete-time signal, $x(n)$, given by:

$$X(\omega) = \sum_{n=-\infty}^{\infty} x(n)e^{-j\omega n} \tag{4.10}$$

where the frequency variable ω is in radians and the $X(\omega)$ is the periodic signal of period 2π and defined as:

$$X(\omega) = |X(\omega)|e^{j\angle X(\omega)} \tag{4.11}$$

The forward Fourier transform of such a discrete-time signal, $x(n)$, is given by:

$$X_{DTFT}(\omega) = X(z)|_{z=e^{j\omega}} = \sum_{n=-\infty}^{\infty} x(n)e^{-j\omega n} \tag{4.12}$$

The inverse transform is given by:

$$x(n) = \frac{1}{2\pi} \int X(\omega)e^{j\omega n} d\omega \tag{4.13}$$

The DTFT is always a periodic function of ω. Among DTFT properties, it includes the following:

Periodic property:

$$x(\omega) = x(\omega + 2\pi k) \tag{4.14}$$

Symmetry property:

$$x(n) \in R, X(-\omega) = X(\omega) \tag{4.15}$$

Time-shifting property:

$$F(x(n - n_k)) = |X(\omega)|e^{-j\omega_n k} \tag{4.16}$$

Frequency shifting property:

$$F(e^{j\omega_k n}x(n)) = X(\omega - \omega_k) \tag{4.17}$$

Linearity property:

$$F(a_1 x(n) + a_2 x(n)) = a_1 X_1(\omega) + a_2 X_2(\omega) \tag{4.18}$$

Parseval's relation property:

$$\sum_{n=-\infty}^{\infty} |x(n)|^2 = \frac{1}{2\pi} \int_{2\pi}^{0} |X(\omega)|^2 d\omega \tag{4.19}$$

DTFT can be used to determine the frequency response of systems characterized by a linear constant coefficient of the difference equation using its time-shifting, linearity, convolution, and modulation properties. In addition, with random signal analysis, by using DFT it is possible to compute a signal's frequency spectrum by deriving the signal Power Spectrum Density (PSD) equivalent to the Fourier transform of the auto-covariance sequence. This is used to extract the information encoded in the frequency, phase, and amplitude of the component sinusoids. Hence, the system can be analyzed in the frequency domain by the representation of each system output signal from a group of cosine waves. DFT is intensively used in digital signal processing because there are some algorithms for its computation. These algorithms are usually referred to as FFT.

4.2.2.1 Discrete System Frequency Response

Consider an input signal to a discrete system characterized by:

$$x(n) = \cos(\omega_n) \tag{4.20}$$

The equivalent series expansion (here input signal ω_n at the nth sample) yields:

$$x(n) = \frac{1}{2}(e^{j\omega_n} + e^{-j\omega_n}) \tag{4.21}$$

With the system $H(s)$, the corresponding output signal is given by:

$$y(n) = H(e^{j\omega_n})x(n) = \frac{1}{2}(H(e^{j\omega_n})e^{j\omega_n} + H(e^{-j\omega_n})e^{-j\omega_n}) \tag{4.22}$$

such that:

$$y(n) = \text{Re}\{H(e^{j\omega_n})x(n)\} = \text{Re}\{|H(e^{j\omega_n})|e^{j\angle H(e^{j\omega_n})}e^{j\omega_n}\} \tag{4.23}$$

Therefore, it can be written as:

$$y(n) = |H(e^{j\omega_n})| \cos(\omega_n + \angle H(e^{j\omega_n})) \tag{4.24}$$

defining:

$$z = re^{j\omega T} \tag{4.25}$$

Using a unit circle $r = 1$ and an angle given by $e^{j\omega}$, the magnitude and phase of the response of the system to an external cosine signal are defined as the magnitude and phase of the system through the function evaluated with respect to this unit circle. This could be rewritten as:

$$H(e^{j\omega_n}) = H(z)|_{z=e^{j\omega T}} \tag{4.26}$$

It can be noted that, with the frequency response, only the steady-state response characteristics are covered, which are given by the angle of poles. For a generic discrete system given by:

$$H(z_0) = K\frac{(z_0 - q_0)(z_0 - q_1)(z_0 - q_2)\ldots}{(z_0 - p_0)(z_0 - p_1)(z_0 - p_2)\ldots} \tag{4.27}$$

The value of $H(z)$ at $z = z_0$ can be determined by combining the contribution of vectors associated with each of the poles and zeros, such that the magnitude is given by the product of their magnitude, so that:

$$|H(z_0)| = K\frac{|(z_0 - q_0)||(z_0 - q_1)||(z_0 - q_2)| \ldots}{|(z_0 - p_0)||(z_0 - p_1)||(z_0 - p_2)| \ldots} \tag{4.28}$$

and the angle is determined by the sum of the angles:

$$\angle H(z_0) = (\angle K + \angle(z_0 - q_0) + \angle(z_0 - q_1) + \ldots) - (\angle(z_0 - p_0) + \angle(z_0 - p_1)\cdots) \tag{4.29}$$

Table 4.2 Frequency response data.

ωT(rad)	ω_ω (rad s^{-1})	$|G_p(e^{j\omega T})|$	$20\log_{10}|G_p(e^{j\omega T})|$ (dB)	$\angle G_p(e^{j\omega T})$ (rad)	$\angle G_p(e^{j\omega T})$ (degree)
0.003					
0.031					
0.314					
3.14					

The Bode diagram for discrete systems consists of both the magnitude and phase responses of a discrete system, and is defined as follows:

$$\begin{cases} Magnitude = |H(z)|_{z=e^{j\omega T}} \\ Phase = \angle H(z)_{z=e^{j\omega T}} for -\pi \leq \omega T \leq \pi \end{cases} \qquad (4.30)$$

4.2.2.2 Sketching Procedure for the Frequency Response of a Discrete System

Given a linear time-invariant system represented by a discrete-time transfer function $G_p(z)$:

1) With $z = e^{j\omega T}$ derive the frequency response $G_p(e^{j\omega T})$ for frequency interval $\left[0\frac{\pi}{T}\right]$, as well as for the bilinear equivalent frequency $\omega_\omega = \frac{2}{T}\tan\frac{\omega T}{2}$.
2) Fill out Table 4.2 and sketch the frequency response of magnitude $|G_p(e^{j\omega T})|$ in decibels and the phase $\angle G_p(e^{j\omega T})$ in degrees over the prewarping frequency ω_ω in rad s^{-1} in semilog scale.
3) Check the frequency response for the discrete-time transfer function by:
   a) verifying that the slope of the low-frequency asymptote determines the system type; and
   b) verifying that the crossover frequency correlates with the bandwidth of the system, and also with the speed of the response.

4.2.2.3 Properties of a Frequency Response

Some key frequency response characteristics are:

1) The open-loop *gain crossover frequency* is the frequency at which the magnitude plot reaches -3 dB.
2) The *PM* is measured at the gain crossover frequency when the phase of the open-loop frequency response becomes $-\pi$ rad ($-180°$). It is the magnitude of the frequency response at the gain crossover frequency.
3) The open-loop *phase crossover frequency* is the frequency at which the phase plot reaches $\pm 180°$.
4) *The gain margin* is measured at the phase crossover frequency where the magnitude of the open-loop frequency response becomes 1 (0 dB). It is the difference between the phase at that frequency and $-\pi$ rad ($-180°$). It is obtained by adding π rad ($180°$) to the phase at that crossover frequency.
5) The GMs and PMs must be positive for the closed-loop system to be stable. The greater their values, the greater the relative stability of the system. Usually, gain and PMs tend to increase and systems tend to become less oscillatory.
6) The PM can be used to predict the damping ratio (ξ) in the closed-loop system, such that PM = $100°\xi$;

7) The PMs, GMs, and predicted bandwidth are useful measures when designing controller gains, and are covered later in this chapter along with the correlation between the PM, the percentage overshoot (PO), and the oscillation of the system.

Example 4.3   *Bode Plot Diagram for Discrete-Time Systems*
Consider an open-loop system given by:

$$H(z) = \frac{0.5416z - 0.4120}{z^2 - 1.931z + 0.943}$$

Using the previous method, Table 4.3 is completed. The Bode diagrams are derived using the method presented and the MATLAB command. The results are displayed in Figures 4.5 and 4.6.

This yields a PM of 55° and GM of 12 dB. Figure 4.6 displays the response derived using Table 4.3 and MATLAB simulation tools. It should be recalled that both the magnitude and phase responses of a discrete system repeat after $2\pi/T$.

Table 4.3 Derived frequency response data.

ωT(rad)	ω_ω (rad s^{-1})	$|G_p(e^{j\omega T})|$	$20\log_{10}|G_p(e^{j\omega T})|$ (dB)	$\angle G_p(e^{j\omega T})$ (rad)	$\angle G_p(e^{j\omega T})$ (degree)
0	0	6.73e+15	316.5530	0	0
0.003	0	712.3541	57.0539	−1.6133	−92.438
0.031	0	63.5567	36.0632	−1.9557	−112.056
0.314	0.0003	2.0357	6.1741	−2.2519	−129.026
3.14	2.3553	0.2455	−12.1992	−3.1408	−179.956

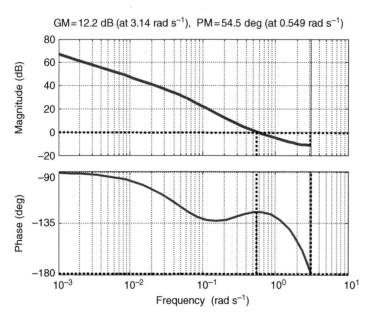

Figure 4.5 Magnitude and phase frequency responses using MATLAB.

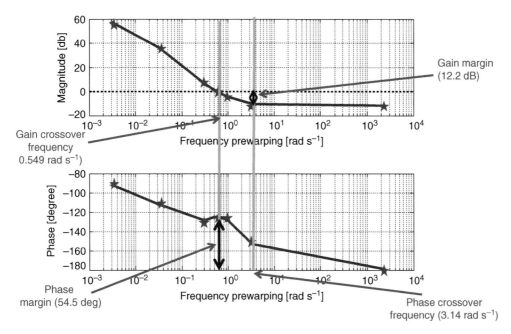

Figure 4.6 Magnitude and phase frequency responses with PM and GM values.

```
clear all;
clear figure;clf;T=1;
Num_sysD=[0.5417 -0.4118];Den_sysD=[1 -1.942 +0.942];
wT=pi*[0 0.001 0.01 0.1 1]
sysD1=tf(Num_sysD,Den_sysD,T);
figure(1);bode(sysD1);grid;margin(sysD1);grid;
i=1;
for angl=[0 0.179908 1.79908 17.9908 28 31.45 50 120 179.908] %in degrees
wt(i)=(3.1415/180)*angl; % in rad
wwa(i)=(2/1)*tan(wt(i)/2); % in rad/sec (pre-warping scaling)
[mag,phase]=bode(sysD1,wt(i));
phas(i)=phase;
db1(i)=20*log10(mag); % magnitude in db
i=i+1;
end;
figure(2);subplot(2,1,1), semilogx(wwa,db1,'*');
hold on; semilogx(wwa,db1);hold off;
grid; ylabel ('magnitude [db]'); xlabel ('frequency pre-warping [rad/sec]');
subplot(2,1,2), semilogx(wwa,phas,'*');hold on; semilogx(wwa,phas);hold off;
grid; ylabel
('phase [degree]'); xlabel ('frequency pre-warping [rad/sec]');
```

4.3 Discrete-Time Controller Specifications

The controller performance requirements result from process operational audit outlining: the time domain (transient and stationary) characteristics of the dynamic response to various

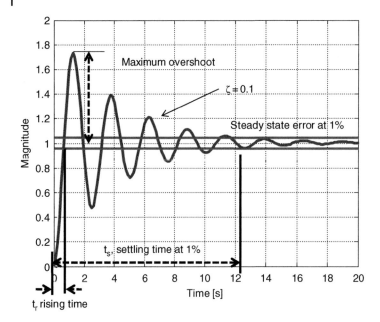

Figure 4.7 Typical percentage second-order oscillatory response.

inputs such as the unit step and the ramp functions. Among those characteristics are the rise time, overshoot, settling time, percentage overshoot, disturbance rejection, steady-state accuracy, and stability. From such auditing in the frequency domain, the process response analysis characteristics lead to PM, GM, and stability system bandwidth. In the following sections. such characterization of process performance is presented, both in the time domain and the frequency response.

4.3.1 Time Domain Specifications

Commonly used transient-based system dynamics have specifications with respect to: (i) the PO; (ii) the rise time; and (iii) the settling time. By considering a continuous second-order underdamped system with no zeros, as illustrated in Figure 4.7, the following criteria can be defined:

1) The *percentage overshoot* (PO), which is dependent only on the damping ratio, ξ, such that:

$$PO = 100 \times e - \xi\pi\sqrt{1 - \xi^2} \approx 100 \times \left(1 - \frac{\xi}{0.6}\right) \tag{4.31}$$

2) The rise *time t_r*, which is the time from 10 to 90%. It indicates the speed of the response. It is strongly dependent on natural frequency, ω_n, such that:

$$t_r \approx \frac{1.8}{\omega_n} \text{ best accuracy when } \xi \approx 0.5 \tag{4.32}$$

3) The settling *time, t_s*, which is the time elapsed until the process response and lies within a desired tolerance of the steady-state value, expressed as a percentage criterion given by:

$$y(t) = 1 - e^{-\xi\omega_n t}\cos(\omega_d t + \phi) \tag{4.33}$$

where damped oscillation frequency is $\omega_d = \omega_n\sqrt{1 - \xi^2}$. The system error transient is contained within the envelope $e^{-\xi\omega_n t}$. For a large value of $\xi\omega_n$, the system error decreases as fast as necessary. For example, for the error to be less than 1% after t_S seconds, then

$$e^{-\xi\omega_n t_s} \leq 0.01 \tag{4.34}$$

and taking the natural log of both sides, which is equivalent to

$$\xi\omega_n \geq \frac{4.6}{t_s} \tag{4.35}$$

multiple time constants can be achieved depending on the specified percentage, such that the settling time within a 2% specification is given by:

$$t_s = \frac{4}{\xi\omega_n} = 4\tau \tag{4.36}$$

while for the 5% criterion it is equivalent to:

$$t_s = \frac{3}{\xi\omega_n} = 3\tau \tag{4.37}$$

4) The system response-time constant, τ, is the time constant of exponential decay of the continuous response given by:

$$\tau = \frac{1}{\xi\omega_n} \tag{4.38}$$

Hence, the assessment criteria for efficient command tracking of the steady-state dynamics response include: (i) the steady-state error constant; (ii) the disturbance rejection; (iii) the sensitivity; and (iv) even the stability.

1) The *steady-state error constant* measures the steady-state accuracy. The relationship between error constants K_v^*, K_p^*, K_a^* and the steady-state errors, e_{ss}, for various command inputs (step, ramp with magnitude A, and parabolic) and, depending on the system type (order), the expected results, are listed in Table 4.4.
2) The *disturbance rejection*, or the capacity to reduce the disturbance effect on the system output signal. By considering the reference input $R(z) = 0$, $W(z)$ being a disturbance input, the error $E(z)$ is given by:

$$E(z) = \frac{G(z)W(z)}{1 + G(z)D(z)} \approx \frac{W(z)}{D(z)} \tag{4.39}$$

Assuming that loop gain $|G(z)D(z)| >> 1$, a high gain in $G(z)$ would not greatly influence the disturbance rejection rather defined by $D(z)$. Also, if $D(z)$ has an integration (pole at $z = 1$), it would result in infinite gain at zero frequency; therefore, zero frequency disturbances (not oscillatory) would be rejected completely, within the limits of the actuating devices.
3) The system's capacity to reduce its *sensitivity* to changes in parameter values or how those changes affect the system output signal. This could be analyzed using the eigenvalue migration plot.

Table 4.4 Error constants and steady-state errors for various system types and command inputs.

	Step input		Ramp input		Parabolic input	
System type	Kp	ess	Kv	ess	Ka	ess
$G(z)D(z)$ no pole at $z=1$	K	$A/(1+K)$	0	∞	0	∞
$G(z)D(z)$ one pole at $z=1$	∞	0	K	A/K	0	∞
$G(z)D(z)$ two poles at $z=1$	∞	0	∞	0	K	A/K

4) The *stability* requirement is given by the PO.
5) The *frequency of oscillations*, ω_d, is the angle of the dominant complex conjugated poles that would be divided by the sampling period. It is given by:

$$\omega_d = \omega_n \sqrt{1 - \xi^2} \tag{4.40}$$

6) The *energy required and gain distribution* from each component, such as commands, amplifiers, actuators, power amplifier, signal conditioning and so on, are not saturated. These are calculated by evaluating the response of the appropriate variables from inputs such as command set point or disturbance.

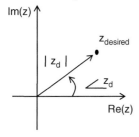

As noted from all the design specifications in the time domain listed here, the key dynamics parameters to be derived are (ω_n, ξ). This allows derivation of the desired dynamics characterization based on the desired roots placement. For example, as illustrated in Figure 4.8, for a second-order process model, derivation consists of selecting a dominant closed-loop in the z-plane corresponding to the desired system dynamics, which would be given by:

$$z_{\text{desired}} = e^{-\omega_n(\xi + j\sqrt{1-\xi^2})T} = e^{-\omega_n\xi T} \angle \pm (\omega_n T\sqrt{1 - \xi^2}) \tag{4.41}$$

Figure 4.8 Complex roots locus phase and magnitude value definition.

4.3.2 Frequency Response Specifications

Commonly encountered dynamics specifications of the amplitude frequency response are: (i) resonant peak; (ii) resonant frequency; (iii) system bandwidth; and (iv) PMs and GMs. By considering a second-order process, the following definitions can be derived:

1) The amplitude ratio and phase are given such that:
   a) the amplitude ratio of the steady-state output $c(t)$ versus the sinusoid input is:

   $$|G(j\omega)| = \frac{|C(j\omega)|}{|R(j\omega)|} \tag{4.42}$$

   b) the phase difference between the steady-state output and the sinusoid input $r(t)$ is:

   $$\varphi(j\omega) = G(j\omega) = \angle C(j\omega) - \angle R(j\omega) \tag{4.43}$$

2) The closed-loop *resonant peak* M_r at the maximum value of $|G(j\omega)|$ is obtained through $\frac{d|G(j\omega)|}{d\omega} = 0$, such that:

   $$M_r = |G(j\omega)|_{\text{max}} = |G(j\omega_r)| = \frac{1}{2\xi\sqrt{1 - 2\xi^2}} \tag{4.44}$$

   for $0 \leq \zeta \leq 0.707$. There is a special case for $\zeta > 0.707$, $\omega_r = 0$, and $M_r = 1$. M_r indicates the relative stability of a stable closed-loop system. A large M_r corresponds to a larger maximum overshoot of the step response and its suitable value range is between: 1.1 and 1.5.

3) The closed-loop *resonant frequency* ω_r is the frequency at which the peak resonance M_r occurs, such that:

   $$\omega_r = \omega_n \sqrt{1 - 2\xi^2} \tag{4.45}$$

   for $0 \leq \zeta \leq \sqrt{2}/2$. From the typical frequency system response analysis, the following points are made:

a) If ζ decreases, then ω_r increases (ω_n is maintained), then M_r increases.

b) If $\zeta = 0$, then ω_r is equal to ω_n, then the system is unstable.

c) If $\zeta = \frac{\sqrt{2}}{2}$, then $\omega_r = 0$, then there is no resonance.

4) The *closed-loop bandwidth* ω_b is the frequency at which $M(j\omega)$ drops to 70.7% (3 dB) of its zero-frequency value; or another definition of ω_b, based on the transfer function $G(s)$, such that:

$$\left| \frac{G(j\omega_b)}{G(0)} \right| = \frac{1}{\sqrt{2}} \tag{4.46}$$

which is equivalent to:

$$\omega_b = \omega_n \sqrt{(1 - 2\xi^2) + \sqrt{2 - 4\xi^2 + 4\xi^4}} = \omega_n \sqrt{\sqrt{(1 + 4\xi^4)} - 2\xi^2} \tag{4.47}$$

The bandwidth ω_b gives an indication of the transient response properties of a control system. For example, it is inversely proportional to the rise time t_r. ω_b also indicates the noise-filtering characteristics and robustness of the system. ω_n is directly proportional to ω_b. Increasing ζ decreases ω_b as well as M_r. In addition, M_r and ω_n are proportional to each other for $0 \le \zeta \le 0.707$.

5) The *Open-loop gain crossover frequency* ω_g is given by:

$$\omega_g = \omega_n = \frac{4.6}{\xi t_s} \tag{4.48}$$

6) The *relative stability* analysis in the frequency domain is defined by gain and PM values such as $\gamma(PM)$, which is the amount of additional lag at the gain crossover frequency ω_g required to bring the system to the verge of instability. At gain crossover frequency, the magnitude of open-loop gain is unity, that is $|G(\omega_g)| = 1$. The $\gamma(PM)$ is defined as the phase of the open-loop transfer function $G(s)$ at the gain crossover frequency ω_g, so that:

$$PM = \gamma = 180° + \angle G(\omega_g) = tg^{-1}\left(2\xi \sqrt{\frac{1}{\sqrt{(1 + 4\xi^4)} - 2\xi^2}} \right) \tag{4.49}$$

where the open-loop frequency ω_g is defined as the lowest frequency at $|G(\omega_g)| = 1$. For unity-feedback systems, $\omega_g \approx \omega_b$, and knowing that yields:

$$G_{\text{closed loop}}(j\omega) = \frac{G(j\omega)}{1 + G(j\omega)} \tag{4.50}$$

Gain Margin, $K_g(GM)$, is the amount of additional gain at phase crossover frequency ω_p that can move the system to the verge of instability. At the phase crossover frequency, the phase angle of the open-loop transfer function, $G(s)$, equals $-180°$; that is, $\angle G(s) = -180°$. Thus, the GM $\omega_g K_g$ is defined as the inverse of the open-loop gain at the phase crossover frequency, ω_p, such that:

$$GM = K_g = \frac{1}{|G(\omega_p)|} = -20 \log|G(j\omega_p)| \tag{4.51}$$

where the frequency ω_p is defined as the lowest frequency with:

$$\angle G(j\omega)|_{\omega_p} = \arg(G(j\omega_p)) = -\pi = -180° \tag{4.52}$$

It is noted that if K_g and γ increase, then M_r decreases. Similarly, if ω_g increases, then ω_r and ω_b decrease. The phase crossover frequency ω_p is also called the ultimate frequency of the control system. Hence, both PM and GM must be positive for a minimum-phase system to be stable. Negative margins indicate instability. For satisfactory performance, the PM should be between 30° and 60° and the GM should be greater than 6 dB. For a first- and second-order system, the GM is always infinity.

7) *Relationships between the frequency and the time domains*
   a) The closed-loop bandwidth, ω_b, and the rise time, t_r, are related such that when ω_b increases, then t_r decreases due to the time scale. For large ω_b, there are more high-frequency components in the process response $c(t)$, making the time response faster.
   b) Resonance peak, M_r and overshoot P.O.(%) are related such that, when M_r increases, then P.O.(%) increases because of the large imbalance of the frequency signals passing to $c(t)$. If K_g and γ decrease, then P.O.(%) increases similarly because of the decrease in K_g and γ, and M_r is given by:

$$P.O.(\%) = 0.16 + 0.4(M_r - 1) \tag{4.53}$$

such that $1.1 \leq M_r \leq 1.8$, and:

$$M_r \approx \frac{1}{\sin \gamma_c} \tag{4.54}$$

The settling time is given by:

$$t_s \approx \frac{k\pi}{\omega_c} \tag{4.55}$$

with $k \approx 2 + 1.5 \left(\dfrac{1}{\sin \gamma_c} - 1 \right) + 2.5 \left(\dfrac{1}{\sin \gamma_c} - 1 \right)$. Thus,

$$35° \leq \gamma \leq 90° \tag{4.56}$$

In addition, the following rules define a typical frequency response plot:

1) The low-frequency region is related to the control accuracy of the system. The higher the negative value of the slope of $20 \log |G(j\omega)|$, the higher the control accuracy of the system. The higher the magnitude of $G(j\omega)$, the smaller the steady-state error e_{ss}.
2) The middle frequency region is related to the transient performance of the system. An increase in ω_b causes a decrease of t_r, and a decrease of K_g and γ causes an increase of P.O.(%). The slope of $20 \log |G(j\omega)|$ should be $-20 \, \text{dB s}^{-1}$.
3) The high-frequency region is mainly related to the ability of the system to reject high-frequency noise. When the magnitude of $20 \log |G(j\omega)|$ is low, the ability of the system to reject high-frequency noise improves.

4.4 Discrete-Time Steady-State Error Analysis

For single input and single output (SISO) system, the error signal is defined by:

$$e(t) = r(t) - y(t) \tag{4.57}$$

where $r(t)$ is the input and $y(t)$ is the output. The error analysis conducted here is only for an SISO system with unity feedback. For digital system analysis, z-transform or the difference equations are often used to represent the input and output in sampled form, $r(kT)$ and $y(kT)$ respectively. Subsequently, the digital error signal can be represented by $e(kT)$. That is,

$$e(kT) = r(kT) - y(kT) \tag{4.58}$$

By using the final-value theorem of the z-transform, the steady-state error at the sampling instants is defined as:

$$e_{ss} = \lim_{k \to \infty} e^*(t) = \lim_{k \to \infty} e^*(kT) = \lim_{z \to 1}(1 - z^{-1})E(z) \tag{4.59}$$

Using $E(z)$ in terms of $R(z)$, and $G_H G_P(z)$, Equation (4.59) is written as:

$$e_{ss} = \lim_{k \to \infty} e(kT) = \lim_{z \to 1}(1 - z^{-1})\frac{R(z)}{1 + G_H G_P(z)} \tag{4.60}$$

Let a transfer function of the process be in the form:

$$G_P(s) = K\frac{(1 + T_a s)(1 + T_b s) \ldots (1 + T_m s)}{s^j(1 + T_1 s)(1 + T_2 s) \ldots (1 + T_n s)} \tag{4.61}$$

Thus, in the z-domain:

$$G_H G_P(s) = (1 - z^{-1})Z\left(\frac{(1 + T_a s)(1 + T_b s) \ldots (1 + T_m s)}{s^j(1 + T_1 s)(1 + T_2 s) \ldots (1 + T_n s)}\right) \tag{4.62}$$

with a given digital system controller $D(z)$, and depending on the input signal type and error constants (the position error and velocity error), the closed-loop steady-state error can be derived as:

1) Position error constant:

$$K_p = \lim_{z \to 1}(z - 1)D(z)G_H G_P(z) \tag{4.63}$$

$$e_{ss} = \frac{1}{1 + K_p} \tag{4.64}$$

2) Velocity error constant:

$$K_v = \lim_{z \to 1}\frac{(1 - z^{-1})(1 + D(z)G_H G_P(z))}{Tz} \tag{4.65}$$

$$e_{ss} = \frac{1}{K_v} \tag{4.66}$$

4.5 Stability Test for Discrete-Time Systems

Section 4.3 describes how to derive the characteristics equation as well as the roots. For continuous-time linear time-invariant systems, the stability condition can be easily analyzed using graphical tools such as the z-plane, Nyquist criterion, root-locus plot, and Bode plot. This can be extended to stability studies of discrete-data systems. In this section, first the stability analysis definition is presented through the bound-input bound-output (BIBO) and the zero-input stability definitions. Then, stability tests based on the bilinear transformation method or the direct stability test methods are outlined. These stability tests can be applied directly to the characteristics equation in the discrete domain by using the unit circle in the z-plane. Among the methods providing the necessary and sufficient conditions for the characteristics equation roots lying inside the unit circle are: (i) the *Schur–Cohn* criterion; and (ii) tabulation-based methods such as the *Blanchard–Jury–Marden* stability criterion or the *Raible–Jury* stability test. Unfortunately, these analytical tests become very tedious for system models captured by equations with orders higher than the second, especially when these equations include unknown parameters. There is no reason to use any of these tests if all the

Table 4.5 Stability conditions and corresponding root values.

Stability condition	Root values
Asymptotically stable or simply stable	$\lvert z_i \rvert < 1$ for all $i = 1,2,\ldots,n$ (all roots inside the unit circle)
Marginally stable or marginally unstable	$\lvert z_i \rvert = 1$ for any simple roots and no $\lvert z_i \rvert > 1$ for $i = 1,2,\ldots,n$ (at least one simple root, no multiplier-order roots on the unit circle, no roots outside the unit circle)
Unstable	$\lvert z_i \rvert > 1$ for any i, or $\lvert z_i \rvert = 1$ for any multiple-order root $i = 1,2,\ldots,n$ (at least one simple root outside the unit circle and at least one multiple-order root on the unit circle)

coefficients of the equation are known. Rather, the execution of a root-finding program can be performed. It is believed that when the characteristics equation has at least one unknown parameter, the bilinear transformation method combined with the *Routh–Hurwitz* stability test is the suitable manual method to determine the stability of a linear discrete-data system.

4.5.1 Bound-Input Bound-Output (BIBO) Stability Definition

Let $u(kT)$, $y(kT)$, and $g(kT)$ be the input, output, and impulse sequence of a linear time-invariant SISO discrete-data system, respectively. With zero initial conditions, the system is said to be BIBO-stable, or simply stable, if its output sequence $y(kT)$ is bounded to a bound input $u(kT)$. Thus, for the system to be BIBO-stable, the following condition must be met:

$$\sum_{k=0}^{\infty} \lvert g(kT) \rvert < \infty \tag{4.67}$$

4.5.2 Zero-Input Stability Definition

For zero-input stability, the output sequence of the system must satisfy the following conditions:

$$\begin{cases} \lvert y(kT) \rvert \le M < \infty \\ \lim_{k \to \infty} \lvert y(kT) \rvert = 0 \end{cases} \tag{4.68}$$

Thus, zero-input stability can also be referred to as *asymptotic stability*. Both the BIBO stability and the zero-input stability of discrete-data systems require that the roots of the characteristics equation lie inside the unit circle in the z-plane. Let the characteristics equation roots of a linear discrete-data time-invariant SISO system be $z = 1, 2, \ldots, n$. The possible stability conditions of the system are summarized in Table 4.5 with respect to the roots of the characteristics equation.

4.5.3 Bilinear Transformation and the Routh–Hurwitz Criterion

Consider a discrete closed-loop process model given by:

$$G_P(z) = \frac{c_0 + c_1 z^{-1} + c_2 z^{-2} \ldots + c_m z^{-m}}{a_0 + a_1 z^{-1} + a_2 z^{-2} \ldots + a_n z^{-n}} \tag{4.69}$$

The corresponding characteristics polynomial is given by:

$$a_0 + a_1 z^{-1} + a_2 z^{-2} + \ldots + a_n z^{-n} = 0 \tag{4.70}$$

It is possible to transform the lines in the z-plane into a circle in the s-plane by transforming Equation (4.70) into the z-plane using the r-transformation given by:

$$z = \frac{1+r}{1-r} \tag{4.71}$$

This would result in:

$$d_0 + d_1 r^1 + d_2 r^2 + \ldots + d_n r^n = 0 \tag{4.72}$$

This bilinear transform method would map the inside of the unit circle of the s-plane into the left half of the r-plane, which would result in an equivalent characteristics polynomial in the form of:

$$b_0 + b_1 s + b_2 s^2 + \ldots + b_n s^n = b_n \prod_{i=1}^{n}(s - r_i) = 0 \tag{4.73}$$

Then apply the Routh stability criterion, given that the system would be stable if all its zeros **Re** $r_i < 0$, such that:

$$
\begin{array}{cccc}
s^n & b_n & b_{n-2} & b_{n-4} \\
s^{n-1} & b_{n-1} & b_{n-3} & b_{n-5} \\
s^{n-2} & e_{n-1} & e_{n-3} & e_{n-5} \\
\ldots & \ldots & \ldots & \ldots \\
s^0 & k_{n-1} & \ldots &
\end{array}
\tag{4.74}
$$

with

$$e_{n-1} = \frac{-1}{b_{n-1}} \begin{vmatrix} b_n & b_{n-2} \\ b_{n-1} & b_{n-3} \end{vmatrix}, \quad e_{n-3} = \frac{-1}{b_{n-1}} \begin{vmatrix} b_n & b_{n-4} \\ b_{n-1} & b_{n-5} \end{vmatrix}, \quad e_{n-2} = \frac{-1}{b_{n-1}} \begin{vmatrix} b_{n-1} & b_{n-3} \\ b_{n-1} & b_{n-3} \end{vmatrix}$$

So, the number of changes in the sign of the first column of the Routh array defines the number of roots of the characteristics equation with a positive real part **Re** $r_i > 0$.

Example 4.4   Consider that the characteristics equation of a discrete-data control system is:

$$z^3 + 0.6z^2 + 0.09z + 0.5 = 0$$

Applying the r-transformation, it becomes:

$$1.9r^3 + 3.81r^2 + 5.01r + 2.19 = 0$$

Routh's tabulation of the last equation is:

$$
\begin{array}{llll}
& r^3 & 1.9 & 5.01 \\
\text{No sign change} & r^2 & 3.81 & 2.19 \\
\text{No sign change} & r^1 & 3.91 & 0 \\
& r^0 & 2.19 &
\end{array}
$$

From the s-plane to the z-plane, the stability boundary is no longer an imaginary axis, but rather it is the unit circle in the z-plane. The system is stable when all poles are located inside the unit circle and it is unstable when any pole is located outside. Since there are no sign changes in the first column of the tabulation, the characteristics equation has any no roots in the right half of the r-plane. This corresponds to having no roots outside the unit circle in the z-plane.

4.5.4 Jury–Marden Stability Test

A discrete system is stable if its impulse response fulfills the additive property (i.e. superposition). This requirement translates into the z-domain requirement that all the poles of the transfer function must lie in the open interior of the unit circle. Let a transfer function be in the form $H(z) = \frac{N(z)}{D(Z)}$, where

$$D(z) = a_D z^D + a_{D-1} z^{D-1} + \ldots + a_1 z^1 + a_0 = 0 \tag{4.75}$$

The array is given by:

	z^0	z^1	z^2	...	z^{D-2}	z^{D-1}	z^D
1	a_0	a_1	a_2	...	a_{D-2}	a_{D-1}	a_D
2	a_D	a_{D-1}	a_{D-2}	...	a_2	a_1	a_0
3	b_0	b_1	b_2	...	b_{D-2}	b_{D-1}	
4	b_{D-1}	b_{D-2}	b_{D-3}	...	b_1	b_0	
5	c_0	c_1	c_2	...	c_{D-2}		
6	c_{D-2}	c_{D-3}	c_{D-4}	...	c_0		
...	...	...	...	...			
$2D-3$	s_0	s_1	s_2	...			

The third row is computed from the first two by:

$$b_0 = \begin{vmatrix} a_0 & a_D \\ a_D & a_0 \end{vmatrix}, b_1 = \begin{vmatrix} a_0 & a_{D-1} \\ a_D & a_1 \end{vmatrix}, b_2 = \begin{vmatrix} a_0 & a_{D-2} \\ a_D & a_2 \end{vmatrix}, \ldots, b_{D-1} = \begin{vmatrix} a_0 & a_1 \\ a_D & a_{D-1} \end{vmatrix}$$

The fourth row is the same set as the third row, except it is in the reverse order. Then the c's are derived using the same approach to derive the b's. This operation could continue until all entries in the tabular array are obtained. In the case of a system of an nth order, each array has $2n-3$ rows. The system is stable if the following conditions are satisfied:

$$D(1) > 0 \quad (-1)^D D(-1) > 0$$

$$\text{(i.e. } D(-1) > 0 \text{ for } n \text{ being an even integer and } D(-1) < 0 \text{ for } n \text{ being odd)} \tag{4.76}$$

$$a_D > |a_0|, |b_0| > |b_{D-1}|, |a_0| > |c_{D-2}|, \ldots, |s_0| > |s_2| \tag{4.77}$$

In summary, when all the coefficients are real, the Jury–Marden stability test consists of satisfying the following necessary conditions for $D(z)$ to have no roots on or outside the unit circle:
If n is an even integer:

$$\begin{cases} D(1) > 0 \\ D(-1) > 0 \end{cases} \tag{4.78}$$

If n is an odd integer:

$$D(-1) < 0 \tag{4.79}$$

$$|a_0| > a_n \tag{4.80}$$

If any stability condition here is violated, then not all of the roots are inside the unit circle and the system would not be stable. These necessary conditions can be checked easily by inspection.

4.5.5 Frequency-Based Stability Analysis

As a rule of thumb, the stability reduction is small and tolerable if the time delay is less than one tenth of the response time of the control system, as it would have been with a continuous-time controller or a controller having a very small sampling time.

4.6 Performance Indices and System Dynamical Analysis

A performance index is a quantitative measure of the dynamic performance of a process based on system specifications. The performance index is always positive or zero and based on merit measures (criteria). Among some commonly encountered performance assessment criteria are: (i) stability; (ii) minimal effect of disturbance; (iii) rapid smooth response to set point changes; (iv) minimal input tracking and zero offset; (v) reduced level of controller effect; and (vi) robustness to plant–model mismatch. A system is considered to be optimally controlled when, after the system parameters are adjusted, the performance index reaches an extremum value (i.e. a positive minimum value).

Some performance indices or quantitative performance measures are the rise time, PO, settling time, and steady-state error, in addition to the overall optimal performance indices such as the integral of the absolute value of error, the integral of the squared error, and so on. Those indices are also used to guide controller designers. Some mathematical formulations of such indices are presented as follows:

ISE (integral of the squared error) Performance Index, defined as:

$$ISE = \int_0^{tf} (e(t))^2 dt \tag{4.81}$$

with the final time, t_f being the final time of measurement. The ISE will be large if the system is highly overdamped or highly underdamped. Meaning that a large error contributes more while a small error contributes less, and there is also a large penalty for a large overshoot, while there is a small penalty for persisting oscillation.

IAE (integral of the absolute error) Performance Index, defined as:

$$IAE = \int_0^{tf} |e(t)| dt \tag{4.82}$$

ITSE (integral of time x square of error) Performance Index, defined as:

$$ITSE = \int_0^{tf} t(e(t))^2 dt \tag{4.83}$$

In comparison to ISE, the inclusion of "t" is used initially to reduce the contribution of a large initial error while maintaining its contribution over a longer period of time.

ITAE (integral of the time of absolute error) Performance Index, defined as:

$$ITAE = \int_0^{tf} t|e(t)| dt \tag{4.84}$$

Here, there should be a large penalty for persisting oscillation, with a small penalty for the initial transient response. Among IAE, ISE, and ITAE, ITAE is the most conservative setting and ISE is the least conservative. Also, common sources of system performance degradation are: (i) changing the process operating conditions (i.e. throughput rate); (ii) changing or maintaining the control settings; (iii) the degradation of the data acquisition (DAQ) and measurement devices; and (iv) process configuration or actuating system degradation.

Exercises and Problems

4.1 Stability analysis in the z-domain

Consider a motor with a printer positioning system driven by a direct current (DC) motor, as depicted in Figure 4.9. Here the motor is approximated as an integration process.

a) Derive the discrete closed-loop printer positioning system.

b) Derive the maximum gain K_c to ensure absolute stability.

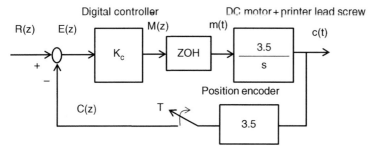

Figure 4.9 Block diagram of a DC motor with a lead screw system.

4.2 a) Derive a discrete-time model of the following functions using backward, forward, and trapezoidal approximation methods

$$G_p(s) = \frac{1}{s^2 + 1}$$

$$G_p(s) = \frac{10(s + 4)}{s^2}$$

b) For the discrete-time model obtained, plot the unit step response for $T = 0.5$ s and perform a comparative analysis of the results obtained.

c) Plot the unit step response for $T = 0.05$ s, $T = 0.5$ s, $T = 2$ s, $T = 12.5$ s.

d) Perform a comparative analysis of the results obtained.

4.3 a) Consider a DC motor transfer function $G_p(s) = \frac{12.5}{s}$, with feedback gain $K_e = 340$ and system amplifier gain $K_a = 3.5$. Assuming a zero-order hold is used:

i) Derive the discrete open-loop transfer function $\frac{C(z)}{E(z)}$.

ii) Derive the discrete closed-loop transfer function $\frac{C(z)}{R(z)}$.

b) Consider a process model given by:

$$G_p(z) = \frac{[0.01 - 0.006(z - 1)] + 0.006(z - 1)^2}{2.5(z - 1)\left(z - e^{-\frac{0.01}{0.006}}\right)}$$

i) Derive the closed-loop gain such that the system response has an overshoot of less than 35% for a unit step.

ii) Determine the system steady-state error to the unit step input.

iii) Determine the closed-loop gain minimizing the integral square error.

4.4 The garage gate angular position $\theta(t)$ (in degrees) is proportional to the applied voltage $v(t)$ and can be modeled as:

$$\frac{d\theta(t)}{dt} = 2.6v(t)$$

The rule used to control (proportional control) the gate position is given by:

$$v(k) = 0.35(\theta^*(k) - \theta(k))$$

in volts, with $\theta^*(k)$ being the command input gate position. The sample period is 5 s.

a) Find the generating function for the gate position that results from the gate being closed using a step input of 840°.

b) Find the generating function for the gate position that results from the gate being closed using a ramp input of 10.5° per s, lasting 210 s.

4.5 A 2D robot arm is attached to a laser machining system in order to cut a circular part using the chordal approximation, as shown in Figure 4.10.

The accuracy is to be 0.002 cm and the radius of the workpiece has to be no more than 10 cm such that the clockwise motion at 600 cm min^{-1} starts at the top of the circle. What are the straight-line interpolation equations for the first three chordal segments of the motion? Consider the case where a hole is to be cut in a workpiece, and the radius of the hole, R, is to be less than 10 cm, and derive the straight-line interpolation equations.

4.6 a) Consider the process illustrated in Figure 4.11.

i) With $K_c = 3.5$ and $\delta = 0.21$, derive the steady-state error in this system for a command ramp input $\theta^*(t) = 200t$.

ii) In order to simplify the system shown in Figure 4.11, it is desired to obtain a simplified model that will make it easier to set the controller gain K_c.

- Plot the frequency response ($T = 0.01$ sec) of each system component.
- If the closed-loop system is not required to respond to frequencies above 1 Hz, draw a new block diagram with simplified blocks wherever possible.
- Plot and compare the open-loop frequency responses of the original system and the simplified system.

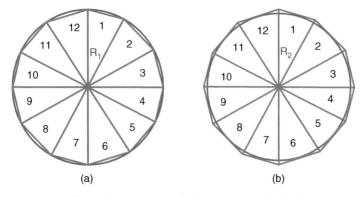

(a)   (b)

Figure 4.10 Chordal approximation for hole cutting with a different radius.

Figure 4.11 Process block diagram.

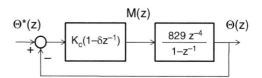

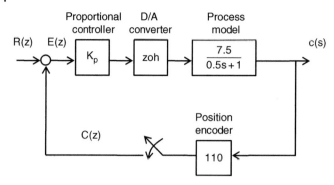

Figure 4.12 Process block diagram.

    b) Perform stability analysis on the following system transfer functions:

        i) $G_p(z) = \dfrac{3(z+1.2)(z+1.2)}{z^2(z+1)(z+0.1)}$

        ii) $G_p(z) = \dfrac{(3z+1)(z+4)}{z(z+2.2)(z+0.9)}$

4.7   a) With $T = 0.01$ s, $\tau = 0.008$ s, and $E = e^{-T/\tau}$, consider a process model given by:

$$G_p(z) = \frac{K(z-E)[T-\tau(z-1)]+\tau(z-1)^2}{2.5(z-1)(z-E)}$$

        i) Sketch the root locus of the process. What is its corresponding frequency response?

        ii) Derive the transient and stationary dynamic characteristics in both time and frequency domains.

    b) Considering a motor-driven positioning system in which the transfer function can be approximated as:

$$G_p(s) = \frac{K_a s}{s^2 + 1}$$

    Plot the discrete frequency response of the position of the system with a sampling period of 200 Hz.

4.8   Consider the linear process system depicted as a block diagram in Figure 4.12.

    a) Derive the parametric characteristics equation of this process.

    b) Assuming that $T = 20$ m s^{-1}, derive the range of values of K_p so that the system is stable.

Bibliography

1 Ackermann, J. (1985). *Sampled Data Control Systems: Analysis and Synthesis, Robust System Design*. Springer-Verlag.

2 Åström, K.J. and Wittenmark, B. (2011). *Computer-Controlled Systems: Theory and Design*. Courier Dover.

3 Bollinger, J.D. and Duffie, N.A. (1989). *Computer Control Machines and Processes.* Addison-Wesley.

4 Brockwell, P.J. and Davis, R.A. (2009). *Time Series: Theory and Methods*, 2e. Springer.

5 Fadali, S. and Visioli, A. (2012). *Digital Control Engineering*, 2e. Academic Press.

6 Isermann, R. (1989). *Digital Control Systems.* Berlin: Springer-Verlag.

7 Jury, E. and Tsypkin, Y.Z. (1971). *On the Theory of Discrete Systems.* Automatica, Elsevier.

8 Powell, G.F.J.D. and Workman, M. (2006). *Digital Control of Dynamic Systems*, 3e. Ellis-Kagle Press.

9 Golnaraghi, F., Kuo, B.C., and Adams, J.A. (2009). *Automatic Control*, 9e. Wiley.

10 Lorenz D.R. (1999). *Advances in Electric Drive Control, IEEE International Conference Electric Machines and Drives*, pp. 9–16.

11 Patterson, D.A. and Hennessy, J.L. (2013). *Computer Organization and Design: The Hardware/Software Interface.* Morgan Kaufmann.

12 Paraskevopoulos, P.N. (1996). *Digital Control Systems.* Prentice Hall.

13 Polderman, J.W. and Willems, J.C. (2013). *Introduction to the Mathematical Theory of Systems and Control.* Springer.

14 Santina, M., Stubberud, A., and Hostetter, G. (1994). *Digital Control System Design.* Oxford University Press.

15 Skogestad, S. and Postlethwaite, I. (2005). *Multivariable Feedback Control: Analysis and Design*, 2e. Wiley.

5

Continuous Digital Controller Design

5.1 Introduction

The choice of digital control of systems and processes aims to ease control implementation, to integrate local actions (e.g. shop-floor) and global decisions (e.g. production planning), as well as to combine logic and feedback control algorithms, and, eventually, to optimize system or process operating performance. Among the design requirements of digital control are the mathematical formulations to capture the time-based behavioristic characteristics of the process; the formal modeling of the control systems instrumentation, including their technological constraints; and the integration and modeling of control systems' instrumentation interfacing operations with process equipment with respect to performance and operating constraints (e.g. time delays, signal-to-noise ratios, etc.).

The design objectives of discrete control systems are similar in principle to those of continuous control systems, so that the resulting controller ensures the system will perform in accordance with some technical specifications (or controller design requirements). These controller specifications can be established through the process benchmarking (in both time and frequency domains) derived from the audit of process dynamics behavior. Based on the formulation of the controller specifications, as covered in Chapter 4, it is possible to develop a control algorithm using approaches such as continuous time design, discrete-time design and direct design. The implementation of the control algorithm has to take into consideration some control instrumentation constraints (e.g. time delays, signal-to-noise ratio, etc.).

In this chapter, root locus and Bode plot techniques are used to design feedback computer control algorithms, such as proportional-integral-derivative (PID) and feedforward, including the digital state feedback controller concept, which is revisited for cases where it is not possible to measure all state variables directly. Comparatively, analyses between classical PID controllers and various state feedback topologies for DC motor speed control are presented. Advanced control algorithms, such as model predictive control suitable for process operations with physical, safety, and performance constraints, are presented along with distributed control, open-loop control, and scalar and vector control design for induction motors.

5.2 Design of Control Algorithms for Continuous Systems and Processes

Process compensation and regulation aim to reduce or remove unwanted behavior due to process disturbances and even nonlinearities. Hence, the system and process control objectives are:

Control Of Mechatronic Systems: Model-Driven Design And Implementation Guidelines,
First Edition. Patrick O.J. Kaltjob.
© 2021 Patrick O.J. Kaltjob. Published 2021 by John Wiley & Sons Ltd.

1) input command tracking or set point tracking; this means having the output to track the reference input closely;
2) disturbance rejection to cancel the output from unwanted disturbance inputs due to model uncertainty or changes in operating conditions;
3) parameter sensitivity, or the process output variability to reduce the effect on the output of variations in process parameters.
4) stability issues.

A generic discrete controller design procedure consists in:

1) defining or characterizing the process boundaries and process variables to be controlled with respect to noise or other disturbances related to the operating range of involved sensing, actuating, and communicating units;
2) estimating the process model parameters and deriving the resulting dynamics characterization in differential equation form, difference equation form, or state space form in the case of multiple inputs or outputs processes;
3) performing an audit of the process dynamics and control system analysis in terms of controllability, stability, robustness, observability, and sensitivity;
4) formulating the controller specifications and the expected process dynamics in terms of (ξ, ω_n), such as:
   a) in the time domain, performance indicators like rise time t_r, time delay D, overshoot PO (%), settling time t_s, steady-state error of the response to the step reference, disturbance inputs e_{ss}, and error constants K_{pp}, K_v;
   b) in the frequency domain, performance indicators like resonance peak M_r, bandwidth of the closed-loop frequency response ω_b, sensitivity functions and stability margins derived from GM, PM.
5) deriving the desired system dynamics characteristics in either:
   a) the discrete-time domain using the root locus located at the desired roots given by: (ξ, ω_n);
   b) the frequency domain using the Bode plot, identified by the desired crossover frequency (ω_g, ω_p);
6) choosing the controller type to be designed accordingly;
7) deriving the sampling time and the corresponding discrete-time process model and the corresponding controller;
8) choosing the design approach of the discrete controller (i.e. algorithm) from:
   a) the *continuous time design* approach, which consists of designing the continuous time controller based on the continuous time process model and then deriving the equivalent discrete control algorithm. Here, the sampling period must be selected according to the suitable open and closed-loop results;
   b) the *discrete-time design* approach, which uses the continuous time process model to derive an equivalent discrete-time process model and then any discrete controller design techniques can be applied. Here, some resulting intersampled behavior problems due to the discrete sampling zeros could be circumvented using an anti-aliasing filter;
   c) the *direct design* approach, which consists of designing the discrete-time controller from the continuous time process model;
9) choosing among the control design techniques classified as follows:
   a) a classical control paradigm based on command input tracking. This uses classical controller design tools like PID family, feedforward, cascade controller types, and tuning. It is applied to the control of industrial processes such as chemical plants;

b) optimal control schemes, such as model predictive control, linear quadratic regula-tor/gain control (LQR/LQG), a Kalman filter, and H_2 control, which have been developed to achieve an optimal performance. It is applied within aerospace explorations projects;

c) a robust control paradigm, such as H_∞ control, which has been developed to handle systems with uncertainties and disturbances with high precision requirements such as missile defense systems;

d) a nonlinear control paradigm, which has been developed to handle high-performance nonlinear systems such as combat aircraft systems;

e) artificial intelligence control schemes, such as knowledge-based control, adaptive con-trol, and neural and fuzzy control, which have been developed to handle systems with unknown models such as human metabolism systems;

10) evaluating the process dynamics performance achieved and comparing them with the per-formance benchmarks and objectives stated in the previous step, including:

a) *steady-state accuracy* with tracking error being due either to changes in the reference input or to onset of disturbances;

b) *transient response* for criteria related to the speed of response, such as settling time t_s or rise time t_r;

c) *disturbance rejection* for a process subjected to disturbance inputs during its operation;

d) *control effort* and *gain distribution,* to avoid process control component saturation;

e) *sensitivity analysis,* to maintain performance despite changes in process parameters.

5.2.1 Direct Design Controller Algorithms

Based on the desired input/output or output/error relationship, the controller transfer func-tion can be directly defined. This design approach is suitable for a process with a complete mathematical, linear difference-based description of its dynamics. Furthermore, it is possible to represent any of their input sample sequences and the subsequent system responses from the corresponding discrete process transfer function. The control algorithm could be derived by shaping the required digital controller, $D(z)$, in the following generic form:

$$D(z) = \frac{1}{G_p(Z)} \left[\frac{G_e(z)}{1 - G_e(Z)} \right] \tag{5.1}$$

where $G_p(z)$ is the discrete process transfer function and the expected system transfer function $G_e(z)$ can be obtained from the desired roots locations for various (ξ, ω_n) at step response. A special feature of digital control is the possibility of designing a controller in order to have the system output capable of tracking the reference input in a finite number of steps, n; that is, in a finite time interval, nT. This is called a deadbeat controller, which is equivalent to placing all the closed-loop system poles at the origin. Thus, the desired closed-loop transfer function under the deadbeat is

$$G_e(z) = \frac{1}{z^n} \tag{5.2}$$

The resulting overshoot is huge, and the control input requires a great effort to achieve such a performance. This makes the deadbeat controller impractical, and it is rarely used in practical situations.

Example 5.1 *Deadbeat Controller Design*
Consider a first-order system with the closed loop given by:

$$\frac{\phi(z)}{\phi^*(z)} = \frac{Az^{-1}}{1 - 2z^{-1}}$$

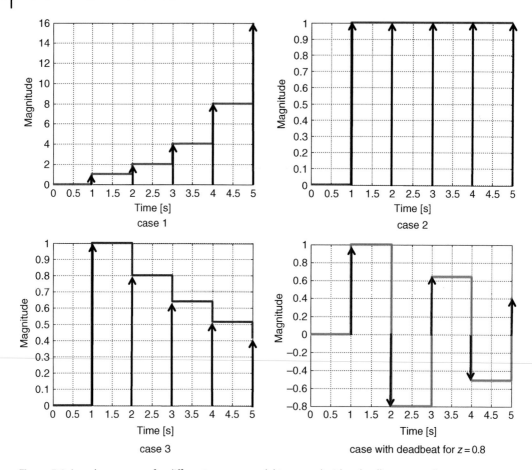

Figure 5.1 Impulse response for different process model types and with a deadbeat controller.

The difference equation can be obtained through direct inverse and yields:

$$\phi(z) = 2\phi(k-1) - A\phi^*(k-1)$$

As shown in Figure 5.1, *case 1*, the system will be unstable when a root is located at +2 and its corresponding transfer is given by:

$$\frac{\phi(z)}{\phi^*(z)} = \frac{Az^{-1}}{1 - 2z^{-1}}$$

As shown in Figure 5.1, *case 2*, the system will be stable but will display an integration type of response when a root is located at +1 and its corresponding transfer is given by:

$$\frac{\phi(z)}{\phi^*(z)} = \frac{Az^{-1}}{1 - z^{-1}}$$

As shown in Figure 5.1, *case 3*, the system will be stable when a root is located at +0.8 and its corresponding transfer is given by:

$$\frac{\phi(z)}{\phi^*(z)} = \frac{Az^{-1}}{1 - 0.8z^{-1}}$$

Table 5.1 summarizes how recursively the system responses are computed for the first four samples and for all cases mentioned previously.

Table 5.1 Table of discrete values of input-output at different sample periods.

| | | | | Without deadbeat | | | With deadbeat | | |
| | | | | Case 1 | Case 2 | Case 3 | | $z + 0.2$ | $z + 0.8$ |
k	Φ^*(k-1)	Φ(k)	Φ(k-1)	Φ(k)	Φ(k)	Φ(k)	$\Phi^*(k-1)$	Φ(k)	Φ(k−1)
0	0	1	0	0	0	0	0	0	0
1	1	0	0	A	A	A	A	A	A
2	0	0	A	2A	A	0.8A	0	−0.2A	−0.8A
3	0	0	2A	4A	A	0.64A	0	(−0.2)2A	(−0.8)2A
4	—	—	—	—	—	—	0	(−0.2)3A	(−0.8)3A

The system displays a *deadbeat response* when a root is located at 0 and its corresponding transfer function is given by:

$$\frac{\phi(z)}{\phi^*(z)} = Az^{-1}$$

In Figure 5.1, *case with deadbeat,* note the oscillatory nature of the response for a deadbeat.

```
a=1;numz=[a];denz=[1 -2];sys_D=tf(numz,denz,1); [y,t]=impulse(sys_D,5);
numz2=[a];denz2=[1 -1];sys_D2=tf(numz2,denz2,1); [y2,t]=impulse(sys_D2,5);
numz3=[a];denz3=[1 -0.8];
sys_D3=tf(numz3,denz3,1); [y3,t]=impulse(sys_D3,5);numz4=[a];denz4=[1 0.8];
sys_D4=tf(numz4,denz4,1); [y4,t]=impulse(sys_D4,5);
figure(1); stairs(t,y,'*r');grid;xlabel('Time(sec)');ylabel('Magnitude');
figure(2),stairs(t,y2,'*b');grid;xlabel('Time(sec)');
ylabel('Magnitude');figure(3);stairs(t,y3,'*g');grid;xlabel('Time(sec)');
ylabel('Magnitude');figure(4); plot(t,y4,'*k');grid;xlabel('Time(sec)');
ylabel('Magnitude');
```

5.2.2 Discrete PID Controller Algorithms

PID control algorithms are the most common control technique, especially when the process model is not available. There are three components, namely P (proportional control), I (integral control), and D (derivative control), with various combinations thereof.

5.2.2.1 Proportional Control Algorithm

A discrete implementation of the discrete proportional control is given by $D(z) = K_p$, making the computer control algorithm (with $u_n = u(k)$):

$$u(k) = K_p e(k) \tag{5.3}$$

where $e(k)$ is the error signal as given in the feedback block diagram. A proportional controller K_p has the effect of reducing the rise time and reducing but never eliminating the steady-state error.

5.2.2.2 Derivative Control Algorithm

The continuous time derivative controller is $u_n = K_d \dot{e}_n$. Thus, the equivalent discrete derivative control algorithm is approximated via the backward difference as:

$$u_n = K_d \frac{e_n - e_{n-1}}{T} \Rightarrow D(z) = K_d \frac{1 - z^{-1}}{T} \tag{5.4}$$

A derivative control (K_d/T) has the effect of increasing the stability of the system, reducing the overshoot, and improving transient response characteristics.

5.2.2.3 Integral Control Algorithm

Typical continuous integral control is given by $u(t) = K_i \int \dot{e}(k)dt$. Thus, the integral computer control algorithm is approximated via the backward difference as:

$$u_n = u_{n-1} + K_i T e_n \Rightarrow D(z) = \frac{K_i T}{1 - z^{-1}} \tag{5.5}$$

An integral control $(K_i T)$ has the effect of eliminating the steady-state error, but it may worsen the transient response characteristics.

5.2.2.4 PI Control Algorithm

The purpose of the PI or *lag compensation* is to improve steady-state accuracy. The traditional design approach to PI or lag compensation is to put the zero of the compensator very close to the dominant pole of the process. The discrete PI controller transfer function is given by:

$$D(z) = K_P + K_i \frac{Tz}{z - 1} = \frac{[(K_P + K_i T)z - K_P]}{(z - 1)} \tag{5.6}$$

This equivalent *lag compensation* increases the low frequency gain without affecting the phase in the crossover region.

5.2.2.5 PD Control Algorithm

The objective of the PD or lead compensation is to improve the closed-loop response's speed and the percentage overshoot (i.e. to improve the transient response), typically making the system respond faster without introducing instability. Thus, the derivative term (D) gives an additional control effort due to the rate of change in the error with respect to the system control input. This improves the response to a sudden change in the system state or reference value. It should be noted that the D term essentially behaves as a high-pass filter on the error signal and makes it more sensitive to noise. The discrete PD controller transfer function is given by:

$$D(z) = K_p + K_d \frac{z - 1}{z} = \frac{(K_p + K_d)z - K_d}{z} \tag{5.7}$$

This equivalent *lead compensation* improves stability margins by adding an extra phase in the crossover region.

5.2.2.6 Classical PID Controller Algorithm

The PID and lead-lag compensation improve both transient and stationary process dynamics characteristics. Taking $K_I = K_i T$ and $K_D = K_d/T$, and using the combined algorithm as found before (or directly using the backward difference approximation for s), the *classical PID controller algorithm* yields:

$$D(z) = \frac{M(z)}{E(z)} = K_p + K_I \frac{z}{z - 1} + K_D \frac{z - 1}{z} = \frac{(K_p + K_D + K_I)z^2 - (K_p + K_D)z}{z(z - 1)} \tag{5.8}$$

Another PID family controller is the *position algorithm of PID*, derived from taking:

$$m = \overline{m} + K_p \left[e_n + K_i T \sum_{k=1}^{n} e_k + \frac{K_d}{T}(e_n - e_{n-1}) \right] \tag{5.9}$$

where $\overline{m}$ is the nominal steady-state value (or bias). Using the z-transform after simplification (approximation of integration term into the limit of $\frac{1}{1-z^{-1}}$), this is equivalent to:

$$M'(z) = M(z) - \overline{M} = K_p \left[1 + K_i T \left(\frac{1}{1-z^{-1}} \right) + \frac{K_d}{T}(1 - z^{-1}) \right] E(z) \tag{5.10}$$

where $m'_n = m_n - \overline{m}$, so the digital controller transfer function yields:

$$D(z) = \frac{M'(z)}{E(z)} = K_p \left[1 + K_i T \left(\frac{1}{1-z^{-1}} \right) + \frac{K_d}{T}(1 - z^{-1}) \right] \tag{5.11}$$

This algorithm yields the value of the controller directly, but this requires memory space to complete the computing summation for the integration component, which also requires the value of $\overline{m}$. Since this is constant, the controller output Δm_n is given by:

$$m'_n - m'_{n-1} = m_n - m_{n-1} = \Delta m_n \tag{5.12}$$

Thus, from Equation (5.9):

$$\Delta m_n = K_p \left[(e_n - e_{n-1}) + K_i T e_n + \frac{K_d}{T}(e_n - 2e_{n-1} - e_{n-2}) \right] \tag{5.13}$$

Taking the z-transform:

$$\Delta M(z) = K_p \left[(1 - z^{-1}) + K_i T + \frac{K_d}{T}(1 - 2z^{-1} - z^{-2}) \right] E(z) \tag{5.14}$$

The discrete transfer function defined as *the velocity algorithm of PID* is:

$$D(z) = \frac{\Delta M(z)}{E(z)} = K_p \left[(1 - z^{-1}) + K_i T + \frac{K_d}{T}(1 - 2z^{-1} - z^{-2}) \right] \tag{5.15}$$

This *velocity algorithm* is suggested to avoid computing the summation and circumventing the bias term of $\overline{m}$ (less prone to reset windup). Please note that multiplying both sides of position algorithm by $(1 - z^{-1})$ is equivalent to the velocity algorithm. Using the trapezoidal approximation for the integral of the PID controller, such as $K_i T \sum_{k=1}^{\Delta t} \frac{(e_k - e_{k-1})}{2}$ instead of $K_i T \sum_{k=1}^{\Delta t} e_k$, another velocity algorithm of PID computer control would yield:

$$D(z) = \frac{\Delta M(z)}{E(z)} = K_p \left[(1 - z^{-1}) + K_i T(1 + z^{-1}) + \frac{K_d}{T}(1 - 2z^{-1} - z^{-2}) \right] E(z) \tag{5.16}$$

Even if it is a more accurate discrete approximation of the integral action, however, this does not necessarily improve the control loop performance. The digital PID controller design challenge is to derive the different settings values (K_p, K_d, K_i, T). It should be noted that a generic form of the PID computer control algorithm combined action can be written as:

$$\Delta m_n = (\Delta m_n)_p + (\Delta m_n)_i + (\Delta m_n)_d = K_0 e_n + K_1 e_{n-1} + K_2 e_{n-2} \tag{5.17}$$

with

$$
\begin{cases}
K_0 = K_p + K_p K_i T + K_p \dfrac{K_d}{T} \\[2mm]
K_1 = -2\dfrac{K_d}{T} - K_p \\[2mm]
K_2 = \dfrac{K_d K_p}{T}
\end{cases}
\tag{5.18}
$$

The *lead-lag compensation or PID controller* jointly uses phase-lag compensation at low frequencies and phase lead compensation at crossover frequency. It is used to meet the design specifications, such as time to peak, percentage overshoot, and steady-state accuracy.

5.2.2.7 Properties of and Some Remarks on PID Controller Algorithms

Each PID form produces the same result but incorporates information in a different manner. For example, it is possible to adjust each term independently using the PID parallel form. The PID controller form depends on design decisions such as how to manipulate the output of the controller.

1) From the position algorithm of PID, the error summation from *integral term* could grow to a very large value, even though the measured variable does not eventually reach the set point ($e_n = 0$). Until the error changes sign, the summation term will not be reduced. This is the case for excessive overshoot of the controlled variables or the saturation of the control effort (manipulated variables). To circumvent this situation, either (i) an upper limit could be placed on the value of the manipulated variable to suspend the summation until the controller moves away from the limit, or (ii) e_{n-1} could be used instead of e_n in the next controller action when saturation occurs. This is called elimination of *windup reset* due to the *integral term* problem. In the case of the velocity algorithm, the discrete controller must be programmed to update Δm_n such that if *continuously monitored*, m_n reaches its limits.
2) If set point change is reached (Δr_n), the resulting control algorithms produce a large and sudden change in the output due to the *derivative term*. This can be avoided by either (i) using the measured variable instead of the error signal, ($e_n = r_n - b_n$), such that for the position algorithm $m_n = \overline{m} + K_p \left[e_n + K_i T \sum\limits_{k=1}^{\frac{\Delta t}{}} e_k + \frac{K_d}{T}(b_n - b_{n-1}) \right]$; or (ii) changing the ramp set point to a new value instead of the step point change, limiting the value of r_n. This is called elimination of the *derivative kick due to a derivative term.*
3) There is also a *saturation of controller performance* from the selection of the sampling period due to a scaling problem. The high sampling frequency can cause a large derivative action such that $\frac{K_p K_d}{T}$, while the small change in error Δe_n can yield a large manipulated controller output, Δm_n (saturation).
4) *The physical realizability requirements of the digital controllers* are: (i) PID with a time delay compensation implemented using the Smith predictor; (ii) the use of dimensionless controller gain; and (iii) the avoidance by the algorithm of the utilization of the future process inputs in deriving its output signal.
5) (K_p, K_d, K_i) gains are dependent on one another, as summarized in Table 5.2. Changing one of these variables can change the effects of the other two. Hence, Table 5.2 should only be used as a reference to derive the values for (K_p, K_d, K_i) gains.

Table 5.2 Effects of PID controller components on process dynamics characteristics.

Closed-loop response	Rise time	Overshoot	Settling time	Stability	Steady-state error
K_p	Decrease	Increase	Small change	Degrade	Decrease
K_i	Decrease	Increase	Increase	Degrade	Eliminate
K_d	Small change	Decrease	Decrease	Improve	Small change

5.2.3 PID Controller Gains Design Using a Frequency Response Technique

The frequency-based design techniques for continuous time systems and discrete-time systems are the same. Here, the process model $G(s)$, combined with the hold element, is equivalent to a discrete time process model $G(z)$. This can be transformed into the ω-domain (bilinear transformation) in order to ensure it is in the sable region of the z-plane. Thus, it is possible to apply the frequency response controller design technique based on the phase and gain margins established in the Laplace domain. In other words, the technique used for the controller design of the continuous system can be directly applied to a compensator design in the ω-domain. Finally, using the bilinear transform, the resulting discrete controller can be derived. The *generic procedure to design a controller $D(\omega)$* consists of:

1) deriving the transient and the stationary performance specifications for the continuous time process dynamics and the controller design requirements, such as: $(PM, GM, e_{ss}, \omega_b, P.O(\%), t_s, t_r, etc \ldots)$;
2) deriving the discrete process transfer function $G(z)$;
3) performing a ω-transformation; $G(\omega)|_{z=\frac{\frac{2}{T}+\omega}{\frac{2}{T}-\omega}} = G(z)$
4) designing a controller $D(\omega)$ using a continuous time technique, as follows:
   a) draw the uncompensated open-loop transfer function, $G(\omega)$ Bode plot;
   b) select a gain crossover frequency for the process based on the requirements for a transient response;
   c) determine the corresponding desired phase margin $\gamma_{desired}(PM_{desired})$;
   d) analyze a Bode plot to conclude on the requirements with respect to the increase or reduction of the phase margin;
   e) using the percentage overshoot, find the corresponding value; then derive the corresponding process phase margin $\gamma(PM)$, using $100\xi \approx Phase\ Margin(PM)$;
   f) determine the phase required to reach the desired phase margin;
   g) determine the frequency at which the phase should be added, using:
   $$T_{settling\_time} = f(\omega_{closed\_loop\_bandwidth})$$
   h) apply the inverse bilinear transform to derive the discrete controller from: $D(z) = D(\omega)|_{\omega=\frac{2z-1}{Tz+1}}$

5.2.3.1 Design Procedure for PID Controller Design

The PID controller deals essentially with the manipulation of the phase of the closed-loop system. Indeed, a D-action of the controller can increase the stability or the speed of response of a system, while I-action of the controller can reduce the steady-state error. Thus, for the implementation of a PID controller, one must first design the derivative controller gain to achieve the desired transient response and stability and then add on an integral controller gain to improve the steady-state response.

PID Discrete Compensator Design Method   Recall that a digital PID controller algorithm is given by:

$$D(z) = K_p + K_i \frac{T}{2}\frac{z+1}{z-1} + K_d \frac{1}{T}(1-z^{-1}) = \frac{K_p + K_i + K_d - (K_p + 2K_d) + K_d}{z^2 - z} \qquad (5.19\text{a})$$

whereas an analog PID controller is defined as:

$$D(z) = K_p + K_i \frac{1}{\omega} + K_d \omega = \frac{K_p \omega + K_i + K_d \omega^2}{\omega} \qquad (5.19\text{b})$$

such that the derivative term $K_d \omega$ is given by $\frac{1}{T}(1-z^{-1}) = \frac{z-1}{Tz}$. Therefore, defining a bilinear transformation such as $\omega = \frac{T}{2}\frac{z-1}{z+1}$, D-action is given by:

$$\left. \frac{z-1}{Tz} \right|_{z = \frac{1+\frac{T}{2}\omega}{1-\frac{T}{2}\omega}} = \frac{\omega}{1+\frac{T}{2}\omega} = \frac{\omega}{1+\frac{\omega}{\frac{2}{T}}} = \frac{\omega}{1+\frac{\omega}{\frac{\omega_s}{\pi}}} \qquad (5.20)$$

Setting $\omega_\omega = \frac{2}{T}\tan\frac{\omega T}{2}$ and the corner frequency at $\omega_\omega = \frac{\omega_s}{\pi} < \frac{\omega_s}{2}$. The derivative controller adjusts the desired phase margin, $\gamma(PM)$, to a desired value, $|\dot{D}(j\omega_\omega)| \to \infty$, when $\omega T = \pi$. From Equation (5.19b), for purpose of PID controller design, the frequency response $D(j\omega_\omega)$ yields:

$$D(j\omega_\omega) = K_p - jK_i\frac{1}{\omega_\omega} + jK_d\omega_\omega = K_p + j\left(-K_i\frac{1}{\omega_\omega} + K_d\omega_\omega\right) = |D(j\omega_\omega)|\angle\theta \qquad (5.21)$$

where:

$$|D(j\omega_\omega)| = \sqrt{K_p^2 + \left(-K_i\frac{1}{\omega_\omega} + K_d\omega_\omega\right)^2} \qquad (5.22)$$

$$\theta = \tan^{-1}\left(\frac{\left(-K_i\frac{1}{\omega_\omega} + K_d\omega_\omega\right)}{K_p}\right) \qquad (5.23)$$

In order to derive K_d, K_p, K_i such that the desired phase margin $\gamma(PM_{desired})$ is achieved at a gain crossover frequency $\omega_{\omega 1}$, the following requirements must be fulfilled:

$$D(j\omega_{\omega 1})G(j\omega_{\omega 1}) = 1\angle -180° + \gamma(PM) \qquad (5.24)$$

for the gain condition:

$$|D(j\omega_{\omega 1})| = \frac{1}{|G(j\omega_{\omega 1})|} \qquad (5.25)$$

for the phase condition:

$$\angle D(j\omega_{\omega 1}) + \angle G(j\omega_{\omega 1}) = \pm 180° + \gamma(PM) \qquad (5.26)$$

Recall:

$$K_p + j\left(K_d\omega_{\omega 1} - K_i\frac{1}{\omega_{\omega 1}}\right) = |D(j\omega_{\omega 1})|e^{j\theta} = \frac{1}{|G(j\omega_{\omega 1})|}(\cos\theta + j\sin\theta) \qquad (5.27)$$

So:

$$K_p = \frac{1}{|G(j\omega_{\omega 1})|}\cos\theta \qquad (5.28)$$

$$K_d\omega_{\omega 1} - K_i\frac{1}{\omega_{\omega 1}} = \frac{1}{|G(j\omega_{\omega 1})|}\sin\theta \qquad (5.29)$$

For a phase-lag compensation design (PI controller), this is equivalent to $\theta < \sin\theta < 0$ and $K_d\omega_{\omega 1} < K_i\frac{1}{\omega_{\omega 1}}$, while for phase lead compensation design (PD controller), it is equivalent to $\theta > 0$, $\sin\theta > 0$ and $K_d\omega_{\omega 1} > K_i\frac{1}{\omega_{\omega 1}}$.

Some remarks on PID controller gains: (i) increasing K_d increases process bandwidth ω_b; (ii) increasing K_i decreases steady-state error e_{ss}, causing, in both cases, a change in the gain margin $K_g(GM)$, while the phase margin $\gamma(PM)$ remains unchanged. From the relationship $\theta = \angle D(j\omega_{\omega 1}) = \pm 180° + \gamma(PM) - \angle G(j\omega_{\omega 1})$, the crossover frequency, $\omega_{\omega 1}$, must be chosen depending on whether the PID controller should be the leading or the lagging phase. This could be determined a trial-and-error process, referring to this condition: $\angle G(j\omega_{\omega 1}) = -180° + \gamma(PM_{desired})$ with $\theta = \angle D(j\omega_{\omega 1}) = \pm 180° + \phi_m - \angle G(j\omega_{\omega 1}) > 0$. Hence, based only on this condition:

1) $\omega_{\omega 1}$ should be chosen where $\angle G(j\omega_{\omega 1}) < \pm 180° + \gamma(PM_{desired})$ in the case where the derivative controller action is expected to be dominant over the integral controller action;
2) $\omega_{\omega 1}$ should be chosen where $\angle G(j\omega_{\omega 1}) > \pm 180° + \gamma(PM_{desired})$ in the case where the integral controller action is expected to be dominant over the derivative controller action.

A PID controller increases the system gain both at low and at high frequency. A large gain at low frequency decreases the steady-state error, e_{ss}, while at high frequency, a large gain shifts the gain crossover frequency to a higher frequency, causing a decrease of phase margin, $\gamma(PM)$, but adds a leading phase up to 90°. A larger gain at high frequency makes the bandwidth wider. This trial-and-error approach could provide multiple solution values for the PID controller gains K_i and K_d.

Discrete PI Controller Design ($\theta < 0$)

$$K_p - jK_i\frac{1}{\omega_{\omega 1}} = |D(j\omega_{\omega 1})|e^{j\theta} = \frac{1}{|G(j\omega_{\omega 1})|}(\cos\theta + j\sin\theta) \tag{5.30}$$

with:

$$K_p = \frac{1}{|G(j\omega_{\omega 1})|}\cos\theta \tag{5.31}$$

$$K_i = -\omega_{\omega 1}\frac{1}{|G(j\omega_{\omega 1})|}\sin\theta \tag{5.32}$$

Discrete PD Controller Design ($\theta > 0$)

$$K_p - jK_d\omega_{\omega 1} = |D(j\omega_{\omega 1})|e^{j\theta} = \frac{1}{|G(j\omega_{\omega 1})|}(\cos\theta + j\sin\theta) \tag{5.33}$$

with:

$$K_p = \frac{1}{|G(j\omega_{\omega 1})|}\cos\theta \tag{5.34}$$

$$K_d = \frac{1}{\omega_{\omega 1}|G(j\omega_{\omega 1})|}\sin\theta \tag{5.35}$$

Exact Solution for Discrete PID Controller Design Recall the digital PID controller given by $D(z) = K_p + K_i\frac{T}{2}\frac{z+1}{z-1} + K_d\frac{1}{T}(1 - z^{-1})$, which is mapped into the ω-domain (using bilinear transformation for derivative terms) as:

$$D(\omega) = K_p + K_i\frac{1}{\omega} + K_d\frac{\omega}{1 + \frac{T}{2}\omega} \tag{5.36}$$

The frequency response is:

$$D(j\omega_\omega) = K_p - jK_i \frac{1}{\omega_\omega} + K_d \frac{j\omega_\omega}{1 + \frac{T}{2}j\omega_\omega} \tag{5.37}$$

At the crossover frequency, $\omega_{\omega 1}$:

$$D(j\omega_{\omega 1}) = K_p - jK_i \frac{1}{\omega_{\omega 1}} + K_d \frac{j\omega_{\omega 1}}{1 + \frac{T}{2}j\omega_{\omega 1}} = \frac{1}{|G(j\omega_{\omega 1})|}(\cos\theta + j\sin\theta) \tag{5.38}$$

The real part is given by:

$$K_p + K_d \frac{\frac{2}{T}\omega_{\omega 1}^2}{\omega_{\omega 1}^2 + \left(\frac{T}{2}\right)^2} = \frac{\cos\theta}{|G(j\omega_{\omega 1})|} \tag{5.39}$$

The imaginary part is given by:

$$-K_i \frac{1}{\omega_{\omega 1}} + K_d \frac{\frac{2}{T}\omega_{\omega 1}^2}{\omega_{\omega 1}^2 + \left(\frac{T}{2}\right)^2} = \frac{\sin\theta}{|G(j\omega_{\omega 1})|} \tag{5.40}$$

Another design technique is based on the first-order compensator in the ω-domain, given by analogy to the s-domain as the compensator transfer function. This yields:

$$D(\omega) = \frac{a_1\omega + a_0}{b_1\omega + 1} = a_0 \frac{1 + \frac{a_1}{a_0}\omega}{b_1\omega + 1} = a_0 \frac{1 + \frac{\omega}{\omega_{\omega 0}}}{1 + \frac{\omega}{\omega_{\omega p}}} \tag{5.41}$$

with zero $\omega_{\omega 0} = {}^{a_0}/a_1$ and pole $\omega_{\omega p} = {}^1/b_1$, which are set to satisfy the controller parameters a_1 and b_1 as:

$$a_1 = \frac{1 - a_0|G(j\omega_{\omega 1})|\cos\theta}{\omega_{\omega 1}|G(j\omega_{\omega 1})|\sin\theta} \tag{5.42}$$

$$b_1 = \frac{\cos\theta - a_0|G(j\omega_{\omega 1})|}{\omega_{\omega 1}\sin\theta} \tag{5.43}$$

where $\theta = \angle D(j\omega_{\omega 1}) = \pm 180° + \gamma(PM) - \angle G(j\omega_{\omega 1})$. Here, the DC gain a_0 can be derived from other criteria, such as the steady-state error:

$$D(\omega) = a_0 \frac{1 + \frac{\frac{2}{T}\frac{z-1}{z+1}}{\omega_{\omega 0}}}{1 + \frac{\frac{2}{T}\frac{z-1}{z+1}}{\omega_{\omega p}}} \tag{5.44}$$

which is equivalent to:

$$D(z) = a_0 \frac{\omega_{\omega 0}\omega_{\omega p} + \omega_{\omega p}\frac{2}{T}\frac{z-1}{z+1}}{\omega_{\omega 0}\omega_{\omega p} + \omega_{\omega 0}\frac{2}{T}\frac{z-1}{z+1}} = a_0 \frac{\omega_{\omega 0}\omega_{\omega p}T(z+1) + 2\omega_{\omega p}(z-1)}{\omega_{\omega 0}\omega_{\omega p}T(z+1) + 2\omega_{\omega 0}(z-1)} = K\frac{(z - z_0)}{(z - z_p)} \tag{5.45}$$

such as:

$$K = a_0 \frac{\omega_{\omega p}(\omega_{\omega 0}T + 2)}{\omega_{\omega 0}(\omega_{\omega b}T + 2)} = a_0 \frac{\omega_{\omega p}\left(\omega_{\omega 0} + \frac{2}{T}\right)}{\omega_{\omega 0}\left(\omega_{\omega b} + \frac{2}{T}\right)}$$

$$z_0 = \frac{\omega_{\omega 0}T - 2}{\omega_{\omega 0}T + 2} = \frac{\omega_{\omega 0} - \frac{2}{T}}{\omega_{\omega 0} + \frac{2}{T}}$$

$$z_p = \frac{\omega_{\omega b}T - 2}{\omega_{\omega b}T + 2} = -\frac{\omega_{\omega b} - \frac{2}{T}}{\omega_{\omega b} + \frac{2}{T}}$$

When $z_0 > z_p > 0$, this is considered a lead compensator; otherwise ($z_0 < z_p$), it is a lag compensator. Therefore, the discrete controller parameters K_i, z_0, z_p of the first-order phase lead or lag compensator can be determined in terms of $a_0, \omega_{\omega_0}, \omega_{\omega b}$ defined in the w-domain. The lag compensator (equivalent to PI controller) aims to improve and reduce the steady-state error by using a pole near the origin. The lead compensator (equivalent to PD controller) improves the transient characteristics. There is also a lag and lead compensator given by:

$$D_c(s) = K_c \frac{(s + z_1)(s + z_2)}{(s + p_1)(s + p_2)} \tag{5.46}$$

with $|p_1| > |z_1| > |z_2| > |p_2| > 0$. Because this allows modification of the transient and stationary system dynamic characteristics, it is similar to the PID controller. Because the latter is the most commonly used industrial application, PID controller design methods are covered throughout this textbook. A simple PID design method would consist of canceling all dominant poles of the system by the controller zeros, then adding zeros and poles to meet the controller design requirements. This is called the pole-zero cancelation. Pole placement is critical and must consider stability issues in the z-domain within trial-and-error procedures.

Example 5.2   PI Controller Design Using a Frequency Response Method
Considering a discrete process model given by:

$$G_p(z) = \frac{14.01}{z^2 - 0.381z}$$

Using a sampling period $T = 0.25$ s, design a controller with the following requirements: PO (%) should be less 5% and near the zero steady-state error. The design of an *equivalent PI* controller is presented as follows:

Step 1. Sketch the Bode plot $G_p(\omega)$ in the ω-domain with respect to the frequency axis of ω and perform a dynamical step response of the uncompensated open-loop system. From the uncompensated Bode plot analysis illustrated in Figure 5.2, the performance characteristics are (i) $K_g(GM) = 20\log_{10}(-22.9$ dB$) = 0.0714$ at phase crossover frequency, where ω_p $2\pi \times 0.879$ Hz $= 5.52$ rad s^{-1} ω_p and $\gamma(PM)$ is not defined and (ii) settling time $t_s = 1.5$ s for an interval ($min; max$ ()(21.38; 22.63)) PO (%) above 100%, where peak $= 22.63$ and peak time $t_p = 1.5$ s.

Step 2. Define the discrete controller specifications. Knowing that it is desired to have:

$$P.O(\%) = 100e^{-\xi\pi/\sqrt{1-\xi}} \approx 100\left(\frac{0.6 - \xi}{0.6}\right) \leq 5\%$$

This is valid if $\xi = 0.5439$. Thus:

$$PM_{desired} = 100\xi = 54.89°$$

Step 3. Derive the controller parameters, such as the frequency $\omega_{\omega 1}$, that meet design specifications for:

$$\angle G(j\omega_{\omega 1}) + \angle D(j\omega_{\omega 1}) = \pm 180° + \gamma(PM_{desired})$$

Uncompensated system GM=22.9 dB (at 5.52 rad s^{-1}), PM= Inf
Compensated system GM=7.81 dB (at 4.45 rad s^{-1}), PM=61.7 deg (at 1.55 rad s^{-1})

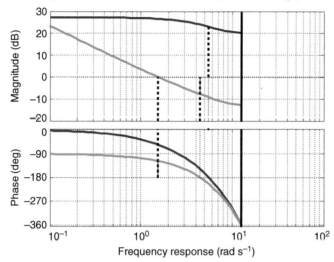

Figure 5.2 Frequency response of uncompensated $G(\omega)$ and compensated closed-loop $D(\omega)G(\omega)$.

However, a key condition is to have:

$$\angle G(j\omega_{\omega 1}) < \pm 180° + \gamma(PM_{desired}) = 180° + 54.39° = 234.4°$$

$$\theta = \angle D(j\omega_{\omega 1}) = \pm 180° + \gamma(PM_{desired}) - \angle G(j\omega_{\omega 1}) = 301.9°$$

The gain values can then be derived graphically or by computing, such as:

$$|G(j\omega_{\omega 1})| = 26.3 dB = 20.65$$

Then, using the formula, the controller gains are given by:

$$K_p = \frac{1}{|G(j\omega_{\omega 1})|} \cos\theta = 0.0256$$

$$K_i = -\omega_{\omega 1} \frac{1}{|G(j\omega_{\omega 1})|} \sin\theta = 0.0761$$

If the performance is not met, choose another $\angle G_p(j\omega_{\omega 1})$ and start the computation again.

$$D(z) = K_p + K_i \frac{T}{2}\frac{z+1}{z-1} = \frac{K_p 2(z-1) + K_i T(z+1)}{2(z-1)} = \frac{(2K_p + K_i T)z - (-K_i T + 2K_p)}{2(z-1)}$$

Step 4. Tune the controller parameters if the controller design specifications are not met. Using these parameters, an overshoot of 14.3% is achieved. Tuning controller parameters include:

$$K_p = 0.0226$$

$$K_i = 0.051$$

Step 5. Validate/check the controller parameter dynamics from the compensated Bode plot and time response. The open-loop transfer function of the compensated system is given by:

$$T(z) = \frac{0.8163z - 0.4052}{2z^3 - 2.762z^2 + 0.762z}$$

Figure 5.3 System unit step response.

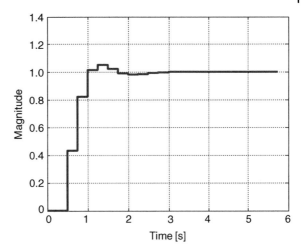

Therefore, the performance of the compensated system has the following characteristics: (i) from the Bode plot $K_g(GM) = 7.81$ dB at phase crossover frequency ω_p of 0.7082 Hz and γ (PM) = 61.7° at the gain crossover frequency ω_g of 0.2467 Hz and (ii) from the resulting step response, the rise time $t_r = 0.5$ s and the settling time $t_s = 1.5$ s, as depicted in Figure 5.3.

```
clear all; clf; Ts=0.25; sys_D = tf([14.01],[1 -0.381 0],0.25);
kp=0.0226;ki=0.0651; cont_numz=[ki*Ts+2*kp ki*Ts-2*kp];
cont_denz=[2 -2];cont_sys_D8=tf(cont_numz,cont_denz,0.25);
cont_sys_D=series(cont_sys_D8,sys_D);
figure(1);bode(sys_D,'b'); margin(sys_D);hold on;
bode(cont_sys_D,'r');hold off;
margin(cont_sys_D);grid;[cl_sys_D]=feedback(sys_D,1);
[cl_cont_sys_D]=feedback(cont_sys_D,1);
[y,t]=step(cl_sys_D);info1=stepinfo(cl_sys_D);
[y1,t1]=step(cl_cont_sys_D);info2=stepinfo(cl_cont_sys_D);
figure(2);stairs(t,y,'r'); figure(2);hold on;stairs(t1,y1,'b');
xlabel('Time[sec]');ylabel('Magnitude');
hold off;grid;axis([0 8 0 1.2]);
%y = db2mag(ydb); to convert from dB in mag values
%ydb = mag2db(y); to convert from mag in dB values
```

Example 5.3   Consider the discrete compensator given by:

$$D(z) = \frac{M(z)}{E(z)} = \frac{0.33 - 0.148z^{-1}}{1 - 0.87z^{-1}}$$

Then, the control algorithm yields:

$$m(k + 1) = 0.87m(k) + 0.33e(k + 1) - 0.148e(k)$$

However, it is neither unfeasible nor impractical to pretend to know in advance the future error values of $e(k+1)$ at the kth sampling time. In the case of infinite zeros, it is suitable to add zeros at -1 in discrete time for each infinite zero.

Remarks on Higher-Order Systems

When the given process has a dynamic order higher than 1 or a PID controller is used, the overall closed-loop transfer function from $H(z)$ will have an order larger than 2, such as:

$$H(z) = \frac{b_m z^m + b_{m-1} z^{m-1} + \ldots + b_1 z + b_0}{z^n + a_{n-1} z^{n-1} + \ldots + a_1 z + a_0} \tag{5.47}$$

where $m \leq n$ and $n > 2$. In this case, the poles of transfer function should be compared to the following desired transfer function:

$$H_{desired}(z) = \frac{*}{(z - \alpha_1) \ldots (z - \alpha_{n-2})(z - z_p)(z - \bar{z}_p)} \tag{5.48}$$

This entails placing all the rest poles close to the origin. Eventually, the dynamics associated with the poles close to the origin will die out very fast and the pair left will dominate the system dynamics. For a first-order system, it is suitable to use a deadbeat controller. For second- or higher-order systems, the impulse system response could be obtained through a finite settling step (FFS), as it is unfeasible to achieve a deadbeat response. However, this method requires an accurate model, allowing the process to achieve a faster response.

Example 5.4   PD Controller Design

A positioning system can be modeled with a sampling period $T = 0.1$ s as follows:

$$G(z) = \frac{0.6z + 0.3}{(z - 0.4)(z - 0.8)}$$

The step response is displayed in Figure 5.4, with $e_{ss} \approx 0.72$, $P.O.(\%) < 46\%$ and a settling time of 1.7 s.

It is desired to design a digital PD controller such that the resulting system output tracks a step reference. It has the following transient characteristics: a settling time less than 1 s with an

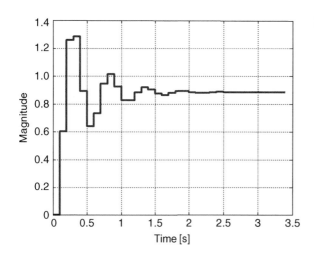

Figure 5.4 Uncompensated step response.

overshoot PO(%) ≤ 15% and a steady-state error e_{ss} ≤ 10% of final value. Converting this model into ω-transform yields:

$$G(z) = G(\omega)\big|_{z=\frac{\frac{2}{T}+\omega}{\frac{2}{T}-\omega}} = \frac{\left(0.6\frac{20+\omega}{20-\omega}+0.3\right)}{\left(\frac{20+\omega}{20-\omega}+0.4\right)\left(\frac{20+\omega}{20-\omega}-0.8\right)} = \frac{0.3\omega+18}{(1.4\omega+12)(1.2\omega+4)}$$

Step 1. From the controller design specifications, the overshoot is related to the damping ratio ξ such that:

$$PO(\%) = 100e^{-\xi\pi/\sqrt{1-\xi}} \approx 100\left(\frac{0.6-\xi}{0.6}\right) \le 15\%$$

This is valid for $\xi = 0.51$. Thus, the desired phase margin is given as $PM_{desired} = 100\xi = 51°$.
Step 2. Sketch the uncompensated Bode plot as illustrated in Figure 5.4(b).
It is possible to derive the open-loop gain crossover frequency from design specifications $t_s = 5$ s such that:

$$\omega_g = \omega_n = \frac{4.6}{\xi t_s} = \frac{4.6}{0.51 \times 5} = 1.8 \text{ rad s}^{-1}$$

with:

$$\omega_\omega = \frac{2}{T}\tan\frac{\omega T}{2}$$

With, as the main criteria:

$$\angle G(j\omega_{\omega1}) < \pm180° + \gamma(PM_{desired}) = -180° + 51° = -129°$$

Step 3. Determine the static (DC) gain a_0 using a steady-state error e_{ss} of 10% such that:

$$e_{ss} = e(\infty) = \frac{1}{1+K_{pv}} = 0.10 \Rightarrow K_{pv} = 9$$

with a PD controller given by:

$$D(z) = K_p + K_d\frac{1}{T}(1-z^{-1})$$

such that the DC gain is given by:

$$D(1)G(1) = K_p(1)\frac{0.6+0.3}{(1-0.4)(1-0.8)} = 9 \Rightarrow K_p = 1.2$$

Step 4. By setting $\angle G(j\omega_{\omega1}) = -148°$, this will correspond to $\omega_{\omega1} = 9.89$ rad s^{-1}, and from the phase condition $\theta = \angle D(j\omega_{\omega1}) = \pm180° + PM = -180 + 51 - (-148) = 19°$.
Step 5. Finally, the PD controller yields:

$$D(z) = \frac{\left(K_p + K_d\frac{1}{T}\right)z - K_d\frac{1}{T}}{z}$$

Using an uncompensated Bode plot, the gain crossover frequency and its corresponding magnitude are given by:

$$\omega_{\omega1} = 9.89 \text{ rad s}^{-1}$$

$$|G(j\omega_{\omega1})| = 0.73 \text{ dB} = 1.087$$

Thus, other gain values will be:

$$K_p = \frac{1}{|G(j\omega_{\omega 1})|}\cos\theta = 0.8698$$

$$K_d = \frac{1}{\omega_{\omega 1}|G(j\omega_{\omega 1})|}\sin\theta = 0.0303$$

With these values, the steady-state error is 12%, while the settling time and the percentage overshoot are roughly 1 s and 41%, respectively. Hence, after tuning, this results in the step response displayed in Figure 5.4(c) for $K_p = 0.6089$, $K_d = 0.0903$, where a settling time is 0.8 s, the percentage overshoot is about 30%, and a steady-state value is 0.81.

In order to further reduce the steady-state error, a strategy of increasing gain K_p could be used. However, such a decision could degrade the percentage overshoot. As such, a tradeoff is required between achieving the steady-state condition and degrading the transient conditions. An alternative strategy would be to add an integral action to reduce or even to eliminate the steady-state error. As such, the controller gain of the integral action could be added into the PD controller to form a PID controller for a new value of θ:

$$K_i - -\omega_{\omega 1}\frac{1}{|G(j\omega_{\omega 1})|}\sin\theta$$

The new controller gain for the D-action would be:

$$K_d = \frac{1}{\omega_{\omega 1}|G_p(j\omega_{\omega 1})|}\sin\theta + K_i\frac{1}{\omega_{\omega 1}{}^2}$$

Once the values were derived (e.g. after tuning $K_p = 0.8689$, $K_d = 0.0303$ and $K_i = 0.4622$), they could be inserted in the generic PID discrete controller equation:

$$D(z) = K_p + K_i\frac{T}{2}\frac{z+1}{z-1} + K_d\frac{1}{T}(1-z^{-1}) = \frac{K_p + K_i + K_d - (K_p + 2K_d) + K_d}{z^2 - z}$$

An alternative approach would be to design an equivalent phase lead compensator, given by:

$$D(z) = K\frac{z-z_c}{z-z_p}$$

which is equivalent to in the ω-domain (by bilinear transformation):

$$D(\omega) = K\frac{\tau\omega + 1}{\alpha\tau\omega + 1}$$

$$\theta = \angle D(j\omega_{\omega 1}) = \pm 180° + \gamma(PM) - \angle G(j\omega_{\omega 1}) = -180° + 51° - (-171°) + 5° = 47°$$

$$\alpha = \frac{1-\sin\theta}{1+\sin\theta} = \frac{1-\sin 47°}{1+\sin 47°} = 0.1552\,rad \text{ and } \tau = \frac{1}{\omega_{gmax}\sqrt{\alpha}} = 0.2538$$

$$D(\omega) = 1.2\frac{0.1552\omega + 1}{0.0394\omega + 1}$$

This could be converted back to the z-domain by using the inverse bilinear transformation:

$$D(z) = D(\omega)|_{\omega = \frac{2z-1}{Tz+1}} = 1.2\frac{0.1552\frac{2}{0.1}\frac{z-1}{z+1} + 1}{0.0394\frac{2}{0.1}\frac{z-1}{z+1} + 1} = \frac{3.725z - 1.325}{1.788z - 0.212}$$

When using the resulting digital lead compensator, the gain crossover frequency is about 9.9 rad s^{-1} and $PM = 180 - 134 = 46°$. Frequency responses of uncompensated and compensated systems are displayed in Figure 5.5(b), while Figure 5.5(c) depicts the system step response.

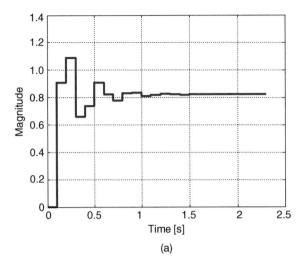

(a)

Uncompensated system GM=7.11 dB (at 16.5 rad s⁻¹), PM=28.6 deg (at 10.4 rad s⁻¹)
Compensated system GM=3.01 dB (at 19.8 rad s⁻¹), PM=13 deg (at 16.3 rad s⁻¹)

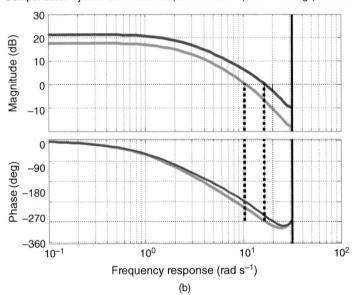

(b)

Figure 5.5 (a) Uncompensated step response with $K_p = 0.8689$ and $K_d = 0.0303$. (b) Frequency response of open-loop uncompensated $G(\omega)$ and compensated system. (c) Step response of closed-loop $D(\omega)G(\omega)$.

```
clear all;clf;Ts=0.1;kp = 0.8698;%kp=0.95;kd = 0.0303;ki=0.4622;
uncomp_sys_D = tf([0.6 0.3],conv([1 -0.4],[1 -0.8]),0.1);
uncomp_sys_cl=feedback(uncomp_sys_D,1);
%pdcont=tf([kd/Ts+kp -kd/Ts],[1 0],0.1);
%pid=tf([kp+ki*(Ts/2)+kd/Ts ki*(Ts/2)-kp-(2*kd)/Ts kd/Ts],[1 -10],0.1);
%%% gain=1.2
pleadcont=tf([3.725 -1.325],[1.788 -0.212],0.1);
```

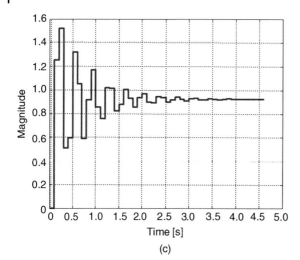

Figure 5.5 (Continued)

(c)

```
%sys_comp=series(uncomp_sys_D,pdcont);%sys_comp=series(uncomp_sys_D,pid);
sys_comp=series(uncomp_sys_D,pleadcont);sys_cl=feedback(sys_comp,1); figure(1);
bode(uncomp_sys_D,'b');margin(uncomp_sys_D);
hold on;bode(sys_comp,'r');margin(sys_comp);
hold off;grid;figure(2); [xc,tc]=step(sys_cl);
figure(2);stairs(tc,xc); xlabel('Time [sec]');
ylabel('Magnitude');
% [xc1,tc1]=step(uncomp_sys_cl); %figure(3);stairs(tc1,xc1);
info1=stepinfo(sys_cl);%axis([0 20 0 1.5]);
grid; %y = db2mag(ydb); to convert from dB in mag values
```

Example 5.5 *Deadbeat and PI Controller System*

Consider the system given by:

$$G_p(z) = \frac{0.00058}{z - 0.942}$$

Step 1. The step response of the uncompensated value is given in Figure 5.6(a).

Step 2. From the dynamical analysis of the step response, the process steady-state error to step input is about $(1-0.01)$. Thus, the controller should be designed for a faster response in one sampling period. In addition, the compensated response should avoid intersampling oscillations. Choose a PI controller to reduce the error such that:

$$D(z) = K_c \left(1 + \frac{T}{\tau_i} \frac{1}{1 - z^{-1}} \right) = \frac{K_c(1 + \tau_i) - K_c \tau_i z^{-1}}{\tau_i(1 - z^{-1})}$$

Step 3. Using the deadbeat control strategy, the closed-loop transfer function is defined such that:

$$T(z) = \frac{D(z)G_p(z)}{1 + D(z)G_p(z)} = z^{-1}$$

Therefore, the equivalent discrete controller is given by:

$$D(z) = \frac{z^{-1}}{(1 - z^{-1})G_p(z)} = \frac{K_c(1 + \tau_i) - K_c \tau_i z^{-1}}{\tau_i(1 - z^{-1})} = \frac{1 - 0.942 z^{-1}}{0.00058(1 - z^{-1})}$$

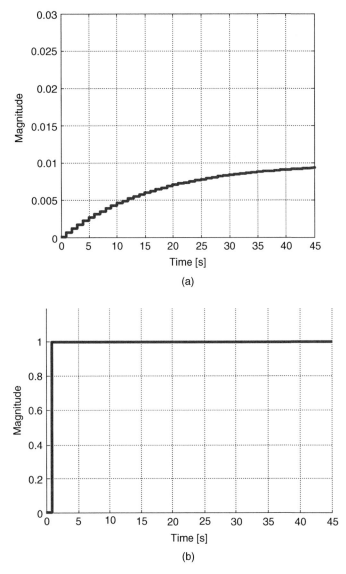

Figure 5.6 (a) Uncompensated system response. (b) Compensated system response.

Step 4. By equivalence, derive the controller parameters such that:

$$D(z) = \frac{K_c(1 + \tau_i) - K_c\tau_i z^{-1}}{\tau_i(1 - z^{-1})} = \frac{1 - 0.942z^{-1}}{0.00058(1 - z^{-1})}$$

By identification, the controller characteristics are given by:

$$K_c = 1.624$$

$$\tau_i = 16.2414$$

The compensated system response is illustrated in Figure 5.6(b).

```
numz=[0.00058];denz=[1 -0.942];uncomp_sys_D=tf(numz,denz,1);
T=300;uncomp_sys_cl=feedback(uncomp_sys_D,1);
[x,t] = step(uncomp_sys_cl);
figure(1);stairs(t,x,'b');xlabel('Time [sec]'); ylabel('Magnitude');
grid;axis([0 45 0 0.03]);pidcont1=tf([1 -0.942],[0.00058 -0.00058],1);
sys_comp1=series(uncomp_sys_D,pidcont1);sys_cl1=feedback(sys_comp1,1);
[xd,td]=step(sys_cl1,T);figure(2);hold on; stairs(td,xd,'b');
xlabel('Time [sec]'); ylabel('Magnitude');grid;axis([0 45 0 1.2]);
```

Example 5.6   *PID Controller Design Using Frequency Response*
Consider a process model given by:

$$G_p(z) = \frac{0.009057z + 0.008194}{z(z^2 - 1.7424z + 0.7408)}$$

The design requirements are as follows: for a sampling time $T = 1$ s, there should be minimal settling time, no steady-state error, $e_{ss} = 0$ for step input, a time delay of 1 s, and a phase margin $PM_{desired} = 60°$ at a frequency $\omega_{\omega 1}$ not defined.

Step 1. Sketch the Bode plot of the uncompensated process $G_p(\omega)$, as illustrated in Figure 5.7(a).
Step 2. Determine the parameters of the PID controller such that:

$$D(z) = K_p + K_i \frac{T}{2} \frac{z+1}{z-1} + K_d \frac{1}{T}(1 - z^{-1})$$

with K_d, K_p, K_i being controller gain and

$$\angle G(j\omega_{\omega 1}) < -180° + \gamma(PM_{desired}) = 180° + 60° = 240°$$

Step 3. From the uncompensated system Bode plot, find a new crossover frequency $\omega_{\omega 1}$ such that $PM_{desired} = 60°$. This is equivalent to choosing $\angle G(j\omega_{\omega 1}) < 240°$. By setting $\angle G(j\omega_{\omega 1}) = -106°$, this will correspond to $\omega_{\omega 1} = 0.151$ rad s^{-1}, and from the phase condition $\theta = \angle D(j\omega_{\omega 1}) = \pm 180° + \gamma(PM_{desired}) - \angle G(j\omega_{\omega 1}) = 180° + 60° + 106° = 346°$
Using the Bode plot, the gain crossover frequency is given by:

$$\omega_{\omega 1} = 0.241$$

$$|G_p(j\omega_{\omega 1})| = -7.28 dB = 0.241$$

Thus, controller gains result in:

$$K_p = \frac{1}{|G(j\omega_{\omega 1})|} \cos\theta = 2.2435$$

$$K_i = -\omega_{\omega 1} \frac{1}{|G(j\omega_{\omega 1})|} \sin\theta = 0.1348$$

$$K_d \omega_{\omega 1} = \frac{1}{|G(j\omega_{\omega 1})|} \sin\theta + K_i \frac{1}{\omega_{\omega 1}} = -0.00008$$

Thus, the discrete PID controller is given by:

$$D(z) = -0.17 + 0.055\frac{(z+1)}{z(z-1)} - 0.0019(1 - z^{-1})$$

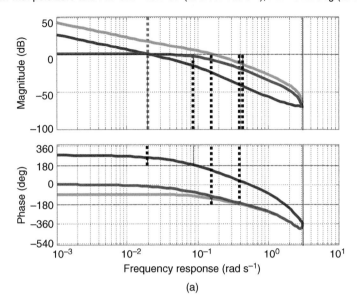

Uncompensated system GM = 20.9 dB (at 0.441 rad s^{-1}), PM = inf
System with PID GM = 11.9 dB (at 0.39 rad s^{-1}), PM = 50.4 deg (at 0.156 rad s^{-1})
System with pole zero and PID GM = 15.7 dB (at 0.087 rad s^{-1}), PM = 67.2 deg (at 0.0197 rad s^{-1})

(a)

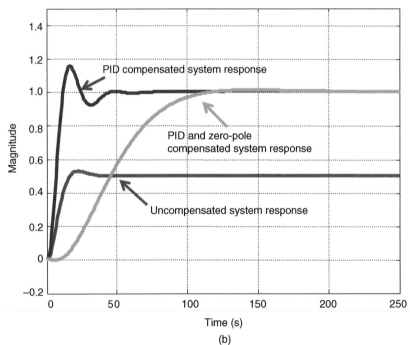

(b)

Figure 5.7 (a) Frequency responses for uncompensated PID with a pole-zero compensated system. (b) Step responses of uncompensated PID with a pole-zero compensated system.

Step 3. Plot the step response of the uncompensated process and compensated process with the PID controller as shown in Figure 5.6.

Step 4. Check the system stability through the determination of system roots. The poles and zeros of the compensated system with PID controller are as follows: for poles, $z = 0$; $z = 1$; $z = 0.9019$; $z = 0.8216$, and for zeros, $z = 1.3846$; $z = -0.9050$; $z = 0.0095$. An undesired zero is $z = -0.9050$.

Step 5. Add the pole to cancel the zero such that:

$$G_{pz_c}(z) = \frac{K_c}{z - 0.9050}$$

Step 6. Apply the pole-zero cancelation with PID controller such that:

$$G_{pz_c}(z) = \frac{K_c(z - 1.3846)(z - 0.0095)}{z^2(z - 0.9019)(z - 0.8216)(z - 1)}$$

Step 7. Through gain K_c tuning, $K_c = 0.7$.

Step 8. The discrete controller transfer function is:

$$G_{pz_c}(z) = \frac{-0.2022z^2 + 0.2818z - 0.00266}{2z^3 - 0.19z^2 - 1.81z}$$

```
clear all; clc;Ts=1;
uncomp_sys_D=tf([0.009057 0.008194],[1 -1.7183 0.7356 0],Ts)
figure(1); bode(uncomp_sys_D,'g');margin(uncomp_sys_D);grid;
uncomp_sys_cl=feedback(uncomp_sys_D,1);
T=700;[x,t] = step(uncomp_sys_cl,T);
figure(2);plot(t,x,'r');hold on;
kp=2.2435;ki=0.1348;kd=-0.00008;
pidcont=tf(parallel(parallel(kp,ki*tf([1 1],[2 -2],1)),kd*tf([1 -1],
[1 0],1)));
sys_comp=series(uncomp_sys_D,pidcont);sys_cl=feedback(sys_comp,1);
pidcont1=tf([-0.2022 0.2818 -0.00266],[2 -0.19 -1.81 0],1);
sys_comp1=series(uncomp_sys_D,pidcont1);
sys_cl1=feedback(sys_comp1,1); figure(1); bode(uncomp_sys_D,'g');
bandwidth(sys_comp);margin(sys_comp);
hold on;bode(sys_comp1,'r');
bandwidth(sys_comp1);margin(sys_comp1);
[xc,tc]=step(sys_cl,T);[xd,td]=step(sys_cl1,T);
figure(2);hold on;plot(tc,xc,'b');hold on;plot(td,xd,'g');
xlabel('Time (sec)');ylabel('Magnitude');axis([0   250 -0.2 1.5]) ;
legend('uncompensated','compensated','cancelation');
```

5.2.4 PID Controller Gains Design Using a Root Locus Technique

The fundamental idea behind the root-locus technique used for the controller design consists of modifying the process characteristics equation in order to change the root locations until the desirable root's location is reached. Among several possible PID controllers, there may be a number of conflicting transient and stationary characteristics, which can lead to a faster but oscillatory and even unstable response. In the following sections, a variety of controller configurations from the PID family are derived using the root locus design technique.

5.2.4.1   Design Procedures

A lead compensator approximates the PD controller by moving the locus to the left. This improves process damping while lowering the rise time and decreasing the transient overshoot. Therefore, it affects the overall transient performance characteristics. In contrast, the lag compensator (similar to PI control) improves steady-state accuracy requirements by moving the locus to the left due to its zero, while its poles adjust the locus to the right. However, such a controller degrades the system stability. Two design approaches are presented: the discrete-time controller and the continuous time controller. The continuous time controller is then converted into a discrete controller using discrete equivalence.

Lead or Lag Continuous-Based Controller Design Approach

Step 1. Audit the performance to benchmark the process dynamics with respect to the transient and the stationary response characteristics, the stability, and even the disturbance rejection (robustness). Among time-based performance indicators, there are the overshoot PO (%), rise time t_r, settling time t_s, and steady-state error e_{ss}. These can be derived in terms of known or selected (ξ, ω_n).

Step 2. Define the expected performance or design specifications (settling time t_s, percentage overshoot, PO (%), etc.) and translate them into the natural frequency (ω_n) and damping ratio (ξ) in order to derive the desired root loci of the compensated system through the equation, for a second-order system:

$$S_d = -\xi\omega_n \pm j\omega_n\sqrt{1 - \xi^2} \tag{5.49}$$

Step 3. Derive the uncompensated transfer function in the Laplace time domain, $G_p(s)$.

Step 4. Sketch the root locus of $G_p(z)$ and determine whether the desired roots can be realized using this uncompensated system (evaluate whether the root locus passes through the desired loci of the dominant root according to the specifications).

Step 5. Select the controller given by:

$$G_c(z) = K\frac{s + z_0}{s + z_p} \tag{5.50}$$

with $(|z_0| < |z_p|)$, in the case of a lead compensator, or $(|z_0| > |z_p|)$ for a lag compensator.

Step 6. If a compensator is required, choose a zero value for the controller, z_0, such as:

$$z_0 = -\xi\omega_n \tag{5.51}$$

In the case where the lead compensator is chosen, place the zero of the phase-lead directly below the desired root location (or to the left of the first two real poles), as shown in Figure 5.8.

Figure 5.8 An *s*-plane showing root structure.

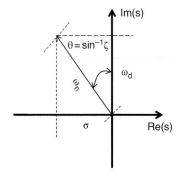

In the case of a lag compensator, set the zero and the pole of controller near the origin such that their difference is given by:

$$|\arg(z_p + s_d) - \arg(z_p + s_d)| < 2°$$

Step 7. Choose the pole of the controller:

$$z_p = \xi\omega_n + \omega_n\sqrt{1-\xi^2}\tan\left(-\varphi \pm \frac{\pi}{2}\right) \tag{5.52}$$

with $\angle G_c(s_d) = \theta_c = 180° - \angle G_p(s_d)$ and $\varphi = \angle(s_d + a)$. Usually, the pole location is such that the total angle at the desired root location is $180°$ (angle condition). Therefore, it is on the compensated root locus.

Step 8. Evaluate the value of the system gain K at the desired root location using the magnitude criteria, as:

$$K = \frac{1}{|G_c(s_d)G_p(s_d).|} \tag{5.53}$$

then calculate the error constant by taking the value of K obtained previously as the new system, then deriving the value of the constant errors for a process $G(s)$ such that:

- for step input: $K_p - \lim_{s\to 0} D(s)G(s)$ and for a given $e_{ss} - \frac{1}{1+K_p}$
- for ramp input: $K_v = \lim_{s\to 0} sD(s)G(s)$ and for a given $e_{ss} = \frac{1}{K_v}$

(repeat if necessary).

Step 9. Construct a closed-loop process step response. If the process dynamics performance requirements are not fulfilled, perform a controller tuning procedure by selecting the controller zero and pole (z_0, z_p) such that the ratio $\frac{z_p}{z_0}$ moves near zero in the case of a lead compensator. In another term, move the zero far away from the origin. This should result in: (i) the zero being placed in the neighborhood of the closed-loop ω_n; that is, $z_0 = \omega_n$; and (ii) the pole being set in the range of $z_p = 5z_0$ to $20z_0$. In the case of lag compensation, where the stationary condition requirements are not met, a tuning procedure would consist into moving the zero of the controller further to the right.

Step 10. Obtain the discrete approximation, $D(z)$, of the derived continuous controller, $G_c(s)$, using the pole-zero mapping technique or another discrete approximation technique.

Lead or Lag Discrete Controller Design Approach Consider a discrete controller in the form of

$$D(z) = \frac{M(z)}{E(z)} = K\frac{z - z_0}{z - z_p} \quad \text{with } z_0 > z_p \tag{5.54}$$

Step 1. Translate the provided design specifications (settling time t_s, percentage overshoot, PO (%), etc.) into natural frequency ω_n and damping ratio ξ in order to derive the desired discrete root loci of the compensated system such that for a given sampling period:

$$z_{desired} = e^{-\omega_n(\xi-j\sqrt{1-\xi^2})T} = e^{-\omega_n\xi T}\angle \pm (\omega_n T\sqrt{1-\xi^2}) \tag{5.55}$$

Step 2. Derive the uncompensated transfer function in the discrete-time domain, $G_p(z)$.

Step 3. Draw the root locus of the uncompensated system and evaluate whether it passes through the desired root loci z_d specified previously. Compute $\angle G_p(z_d)$.

Step 4. If a lead compensator is required, the zero value of the controller, z_o, could be added directly under the dominant pole along the real axis or to the left of the first two poles (near the origin for the higher-order system). If a lag compensator is required: (i) derive $\angle D(z) = \theta = \pm n180° - G_p(z)$, where n is an odd number; (ii) set z_0 such that $\angle D(z_0) > D(z)$ and does not impair closed-loop stability; and (iii) check the value by computing the phase angle of the compensator zero at the desired closed-loop pole location.

Step 5. Derive the pole of the controller, z_p, using the angle condition (i.e. the sum of the angle contributions of poles and zeros from this dominant pole, which should be equal to 180°). Here, the distance from the desired closed-loop pole to the compensator pole is the required phase angle for the compensator pole.

Step 6. Evaluate the value of the system gain K at the desired pole using the magnitude condition and derive the error constant. Recall that the steady-state characteristics are given by the position error $K_p = \lim_{k \to 1} zD(z)G(z)$ and the velocity error constant $K_v = \lim_{k \to 1} \frac{(z-1)(1+D(z)G(z))}{Tz}$. Thus, for a step input, they are given by $e_{ss} = \frac{1}{1+K_p}$, while for a ramp input, they are given by $e_{ss} = \frac{1}{K_v}$.

Step 7. If the design requirements are not fulfilled, the controller tuning procedure should be performed by choosing the controller pole and zero (z_p, z_0) so that the ratio $\frac{z_p}{z_o}$ moves near zero. This will result in the zero moving far away from the origin. It should be noted that the condition of $(|z_p| < |z_o| < 1)$ should be checked to ensure system stability.

Example 5.7   *Controller Design Using Root Locus*
Consider a discrete process model given by:

$$G_p(z) = \frac{K(z+1)}{(z-1)^2}$$

with $T = \sqrt{2}$ s. Consider a digital controller algorithm given by:

$$D(z) = \frac{(z-a)}{(z-b)}$$

The uncompensated root locus plot is depicted in Figure 5.9(a).

It is desired to cancel one pole of the open-loop process transfer function $G_p(z)$ that lies on the positive real axis of the z-plane by $(z-a)$. Select $(z-b)$ such that the compensated process gives a set of complex roots at a desired pole within the unit circle on the z-plane. By setting $a = 1$ to cancel one pole and $b = 0.2$, then:

$$D(z)G_p(z) = \frac{K(z+1)}{(z-1)(z-0.2)}$$

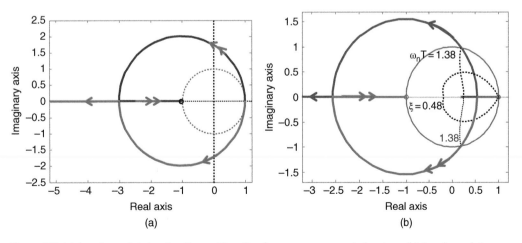

Figure 5.9 (a) A z-plane plot showing the root location for an uncompensated system. (b) A z-plane plot showing the root location for a compensated system.

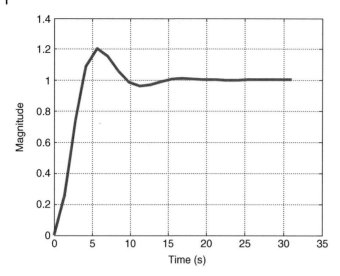

Figure 5.10 Step response of a compensated system.

Derive the entry point using $F(\sigma)$:

$$1 + D(z)G_p(z) = F(\sigma) = \frac{(\sigma - 1)(\sigma - 0.2) + K(\sigma + 1)}{(\sigma - 1)(\sigma - 0.2)}$$

Then, $\frac{dF(\sigma)}{d\sigma} = 0$, so that $\sigma = 2.5$ as an entry point.

From the root locus plot depicted in Figure 5.9(b), it is possible to select $K = 0.25$ in order to achieve an overshoot of 20% and a settling time of 8.5 s, as shown in Figure 5.10. Different results can be achieved if $a = 1$ and $b = -0.98$.

$$D(z)G_p(z) = \frac{K(z + 1)}{(z - 1)(z + 0.98)} \approx \frac{K}{(z - 1)}$$

Design Procedures for Lead-Lag Compensation

The lead-lag compensator is similar to the PID controller. Its discrete controller design is in the form of:

$$G_p(z) = \frac{M(s)}{E(s)} = K_c \frac{(s + z_1)(s + z_2)}{(s + p_1)(s + p_2)} = K_c \frac{\left(s + \frac{1}{T_1}\right)\left(s + \frac{1}{T_2}\right)}{\left(s + \frac{\gamma}{T_1}\right)\left(s + \frac{1}{\beta T_2}\right)} \tag{5.56}$$

with $|p_1| > |z_1| > |z_2| > |p_2| > 0$ and K_c being the controller gain, and where $\gamma > 1$ and $\beta > 1$. The continuous time controller-based design approach procedure consists of deriving $K_c, \gamma, \beta, T_1, T_2$. For the case of $\gamma \neq \beta$, the following design steps here can be taken:

Step 1. From given performance specifications (settling time t_s, percentage overshoot PO (%) ...), derive the corresponding natural frequency (ω_n) and damping ratio (ξ) desired root of closed-loop poles through:

$$s_d = -\xi\omega_n \pm j\omega_n\sqrt{1 - 2\xi^2}$$

Step 2. Derive the uncompensated open-loop transfer function in the discrete-time domain, $G_p(s)$.

Step 3. Using the uncompensated open-loop transfer function $G_p(s)$, determine the angle deficiency φ, if the dominant closed-loop poles are closed to the desired root's location. The phase lead portion of the lag-lead compensator must contribute to this angle φ, such that:

$$\varphi = \angle G_p(s_d)$$

Step 4. Choose T_2 at a sufficiently large value so that the magnitude of the lag portion is almost equal to unity, then choose T_2 such that:

$$\frac{\left|s_{d1} + \frac{1}{T_2}\right|}{\left|s_{d1} + \frac{1}{\beta T_2}\right|} = 1 \tag{5.57}$$

where s_{d1} is one of the dominant closed-loop poles. Choose the values of T_1 and γ such that:

$$\angle \frac{s_{d1} + \frac{1}{T_1}}{s_{d1} + \frac{\gamma}{T_1}} = \varphi \tag{5.58}$$

then determine K_c such that:

$$\left|K_c \frac{s_{d1} + \frac{1}{T_1}}{s_{d1} + \frac{\gamma}{T_1}} G_p(s_{d1})\right| = \varphi \tag{5.59}$$

Step 5. If the velocity error constant K_v is specified, determine the value of β that satisfies the requirement for K_v so that:

$$K_C = \lim_{s \leftarrow 0} sK_c \frac{\beta}{\gamma} G_p(s) \tag{5.60}$$

Remarks on Compensation Controller Properties

The discrete pole placement (or root locus technique) is not recommended for designing PID controllers because the dynamic response is not uniquely determined by the closed-loop pole locations (the value of transfer function zeros is also important). Hence, post-performance evaluation of controller candidates is highly recommended to select the best ones among those suitable.

Common Guidelines for PID Family Control Loops

In the case of *flow and liquid pressure control*, such a process displays the combined characteristics of fast response and no time delay, as well as small high-frequency noise. It is suitable to use a PI controller with intermediate controller gain; that is, $0.5 < K_c < 0.7$ and $0.2 < \tau_I < 0.3$ min. In the case of *liquid level control*, such an integrating process displays an output signal with noise due to the liquid turbulence flow. It is suitable to use a high-gain PI controller by increasing the gain K_c, so the oscillation magnitude will decrease. Otherwise, it may be advisable to use either an averaging control in order to dampen the fluctuation, such as a PI controller with $K_c = < \frac{100\%}{\Delta h}$ and $\tau_I = \frac{4V}{K_c Q_{max}} (\Delta h \cong \min(h_{max} - h_{1p}, h_{1p} - h_{min})$ or an error-squared controller with refined tuning. In the case of *pressure control*, such a process displays some fast and self-regulating dynamic characteristics. Here, it is suitable to use a PI controller with small integral gain (large reset time). In the case of *temperature control*, such a process displays a slow response characteristic of time delay. The PID controller can be used for faster system response. In the case of *control of composition process control*, such a process displays some characteristics similar to temperature control, usually with larger noise and more time delay. However, the derivative gain should be small to limit its action. The temperature and composition controls are usually implemented using advanced control strategies such as model predictive control (MPC), due the difficulty of control.

5.2.5 Feedforward Control Methods

Feedforward control design techniques provide the effective elimination of undesirable responses to various types of process inputs, such as: (i) process disturbance; (ii) reduction of error in process output through command input; and (iii) modification of the manipulated input. Usually, there are three types of feedforward control designs: (i) the feedforward control of disturbance; (ii) open-loop feedforward; and (iii) modified command input feedforward. The principle is to feed the disturbance or the command input signal forward through a feedforward controller and to combine it with manipulated input generated by a feedback controller. This method is especially applicable when the process disturbance is measurable (i.e. dominant process disturbance [loads] signals), meaning its effect on the process is predictable and the controller design is feasible. In the case of command input feedforward, this consists in precomputing a command input in order to have the system track a command trajectory. Figure 5.11 depicts the typical block diagram of the process with a command input feedforward loop.

5.2.5.1 Command Input Feedforward Control Algorithm
During the execution of a command input trajectory, the position errors are large, so the feedback control actions are considerable. The actual velocity and acceleration (including the actuator force) may therefore be much larger than planned. This leads to undesired and even dangerous deviations from the planned trajectory, and damage to the actuator or even large positioning errors when arriving at the desired endpoint; thus, the dynamical state of the controlled system is not settled. In order to circumvent this, a command input feedforward should be added, with the objective of modifying the manipulated input to the process. This is expected to improve dynamics performance by reducing or eliminating errors between the command input and the process output. Figure 5.11 illustrates a generic configuration of command input feedforward control. Thus, the objective command input feedforward is:

$$\frac{c(z)}{r(z)} = 1 \tag{5.61}$$

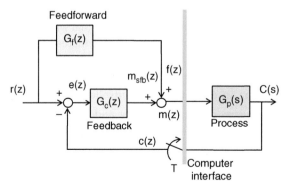

Figure 5.11 Generic command input feedforward block diagram.

It can be observed from Figure 5.11 that the system transfer function is:

$$\frac{c(z)}{r(z)} = \left(1 + \frac{G_f(z)}{G_C(z)}\right)\left(\frac{G_C(z)G_P(z)}{1 + G_C(z)G_P(z)}\right) \tag{5.62}$$

Therefore, $G_f(z)$ could be chosen such that:

$$G_f(z) = \frac{1}{G_P(z)} \tag{5.63}$$

Thus, as expected:

$$r(z) \approx c(z) \tag{5.64}$$

and the effect of command feedforward control is $\begin{cases} e_{ss} = 0 \\ m_{ss} = 0 \end{cases}$.

$G_f(z)$ is similar to implementing the open-loop control of the process output $c(z)$. The feed-forward control output is not affected by $c(z)$, thus the feedforward control scheme is essentially an open-loop control. The accuracy of control depends on the accuracy of the process model. Any inaccuracy is compensated for by the closed-loop controller $G_c(z)$. The implementation of command feedforward control can be simplified using a modified block diagram, as shown in Figure 5.12.

The modified command input to the feedback controller is given by:

$$r'(z) = \left(1 + \frac{G_f(z)}{G_c(z)}\right)r(z) \tag{5.65}$$

This requires that the entire sequence of command inputs $r(z)$ is known in advance. Hence, $r'(z)$ rather than $r(z)$ is known in advance, allowing better elimination of system errors. This is done by precalculating and storing the entire modified command sequence $r'(z)$, rather than using $r(z)$ as the command input for each sample period. This approach speeds up computations, as it eliminates the need to calculate $f(z)$ at each sample period. Regardless of feedback controller $G_c(z)$, a feedforward controller $G_f(z)$ is designed in order to eliminate systems errors. Hence, a lower performance feedback controller would not need feedforward control. Large step inputs should be avoided when using feedforward control because overly large manipulations may be needed to achieve the required response.

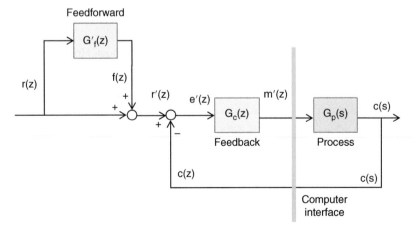

Figure 5.12 Modified command input feedforward block diagram.

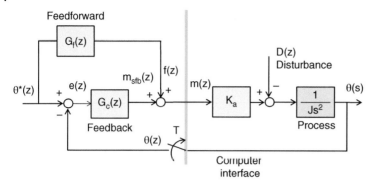

Figure 5.13 Block diagram of a DC motor with feedforward control.

Example 5.8 Consider a simple motor given by a double integration process, as shown in Figure 5.13, such that:

$$G_p(s) = \frac{1}{Js^2}$$

The discrete process transfer function is given by:

$$\frac{\theta(z)}{m(z)} = G_p(z) = \frac{K_a}{J}\frac{T^2}{2}\frac{z^{-1} + z^{-2}}{(1 - z^{-1})^2}$$

Thus:

$$f(z) = G_f(z)\theta^*(z) = \frac{1}{G_p(z)}\theta^*(z) = \frac{m(z)}{\theta(z)}\theta^*(z) = \theta^*(z)\frac{J}{K_a}\frac{2}{T^2}\frac{(1 - z^{-1})^2}{z^{-1} + z^{-2}}$$

$$= \theta^*(z)\frac{J}{K_a}\frac{2}{T^2}\frac{(z - 1)^2}{(z + 1)}$$

By inversion, this equation yields:

$$\frac{f(n) + f(n-1)}{2} = \frac{J}{K_a}\frac{\dfrac{\theta^*(n) - \theta^*(n-1)}{T} - \dfrac{\theta^*(n-1) - \theta^*(n-2)}{T}}{T} = \frac{J}{K_a}\frac{\omega^*(n) - \omega^*(n-1)}{T}$$

meaning that the average input torque is the average acceleration torque. Alternatively:

$$f(n) = -f(n-1) + \frac{J}{K_a}\frac{\dfrac{\theta^*(n+1) - \theta^*(n)}{T} - \dfrac{\theta^*(n+1) - \theta^*(n)}{T}}{T/2}$$

This is realizable if the command trajectory θ^* is feasible and known a priori. However, the negative sign on the value of $f(n-1)$ causes some force oscillations if the average acceleration has a zero mean. A position ramp command input trajectory for the closed-loop motor system is given by:

$$\theta^*(z) = \theta_{slope}\frac{Tz^{-1}}{(1 - z^{-1})^2} = \omega^*\frac{Tz^{-1}}{(1 - z^{-1})^2}$$

thus:

$$f(z) = \theta^*(z)\frac{J}{K_a}\frac{2}{T^2}\frac{(z - 1)^2}{(z + 1)} = \omega^*\frac{Tz^{-1}}{(1 - z^{-1})^2}\frac{J}{K_a}\frac{2}{T^2}\frac{(1 - z^{-1})^2}{(z^{-1} + z^{-2})} = \omega^*\frac{J}{K_a}\frac{2}{T}\frac{1}{(1 + z^{-1})}$$

With the initial impulse, this results in an average input torque given by:

$$\frac{f(n) + f(n-1)}{2} = 0 + \omega^*\frac{J}{K_a}\frac{1}{T}\delta(0)$$

This command feedforward can be rewritten as:

$$f(n) = -f(n-1) + \omega^* \frac{J}{K_a} \frac{2}{T} \delta(0)$$

If:

$$\begin{cases} m_{sfb\_ss} = 0 \\ e_{ss} = 0 \end{cases}$$

then:

$$m(z) = f(z)$$

From the previous equation:

$$\frac{\theta(z)}{m(z)} = \frac{\theta(z)}{f(z)} = \frac{K_a}{J} \frac{T^2}{2} \frac{(z^{-1} + z^{-2})}{(1 - z^{-1})^2}$$

Therefore, the response can be derived from the algorithm given by:

$$\theta(n) = 2\theta(n-1) - \theta(n-2) + \frac{K_a}{J} \frac{T^2}{2} (f(n-1) + f(n-2)) :$$

Table 5.3 summarizes the discrete control and process values for the first four sequences. Figure 5.14(a) and (b) illustrates the manipulated inputs and the desired and achieved positions.

It should be noticed that the manipulated input is a square signal wave of amplitude $(\frac{\hat{J}\omega^*}{\hat{K}_a} \frac{2}{T})$. However, this is unacceptable, even though it produces the correct mean value of command input. Indeed, the velocity profile achieves the correct mean velocity and the correct position at each sample instant, but with poor tracking. In order to achieve better position tracking where tracking errors are tolerated and a trajectory is fixed, the following procedure should be adopted: (i) solve for continuous feedforward command input $f(s)$ and (ii) develop a discrete-time nearest equivalent input sequence form, $f(k)$. An impulse will be required to achieve this possible physical input (nearest approximation). The control strategy consists of deriving the required velocity of the zero error, providing the correct value during the first interval, and turning off this input after the initial pulse. This means that if $\omega^* = \frac{\Delta\theta^*}{T}$ is the

Table 5.3 Derived sequence values.

n	$f(n-2)$	$f(n-1)$	$f(n)$	$\theta(n-2)$	$\theta(n-1)$	$\theta(n)$
0	0	0	$\frac{\hat{J}}{\hat{K}_a} \frac{\omega^*}{T}$	0	0	0
1	0	$\frac{\hat{J}}{\hat{K}_a} \frac{\omega^*}{T}$	$-\frac{\hat{J}}{\hat{K}_a} \frac{\omega^*}{T}$	0	0	$\frac{\omega^* T}{2}$
2	$\frac{\hat{J}}{\hat{K}_a} \frac{\omega^*}{T}$	$-\frac{\hat{J}}{\hat{K}_a} \frac{\omega^*}{T}$	$\frac{\hat{J}}{\hat{K}_a} \frac{\omega^*}{T}$	0	$\frac{\omega^* T}{2}$	$\omega^* T$
3	$-\frac{2\hat{J}}{\hat{K}_a} \frac{\omega^*}{T}$	$\frac{\hat{J}}{\hat{K}_a} \frac{\omega^*}{T}$	$-\frac{\hat{J}}{\hat{K}_a} \frac{\omega^*}{T}$	$\frac{\omega^* T}{2}$	$\omega^* T$	$\frac{3\omega^* T}{2}$
4	$\frac{\hat{J}}{\hat{K}_a} \frac{\omega^*}{T}$	$-\frac{\hat{J}}{\hat{K}_a} \frac{\omega^*}{T}$	$\frac{\hat{J}}{\hat{K}_a} \frac{\omega^*}{T}$	$\omega^* T$	$\frac{3\omega^* T}{2}$	$2\omega^* T$

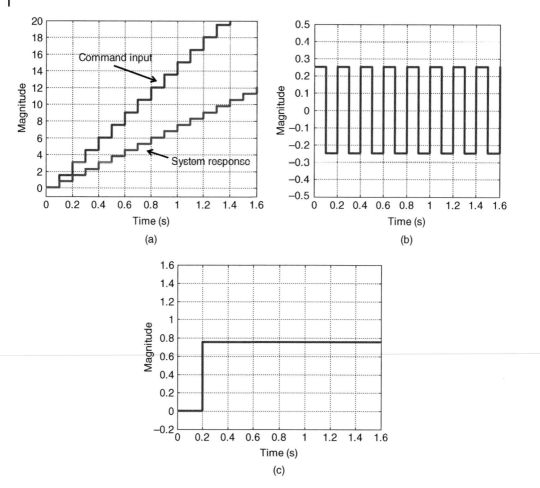

Figure 5.14 (a) Desired ramp command. (b) Required manipulated input. (c) Velocity achieved.

desired ramp rate, the command input feedforward torque $f(k)$ will be:

$$f(0)\frac{K_a}{J}T = \frac{\Delta\theta^*}{T}$$

The system velocity response is illustrated in Figure 5.14(c).
This is equivalent to:

$$\begin{cases} f(0) = \dfrac{J}{K_a}\dfrac{\Delta\theta^*}{T^2} \\ f(k>0) = 0 \end{cases}$$

In this case, it should be noted that a position error has occurred and is never corrected. However, it results in an approximation of the velocity trajectory with near-zero tracking error. This is based on the plant response equation given by:

$$\theta(k) = 2\theta(k-1) - \theta(k-2) + \frac{K_a}{J}\frac{T^2}{2}(f(k-1)+f(k-2))$$

Table 5.4 Sequence values of system with input feedforward.

n	$f(n-2)$	$f(n-1)$	$f(n)$	$\theta(n-2)$	$\theta(n-1)$	$\theta(n)$
0	0	0	$\dfrac{\hat{J}}{\hat{K}_a}\dfrac{\Delta\theta^*}{T^2}$	0	0	0
1	0	$\dfrac{\hat{J}}{\hat{K}_a}\dfrac{\Delta\theta^*}{T^2}$	0	0	0	$\dfrac{\Delta\theta^*}{2}$
2	$\dfrac{\hat{J}}{\hat{K}_a}\dfrac{\Delta\theta^*}{T^2}$	0	0	0	$\dfrac{\Delta\theta^*}{2}$	$\dfrac{3\Delta\theta^*}{2}$
3	0	0	0	$\dfrac{\Delta\theta^*}{2}$	$\dfrac{3\Delta\theta^*}{2}$	$\dfrac{5\Delta\theta^*}{2}$
4	0	0	0	$\dfrac{3\Delta\theta^*}{2}$	$\dfrac{5\Delta\theta^*}{2}$	$\dfrac{7\Delta\theta^*}{2}$

Table 5.4 summarizes the discrete control and process values for the first four sequences. Figure 5.15(a)–(c) illustrates the manipulated inputs, desired and achieved position, and even velocity. It could be noticed that the zero-velocity error is achieved, but the position error is never corrected; rather, it is finite. Another procedure to achieve realizable tracking with zero steady-state error is to use a command input trajectory with no instantaneous energy changes. For an error at time $k = 0$, and where the position value at the second sampling period is $[2\omega^* T]$, the velocity at $k = 1T$ is $[\omega_1]$ for the first two samples, such that:

$$0.5\omega_1 T + 0.5(\omega_1 - \omega^*)T + \omega^* T = 2\omega^* T$$

Thus, an optimal solution would be to have $\omega_1 = 1.5\omega^*$. Meaning that:

$$\begin{cases} f(0) = \dfrac{3}{2}\dfrac{J}{K_a}\dfrac{\Delta\theta^*}{T^2} \\ f(k > 0) = -\dfrac{1}{2}\dfrac{J}{K_a}\dfrac{\Delta\theta^*}{T^2} \end{cases}$$

Thus, the process response equation is given by:

$$\theta(k) = 2\theta(k-1) - \theta(k-2) + \frac{K_a}{J}\frac{T^2}{2}(f(k-1) + f(k-2))$$

Table 5.5 summarizes the discrete control and process values for the first four sequences. It should be noticed that the zero-velocity error and the position error are never corrected, but are reduced. In the case of a parabolic input, the command trajectory is given by:

$$\theta^*(t) = \frac{1}{2}\omega^*(t)t^2$$

such that:

$$\theta^*(z) = \omega^* \frac{T(z^{-1} + z^{-2})}{(1 - z^{-1})^3}$$

Recall that:

$$f(z) = \theta^*(z)\frac{m(z)}{\theta(z)} = \omega^* \frac{T^2}{2}\frac{(z^{-1} + z^{-2})}{(1 - z^{-1})^3}\frac{J}{K_a}\frac{2}{T^2}\frac{(1 - z^{-1})^2}{(z^{-1} + z^{-2})} = \omega^* \frac{J}{K_a}\frac{1}{(1 - z^{-1})}$$

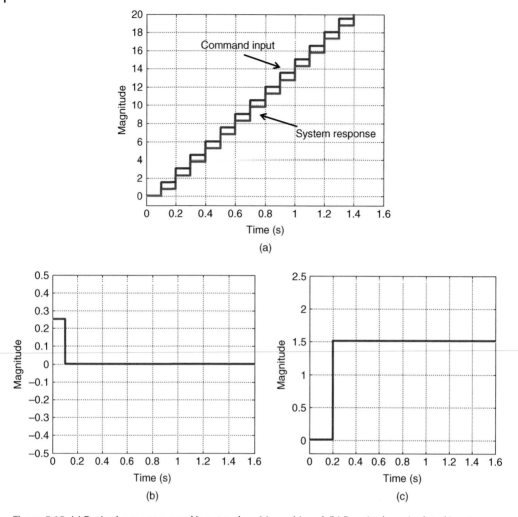

Figure 5.15 (a) Desired ramp command input and position achieved. (b) Required manipulated input. (c) Velocity achieved.

Table 5.5 Sequence values of system with adjusted input feedforward.

n	$f(n-2)$	$f(n-1)$	$f(n)$	$\theta(n-2)$	$\theta(n-1)$	$\theta(n)$
0	0	0	$\dfrac{3}{2}\dfrac{\hat{J}}{\hat{K}_a}\dfrac{\Delta\theta^*}{T^2}$	0	0	0
1	0	$\dfrac{3}{2}\dfrac{\hat{J}}{\hat{K}_a}\dfrac{\Delta\theta^*}{T^2}$	$-\dfrac{1}{2}\dfrac{\hat{J}}{\hat{K}_a}\dfrac{\Delta\theta^*}{T^2}$	0	0	$\dfrac{3\Delta\theta^*}{4}$
2	$\dfrac{3}{2}\dfrac{\hat{J}}{\hat{K}_a}\dfrac{\Delta\theta^*}{T^2}$	$-\dfrac{1}{2}\dfrac{\hat{J}}{\hat{K}_a}\dfrac{\Delta\theta^*}{T^2}$	0	0	$\dfrac{3\Delta\theta^*}{4}$	$2\Delta\theta^*$
3	$-\dfrac{1}{2}\dfrac{\hat{J}}{\hat{K}_a}\dfrac{\Delta\theta^*}{T^2}$	0	0	$\dfrac{3\Delta\theta^*}{4}$	$2\Delta\theta^*$	$3\Delta\theta^*$
4	0	0	0	$2\Delta\theta^*$	$3\Delta\theta^*$	$4\Delta\theta^*$

Inverting the result with the initial impulse yields:

$$f(k) = f(k-1) + \omega^* \frac{J}{K_a} \delta(0)$$

It could be noted that a step manipulation input is feasible. In conclusion, note that the algebraic inverse solution should not be applied without checking the feasibility of the manipulated input result. The generalized procedure to obtain a solution for command input feedforward could be as follows:

1) solve for the continuous command input feedforward $f(t)$;
2) check if $f(t)$ is unfeasible (when several impulse inputs are required);
3) check if the command trajectory θ^* is fixed;
4) then develop the discrete-time equivalent $f(k)$ as the input sequence form, with some tracking errors;
5) otherwise, create a feasible trajectory $\theta^*(k)$.

Example 5.9 Consider the robot arm modeled by block diagram in Figure 5.16. Assume it is desired to eliminate the following errors that occur as a result of the command input $r(z)$. The error transfer function is given by:

$$\frac{e(z)}{r(z)} = \frac{1}{1 + 140 \frac{0.11 Tz^{-1}}{1 - z^{-1}}} = \frac{1 - z^{-1}}{1 - (1 - 15.4T)z}$$

For ramp input of $0.1\ \mathrm{m\,s^{-1}}$,

$$r(z) = \frac{0.1 Tz^{-1}}{(1 - z^{-1})^2} \delta(0)$$

the final value of the error is:

$$e_\infty = \lim_{z^{-1} \to 1} (1 - z^{-1}) \left[\frac{1 - z^{-1}}{1 - (1 - 15.4T)z^{-1}} \right] \frac{0.1 Tz^{-1}}{(1 - z^{-1})^2} = 6.5\ \mathrm{mm}$$

These errors can be eliminated by using feedforward control, as illustrated Figure 5.16. As such, the required feedforward controller is:

$$G_f(z) = \frac{1}{\frac{0.11 Tz^{-1}}{1 - z^{-1}}} = \frac{1 - z^{-1}}{0.11 Tz^{-1}}$$

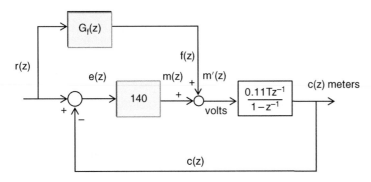

Figure 5.16 Control feedforward of a robot arm.

The corresponding feedforward control equation is:

$$f(k-1) = \frac{1}{0.11T}(r(k) - r(k-1))$$

or:

$$f(k) = \frac{1}{0.11T}(r(k+1) - r(k))$$

Generally, this result is unrealizable because it requires future knowledge of $r(k+1)$; however, in many applications of robot arms, motion commands are stored in a database. In this case, the entire sequence of commands $r(k)$ is known and the foregoing modified feedforward control is used, such that the modified command sequence $r'(k)$ can be precalculated as:

$$r'(z) = \left(1 + \frac{\frac{1-z^{-1}}{0.11Tz^{-1}}}{140}\right)r(z)$$

or:

$$r'(k) = \frac{1}{15.4T}r(k+1) + \left(1 - \frac{1}{15.4T}\right)r(k)$$

In other applications of robot arms, the command $r(k)$ must be generated in real time using outputs from the task-oriented sensors on the robot. In this case, a modified feedforward equation that is realizable must be used. Considering that

$$r(k+1) - r(k) \approx r(k) - r(k-1)$$

for most of the time the robot is operating, a logical choice is:

$$f(k) = \frac{1}{0.11T}r(k) - \frac{1}{0.11T}r(k-1)$$

and:

$$G_f(z) = \frac{1 - z^{-1}}{0.11T}$$

With this feedforward control transfer function, the error transfer function can be determined to be:

$$\frac{e(z)}{r(z)} = \frac{(1 - z^{-1})^2}{1 - (1 - 15.4T)z}$$

and the final value of the error is:

$$e_\infty = \lim_{z^{-1} \to 1}(1 - z^{-1})\left[\frac{(1 - z^{-1})^2}{1 - (1 - 15.4T)z^{-1}}\right]\frac{0.1Tz^{-1}}{(1 - z^{-1})^2} = 0$$

Therefore, the modified command feedforward controller is realizable and eliminates the steady-state errors under discussion.

5.2.5.2 Disturbance Feedforward Control Algorithm

The concept of feedforward control of disturbances is illustrated in Figure 5.17. If a disturbance input $d(z)$ can be sampled by a computing unit, the feedforward controller transfer function $G_f(z)$ can be used to eliminate or at least reduce the effects of the disturbances on the system. If $F(z)$ is the discrete-time disturbance input, then process input $m'(z)$ is given by:

$$m'(z) = m(z) + G_f(z)d(z) + F(z)d(z) \tag{5.66}$$

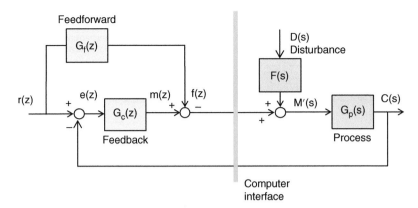

Figure 5.17 Feedforward control of disturbances.

Clearly, the most desirable situation is to design the feedforward control such that:

$$f(z) = -F(z)d(z) \tag{5.67}$$

meaning:

$$G_f(z) = -F(z) \tag{5.68}$$

In this case, then:

$$m(z) = m'(z) \tag{5.69}$$

Although Equation (5.67) is analytically straightforward, practical shortcomings arise. The computer must treat the disturbance signal as a sampled variable, even though the disturbance changes continuously and the process responds continuously to changing disturbances. Calculations using $G_f(z)$ are based on a discrete-time analysis, treating the disturbance as a series of step inputs entering the process at sample times. Ideally, the feedforward control algorithm, $G_f(z)$, should completely compensate for the disturbance. But, in real situations, this goal can only be approximated. The principle of feedforward control action must therefore be combined with feedback control, as illustrated in Figure 5.17. If the feedforward control is successful, the portion of the disturbance input remaining uncompensated for is small, eliminating the disturbance as the major factor affecting system performance.

5.3 Modern Control Topologies

5.3.1 State Feedback PID Control Algorithms

Industrial DC motor-control algorithms are designed around control topologies such as: (i) an average velocity loop cascaded with a position loop; (ii) multiple state-variable closed loops in parallel; or (iii) cascades of current, speed, and position control loops. In general, state-variable controllers allow separation of command tracking issues from disturbance rejection issues. For example, in order to provide the zero-tracking error, a state feedforward approach could be added to the state feedback controller. Hence, in addition to improving the tracking, this approach could maintain the stability and disturbance rejection capability. Advanced control strategies allow estimation of the speed and torque using a reduced-order observer method. However, their implementation through a torque feedforward control scheme for the dynamic load changes poses torque accuracy and torque dynamics predictability challenges.

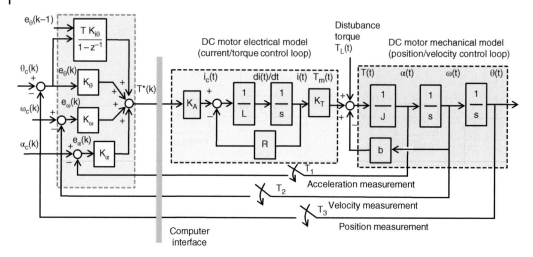

Figure 5.18 Typical PID-based acceleration, velocity, and position state feedback.

Typical *PID-based acceleration, velocity, and position state feedback* is shown in Figure 5.18. Such DC motor control topology offers the ability to maintain the zero steady-state error to a step change in command input, while allowing undesirable speed overshoot from a sudden change in load torque. In addition, the process dynamics is highly sensitive to state controller gains. Also, the full command vector (i.e. position, velocity, and acceleration) offers better command tracking, but for most industrial applications the acceleration is not available or is costly. However, in this topology, the torque or the current control effort is bounded either by (i) limiting the motor operation when the current exceeds a threshold value or (ii) combining the speed control loop with the current loop in order to regulate the motor speed around a pre-set value. Despite providing good stiffness to any disturbance inputs, this state-variable PID control topology is error-driven because it has some dynamic tracking errors (i.e. the integral-of-position-error state).

Some other state-variable loops are used in industrial control applications, such as a PID position controller into which velocity and position commands are fed. In that regard, see Figure 5.19 for an equivalent PID control topology that avoids this drawback. Here, while using

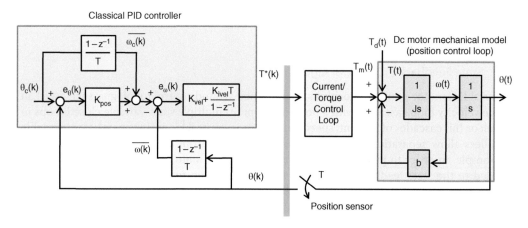

Figure 5.19 Classical (industrial) PID motion controller with velocity and position loops.

$K_A \approx R/K_t$, this PID topology has a cascade PI controller for the velocity loop and a P controller in the position loop. In addition, the velocity input signal is used as a feedforward term that feeds for the command ahead of the integral, making the system more responsive to command changes. Also, this topology requires several synchronized sensors (position, velocity, and acceleration).

Considering that an optical encoder is used as a position-sensing device, while the velocity-sensing device is replaced by an average velocity (denoted $avg\_\omega_c(z)$, derived by numerically differentiating the position measurement), with $a = b/J$ and using the table of equivalence, the discrete-time model of the DC motor mechanical component will result in:

$$G_p(z) = \frac{1}{b}\left[\frac{(aT - 1 + e^{-aT})z + (1 - e^{-aT} - aTe^{-aT})}{a(z - 1)(z - e^{-aT})} \right]$$

(5.70a)

The closed loop yields:

$$\frac{\theta(z)}{\theta_c(z)} = \frac{\left(K_{pos} + \frac{1 - z^{-1}}{T}\right)\left(K_{vel} + \frac{K_{ivel}T}{1 - z^{-1}}\right)G_p(z)}{1 + \left(K_{pos} + \frac{1 - z^{-1}}{T}\right)^2\left(K_{vel} + \frac{K_{ivel}T}{1 - z^{-1}}\right)G_p(z)}$$

(5.70b)

Furthermore, the pulse width modulation (PWM), which is commonly used to control the voltage, generates signals that cause ripples in the motor torque. Therefore, in addition to the conventional PI controller, a proportional term has been proposed in the motor feedback path to eliminate these ripples, as illustrated in Figure 5.20. All of these state feedback PID control topologies aim to be expressed as the state-variable motion controller of a DC motor. Hence, in the examination of those control topologies, it could be demonstrated that they are mathematically identical with the gain relationships. In order to improve the tracking and better reject the disturbance, the control topology with the zero-error tracking state variable combined with the command feedforward, illustrated in Figure 5.21, should be used. From the generic PID control topology, the desired torque command input yields:

$$T^*(z) = M(z) = K_c(\theta_c(z) - \theta(z))\left[1 + K_iT + \frac{K_d}{T}(1 - z^{-1})\right]$$

(5.71)

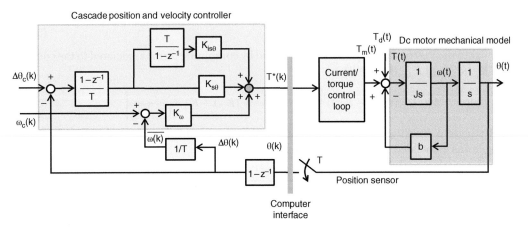

Figure 5.20 PID-based state-position cascade control topology for a DC motor.

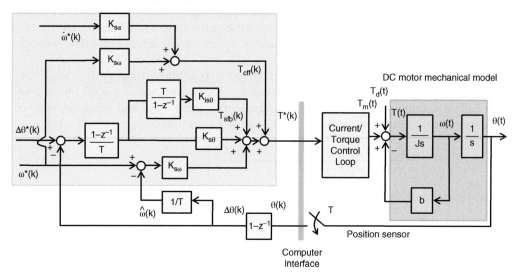

Figure 5.21 Zero-error tracking state-variable motion controller and command feedforward.

In the case of the topology for state feedback with acceleration loop and integrated position loop, this yields:

$$T^*(z) = M(z) = K_c(\theta_c(z) - \theta(z)) \left[\frac{K_{i\theta}T}{(1-z^{-1})} + K_\theta + K_\omega \frac{(1-z^{-1})}{T} + K_\alpha \frac{(1-z^{-1})^2}{T^2} \right] \quad (5.72)$$

Assuming that:

$$avg\_\omega_c(z) = \frac{(1-z^{-1})}{T}\theta_c(z)$$

$$\overline{\alpha_c(z)} = \frac{(1-z^{-1})}{T}avg\_\omega_c(z) = \frac{(1-z^{-1})^2}{T}\theta_c(z)$$

with the classical PID control topology:

$$T^*(z) = M(z) = K_c(\theta_c(z) - \theta(z)) \left[K_{vel}\frac{(1-z^{-1})}{T} + (K_{pos}K_{vel} + K_{ivel}) + K_{pos}K_{ivel}\frac{T}{(1-z^{-1})} \right]$$
$$(5.73)$$

Similar to the PID cascade control topology, and by combining terms associated to the control gains K_ω, $K_{s\theta}$, and K_{iso}, this yields:

$$T^*(z) = M(z) = (\theta^*(z) - \theta(z)) \left[K_\omega \frac{(1-z^{-1})}{T} + (K_{s\theta}) + K_{si\theta}\frac{T}{(1-z^{-1})} \right] \quad (5.74)$$

With the zero-tracking error and command feedforward control topology, as well as by combining terms associated to the control gains, this yields:

$$T^*(z) = M(z) + T_{cff}(z) = M(z) + \left[J\frac{(1-z^{-1})^2}{T^2} + b\frac{(1-z^{-1})}{T} \right]\theta_c(z) \quad (5.75)$$

In Figure 5.21, it should be recalled that $K_{s\omega} \approx b$ and $K_{s\alpha} \approx J$. By comparing controller terms, it could be noted that the physical interpretations of state feedback gains of the physical variables are:

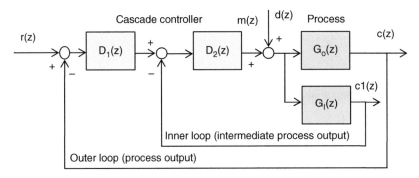

Figure 5.22 Block diagram for generic cascade control topology.

for $K_{s\omega}$ in N.m.sec.rad^{-1}

$\qquad K_{s\omega} = K_{vel}$; that is, the damping gain

for $K_{s\theta}$ in N.m.rad^{-1}

$\qquad K_{s\theta} = K_{pos}K_{vel} + K_{ivel}$; that is, the stiffness gain

for $K_{is\theta}$ in N.m.sec.$^{-1}$.rad^{-1}

$\qquad K_{is\theta} = K_{pos}K_{ivel}$; that is, the integrated stiffness gain

Cascade Controller Design Procedure The cascade control topology requires a process with some intermediate system variables. As illustrated in Figure 5.22, the disturbances affect the interme-diate process output as well as the primary process output. Indeed, some processes often occur in series, with a load distribution between the process variables from the inner and the outer loops. Hence, the cascade control topology aims to limit the effect of the disturbances feeding the secondary variable on the primary output, as well as the secondary process gain variations due to changes in operating points.

The cascade control topology consists of a set of master-slave controllers, where the outer one (master) closes the loop with the main process control variable and the inner one (slave) closes the loop between the master output as a set point and the measurement of the inner process variable. Thus, their resulting discrete transfer functions can be derived individually for each process output. The intermediate output is related to the manipulated input through the inner-loop discrete transfer function, $G_I(z)$, whereas the complete process can be represented by the outer-loop discrete transfer function, $G_o(z)$. Here, the inner process is usually faster than the outer one. Hence, the intermediate variable allows the attenuation of the inner loads or the inner process nonlinearities.

The design of cascade controllers is similar to the case with single-loop controller design. This means that first the slave controller then the master controller are successively designed. There are two controller design specifications: (i) for the command input response; and (ii) for the disturbance input response. Thus, from Figure 5.22, the overall discrete transfer function for the closed loop becomes:

$$\frac{c(z)}{r(z)} = \frac{G_0(z)D_1(z)D_2(z)}{1 + G_0(z)D_1(z)D_2(z) + D_2(z)G_I(z)} \qquad (5.76)$$

Using the linear superposition principle, the outer-loop controller is given by:

$$D_1(z) = \frac{G_e(z)G_1(z)}{G_0(z)[1 - G_e(z)] - G_d(z)} \qquad (5.77)$$

while the inner-loop controller is given by:

$$D_2(z) = \frac{G_0(z)[1 - G_e(z)] - G_d(z)}{G_d(z)G_I(z)} \tag{5.78}$$

with $G_e(z)$ being the expected transfer function derived from the controller design specifications and $G_d(z)$ being the disturbance discrete closed-loop transfer function, given by:

$$G_d(z) = \frac{G_0(z)}{1 + D_1(z)D_2(z)G_0(z) + D_2(z)G_1(z)} \tag{5.79}$$

Example 5.10 *Cross-Coupled Discrete-Time Model of a DC Motor*

Consider the DC motor block diagram illustrated by Figure 5.23. Here, it is desired to develop the corresponding discrete time DC motor model through block diagram manipulation. From the DC motor-control topology, two control loops can be identified: the inner current loop and the outer velocity loop. Figure 5.24(a) and (b) illustrates these equivalent loops in block diagram form.

Through the discretization of the closed-loop transfer function, it is possible to derive these discrete transfer functions such that:

$$\frac{\omega(z)}{V_a(z)} = \frac{b_{\omega 1}z^{-1} + b_{\omega 2}z^{-2}}{1 - b_1 z^{-1} - b_2 z^{-2}}$$

$$\frac{I(z)}{V_a(z)} = \frac{b_{i1}z^{-1} + b_{i2}z^{-2}}{1 - b_1 z^{-1} - b_2 z^{-2}}$$

$$\frac{\theta(z)}{V_a(z)} = \frac{b_{\theta_1}z^{-1} + b_{\theta_2}z^{-2} + b_{\theta_3}z^{-3}}{1 - b_1 z^{-1} - b_2 z^{-2}}$$

Figure 5.25(a) depicts the equivalent open-loop transfer function, while Figure 5.25(b) depicts the equivalent open-loop and cascade transfer function. An equivalent block diagram with

Figure 5.23 DC motor model block diagram.

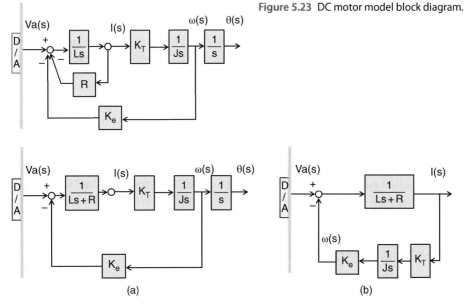

(a)   (b)

Figure 5.24 (a) Velocity control loop. (b) Current loop.

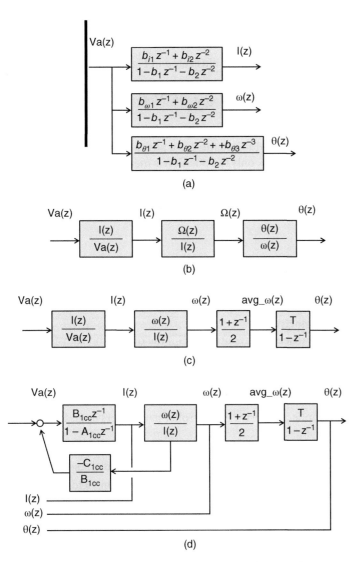

Figure 5.25 (a) Equivalent input voltage to current, velocity, and position transfer functions. (b) Decomposed block diagram of input voltage to motor shaft position. (c) Block diagram of input voltage to motor shaft position derived from average velocity. (d) Cascade current, velocity, and position control loops.

average velocity is presented in Figure 5.25(c), with the relationship between the intermediate variables given by:

$$\frac{\omega(z)}{I(z)} = \frac{\omega(z)V_a(z)}{V_a I(z)} = \frac{b_{\omega1}z^{-1} + b_{\omega2}z^{-2}}{b_{i1}z^{-1} + b_{i2}z^{-2}}$$

This could be rewritten in the form:

$$i(k) = A_{cc}i(k-1) + B_{1cc}V_a(k-1) + C_{1cc}\omega(k-1)$$

This is equivalent to:

$$I(z) = \frac{B_{1cc}z^{-1}}{1 - A_{cc}}\left(V_a(z) - \frac{C_{1cc}}{B_{1cc}}\omega(z)\right)$$

$$I(z) = \frac{B_{1cc}\,z^{-1}}{1 - A_{cc}}\left(V_a(z) - \frac{C_{1cc}}{B_{1cc}}\,\omega(z)\right)$$

The equivalent cascade closed-loop block diagram is illustrated in Figure 5.25(d).

Example 5.11 *Cascade Control and State Control of a DC Motor Using Only Position Feedback*

Consider a cascade controller for DC motor motion where there is a PI controller velocity inner loop and a P controller position outer loop. In addition, there is a position command input and a feedforward command input on velocity. Here, only the position is measured and sampled, as illustrated in Figure 5.26.

From Figure 5.26, the position model is given by:

$$G_p(z) = \frac{K_A}{J}\frac{T^2}{2}\frac{(z^{-1} + z^{-2})}{(1 - z^{-1})^2}$$

and the average velocity model results in:

$$G_{\bar{v}}(z) = G_p(z)\frac{(1 - z^{-1})}{T} = \frac{K_A}{J}\frac{T^2}{2}\frac{(z^{-1} + z^{-2})}{(1 - z^{-1})}$$

Step 1. Tune the inner PI loop on the velocity.

Using the estimated velocity and the discrete PI controller G_{pi}, the open-loop transfer function results in:

$$G_{pi}(z)G_{\bar{v}}(z) = K_c\frac{z - \delta_c}{z - 1}\frac{K_A}{J}\frac{T^2}{2}\frac{(z^2 + z)}{z(z - 1)}$$

Then, the same procedure can be applied to derive PI controller parameters δ_c and K_c, based on the expected dynamic characteristics.

Step 2. Tune outer loop on position.

Consider the average velocity control closed loop $G_{\bar{v}}(z) = \frac{avg\_\omega(z)}{avg\_\omega_c(z)}$. From the resulting position loop of cascaded control of DC motor, the open-loop transfer function yields:

$$G_{pi}(z)G_{\bar{v}}(z) = K_c\frac{z - \delta_c}{z - 1}\frac{K_A}{J}\frac{T^2}{2}\frac{(z^2 + z)}{z(z - 1)}$$

$$K_{pos}\frac{avg\_\omega(z)}{avg\_\omega_c(z)}\frac{T}{1 - z^{-1}}$$

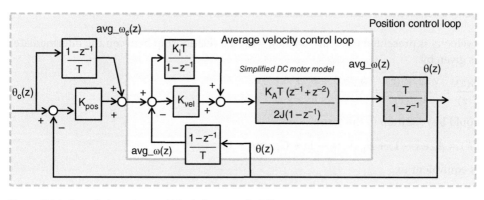

Figure 5.26 Cascade-based control block diagram of a DC motor.

Hence, the position closed-loop transfer function yields:

$$\frac{\theta(z)}{\theta_c(z)} = \frac{avg\_\omega(z)\frac{T}{1-z^{-1}} + K_{pos}\frac{avg\_\omega(z)}{avg\_\omega_c(z)}\frac{T}{1-z^{-1}}}{1 + K_{pos}\frac{avg\_\omega(z)}{avg\_\omega_c(z)}\frac{T}{1-z^{-1}}}$$

Here, using the magnitude condition and based on the controller design requirements, it is possible to derive K_{pos}.

Example 5.12 *State-Variable DC Motor Motion Controller Design with Separated Poles*
Consider a DC motor with state-variable controller ($K_{i\theta}$, $K_{s\theta}$, K_ω) for a position, velocity, and average velocity controller. Consider the DC motor parameters given by $J = 0.01$ Kg.m^2 and $K_a = 100$ N.m V^{-1}, as illustrated in Figure 5.27. It is required to derive the damping K_ω, stiffness $K_{s\theta}$, and integrated stiffness $K_{i\theta}$ gains, which will achieve four real closed-loop roots that are approximately equivalent to break frequencies of 0.2, 0.06, 0.017, and 0.012*500 Hz. There are two possible sample frequencies, $f_s = 500$ Hz and $f_s = 1000$ Hz.

From block diagram manipulation, Figure 5.28 illustrates the equivalent closed-loop cascade control of a DC motor. Hence, from Figure 5.28, the position model is given by:

$$G_p(z) = \frac{K_A}{J}\frac{T^2}{2}\frac{(z^{-1} + z^{-2})}{(1 - z^{-1})^2}$$

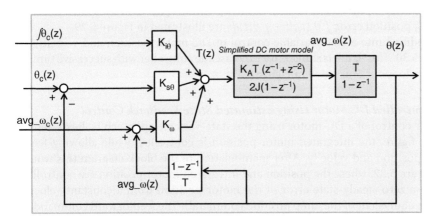

Figure 5.27 Active PID-based state feedback control block diagram of a DC motor.

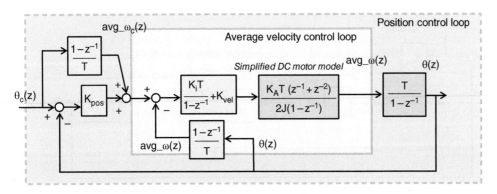

Figure 5.28 Modified position loop of cascaded control of a DC motor.

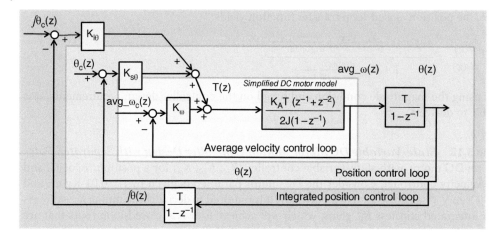

Figure 5.29 Average velocity and position-based state feedback control of a DC motor.

recalling that the model of the DC motor average velocity is given by:

$$G_{\bar{v}}(z) = G_p(z)\frac{(1-z^{-1})}{T} = \frac{K_A}{J}\frac{T^2}{2}\frac{(z^{-1}+z^{-2})}{(1-z^{-1})}$$

State command velocity inputs on all controlled states position $\theta_c(t)$, velocity $\omega_c(t)$, and accumulated (integrated) position error $\int \theta_c(t)dt - \int \theta(t)dt$ are illustrated in Figure 5.29.

This could be modified into a state-variable control of DC motor position and velocity, as illustrated in Figure 5.30. This figure is similar to a PID cascade controller with successive tuning velocity and position loops as illustrated in Figure 5.29.

Example 5.13 *Controlled DC Motor Using Estimated State Feedback Control*
A block diagram for control of a DC motor using the state-variable approach is illustrated in Figure 5.31. In this figure, the integrated motor position is controlled, while allowing feed-forward control of position and velocity. After manipulation of the block diagram this would be equivalent to Figure 5.32, where the position and integrated motor position are controlled in order to achieve a zero steady-state error of the motor position from a constant velocity command input (i.e. equivalent of the ramp position command). The feedforward command of

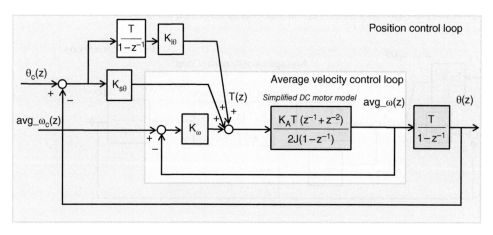

Figure 5.30 Modified average velocity and position-based state feedback control of a DC motor.

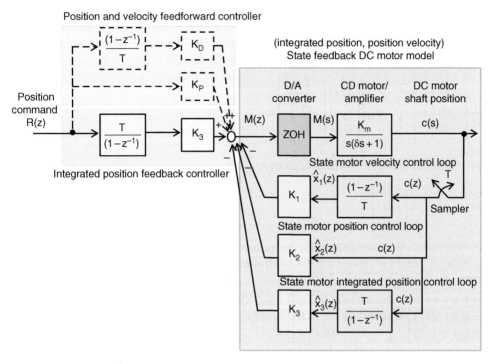

Figure 5.31 Position-based PID state feedback control of a DC motor.

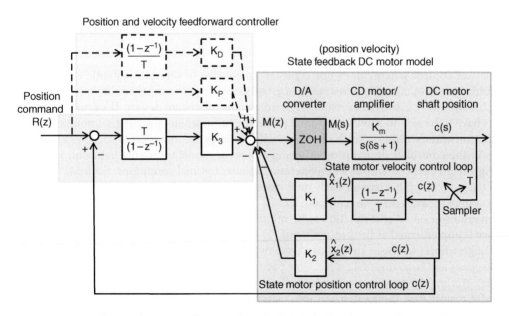

Figure 5.32 Modified block diagram of position-based PID state feedback control of a DC motor.

the position and velocity is added. Here, the state-variable estimates $\hat{x}_1(nT), \hat{x}_2(nT)$, and $\hat{x}_3(nT)$ of the state variables $x_1(nT), x_2(nT)$, and $x_3(nT)$ are derived, corresponding to the motor velocity, motor position, and integrated motor position, respectively. Thus, a state estimator could be used instead of the moving averaging motor position, $c(t)$. Therefore, the state equation for the DC motor system is:

$$\frac{d}{dt}\begin{bmatrix} x_1(t) \\ x_2(t) \\ x_3(t) \end{bmatrix} = \begin{bmatrix} \frac{-1}{\delta} & 0 & 0 \\ 1 & 0 & 0 \\ 0 & 1 & 0 \end{bmatrix}\begin{bmatrix} x_1(t) \\ x_2(t) \\ x_3(t) \end{bmatrix} + \begin{bmatrix} \frac{K_m}{\delta} \\ 0 \\ 0 \end{bmatrix} m(t)$$

and the DC motor position equation is given as:

$$c(t) = \begin{bmatrix} 0 & 1 & 0 \end{bmatrix}\begin{bmatrix} x_1(t) \\ x_2(t) \\ x_3(t) \end{bmatrix}$$

with the time constant of the DC motor being 0.0028 s and the gain being 63 rad s^{-1} V^{-1}. A sample period of 0.002 s is chosen for the controller. The discrete state equations can thus be derived as:

$$\begin{bmatrix} x_1((n+1)T) \\ x_2((n+1)T) \\ x_3((n+1)T) \end{bmatrix} = \begin{bmatrix} 0.48954 & 0 & 0 \\ 0.00143 & 1 & 0 \\ 0 & 0.002 & 1 \end{bmatrix}\begin{bmatrix} x_1(nT) \\ x_2(nT) \\ x_3(nT) \end{bmatrix} + \begin{bmatrix} 32.15888 \\ 0.03596 \\ 0.00003 \end{bmatrix} m(nT)$$

and the DC motor position equation is equivalent to:

$$c(nT) = \begin{bmatrix} 0 & 1 & 0 \end{bmatrix}\begin{bmatrix} x_1(nT) \\ x_2(nT) \\ x_3(nT) \end{bmatrix}$$

5.3.2 MPC Algorithms

The classical control scheme does not consider knowledge of physical systemic constraints (e.g. motor speed limits), safety constraints (e.g. temperature/pressure sensor limits), or performance constraints (e.g. system overshoot) regarding control system design. Thus, such control systems have to be able to show suboptimal plant operation when operating set points are far from the constraints. Otherwise, the MPC paradigm allows the inclusion of such system constraints in the controller design process, using a process model to predict behavior, without violating any process constraints when operating at near optimal conditions. Such MPC algorithms use a receding horizon at each time instant (prediction horizon), during which they receive system measurements to compute a finite horizon control sequence based on the process constraint-oriented cost function to be minimized. The first part of the resulting optimal sequence is implemented as the process input.

MPC algorithms have been applied to a large number of MIMO systems with changing control objectives and large time delays (i.e. slow pace with sampling period from a few seconds up to hours). Among the processes are industrial petrochemical and pulp and paper applications. Successful results ease their extension into classical systems such as car engine control and voltage regulation in electric power generation and distribution networks. However, the speed of execution of their optimal control algorithm (MPC) bounds the finite prediction horizon chosen.

A typical MPC algorithm involves solving an optimization problem at each time instant, k (prediction horizon), by predicting the system performance using a process model (linear or nonlinear, discrete or continuous time) to optimize future system inputs. This is done by computing each time a new control vector, u_k, is fed to the system over the prediction horizon, while considering process constraints such as the amplitude and rate of change of the input signal. This optimization method is done recursively to reduce the effect of uncertainty on the process. Key components of the MPC algorithm include:

1) A *control objective*, J_k (cost function), being the difference between the future system outputs $y_{k+1/L}$ and the future system reference input $r_{k+1/L}$, based on control inputs u_k being costly over a prediction horizon L. This objective function aims to be minimized with respect to the future control vectors, $u_{k+1/L}$. Hence, the typical cost function formulation is given by:

$$J_k = \sum_{=1}^{L} (y_{k+i} - r_{k+i})^T Q(y_{k+i} - r_{k+i}) + u_{k+i}^T P_i u_{k+i-1} + \Delta u_{k+i-1}^T R \Delta u_{k+i-1} \tag{5.80}$$

with $Q_i \in \mathbb{R}^{m \times m}$, $P_i \in \mathbb{R}^{r \times r}$, $R_i \in \mathbb{R}^{r \times r}$ being process-dependent symmetric and positive semidefinite diagonal weighting matrices specified by the designer. The weighting matrices are almost always chosen as constant (time-invariant) over the prediction horizon L (i.e. $Q_1 = Q_2 = \ldots = Q_L$, $P_1 = P_2 = \ldots = P_L$ and $R_1 = R_2 = \ldots = R_L$), while at steady-state, $y = r$. In the matrix form, the cost function formulation could be given as:

$$J_k = (y_{k+1/L} - r_{k+1/L})^T Q(y_{k+1/L} - r_{k+1/L}) + u^T_{k+1/L} P u_{k/L} + \Delta u^T_{k/L} R \Delta u_{k/L} \tag{5.81}$$

Hence, MPC can be considered to be a linear quadratic programming (QP) problem where:

$$Q \in \mathbb{R}^{Lm \times Lm}, P \in \mathbb{R}^{lr \times Lr}, R \in \mathbb{R}^{Lr \times Lr},$$

such that the control problem yields:

$$u_{k/L} = \arg \min_{u_{k/L}} J_k(u_{k/L}) \tag{5.82}$$

2) Definitions of some *process operating constraints*, such as the signal input amplitude constraints that might be active during transients (e.g. the electrovalve saturation) and the constraints on the rate of change of the input signal. These constraints could be classified as either hard (i.e. they must be satisfied at all times or the problem is unfeasible) or soft (i.e. they can be violated to avoid problem infeasibility, such as by removing some constraints until the optimization is feasible). They are used as avoidance case strategies (e.g. anti-windup strategies of the dynamic controller or constrained control law). A typical definition of these constraints includes absolute $\min u_{value} \leq u_k \leq \max u_{value}$ or rate $\min \Delta u_{value} \leq u_k - u_{k-1} \leq \max \Delta u_{value}$. There are also state constraints (active during transients, such as aircraft stall speed, which is active during steady-state, and process economic constraints) that can be defined as $\min x_{value} \leq x_k \leq \max x_{value}$.

3) A *system predictive model* (PM) derived from physics-based or stochastic modeling methods, used to describe the relationship between future control inputs over the output behavior of the system. This is usually done in state space matrix form, so that, for a deterministic linear dynamic process model in state space form, it is given by:

$$u_{k+1} = Ax_k + Bu_k$$
$$y_k = Dx_k + Eu_k \tag{5.83}$$

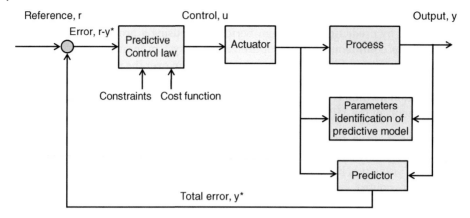

Figure 5.33 Generic MPC control topology.

A PM could be set in the standard form:

$$y_{k+1/L} = F_L u_{k/L} + p_L \tag{5.84}$$

with L being the prediction horizon, $F_L \in \mathbb{R}^{Lm}$ the constant parameters matrix derived from the process model, and $p_L \in \mathbb{R}^{Lm}$ the vector dependent of a number of inputs and outputs older than the time k, as well as model parameters such that:

$$F_L = [\theta_L B H_L^d]$$

$$p_L = O_L A x_k \tag{5.85}$$

The block diagram depicted in Figure 5.33 summarizes some key components of the MPC control process, such as the model parameters estimation using the least-square technique, the predictor, and the MPC algorithm.

Example 5.14 *Blood Glucose Control of Diabetes Type 1 Patient*
Regulation of blood glucose in Type 1 diabetics using an MPC based on parametric insulin-glucose metabolism dynamics modeling is given by the following differential equations:

$$\frac{dI(t)}{dt} = -\gamma_1 I(t) + \beta \max[G(t) - \theta_I, 0] + D_I(t)$$

$$\frac{dN(t)}{dt} = -\gamma_N N(t) + \alpha \max[\theta_N - G(t), 0]$$

$$\frac{dX(t)}{dt} = -p_2 X(t) + p_3 I(t)$$

$$\frac{dG_I(t)}{dt} = -p_1 G_1(t) + X(t)G(t)$$

$$\frac{dG_N(t)}{dt} = -p_4 G_N(t) + p_5 N(t)$$

$$G(t) = G_b(t) + G_I(t) + G_N(t) + D_G(t)$$

where $I(t)$ is the plasma insulin, $N(t)$ the plasma glucagon, and $X(t)$ the insulin action. Based on Sorensen's model, $G_b(t)$ is the deviation of blood glucose from the basal value, $G_I(t)$ is the deviation of the blood glucose from the glucagon action, G is the concentration of blood glucose, D_I is the intravenous insulin input, and $D_G(t)$ is the glucose disturbance (e.g. stress, diet, physical constraints, etc.). The glucose disturbance, D_G, is defined as the effect of three daily meals (breakfast, lunch, and dinner), represented by three gamma-like functions of the form $(t^3 \cdot e^{-0.4t})$, for a 24-hour period, with values of $(40, 50, 60 \text{ mg dl}^{-1})$ for about 80 minutes randomly but uniformly distributed within specified periods. In addition, considering that the daily activity of glucose can be approximated as a sinusoidal signal with a period of 8 hours, an amplitude of 10 mg/dl, and uniformly distributed phase with $[0, \pi]$, then, besides the description of the insulin-glucose metabolism, in order to apply MPC, it is required to predict accurately the future values of blood glucose concentration. At every discrete-time instant, the prediction of the future values of the glucose disturbance using the AR (auto regressive) PM can be given by:

$$D_G(n) = D\alpha + \omega(n)$$

$$D = [D_G(n-1)D_G(n-2) \dots D_G(n-K)]$$

$$\alpha = [\alpha_1 \alpha_2 \dots \alpha_k]$$

where α is the vector of coefficients of the AR model, $\omega(n)$ is an unknown independent process sequence, and K is the order of the AR model. At a discrete-time instant n, the prediction task consists of the estimation of the coefficient vector α. This should allow us to estimate the future values of glucose concentration. Using the least-square estimation method, the coefficients can be derived from:

$$\alpha = [A^T A]^{-1} A^T b$$

where A is a matrix representing all D vectors and b is the vectors of disturbance. Hence, from a model capturing insulin-glucose metabolism dynamics for all the past insulin inputs and the estimation of the future values of glucose disturbance, the MPC at every time instant n enables us to determine the control input value $U(t)$ by minimizing the following cost function:

$$J(n) = [G(n+p/n) - R^T] \cdot \Gamma D_G(n-2) \dots D_G(n-K)]$$

where $G(n+p/n)$ is the vector of predicted glucose values over a future horizon of p, R is the target output value (set equal to G_b), and Γ_y is a diagonal matrix of the weighting coefficients with respect to the near-future predictions. These measures are here to prevent hypoglycemia.

5.3.3 Open-Loop Position Control Using Stepping Motors

In cases where the process behavior is known in advance and free of disturbance, the open-loop control algorithm can be used. The stepping motor uses the open-loop control algorithm because the number of steps per revolution is known. The open-loop control of a stepping motor consists of generating a pulse sequence train that will excite the stator phases. This causes the motor to step after each pulse, so the pulse train command input frequency defines the motor velocity as well as the motor torque. Another utilization of the pulse is to define the direction of the motion (clockwise or counterclockwise). Hence, the command generates two input conditions (direction and enable step) that allow rotation of stator phases. The generation of the pulse trains to be fed into the stepping motor controller is based on the acceleration, deceleration, and frequency limits of the stepping motor. Several methods are used to generate pulse trains, including (i) pulse rate multiplication; and (ii) digital differential analysis (DDA). Figure 5.34 depicts the components of the open-loop control of stepping motor, showing: (i) the command generator; (ii) the command translator; and (iii) the stepping motor.

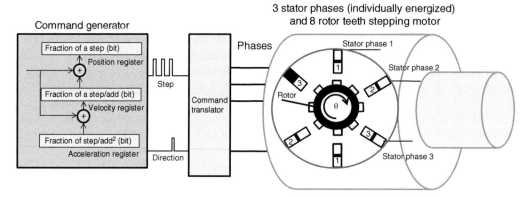

Figure 5.34 Generic structure of a stepping motor structure.

The command generator has three registers (position, velocity, and acceleration), generating a clock-synchronized pulse train.

After the fraction of the position register accumulates to 1, a pulse command signal is generated and a step pulse is sent to the motor command translator. Once the pulse train is generated, the pulse signal activates or deactivates the stator phase, which in turn forces the rotor teeth to be aligned with the stator position. The translator converts this pulse train into a voltage signal for each stator phase of the stepping motor, which is either energized or de-energized depending on the fed signal level. The motor step size is given by:

$$stepsize = \frac{spacing\ of\ rotor\ teeth}{number\ of\ stator\ phases} \tag{5.86}$$

The rotor teeth align when the activated phase is completed. Indeed, when the fraction of position register accumulates to 1, the command generator sends a step pulse to the motor command translator.

Example 5.15 It is desired to achieve a constant velocity of 750 rpm with a stepping motor resolution of 200 steps per rev and a computer clock 10 000 Hz (adds s^{-1}). As such, the velocity is given by:

$$\frac{d\theta(t)}{dt} = \frac{750 rev}{min \cdot \dfrac{min}{60\ s \underbrace{\dfrac{200\ steps}{rev}}_{} = \dfrac{2500\ step}{s}}}$$

which is equivalent to:

$$V = \frac{2500\ steps}{s} \bigg/ \; s\,10\ 000\ adds\,\frac{1\ step}{4\ adds}$$

Therefore, with a 12-bit register, the velocity register has a command input signal given as $0.0100000000_2 = 1/4$. All register values are captured in Figure 5.35.

Based on this constant velocity value, the position register is set to 0.01000000000 after the first addition of the value from the velocity register, but there is no step signal, $\theta(T)$, to be sent to the translator. A second addition of the velocity register value V in the position register P is given by:

$$0.01000000000 + 0.01000000000 = 0.10000000000$$

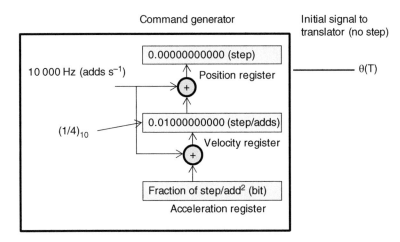

Figure 5.35 Initial *P* and *V* register values.

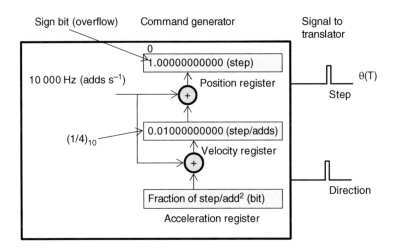

Figure 5.36 *P* and *V* register values after four additions.

but there is no step signal, $\theta(T)$, to be sent. A third operation of addition of the velocity register value results in the position register given by:

$$0.10000000000 + 0.01000000000 = 0.11000000000$$

but there is no step signal, $\theta(T)$, to be sent. A fourth operation of addition of the velocity register value, results in the position register given by:

$$0.11000000000 + 0.01000000000 = 1.00000000000$$

There is a step signal, $\theta(T)$ given by a sign bit or the first bit to be sent to the command translator, and the position register is reset as illustrated in Figure 5.36.

Typical applications of open-loop stepping motors include 2D and 3D printer motion control, as well as the command generation of multiple-axis robot joint angles by derivation of individual robot joint sequence angles such as those shown in Table 5.6.

Table 5.6 Typical recorded motion data of a multi-axis robot.

n	P_1	P_2	V_1	V_2	A_1	A_2
0						
1						
2						
3						

5.4 Induction Motor Controller Design

For all electrical motors, a key challenge has been to change the motor speed without changing the torque. This can be done (i) by changing the motor design (i.e. by changing the number of pair poles); (ii) by varying the magnitude of the supplied voltage; or (iii) by varying the frequency of the voltage supplied. Induction control strategies have either (i) used the scalar control volts per hertz fed into the inverter driver or (ii) used vector control flux or current oriented with the voltage fed into the inverter drive. Vector control implementation requires the identification of the field flux of the motor, while scalar control requires maintaining a constant magnitude and phasing relationship between the voltage and the frequency.

5.4.1 Scalar Control (*V/f* Control)

The scalar *V/f* controller design technique consists of feeding the induction motor windings with a three-phase sinusoidal voltage whose amplitudes are proportional to the frequency. The term *scalar* denotes that the magnitudes of the input variables (frequency and voltage) are controlled. Indeed, with scalar control, by varying the voltage and frequency parameters simultaneously, the alignment of both their magnitude and phase signal components is ensured, such that:

$$\frac{V_{RMS}}{\omega_e} = \frac{V_{RMS}}{4.44 N f \varepsilon} = \psi_s \tag{5.87}$$

with V_{RMS} being the induced voltage in the motor, f the frequency of supplied voltage, N the number of turns, ψ_s the magnetic flux linkage in the stator, and ε the constant of the motor coil. This is usually done without any feedback devices (i.e. an open-loop induction motor speed control with low dynamics requirements). This scalar *V/f* control method is used to provide speed variation but does not handle transient conditions. Hence, this control strategy is valid only during the steady state. Based on the constant volts per hertz principle, scalar control is applied over a sinusoidal steady-state model of the induction motor (i.e. the magnitude of the stator voltage). For a constant ratio, the stator flux remains constant and the motor torque depends only on the slip frequency. The motor speed equation of the induction motor is given by:

$$T_{air\_gap} = \frac{3}{2\omega_n} I_r \frac{R_r}{s} \tag{5.88}$$

with T_{air_gap} being the motor torque in the air gap, ω_n being the mechanical angular speed, I_r being the rotor current, R_r being the resistance rotor, and s being the slip. This method is suitable for applications without position control requirements or high accuracy of speed control. There are two types of scalar control schemes.

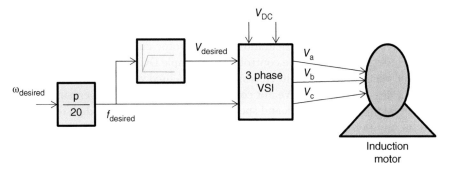

Figure 5.37 Generic open-loop scalar control of an induction motor.

5.4.1.1   Open-Loop Scalar Control

Open-loop scalar control is only suitable for applications requiring speed control. The corresponding desired torque can be found at the minimum of the steady-state operating condition. A typical induction motor open-loop scalar control block diagram is illustrated in Figure 5.37.

5.4.1.2   Closed-Loop Scalar Control (Slip Control)

Here, the speed and the motor torque are concurrently controlled using a PID controller. The uncontrolled magnetic flux is a disadvantage of this scalar controller, due to its inability to capture or modify the transient induction motor behavior. A schematic is illustrated in Figure 5.38.

5.4.2   Vector Control

The difficulty in controlling the speed of an induction motor is due to the lack of a linear relationship between the motor current and the resulting torque. Vector control aims to decouple the complex stator current into two components (i_{ds}, i_{qs}): one for motor flux level or direct current $i_{ds}(t)$, the other for motor torque or quadrature current $i_{qs}(t)$ in the motor stator. This eases the decoupled control of the flux and electromagnetic torque. It is achieved by transforming a sinusoidal quantity into a constant value quantity at the same frequency, giving a new reference frame. In other words, it translates the coordinates from the fixed-reference stator frame to the synchronous rotating frame.

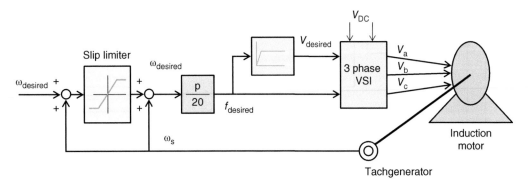

Figure 5.38 Generic closed-loop scalar control of an induction motor.

Typical vector control schemes on an asynchronous motor, which are based on decoupling strategies, include: (i) stator flux field orientation; (ii) rotor flux field orientation; (iii) air-gap flux field orientation; (iv) stator current orientation; and (v) rotor current orientation. This decoupling allows the induction motor to be controlled as a simple DC motor. There are three commonly encountered vector control architectures: (i) field-oriented vector control; (ii) direct torque control; and (iii) direct self-control. Vector control provides performance enhancement in terms of starting, braking, speed reversal, and speed variation. Furthermore, digital control of an induction motor allows the estimation of electrical machine parameters in a sensorless control scheme through rotor speed measurement, slip frequency estimation, flux estimation, and flux vector control.

5.4.2.1 Direct Torque Control

Vector control provides performance enhancement in terms of starting, braking, speed reversal, and speed variation. Furthermore, digital control of the induction motor requires the estimation of electric machine parameters in a sensorless control scheme through rotor speed measurement, slip frequency estimation, flux estimation, or flux vector control.

In the case of the *direct torque control*, the design procedure consists of: (i) the modeling of current and a voltage space vector electrical machine; (ii) the transformation of three-phase speed and a time-dependent system into a two-coordinate time-invariant system; and (iii) an effective PWM pattern, through generation of pulse signals to be supplied to the electrical machine with desired phase voltage. Figure 5.39 depicts a block diagram of direct vector control of the induction motor.

The vector control equation is given by:

$$T_e = \frac{3}{2}\psi_r i_r \sin \varphi = -\frac{3}{2}\overline{\omega}_r \times \overline{i}_r \qquad (5.89)$$

with T_e being the electromagnetic torque of the motor, ψ_r the magnetic flux in the rotor, and i_r the current in the motor. Using the gamma model of the induction motor, the uncoupled

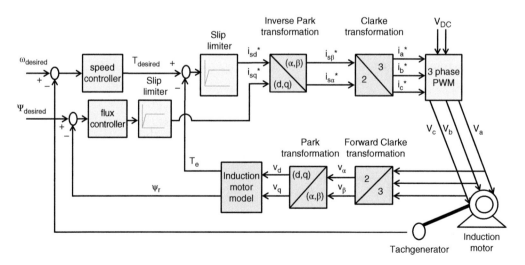

Figure 5.39 Basic structure of indirect oriented vector direct control without flux calculation.

Figure 5.40 Gamma model of an induction motor.

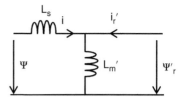

equation is:

$$
\begin{cases}
\dfrac{di_{sd}(t)}{dt} = \dfrac{1}{L_{as}}\left(v_{sq} - i_{sq}R_s - \dfrac{d\psi_{desired}}{dt}\right) + \psi_k i_{sd} \\[2ex]
\dfrac{di_{sd}(t)}{dt} = \dfrac{1}{L_{as}}(v_{sq} - i_{sq}R_s - \omega_k \psi_{desired}) - \psi_k i_{sd} \\[2ex]
\psi_{desired} + \tau\dfrac{d\psi_{desired}}{dt} = L_m i_{sd} \\[2ex]
\omega_k - \omega_s = \psi_s = \dfrac{R_s i_{sq}}{\psi_{desired}}
\end{cases}
\tag{5.90}
$$

Thus, the decoupled torque yields:

$$
T = \frac{3}{2}p\psi_{rd}i_{sd} \tag{5.91}
$$

and, decoupled, the speed is given by:

$$
\frac{d\omega_r(t)}{dt} = \frac{3}{2}\frac{p}{J}\psi_{rs}i_{sq} - \frac{m_l}{J} - \frac{B\omega(t)}{J} \tag{5.92}
$$

with i_{sd}, i_{sq} being the *(d,q)* component of the stator current; L_{as} the inductance of the stator without mutual inductance; v_{sd}, v_{sq} the *(d,q)* component of the stator voltage; ω_k, ω_r the synchronous and rotor angular velocity; ψ_{rs} the component of the rotor magnetic flux linkage; R and R_s the resistance of the stator and rotor; p the pole pair of the motor; B the friction constant; T_r the shaft load; and τ the rotor time constant. Figure 5.40 illustrates the gamma model type of induction motor.

Tests to derive motor parameters include the (i) direct current test; (ii) no load test; and (iii) locked rotor test. Thus, a vector control algorithm is as follows:

1) Measure the motor quantities, phase voltages, and stator phase currents i_a, i_b, i_c. Then, if only the values of i_a and i_b are measured, use $i_a + i_b + i_c = 0$.
2) Transform the set of these three-phase currents into a two-axis (α,β) system. From this conversion, the variables i_α and i_β result from the measured i_a, i_b, i_c values, where i_α and i_β are time-varying quadrature current values. This is known as the Clarke transformation.
3) Calculate the rotor flux space vector magnitude and position angle (its orientation).
4) Rotate the two-axis coordinate system so that it is in alignment with the rotor flux.
5) Use the transformation angle calculated at the last iteration of the control loop.
6) Transform the stator currents to the *d-q* coordinate system using a Park transformation. This conversion provides the i_d and i_q variables from i_α and i_β.
7a) Generate the flux error signal using the reference flux and the estimated flux value. Then, the stator current torque (i_{sq})- and flux (i_{sd})-producing components can be separately controlled.
7b) Compute the output stator voltage space vector using a decoupling block.

8) Use a PI controller to calculate i_d^* and then i_q^* using this error signal.
9a) Convert i_d^* and i_q^* to a set of three-phase currents to produce i_a^*, i_b^*, i_c^*. Here, the stator voltage space vector is transformed by an inverse Park transformation back from the d-q coordinate system to the two-phase system fixed with the stator.
9b) Using the space vector modulation, generate the output three-phase voltage.
10) Compare i_a^*, i_b^*, i_c^* and i_a, i_b, i_c using a hysteresis comparator to generate inverter gate signals.

Park transformation is a translation of the a, b, and c phase variables into d-q components of the rotor flux rotating field reference frame.

Inverse Park transformation is the conversion of the d-q component of the rotor flux rotating field reference frame into a, b, and c phase variables.

i_{qs} calculation consists of computing the rotor flux and torque reference to derive the stator current quadrature component and is required to produce the electromagnetic torque on the motor's shaft.

Flux PI is the estimated rotor flux; the reference rotor flux is the input to a proportional integrator. This flux is applied to the motor and is used to compute the stator current or direct component required to produce the required rotor flux in the machine.

A *current regulator* is a current controller with adjustable hysteresis bandwidth. A modulation technique is used in the current regulator. Hysteresis modulation is a feedback control method where the motor current tracks the reference current within a hysteresis band. The operation principle of hysteresis modulation involves the controller that generates a sinusoidal reference current of desired magnitude frequency, which is compared to the actual motor line current.

5.4.2.2 Speed Control of AC Motors

The speed of AC motors depends on three variables: (i) the fixed number of stator winding sets (known as poles), defining the base speed; (ii) the frequency of the AC voltage or current supply; and (iii) the load torque, defining the slip. Induction motor speed can be varied through:

1) a variable-frequency drive, enabling the conversion of a fixed-frequency AC source into a variable-frequency AC signal. This is implemented using an AC-to-DC converter such as a variable voltage inverter, PWM inverter, current controlled inverter, or even cycloconverter;
2) a variation of the rotor circuit resistance by inserting an AC chopper with a high frequency and a variable ON/OFF time;
3) variation of the stator voltage by using AC regulators;
4) injection of slip frequency EMFs into the rotor circuit;
5) a modification of the AC motor design by changing the number of stator coil poles.

Based on the converter type and the AC motor type, AC motor drives can be classified into four categories: (i) thyristor-based voltage drives, capable of bi-state force switching from gate signal conditions and suitable for controlling an induction motor; (ii) transistor-based volts per hertz (V/f) and vector PWM/cycloconverters, used to control an induction motor; (iii) transistor-based drives with natural commutating states (ON/OFF) capability, such as bridge commutated drives, suitable for controlling a synchronous motor; and (iv) PWM voltage vector drives, used to control a permanent magnet AC motor. In the frequency-control strategy using a PWM, the supply DC voltage and current are smoothed and fed into a PWM inverter to produce a controlled variable magnitude and frequency AC voltage signal. The cycloconverter (AC/AC) requires a large number of thyristors and is appropriate for induction motors and synchronous motors.

Table 5.7 Some motor drives and their corresponding electric motors.

Motor drive types	Drives	Motor types
Thyristor-based	Six-pulse bridge current	Synchronous motor
	Cycloconverter	Induction motor Synchronous motor
	Thyristor voltage controller	Induction motor
Transistor-based	Matrix converter	Induction motor
	Current PWM	Induction motor
	Voltage PWM	Induction motor Permanent magnet motor

In summary, AC motor-control methods consist of either (i) scalar control through the regulation of the command input voltage magnitude or (ii) vector control through the regulation of both command input magnitude and phase. The AC motor speed is controlled using the inverter, which can alter the AC power frequency based on the PWM. Typical industrial applications of AC motors include ball mills, rotary cement kilns, large crushers, mine winders, and mine hoists. Table 5.7 summarizes some motor drives used in electric motor control.

5.4.2.3 Speed Control of DC Motors
Industrial process requirements for wide-range speed variation of electrical motors can be fulfilled with DC motors using variable-speed drives. Speed drives have the ability to match speed and torque with process requirements independent of the load. For DC motors with negligible armature inductance and without load, the induced voltage $E_b(t)$ is given by:

$$E_b(t) = -R_a I_a(t) + V_a(t) = K\phi\omega(t) \tag{5.93}$$

This can be rewritten at steady-state or rated speed as:

$$\omega(t) = \frac{-R_a i_a(t) - V_a(t)}{K\phi} \tag{5.94}$$

Based on Equation (5.94), at the steady state, the speed $\omega(t)$ can be controlled by using any of the following methods: (i) armature voltage control (by varying $V_a(t)$); (ii) field control (by varying ϕ); or (iii) armature resistance control (by varying R_a). Hence, solid state circuits are used to modify DC motor torque-speed characteristics by adjusting either the armature voltage $V_a(t)$ (by using thyristor-based circuits in an AC-to-DC converter for an AC voltage supply or chopper circuits for a DC voltage supply), the field current $i_f(t)$ (by using thyristor-based circuits to supply additional field winding), or both.

Armature Voltage Control, V_a(t) This method of speed control is suitable for separately excited DC motors. Here, R_a and ϕ are kept constant and DC drives aim to keep the constant-rated motor torque at any speed as long as it is lower than the rated motor speed. In normal operation, the voltage drop across the armature resistance is small compared to $E_b(t)$, such that:

$$E_b(t) = V_a(t) \approx K\phi\omega(t) \tag{5.95}$$

For constant flux, ϕ, the speed, $\omega(t)$, changes linearly with $V_a(t)$, as shown in Figure 5.41. Such variable-speed drives can be built either by (i) using thyristors through the variation of the firing phase angle from bridge configuration (e.g. a full-wave 12-pulse bridge, a full-wave six-pulse bridge, or a half-wave three-pulse bridge) relative to the AC supply voltage, as illustrated in Figure 5.42, or (ii) using the variation of the ON and OFF time ratio (duty cycle) of the

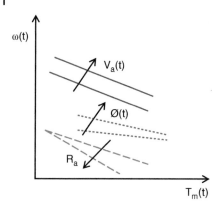

Figure 5.41 Torque-speed curves for varying $R_a(t), \phi, V_a(t)$.

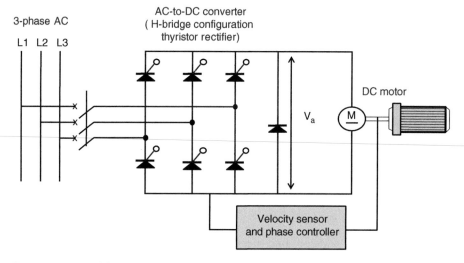

Figure 5.42 Typical thyristor-controlled DC drive.

DC voltage supply from chopper circuits or switch-mode converters (such as Jones, Morgan, oscillation, etc.). Those choppers transform a constant magnitude of input DC signal into a variable magnitude of DC signal by inverting it using a DC-to-AC converter, then passing it through an AC transformer, and finally rectifying it through an AC-to-DC converter circuitry. Among other design alternatives, there are: (i) rectifiers (a combination of bi-stable devices like diodes and thyristors for phase control) to convert a single or three-phase AC signal (either voltage, current, or frequency) into a constant and smooth DC voltage signal and pass it through a chopper circuit to obtain an adjustable mean DC voltage; and (ii) Ward Leonard drives (separately excited DC generators), where three-phase AC voltage is supplied to a fixed-speed induction motor. These devices are suitable for shunt-wound DC motors.

Field Control (ϕ) In this method of speed control, R_a and $V_a(t)$ remain constant. Here, the variation of the motor flux can be derived by setting a variable resistance R_S in series with the field winding for shunt-wound and compound-wound DC motors. This is suitable for a series DC motor. Hence, recalling that R_a is proportional to $1/\phi$ and assuming that magnetic linearity ϕ is proportional to the field current $i_f(t)$, which is inversely proportional to $\omega(t)$, Equation (5.93) leads to:

$$\omega(t) \alpha \frac{V_a(t) - i_a(t)R_a}{i_f(t)} \tag{5.96}$$

Figure 5.43 Variation of speed with an external armature resistance.

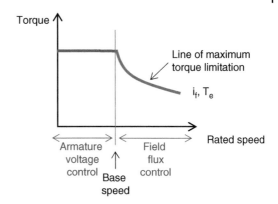

This field current could be varied using an adjustable rheostat in the field circuit. The resulting torque-speed characteristics are illustrated in Figure 5.43. With field control, the applied armature voltage $V_a(t)$ is maintained constant, as shown in Figure 5.43. The reduction of the flux increases the speed beyond approximately three or four times the rated speed and decreases the torque proportionally.

Armature Resistance Control (Rₐ) As the actual resistance of the armature winding is fixed for a given motor, the voltage across the armature can be varied by inserting a variable resistance, R_s, in series with the armature circuit. From torque-speed characteristics, it is known that:

$$T_m(t) \, \alpha \, \frac{K\phi}{R_a}(V_a(t) - K\phi\omega(t)) \tag{5.97}$$

The variation of speed with respect to change in this external resistance is illustrated in Figure 5.41. This control method achieves smooth control of speed while maintaining $i_f(t)$ constant.

All three methods described here can be used to control the speed of DC shunt motors. Except for the commonly used armature resistance control method, they could also be applied for a DC series motor.

Exercises and Problems

5.1 a) For PD controller design using a root locus design technique, between stationary and transient design requirements, in which is it critical to determine the placement of the zero?

b) Answer true or false to the following statements:
    i) phase-lag compensation increases the error constant;
    ii) steady-state errors are improved by increasing the gain from the Bode diagram;
    iii) decreases in the phase and gain margins make the system more stable;
    iv) overshoot is reduced by decreasing the phase margin;
    v) rise time is reduced by increasing the closed-loop system's bandwidth.

c) Draw system block diagrams to illustrate each of the following feedforward controllers:
    i) feedforward control of disturbance;
    ii) open-loop feedforward;
    iii) modified command input.

Table 5.8 PID compensation effects on system response characteristics.

Response	Rise time	Overshoot	Settling time	Steady-state error	Stability
	Decrease		Not defined	Decrease	
		Increase	Increase		Degrade
	Not defined	Decrease		Not defined	Improve

d) Complete Table 5.8.
e) Describe the subsequent system behavior for various sampling frequencies (shorter and longer) in terms of: (i) the speed of response; (ii) stability; and (iii) disturbance rejection.

5.2 Recall the third-order differential equation model of DC motor shaft position $\theta(t)$, with inputs being the developed motor torque $T(t)$ and the table attached to the motor shaft as the motor torque disturbance $T_L(t)$, which are related to the corresponding motor velocity by:

$$J\frac{d\omega(t)}{dt} = T(t) + T_L(t)$$

a) Using a zero-order hold and assuming that the position command input is zero, derive the discrete response from the motor shaft position $\theta(t)$ to a unit step torque disturbance $T_L(t)$ input.
b) Derive the equivalent-discrete closed-loop transfer function for a system with a proportional position controller, K_P.
c) Plot the discrete motor shaft position response to a unit step in a load torque for proportional gains that have the following values: 0.2; 1.2; 3.5. Comment on the results obtained.

5.3 Consider a robot gripper that picks up a box brought by a conveyor and places it in a labeling machine as shown in Figure 5.44. It takes 0.02 s for data logging and processing of the output signal from the vision sensor in order to determine the relative position of the robot gripper. With a sample period 0.05 s, the arm-positioning drives can be approximated as integrations with gains K_x, K_y.
Derive the discrete-time closed-loop transfer functions for robot positioning along the x and y axes. Then, assuming the proportional controller gains are known, determine the steady-state error along the x and y axes when the proportional controller gains are P_x, P_y, respectively, for the box moving on the conveyor at a constant velocity $v(t)$ in the negative x direction. (Hint: the trajectory motion profile is well defined such as that during the synchronized conveyor-robot arm pick up and drop at a height of 0.4 m and spacing between two successive boxes of 0.3 m within 6 sec.)

5.4 For a step input of 10 V, a motor provides a pure integration like velocity response and reaches the steady-state speed of 525 revolutions per minutes after 12 sec, with a motor shaft with a radius 0.15 m.
a) Derive the discrete model of the motor position as a function of voltage command input.
b) Design a proportional controller gain for the motor position loop that would achieve a closed-loop bandwidth of 25 Hz with a sample frequency of 100 Hz.

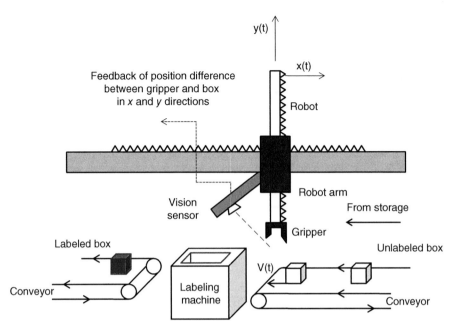

Figure 5.44 Robot for labeling a box in a supply-chain system.

c) Plot the response of the closed-loop system to a 10 V step input. What is the approximate damping ratio? Is this response likely to be acceptable?

d) Tune the proportional controller gain using the model found in part (a) to obtain a closed-loop system damping ratio of 0.7. What is the natural frequency? Compare this with the bandwidth in part (b) and the results obtained.

5.5 As depicted in Figure 5.45, consider a block diagram of an elevator motion system. When the current command input $r(z)$ is applied to a process as shown in Figure 5.46(a) and (b), the process response is described by the following discrete-time transfer

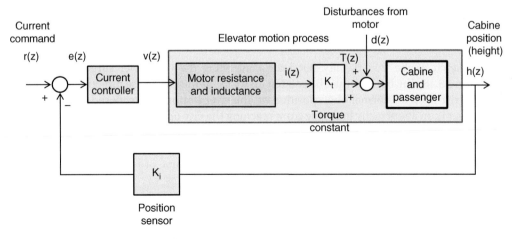

Figure 5.45 Elevator motion block diagram.

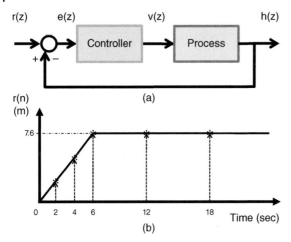

r(z)   e(z)   v(z)   h(z)   **Figure 5.46** (a) Process block diagram. (b)
Command input.

(a)

r(n)
(m)

(b)

function:

$$\frac{h(z)}{v(z)} = \frac{4(1-\alpha)(1-0.5z^{-1})z^{-1}}{(1-\alpha z^{-1})(1-z^{-1}) + 2(1-\alpha)z^{-1}(1-0.5z^{-1})}$$

when the sample period is 2 s and the discrete transfer function of the controller is:

$$\frac{v(z)}{e(z)} = \frac{1-0.5z^{-1}}{1-z^{-1}}$$

a) In order to achieve smooth and precise motion, what type of controller should be selected?
b) Using the discrete controller given above, derive the closed-loop transfer function $c(z) = r(z)$.
c) Plot the unit step response for stablilizing $\alpha = 0.5$.
d) Analyze the transition and steady-state system dynamic characteristics by estimating from the system response the damping ratio, response time constant, and steady-state response. How could the selected controller improve these characteristics?

5.6   Consider the automatic vehicle speed control system captured in the block diagram in Figure 5.47. It is desired to achieve an overall system with dynamics characteristics of time constant $\tau = 5$ s and damping ratio $\zeta < 0.6$ and no steady-state error. Thus, using the design procedure outlined previously, derive the appropriate PD control algorithm.

5.7   A sketch of the Venetia city floodgate controlling the tidal river surge is shown in Figure 5.48. The barrier must be rotated up to 55° in the clockwise direction in order to move it from its normal open position on the river bottom to its closed position. This prevents an abnormally high tide from reaching the city upstream. The barrier is driven by an electric motor, and the barrier position $\theta(t)$ in degrees is digitally controlled. The controller regulates the values of the applied voltages $v(t)$ to be sent to the barrier drive system. The angular speed of the barrier has a first order differential equation relationship $v(t)$, and the barrier motion can be modeled as:

$$\frac{d\theta(t)}{dt} + 2.1\,\theta(t) = 3.5v(t)$$

in degrees per minute. With a sampling period of 2 min:

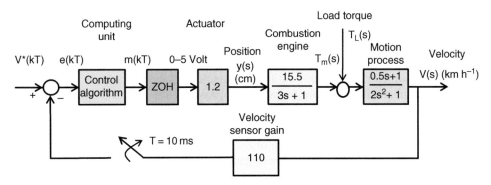

Figure 5.47 Automatic vehicle speed control system.

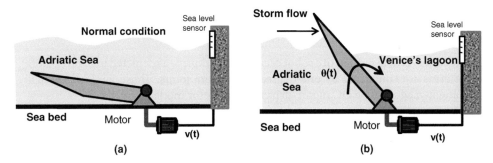

Figure 5.48 Flood tide control gate for (a) a barrier-open position and (b) a barrier-closed position.

a) Develop the mathematical model of a discrete time command input signal for a barrier angular position going from $25°$ to $45°$ in 14 min.
b) If a proportional controller is to be used to control the system, determine the maximum controller gain K_p that results in a closed-loop step response with no oscillation.
c) Using the input signals just derived plot the controlled system responses $\theta(n)$ and calculate the tracking characteristics, especially regarding the steady-state error.
d) In order to analyze the control effort required for the command input, plot the system-manipulated output voltages $m(n)/v(n)$.

5.8 With $T = 0.01$ s; consider a process model given by:

$$G_m(s) = \frac{K(z - 0.3)(-0.01 - 0.001z) + 0.001(z - 1)^2}{(z - 1)(z - 0.3)}$$

a) Sketch its equivalent root locus and find K such that overshoot is less than 25% for a unit step.
b) Determine the steady-state error for a unit step command input.
c) Determine K to minimize the integral square error.

5.9 Design a discrete controller using the roots-locus technique of a double-integration model of a DC motor given by:

$$G(s) = \frac{3.2}{s^2}$$

It is desired to achieve a DC motor operating performance characterization given by $\omega_n = 0.5$ rad s^{-1} and $\zeta = 0.7$.

a) Sketch the root locus.
b) Derive the continuous controller and the equivalent discrete controller.
c) Write the control algorithm and the system response to the step command input.

5.10 Consider the motor-driven positioning system depicted in block diagram in Figure 5.49, where the motor/amplifier transfer function is approximated to be:

$$G_m(s) = K_a$$

a) Plot the discrete frequency response of the position of the system with a sampling period of 200 Hz, a controller proportional gain $K_c = 11.5$, a position encoder gain $K_f = 110$, and lead screw ratio of $K_s = 2.145$.
b) Repeat the plot in part (a), using manually frequency response sketching method describes in Section 4.4. and compare results.
c) Design the controller to have a dominant closed system time constant of 0.4 sec. (By varying the gain).

5.11 Consider the vertical motion of a robot carrying a workpiece as depicted in Figure 5.50(a). Robot arms are connected by one-degree-of-freedom joints. Those joints are actually permanent-magnet-type DC motor-driven. Each motor model is given by:

$$G_p(s) = \frac{\theta(s)}{V(s)} = \frac{K}{(\tau s + 1)}$$

a) Assuming the weight of the robot arm is negligible and the load weight is $m = 0.3\,Kg$, derive the dynamics model of the robot arm for each joint. With $L_1 = 0.8\,m$; $L_2 = 0.4\,m$; $H_1 = 0.35\,m$; $J_{1,2} = 0.002\,\text{N.m.s}^2\text{.rad}^{-1}$; $B_{1,2} = 0.009\,\text{N.m.s.rad}^{-1}$; $K_{1,2} = 0.012\,\text{N.m.rad}^{-1}$; $R_{a\,1,2} = 2.4\,\Omega$; $L_{a\,1,2} = 1.23\,\text{mH}$; $K_{m\,1,2} = 102\,\text{N.m.A}^{-1}$; $K_{m\,1,2} = 102\,\text{N.m.A}^{-1}$; Other parameters are negligible.
b) Considering a sampling time of 5 s, derive the corresponding digital system model.
c) Using Figure 5.50(b) and (c), derive equivalent discrete position command input signals $\theta_1(n), \theta_2(n)$.
d) Plot the system response to $\theta_1(n), \theta_2(n)$ using position proportional gain values of 1.2 and 3.5. Comment on the stationary and transient characteristics of the response.
e) Regarding any robot arm-positioning error, discuss how the command feedforward controller can be adjusted to compensate for steady-state error.

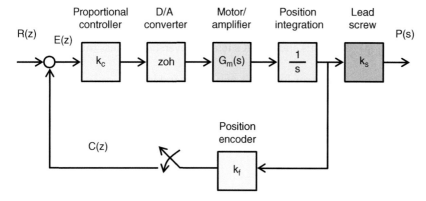

Figure 5.49 Position system block diagram.

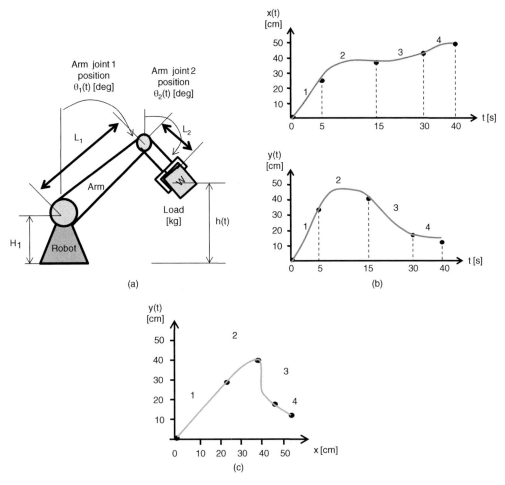

Figure 5.50 (a) Robot arm. (b) Robot arm joint trajectory. (c) Combined robot arm joint trajectory.

5.12 Consider the DC motor block diagram illustrated in Figure 5.51, with:

$$G_c(s) = K_{pi}\frac{(1 - \delta_c z^{-1})}{(1 - z^{-1})}$$

and sampling 10 times faster than the system time constant.

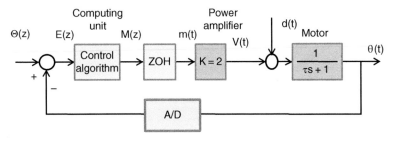

Figure 5.51 Discrete DC motor block diagram.

a) Design a controller (specify K_{pi}, δ_c) for which the dominant closed-loop pole is at 0.1 (slightly slower than the deadbeat dynamics).

b) For a unit step disturbance input occurring at $t = 0$ where the system is at rest $\theta^*(0) = \theta(0)$, compute and plot the error $e(k)$, the manipulated $m(k)$, and the disturbance system response $\theta(k)$ at the first four samples after the step input disturbance.

5.13 Consider a process given by:

$$G_p(s) = \frac{1805}{s(0.25s + 1)}$$

a) Sketch the uncompensated system Bode plot of $G_p(s)$. What is the current phase margin?

b) Using a frequency response design technique, find a discrete-time controller $D(z)$ allowing a phase margin $\gamma(PM_{desired})$ of $45°$ at gain crossover frequency ω_g of 125 rad s^{-1} (for $T = 1$ ms).

c) Plot the compensated system Bode plot.

5.14 The open-loop transfer function of DC motor position over input voltage whose parameters are listed Table 5.9 is given by:

$$\frac{\theta(z)}{V(s)} = \frac{K_e s}{(Js + B)(Ls + R)}$$

The controller design requirements include settling time $t_s \leq 0.04$ s, percentage overshoot PO < 16%, and steady-state error $e_{ss} = 0$ from a command step input.

a) Derive the sampling period T for a discrete-time open-loop transfer function that results in:

$$\frac{\theta(z)}{V(z)} = \frac{0.001039z + 0.002}{z^2 - 1.492z + 0.477}$$

b) Design a pole-zero and an integral controller using the root locus technique.

c) Sketch the responses respectively to unit step command and unit step disturbance inputs. Check the response to a unit step disturbance $\frac{\theta(z)}{T_d(z)}$. Discuss on system disturbance rejection compliance.

5.15 Consider a process model given by:

$$G_p(z) = \frac{0.00158}{z - 0.822}$$

Design a deadbeat controller using the following steps:

Table 5.9 DC motor parameters.

Parameters	Values
R	$4\,\Omega$
L	$2.75 \cdot 10^{-6}$ H
$K_e = K_t$	0.0274 N.m A^{-1}
J	$3.2284 \cdot 10^{-6}$ kg m^2 s^{-1}
B	$3.5077 \cdot 10^{-6}$ N.m s^{-1}

a) Plot the step response of the uncompensated system and analyze the system transient and stationary response characteristics.
b) Add a PI controller to reduce the error.
c) Find the closed-loop transfer function with deadbeat controller.
d) By equivalence, derive the controller parameters.

5.16 Consider a process given by:

$$G_p(s) = \frac{2}{s(s+2)}$$

a) With $T = 1$ s and a zero-order hold used as converter, derive the equivalent discrete-time process transfer function.
b) Using zero-pole cancelation for a PI controller, formulate a discrete controller that allows an overshoot below 4% to be obtained.
c) Plot the discrete step response of the system.

5.17 Blood purification based on a dialysis technique aims to remove extracorporeal waste products such as creatinine and several harmful microorganisms, including bacteria, fungi, viruses, and parasites. Consider a temperature-based blood treatment as illustrated in Figure 5.52.

By reducing the number of infected cells, the blood treatment device could make a difference for severely affected malaria or drug-resistant patients. For example, in the case of malaria, the plasmodium parasites move human red blood cells to consume the hemoglobin protein while merging with the magnetic-based hemozoin. Due to their magnetic properties, these malaria-infected cells can be trapped in the magnetic nanoparticle-based blood filter. In order to ease this filtering, a stage for the pre-heating

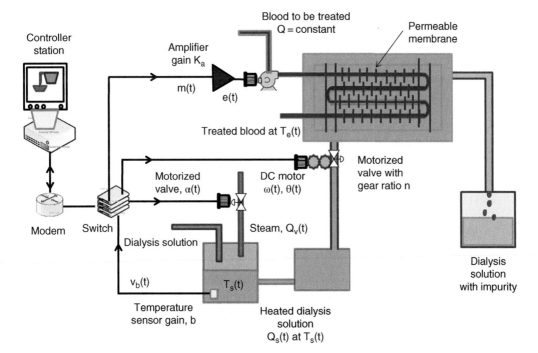

Figure 5.52 Temperature-based blood treatment system.

Table 5.10 Some key parameters and variables of a blood treatment system.

Parameters	Description and values	Variables	Description and units
Q	Liquid inflow rate ($10\,\text{kg s}^{-1}$)	$Q_V(t)$	Input flow rate of the steam in the tank ($\text{m}^3\,\text{s}^{-1}$)
C	Heat coefficient of blood ($2.103\,\text{J (kg}\,°\text{C})^{-1}$)	$Q_i(t)$	Input flow rate of the blood to be heated by the steam in the tank ($\text{m}^3\,\text{s}^{-1}$)
K_1	Valve flow rate constant ($10\,\text{m}^3\,\text{s}^{-1}\,\text{rad}^{-1}$)	$T_e(t)$	Temperature of the inflow blood to be heated ($°\text{C}$)
K_2	Steam energy transfer constant ($105\,\text{J m}^{-3}$)	$Q_o(t)$	Output flow rate of the blood to be heated ($\text{m}^3\,\text{s}^{-1}$)
K	Torque constant ($0.4\,\text{N.m A}^{-1}$)	$T_S(t)$	Temperature of the outflow blood to be heated ($°\text{C}$)
R_a	Armature resistance ($160\,\Omega$)	$\alpha(t)$	Angular position of the steam valve (motorized valve) (rad)
J	Motor shaft inertia ($10^{-3}\,\text{kg m}^2$)	$v_b(t)$	Thermosensor voltage (V)
n	Reduction gear (250)	$\omega(t)$	Motor angular velocity (rad s^{-1})
K_b	Thermo-electrical sensor coefficient ($0.01\,\text{V}\,°\text{C}^{-1}$)		
M	Retained blood mass (40 kg)		

of human blood is required, utilizing a system as depicted in Figure 5.52. The steam injected by a motorized valve is used to heat up the processed blood.

a) For an efficient blood treatment, it is desired to maintain blood temperature at a specific level. Develop a model of this process relating blood temperature $T_S(t)$ to valve angular position $\alpha(t)$, as depicted in Figure 5.49, using the variables and parameters described in Table 5.10.

b) Develop an equivalent discrete model with a sampling period of 10 ms.

c) Plot the system response of the unit step change of blood temperature using a discrete PI controller given as $K_p = 7.8$ and $K_i = 2.1$. Tune the PI gains to derive a performance such that $e_{ss} = 0$ and PO = 5%.

5.18 A process described mathematically by a pure integration so that the discrete transfer function is given by:

$$G_p(z) = \frac{3.5Tz^{-1}}{1 - z^{-1}}$$

and the controller discrete transfer function chosen is:

$G_c(z) = \frac{1}{3.2T}$

with a sampling period $T = 0.05$ s.

a) Plot the system response to a step input $r(t) = R$ for $t \geq 0$. Plot $r(k), e(k), m(k), c(k)$, and $c(t)$ versus t/T.

b) Plot the response to a ramp input $r(t) = 2.5t$ for $t \geq 0$. What is the steady-state error in the system $e(k)$ when the command is a ramp?

c) Derive a controller when the system is required to follow a ramp input $r(t) = 2.5t$.

d) Plot the system response to the step input with the controller derived in (c).

e) Compare the response obtained in (c) with that in (b). Comment on the results obtained with respect to step input signals.

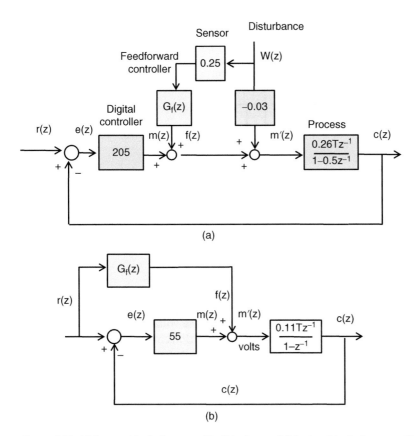

Figure 5.53 (a) System block diagram with disturbance. (b) System block diagram with command feedforward.

5.19 a) A control system schematic diagram is shown in Figure 5.53(a). The step input load disturbance signal $W(z)$ can be modeled as:

$$W(z) = \frac{5}{1 - z^{-1}}\delta(0)$$

    i) Plot the system response to this step disturbance input. Derive the transient and stationary characteristics.

    ii) Derive the feedforward disturbance controller to ensure that there are no disturbances when only the step disturbance acts as an input to the system.

    iii) Analyze the system stability before and after the disturbance rejection. How would the system work if a PI controller were used instead.?

  b) A control system schematic diagram can be modeled as a block diagram, as depicted in Figure 5.53(b).

    i) Use this diagram to derive the error transfer function.

    ii) Find the steady-state error.

    iii) Design the feedforward controller algorithm, if it is feasible. If not, modify accordingly.

    iv) Evaluate the steady-state error with the feedforward controller.

5.20 Consider the airplane aileron servo-to-bank angle sensor that is depicted through block diagram in Figure 5.54. This motor-driven gyro consists of a sensitive motor

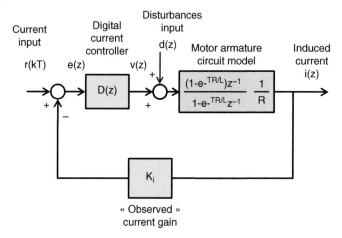

Figure 5.54 Block diagram for control of motor current.

current-driven arm positioned over the top of a gasoline surface so that any variation of airplane sideslips (bank angle) will produce a disturbance voltage inside the current of the motor. Such an event results from the frequent vibrations occurring during airplane motion after any discontinuous contact between the wind and the airplane duriing autopilot driving conditions. The current should be controlled to produce an accurate measurement, as depicted in Figure 5.54. Using $R = 24\ \Omega$, $L = 156$ mH, and $K_I = 17.5$, With a sampling period of 50 ms, design a PI current control algorithm to ensure an overall performance for the steady-state response in 0.5 s step disturbance inputs while approaching a steady-state error below 0.02 with a percentage overshoot less than 10%.

5.21 At regular intervals, the solar panels shown in Figure 5.55 are reoriented to maximize solar energy exposure for the accumulator charging process. This reorientation can be treated as applied thrust forces $T_{F(t)}$ with an effect on the x- and y-axis spinning velocity $\omega_1(t)$ and $\omega_2(t)$ and on the satellite.
   a) Given satellite inertias J_x and J_y, respectively, along the x- and y-axis, respectively, develop the equation of motion of the satellite. Deduce the equivalent discrete-time model.
   b) Draw a block diagram for the system with a closed-loop control showing the command position disturbance input.
   c) Sketch the velocity and position profiles required for the satellite to rotate by 17° and −4° along the x- and y-axes, respectively. Use a sample period of 10 ms.
   d) Show how a feedforward control can be added to nullify the effect of reorientation.

5.22 Consider a robot positioning arm as shown in the block diagram in Figure 5.56 and with the physical configuration of the robot described in Exercise 5.11.
   a) Design a proportional controller so that the system can ensure a percentage overshoot of less than 35% with a settling time better than 2.1 s for a unit step command input. (Hint: $K_t = 0.875$). This is illustrated aside.
   b) Add a command feedforward controller to the block diagram in Figure 5.56(b) to compensate for static errors in the arm position caused by the load.
   c) Plot the position responses obtained with and without the feedforward controller.

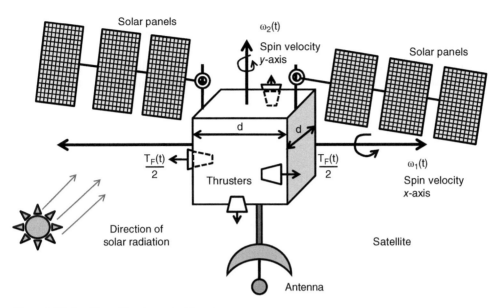

Figure 5.55 Satellite with spin control thrusters.

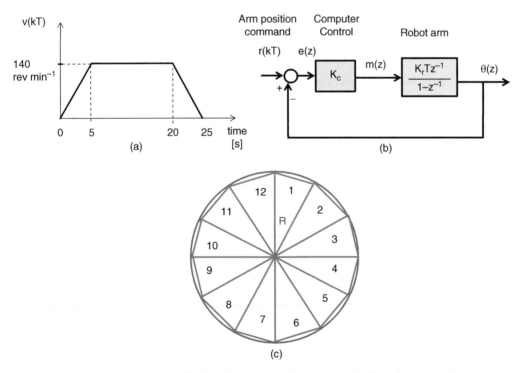

Figure 5.56 (a) Velocity/position profile of a robot arm. (b) Block for control of the robot arm. (c) Robot arm *x-y* trajectory.

d) Derive the equivalent discrete command position input from the velocity profile depicted in Figure 5.56(a) as well as robot motion trajectory illustrated in Figure 5.56(c). All initial conditions are zero. Then, sketch the system response using the controller design in part (b).

e) It is desired to perform a circle with radius R as shown in Figure 5.56(c). Formulate the two-axis corresponding command input for the robot arm-positioning system.

5.23 For the system depicted in Figure 5.57, using the root locus design technique, design a process that enables achievement of $\omega_n = 24$ rad s^{-1} and a damping ratio of $\zeta = 0.7$.

5.24 Consider a tool offset control system for a lathe machining tool where each workpiece produced is measured in real time by a laser measuring system. If an error in a workpiece dimension is detected, a correction is applied before the next part is machined so that errors in the previous part are eliminated in subsequent parts. In short, this tool positioning (cutting width differential) offset control system is used to track and compensate for a constant rate of tool wear on the lathe, as illustrated in Figure 5.58 and summarized in block diagram form in Figure 5.59.

a) In a study of the relationship between $L(kT)$ and $\Delta z(kT) + z(kT)$ for this case, derive some PD controller settings in order to ensure a P.O. less than 12% and a response time constant of 0.5 s for a laser-based sampling period of 0.1 s. Fill out Table 5.11 for the first five data points, while considering a step-like estimated resistive torque given by: $T_r(kT) = 1.1L((k)T)$. Plot the subsequent system response.

b) What conclusions can be drawn about the effectiveness of the controller in compensating for exponential disturbances? (Note that wear rate after several parts is probably more rapid than in reality.)

5.25 A 360-steps-per-revolution stepping motor is controlled using an 8-bit, 1 MHz DDA-based microcontroller algorithm. Consider the velocity profile given in Table 5.12. Initially, the velocity register is $(15 \times 2^{-7} = 0.1171875$ in base 10), while the position register is zero.

a) Based on the velocity profile in Table 5.12, fill this table by deriving the velocity of the motor in revolutions per minute. Obtain subsequently the system velocity resolution.

b) Based on the velocity profile in filled Table 5.12, determine the velocity range $\left[\text{min.}value, \text{ max.}value\right]$.

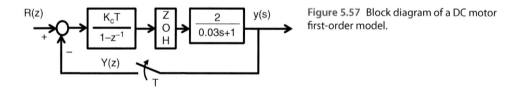

R(z)

$$\frac{K_c T}{1-z^{-1}}$$

ZOH

$$\frac{2}{0.03s+1}$$

y(s)

Y(z)

T

Figure 5.57 Block diagram of a DC motor first-order model.

Table 5.11 Sequence values for the PD-based width cutting control of lathe machining.

kT	$C_w(kT)$	$z(kT)$	$T_r(kT)$	$W(kT)$	$E(kT)$	$\Delta Z(kT+1)$
0						

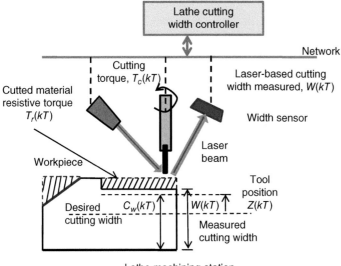

Figure 5.58 Lathe machine and laser-based measuring system.

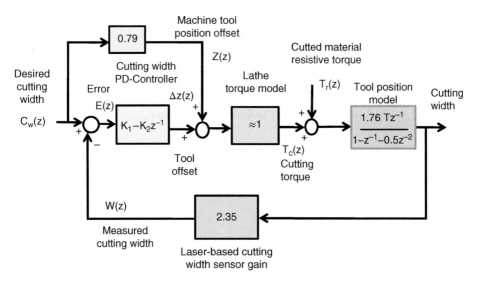

Figure 5.59 Schematic block diagram of a laser-based tool offset controller.

5.26 Consider the stepping motor controlled by an 8-bit, 1 MHz so that its velocity data summarized Table 5.13 with 1000 Hz, 8-bit register for position, velocity and acceleration.

    a) Derive the acceleration in step/add^2 between times 0 and 0.2 sec required to accelerate the motor at a rate of 104.2 rad^{-1} s^2 as well as the number of clock periods are required to achieve the desired speed of 250 rpm?

    b) Between times 0.2 sec and 0.5 sec, derive the number of clock periods that are required when the motor operates during the 0.3 sec of constant velocity interval.

    c) Derive the acceleration and the number of clock between times 0.5 sec and 0.9 sec in order to decelerate the motor to a stop at a rate of 104.2 rad^{-1} s^2.

Table 5.12 With $A = 0.0001111_2$ steps per adds

Time	Velocity (step/adds)	Velocity (rev/min)	Position (resolution)
0	A		360 steps/rev
1	1.4 A		360 steps/rev
3	1.2 A		360 steps/rev
4	0.8 A		360 steps/rev
5	0 A		360 steps/rev

Table 5.13 Velocity and position profiles data

Time	Velocity (rev/min)	Motion phase	Position (resolution)
0 sec	0	Acceleration	360 steps/rev
0.2 sec	250	Constant	360 steps/rev
0.5 sec	250	Deceleration	360 steps/rev
0.9 sec	0	Rest	360 steps/rev
			360 steps/rev

d) Determine the number of revolutions required by the motor to cover those phases. What is the total number of clock periods required for 120 revolutions during constant velocity phase?

e) Write the subsequent control algorithm for this stepping motor.

5.27 Consider the schematic three-bladed horizontal-axis wind turbine depicted in Figure 5.60. The rotor speed must be varied as the wind speed varies, hence the rotor speed can be changed by controlling the generator torque. Consider a typical wind-speed variation with height, which can be expressed as:

$$w(t) = w_{hub}\left(1 + \frac{z}{h}\right)^m$$

where z is the height above the hub, w_{hub} is the wind speed at hub-height h, and m is the power law wind-shear coefficient.

a) Derive a simplified model of three-bladed horizontal-axis wind turbines, considering the tower flexibility, rotor teetering, no-blade flexibility (pitch angle variation negligible), and a variable generator speed. Use the following variables: perturbed rotor speed, $x_1(t) = \frac{\delta\theta_1(t)}{dt}$; perturbed drive-train torsional spring force, $x_2(t) = T_e(t) = K_d(\delta\theta_1(t) - \delta\theta_3(t))$; perturbed generator speed, $x_3(t) = \frac{\delta\theta_3(t)}{dt}$ J_T and J_G inertia moments of the wind turbine rotor and the generator (kg mm); T_T, T_G wind turbine aerodynamic and generator electromagnetic torques, (Nm); θ_T, θ_G angular positions of the rotor and the generator speed (rad s^{-1}). D, K are the damping and the stiffness (N.ms rad^{-1}), (Nm.rad^{-1}), respectively.

b) If a synchronous generator is used, describe the key components of its voltage control with regard to speed.

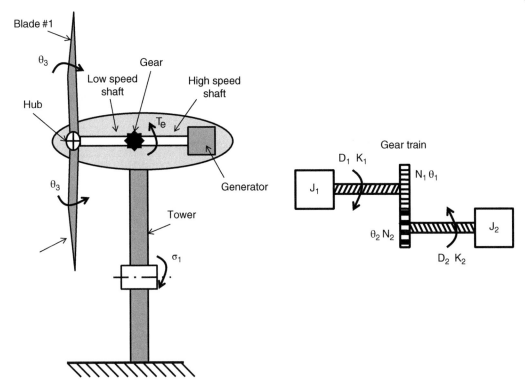

Figure 5.60 Wind turbine control motion.

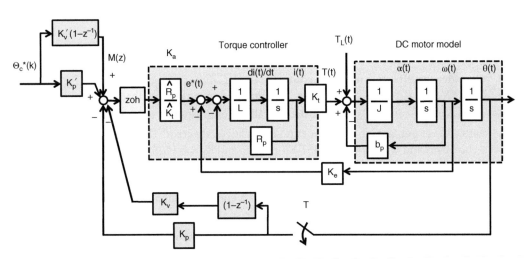

Figure 5.61 DC motor-amplifier position control.

5.28 Revisit the position controller of the DC motor illustrated in Figure 5.61. Here, the velocity is synthesized by digital differentiation. The position encoder $\theta(t)$ is sampled every 0.002 s, the amplifier gain $K_a = \frac{\hat{R}_p}{\hat{K}_t} = 119$ V V^{-1}, the torque constant $K_t = 0.042$ N A^{-1}, the inductance $L = 0.016$ Henrys, the rotor armature resistance $R_p = 1.67$ Ω, the inertia

$J = 0.00036.6$ N.m.s^2/rad^{-1}, the viscous friction (damping) is negligible, and the back EMF is $K_e = 0.0432$ V s^{-1} rad^{-1}.

a) From the block diagram manipulation, derive the discrete transfer function $\theta(z)/M(z)$.

b) Set controller gains K'_p and K'_V for position and velocity feedforward, respectively, to zero. Design an inner velocity gain so that K'_V will result in system-dominated roots that correspond to a damping ratio $\xi = 0.7$. Approximate the corresponding natural frequency.

c) Design a controller to reject torque disturbance in two samples. Determine the value of sample period required to limit the unit step disturbance within four sample periods.

d) By setting $K'_p = 1$, where K'_V is the value found in (b), plot the system response to a step change command input. Discuss your results.

e) Design K'_p so that it results in a system with a dominant root corresponding to a time constant of 0.053 s.

f) Using the values of K'_p and K'_V found previously, plot the response of the motor position to a step command input.

5.29 Consider a first-order system process model given by:

$$G_p(z) = \frac{(1 - \delta_c)}{z - \delta_p}$$

In order to elaborate the design procedure of a PI control algorithm given by:

$$D(z) = K_p + \frac{K_i T}{1 - z^{-1}}$$

a) Derive the closed-loop transfer function.

b) Derive the value of K_i, K_p, T using the pole-zero cancelation associated to the PI controller for a chosen system dynamic scenario.

5.30 Consider the PID state feedback control of the DC motor system depicted in Figure 5.62. With K_i, K_V, K_p, derive the state velocity and position feedback controller gains $K_V = K_V' - \frac{1}{K_p}$ and K_C such that the overall system responds to a unit step disturbance with a lower percentage overshoot and no steady-state error. The sample period is given to be 10 ms and with J = 0.0000431 Kg·m^2/sec and K$_p$ = 0.1425 N.m/A.

5.31 Consider a DC motor system with the position and velocity feedback control topology shown in Figure 5.63. It is desired to design a controller such that the closed-loop system has a bandwidth of $\omega_{closedloop} = \frac{1}{\tau_{cl}} = \frac{K_V' K_p}{\sigma_p} = 16 rad/s$. Consider using $K_{closedloop} = \frac{1}{K_V'}$ from the desired characteristic equation pole τ_{cl} and a sampling period of 10 ms.

5.32 It is desired to use a 12-bit microcontroller to control the resolution of a 200-steps-per-revolution stepping motor with a computer clock 10 000 Hz (adds s^{-1}) at a constant acceleration of 500 rev s^{-2}. Determine the number of additional operations before the step signal will be sent to the command translator. Compute the position, the velocity, and the acceleration register values for the first four additions.

5.33 a) Give the difference between: (i) the control horizon; (ii) the prediction horizon; and (iii) the control horizon.

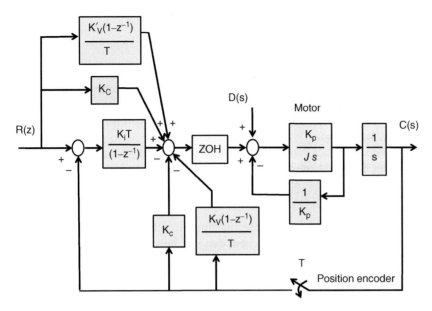

Figure 5.62 Block diagram of a motor-control system.

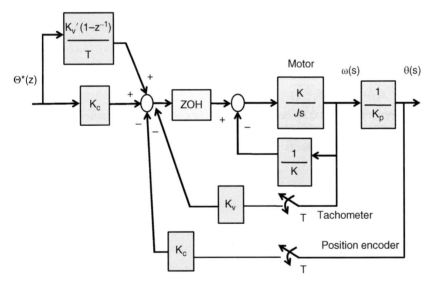

Figure 5.63 DC motor state feedback control topology.

b) In the following systems exhibiting unusual dynamic behavior, define the receding horizon control strategy (finite future horizon of N steps) for the MPC algorithm:

   i) Tracking control for an automatic of steering driverless vehicle moving in busy road.

   ii) Automatic plane altitude positioning system such that lowering altitude requires moving up before a move down motion.

   iii) Aerial navigation of unmanned robot craft during a cave exploration.

   iv) Thermal-based trajectory tracking for a missile defense system.

   v) Wort temperature changes due to conversion of yeast into alcohol in a fermentation tank.

5.34 Consider a discrete process transfer function given by:

$$G_p(z) = \frac{2.5(1 - \delta_p)z^{-1}}{1 - \delta_p z^{-1}}$$

Note the cancelation of the pole in the process by zero in the numerator. If the desired closed-loop bandwidth is 50 rad s^{-1}, the desired closed-loop time constant is $\tau_{dr} = 0.02$ s. Thus, a sample of $T = 0.01$ s can be chosen. Derive the discrete controller for a process and zero-order hold, with $\delta_p = e^{-T/0.03}$.

5.35 Consider a first-order discrete-time process transfer function given by:

$$G(z) = \frac{2(1 - e^{-T/0.04})z^{-1}}{1 - e^{-T/0.04}z^{-1}}$$

Using the roots-locus and frequency response, the aim is to achieve a closed-loop bandwidth of about 50 rad s^{-1}. The sampling frequency should be at least 10 times the bandwidth.

a) Plot the uncompensated system frequency response.
b) Using the pole-zero cancelation method, derive the corresponding digital controller.
c) Plot the compensated system frequency response and sketch the root locus. Comment on the results achieved.

5.36 a) If not specified (directly or indirectly), use a sampling period of 100 ms in all the following exercises.
   Consider a process model given by:

$$G_C(s) = \frac{1}{s(s + 1)(0.32s + 1)}$$

   With a sampling period of 0.05 s, design a phase-lag compensator, ensuring a desired phase margin of 55° for zero steady-state error.
b) Consider a process model given as:

$$G_C(s) = \frac{1.1}{2.6s(s + 1)}$$

   Design a lead compensator that could achieve PO (%) > 0.5 (so $\zeta \leq 0.5975$) and $\omega_n \approx 7$ rad s^{-1}.
c) Consider a process given by:

$$G_P(s) = \frac{2.5}{s^2}$$

   Design a controller with the following design requirements: settling time $t_s = 1.2$ s and damping ratio $\zeta > 0.4$. The controller, with $p_c > z_c > 0$, is given by:

$$G_C(s) = K\frac{s + z_c}{(s + p_c)}$$

d) Consider a process model given by:

$$G(s) = \frac{K}{s(s + 1)}$$

   Design a discrete controller enabling a display of a dynamic closed-loop response with characteristics such that the damping ratio is $\zeta > 0.5$ and $\omega_n > 0.9$ rad s^{-1} (or a settling time of about 8.8 s).

e) Design a lag compensator in order to have $K_v \geq 100$ for a process given by:

$$D(s)G(s) = 127 \frac{s + 5.4}{(s + 20)} \frac{1}{s(s + 1)}$$

f) Consider a process model given by:

$$G(s) = \frac{10}{s(s + 4)}$$

The desired process dynamics performance characteristics are PO $\approx 5\%$ and $t_s = 1.2$ s (2% criterion) for a step input.

g) Consider a motor plus amplifier model given by:

$$G(s) = \frac{1}{s(s + 10)(s + 5)}$$

Design a digital controller with the following specifications: 7.5% overshoot and minimum settling time $t_s = 1.2$ s (2% criterion), with a chosen $T_s = 0.01$ s.

5.37 In a distillation tower, distillation process consists of raising the temperature of the inflow crude oil from ambient temperature to approximately 400°C. The burner providing stream outlet temperature is regulated by the gas flow rate to the furnace, along a reboiler and a total condenser. Consider a continuous distillation process through a tower with five stages, designed to separate crude oil made of benzene, toluene, and p-xylenes, as shown in Figure 5.64.

This distillation tower has a 6-m diameter over 0.07 m weir height, with trays spaced apart. The nominal liquid depths are 0.67 and 1.4875 m in the horizontal reflux drum and sump, respectively. The distillation tower operates at nominal steady-state conditions given by: (i) a feed stream containing 30% benzene, 40% toluene, and 30% oxylenes; (ii) a feed flow rate of 500 kmol h^{-1}; and (iii) distillate containing 95% benzene; during the distillation, 95% benzene is recovered in the feed for 1.7% of the benzene impurity at the bottom.

a) Draw the closed-loop control diagram of the distillation process.
b) Develop the predictive model (PM) of this process, where the process inputs are the condenser duty (W), the reboiler duty (W), the reflux mass flow rate (kg h^{-1}), the distillate mass flow rate (stream #2), the bottoms mass flow rate (kg h^{-1} stream #3), and the feed molar flow rate (kmol h^{-1} stream #1). Similarly, the process outputs are the tower pressure (stage 1), the reflux drum liquid level (m), the sump liquid level (m), the mass fraction toluene in the distillate, and the mass fraction benzene at the bottom.
c) Define the cost function with respect to the following control objectives: (i) maintain a constant tower pressure; (ii) maintain 4.6% toluene (or 96.6% benzene) in the distillate; (iii) maintain 2.1% benzene in the distillation tower bottom; and (iv) maintain the sump and reflux drum in the allowable operating ranges.

5.38 Reconsider a controlled system with a PI controller as illustrated in Figure 5.65. Design a controller allowing a system that settles down to the target reference in two steps; that is, $2 \times 0.6 = 1.2 = 2$ seconds. Plot the compensated system response $y(z)$ to unit step input as well as $u(z)$. What are the control effort required and the percentage overshoot achieved?

5.39 Consider the system depicted in Figure 5.66. Design a discrete-time PD controller based on the requirements of time constant $\tau = 2$ s and damping ratio $\xi > 0.4$.

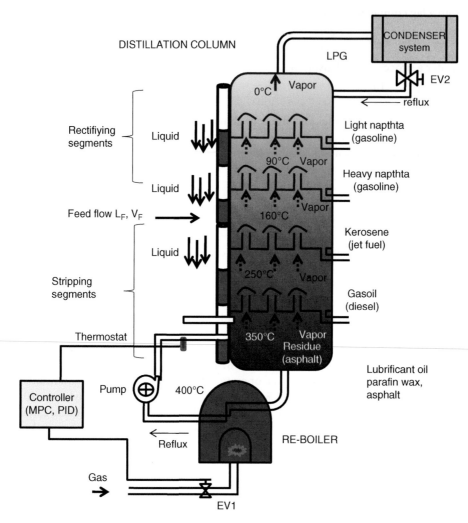

DISTILLATION COLUMN

Figure 5.64 Oil distillation column and reboiler.

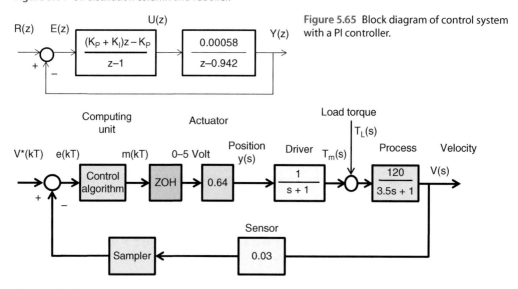

Figure 5.65 Block diagram of control system with a PI controller.

Figure 5.66 Vehicle speed control.

5.40 Consider the system depicted in Figure 2.7.
   a) Develop a discrete time transfer function model of inside incubator temperature using the heating element temperature as the manipulated input and the infant body as the disturbance input.
   b) Sketch a block diagram for this system in which the incubator temperature is explicitly in state feedback form.
   c) Design a discrete-time PI controller to regulate the temperature based on the requirements of time constant $\tau = 10$ s and steady-state error less than 0.05 (use a sampling time = 0.2 s).

5.41 Consider the hybrid vehicle system depicted in Section 2.3.3 and Figures 2.4 and 2.5.
   a) Generate and derive the discrete model of a non-planar road profile for use by the HEV.
   b) Propose a generic operating strategy and subsequent control algorithm for minimizing fuel consumption during hybrid operating mode.

Bibliography

1 Ackermann, J. (1985). *Sampled Data Control Systems: Analysis and Synthesis, Robust System Design*. Springer Verlag.
2 Åström, K.J. and Wittenmark, B. (2011). *Computer-Controlled Systems: Theory and Design*. Courier Dover.
3 Bistritz, Y. (1984). Direct bilinear Routh stability criteria for discrete systems. *Systems and Control Letters* 4: 265–271.
4 Bollinger, J.D. and Duffie, N.A. (1989). *Computer Control Machines and Processes*. Addison-Wesley.
5 Bistritz, Y. (1984). Direct bilinear Routh stability criteria for discrete systems. *Systems and Control Letters* 4: 265–271.
6 Brockwell, P.J. and Davis, R.A. (2009). *Time Series: Theory and Methods*, 2e. Springer.
7 Camacho, E. and Bordons, C. (2004). *Model Predictive Control*. Springer.
8 deSilva, W. (2010). *Mechatronics: A Foundation Course*. CRC Press.
9 Fadali, S. and Visioli, A. (2012). *Digital Control Engineering*, 2e. Academic Press.
10 Franklin, G.F., Powell, J., and Workman, M. (2006). *Digital Control of Dynamic Systems*, 3e. Ellis-Kagle Press.
11 Golnaraghi, F., Kuo, B.C., and Adams, J.A. (2009). *Automatic Control*, 9e. Wiley.
12 Herman, S. (2013). *Industrial Motor*, 7e. Cengage Learning.
13 Isermann, R. (1989). *Digital Control Systems*. Berlin: Springer-Verlag.
14 Jury, E.I. and Tsypkin, Y.Z. (1971). On the theory of discrete systems. *Automatica* 7 (1): 89–107.
15 Katz, P. (1981). *Digital Control Using Microprocessors*. Prentice-Hall.
16 Kothare, M.V., Campo, P.J., Morari, M., and Nett, C.N. (1994). A unified framework for the study of anti-windup designs. *Automatica* 30 (12): 1869–1883.
17 Kuo, B. (1995). *Digital Control Systems*. Oxford University Press.
18 Leigh, J.R. (2006). *Applied Digital Control: Theory, Design and Implementation*, 2e. Dover.
19 Lorenz D.R. (1999). Advances in Electric Drive Control, IEEE International Conference Electric Machines and Drives, pp. 9–16.

20 Luenberger, D.G. (1979). *Introduction to Dynamic Systems: Theory, Models, and Applications*. Wiley.

21 Maciejowski, J.M. (2002). *Predictive Control with Constraints*. Prentice-Hall.

22 Markakis M., Mitsis, G., Papavassilopoulos, G., and Marmarelis, V. (2008). Model predictive control of blood glucose in Type 1 Diabetes: The principal dynamic modes approach, 30th Annual International EMBS Conference, Canada.

23 Paraskevopoulos, P.N. (2001). *Modern Control Engineering*. Marcel Dekker.

24 Ogata, K. (1995). *Discrete-Time Control Systems*, 2e. Prentice-Hall.

25 Patterson, D.A. and Hennessy, J.L. (2013). *Computer Organization and Design: The Hardware/Software Interface*. Morgan Kaufmann.

26 Paraskevopoulos, P.N. (1996). *Digital Control Systems*. Prentice-Hall.

27 Parker, R.S., Doyle, F.J., and Peppas, N.A. (1999). A model-based algorithm for blood glucose control in Type I diabetic patients. *IEEE Transactions on Biomedical Engineering* 46 (2): 148–157.

28 Phillips, C.L. and Nagle, H.T. Jr. (1984). *Digital Control System Analysis and Design*. Prentice-Hall.

29 Polderman, J.W. and Willems, J.C. (2013). *Introduction to the Mathematical Theory of Systems and Control*. Springer.

30 Santina, M., Stubberud, A., and Hostetter, G. (1994). *Digital Control System Design*. Oxford University Press.

31 Skogestad, S. and Postlethwaite, I. (2005). *Multivariable Feedback Control: Analysis and Design*, 2e. Wiley.

32 Sorensen, J. (1985). A physiological model of glucose metabolism in man and its use to design and assess improved insulin therapies for diabetes, PhD Thesis, MIT.

33 Spong, M.W. and Vidyasagar, M. (2008). *Robot Dynamics and Control*. Wiley.

34 Stol, K. and Balas, M.J. (2001). Full-state feedback control of a variable-speed wind turbine: a comparison of periodic and constant gains. *Journal of Solar Energy Engineering*.

35 Visioli, A. (2006). *Practical PID Control*. Springer.

6

Boolean-Based Modeling and Logic Controller Design

6.1   Introduction

Mechatronic systems are operated by turning ON and OFF switches, motor starters, valves, and other devices in response to operating conditions or as a function of time. Such systems are referred to as discrete event requiring logic-based control systems. Hence, binary values such as 1 or 0, ON or OFF, and open or closed define the switched conditions of inputs (e.g. limit switches, relay contacts, push buttons), dictating one of the two states for the output of mechatronic system devices (e.g. light ON or OFF, +V and 0 V input voltage levels for motor) through logic control systems. Thus, a system operating sequence can be defined directly by successive output states of the devices involved. As such, Boolean algebra is the mathematical foundation for system behavioristic modeling. Such formal modeling requires a functional description and analysis of the operating sequences to ensure the execution of predictable operational sequences. For a system with a single cycle of operation, modeling consists of converting the operating sequences into the corresponding truth table or process-switching sequence table. For the modeling of a system with several cycles of operations and concurrent operating sequences, state diagrams can be used. In the case of multiple cycles of operations and parallel operating sequences, a petri net or state function chart (SFC)-based modeling is preferred. Hence, from these formal models, it is possible to derive the Boolean functions relating system state outputs to state input transition conditions.

Furthermore, the sequential logic controller converts switching input conditions into energized or de-energized state outputs for each device. Hence, in logic controller design, the challenge is to determine the logical linkages, always validating any combination of switching input conditions with changes in state outputs. This can be achieved by solving the system operating Boolean-based models.

The main objective of this chapter is to lay out a methodology for Boolean-based modeling and the design of logic controllers that enables input/output (I/O) Boolean functions to be solved. First, Boolean algebra is reviewed, along with combinatorial and sequential logic tools. Then, some logic controller design methods based on functional analysis and various formal process modeling techniques (e.g. switching theory, state diagram, SFC) are presented. Thus, a Boolean-based logic controller is implemented by using solid-state electronic devices. This is covered along with the procedure to develop schematic electrical wiring diagrams for automation design projects. Finally, automation applications for biomedical, production systems, transportation, and handling systems are described to illustrate Boolean-based modeling and design of logic controller algorithms and circuitries.

Control Of Mechatronic Systems: Model-Driven Design And Implementation Guidelines,
First Edition. Patrick O.J. Kaltjob.
© 2021 Patrick O.J. Kaltjob. Published 2021 by John Wiley & Sons Ltd.

6.2 Generic Boolean-Based Modeling Methodology

Discrete event systems have switching system outputs due to a change in the set of binary input variables (binary level 1 or logic 1 is often referred to as TRUE, ON, or HIGH, while binary 0 or logic 0 is referred to as FALSE, OFF, or LOW), sometimes in a specific timely order. The behavioral characteristics of discrete event systems can be captured using a binary-valued condition concept. Discrete events occurring in a given order are called sequential events, while those occurring randomly based on a defined input/output relationship are called combinational events. Thus, discrete event system dynamical modeling describes time-based evolution of Boolean variable values through the powerful Boolean algebra mathematical foundation (including Boolean differential calculus). The resulting Boolean-based dynamical model consists of deriving the Boolean function relating any possible output variable changes with respect to input variable changes. Consequently, the logic controller of a discrete event system aims to ensure the execution of a number of discrete event process operations by solving those Boolean equations.

Thus, the design of a logic controller requires establishing the model of a combination of discrete event system state outputs based on a set of system inputs and present system states. Subsequent sections present some techniques for designing a logic controller that enables the execution of repetitive process operations, while circumventing the time-consuming trial-and-error approach that is commonly used. A design methodology ensuring the automatic execution of process operations by logic controllers is summarized in Table 6.1.

6.2.1 System Operation Description and Functional Analysis

The functional analysis aims to characterize all system operating cycles by decomposing the system operation execution into major operating tasks or functions and identifying their activation and deactivation conditions. It is then possible to establish the chronological order of the process operations execution, also called process sequencing. This is achieved by gathering overall process information, such as present system state conditions (e.g. initial condition, condition of action termination, operating time constraints), input transition conditions (level or push type), and even activated (next) system state outputs. The procedure to perform a functional analysis of a discrete event system can be summarized as follows:

1) Perform the functional decomposition of system operations using a functional analysis system technique (FAST) or structural analysis and design technique (SADT) that consists of breaking down some process operating functions into the primary and secondary levels. The FAST is presented in the Section 6.5.3. With the SADT, a function is defined as the process operation to be carried out, while system operations are lists of actions to be performed by system actuating devices. Hence, in the function table (FT), for each action, there is an activation condition required prior to its execution (precondition). Information on all actions assigned to each function is presented in chronological order in the FT, as depicted in Table 6.2.

2) Once each function has been structured in a chronological order of involved actions, the table can be converted into a dependency chart (DC) to present these functions as a sequence. Here, DC illustrates the relation over time between the different functions through rectangles, as depicted in Figure 6.1. The sequential execution of functions is illustrated by arrows, which could describe either a parallel execution or an alternative execution of functions. DC and FT should be refined while considering all design specifications. Subsequently, FT and DC can be transformed into SFC, as presented later in this chapter.

Table 6.1 Step-by-step logic controller design methodology.

Design steps	Item description
Mechatronic system operating and functional analysis	
1) List all the devices involved in the system operating behavior.	Listing of processes (analog/digital) related to the operating sequence execution, such as the startup/shutdown of actuating, sensing devices, as well as inputs/outputs from the man–machine interface (e.g. AutoPB, Stop PB, mode indicator lamp, top LS), with their technical specifications.
2) Identify all device operational conditions and constraints, operating threshold values, and initial conditions.	Bidirectional motor, n-multispeed drive requiring n contactors, etc. Operating cycle conditions: simultaneous or concurrent operating sequences, level-type or pulse-type switch operating mode, etc. Use a symbol for each I/O device.
3) Classify them into analog/ digital, input/output system variables, and (level or pulse) types.	Inputs: e.g. push buttons, limit switches. Outputs: e.g. solenoids, lights, motors.
4) Draw a piping and instrumentation diagram (P&ID) and process flow diagram (P&FD).	If possible.
5) Develop functional decomposition and analysis.	Using decomposition methods (e.g. FAST, SADT), organize each process combinational or sequential event in the equivalent process operation and input transition conditions.
6) Derive each activation order of process operations (operating cycle) for a group of sequential events or for a combination of all possible input conditions for an events group.	For all sequential events, regroup all correlated process operations into successive sequences (operating cycle). List each sequence with the system outputs and inputs causing the transition between sequences. For all combinational events, summarize within the truth table all conditions of activation for each process operation (output). For remaining conditions with any activation, mark them as undetermined.
7) For each operating cycle: identify and describe the type of system operating cycle.	Single operational cycle. Multiple operational cycle: mutually exclusive or concurrent. Cyclic or acyclic (auto/manual/semi-automatic). Synchronization of the system operations (time-triggered, event-triggered, counter-triggered).
8) Perform a system output activation analysis: either on each sequence for each operating cycle or on each combination of input conditions for each group of combinational event.	Describe the initial condition of each operating cycle (activation). Identify duplicate, priority (emergency) sequences. Identify the system operating cycle deactivation condition.

(Continued)

Table 6.1 (Continued)

Design steps	Item description	
9) Perform a safety analysis of process operational execution.	Identify the potential defaults and develop interlocks to maintain the normal system operating in case of their occurrence; specify other safety constraints and remedy actions. Identify process hardware and software operational defaults.	
10) Derive logic operating sequences for each process operation, including safety measures in diagram format.	Develop interlocks and forcing action designs. Construct either truth, sequence, stable transition tables or state diagrams containing information on all inputs and the corresponding process outputs activated, as well as the current/next state for each operating sequence.	
11) Develop a process start-and-stop mode graphical analysis.	Classify within a coordinated cycle hierarchy, all cycles of process operations in a normal automatic production cycle, a manual, semi-automatic, or a maintenance cycle, a safety cycle, etc.	
Formal modeling of a discrete event system		
12) Sequential or combinational formal modeling of a discrete event system.	In the case of a single operating cycle, construct a truth table (combinational event) or a sequence table (sequential event).	From the sequence table, identify and separate each binary process output activation. Output[1]/deactivation Output[0]; for the input combinations within the truth table, use K-maps.
	In the case of multiple operating cycles with mutually exclusive sequences, construct the corresponding state transition diagram (sequentially structured event) or construct several truth tables (combinational event).	Derive an I/O Boolean equation for each state (even of state output) from the state transition diagram; from input combinations within the truth table, use K-maps and then combine the resulting state outputs with POS and SOP methods.
	In the case of multiple concurrent exclusive operating cycles, construct a state function chart.	
Sequential logic controller circuit design		
13) Select solid-state devices (latches, flips-flops).		
14) From the truth table or the sequence table, derive the state table associated with the logic devices selected.		
15) Use K-maps to derive I/O Boolean functions for device inputs.		
16) Simplify and check the consistency of I/O Boolean functions.	Using a timing diagram.	

(Continued)

Table 6.1 (Continued)

Design steps	Item description
Wiring diagram and logic controller programming languages	
17) Develop the wiring diagram between the sensing/actuating devices, as well as HMI and the logic programmable controller unit.	Electrical power supply diagram. I/O control unit wiring diagram (in/out of control unit).
18) Set a table of mnemonics for the I/O system variables based on the physical addresses in the programming environment.	Use a wiring diagram to assign addresses for system variables. This table should have information on the types of variables (analog/digital), address of I/O, range of values, etc.
19) Choose the programming language and develop a logic controller application.	Convert I/O Boolean equations into a classical programming language such as Ladder, Function Block, Sequential Function Chart, Instruction List (IL), structured text (ST), or even assembly language.
20) Apply the first scan subprogram.	
21) Size and select the computing hardware (control unit) and the detection/sensing equipment.	
22) Verify the cycles of system operation and check if the logic meets the safety specifications.	Use sequential logic software analysis tools and timing diagrams for all input transition conditions and system outputs.
Automation project documentation	
23) Establish wiring and electrical connection diagrams integrated into the control unit schematics with physical addresses.	Compile a P&I diagram, PF diagram, and electrical wiring diagram.
24) Documentation.	Provide a detailed system configuration; I/O wiring connection diagram; I/O address assignments; internal storage address assignments; storage register assignments; variable declaration; control program printout; and stored control program.

Table 6.2 Module component from FT using SADT method.

Function (from module)	Activation condition	Activation device	Action description (event)	Execution device
Action 1				

3) Perform the safety analysis of the process operations execution from expected defaults and specify other safety constraints (e.g. max. operating values). Then, for each expected system hardware and software operational default, develop interlocks and design-forcing actions to restrict the execution of some process operations (by adding process inputs and outputs as well as process states). For each process operation with safety constraints, the FT and DC should be refined according to newly identified process operating sequences, as well as input transitions.

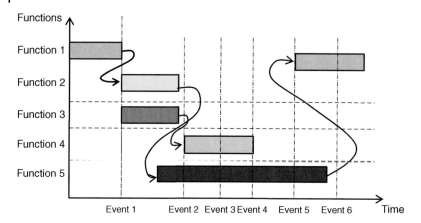

Figure 6.1 Dependency charts showing a sequence of process events.

4) Derive all process operating sequences, including those related to safety measures, for each process operating cycle. This should lead to the gathering together of all input conditions and their corresponding system outputs for each process operating sequence.
5) Present information representing the characterization of the process operations in a truth table (combinational structured event), a sequence table (sequentially structured event), a state transition table, or even an excitation table or a state transition diagram.

6.2.2 Combinatorial and Sequential Logic Systems

Considering each state of a logic system as an event, logic systems can be described as discrete event systems evolving from changes within input conditions. Hence, the logic relationship between switched input conditions and system output states can be either combinatorial or sequential. This could be captured in the form of Boolean functions resulting from the analysis of their corresponding truth table and sequence table, as well as from the switching theory. In the case of combinatorial logic systems, the output is only a function of the current input. Examples of combinational logic systems are solid-state electronics devices such as gates, decoders, and multiplexers. In contrast, sequential logic systems derive their output from past inputs as well as current ones, implying that memory is required. Typical sequential logic systems are latches (e.g. S-R, S-R Latch with Enable, D, etc.) that can change states at any time due to input changes, or flip-flops (e.g. Edge-Triggered D Flip-Flop, Edge-Triggered S-R, Edge-Triggered J-K, T Flip-Flop, etc.) that change states only when a clock edge is applied. Latches and flip-flops are summarized in Appendix D. The formal modeling tool for logic systems is Boolean algebra, with variables having a value of 1 or 0 and using logic gates AND($\cdot$), OR($+$), NOT, EOR($+$), NAND, or NOR to describe the logical links between input conditions and past or current state outputs. In short, the design of combinational or sequential logic systems would consist of deriving I/O Boolean functions representing the operating combination or sequence. In the case of combinational discrete event systems with few inputs (i.e. fewer than three inputs), the I/O Boolean relationship can be derived using the following steps: (i) construct the truth table; (ii) derive the stable transition table; and (iii) convert it into Karnaugh maps (K-maps) by using selected solid-state devices such as flip-flops or latches to derive the Boolean output function.

The formal modeling of such discrete event systems consists of capturing the logic relationship between all input transition conditions causing the activation of output variables in terms of Boolean functions. Hence, among formal modeling methods, there are: (i) the truth table and K-maps; (ii) the switching theory in sequence table analysis; (iii) the state diagram, which is a graphical representation of process operations based on expected input transitions; and (iv) the sequential function chart. The resulting I/O Boolean functions can be implemented using hundreds of relays or logic programmable devices, in the same way that a general-purpose computer is used to solve algebraic equations. This formalism is captured using input and output condition variable events, as well as system operating states. These formal modeling methods are presented in the following subsections.

6.2.2.1 Combinational Modeling Tools: Truth Table, SOP, Product of Sums (POS), K-Maps

Sum of products (SOP) is a method used to derive I/O Boolean relationships for combinational logic systems by using their corresponding truth table. Here, for each ith row of the truth table, the logic input combination activating any output is given by:

$$Product(i) = \prod inputs \tag{6.1}$$

This could be implemented by connecting all involved relays in series (AND gate). Hence, the SOP method defines the Boolean relationship between all input combinations activating a jth output by:

$$Output(j) = SOP = \sum_j Product(i) \tag{6.2}$$

The hardwired implementation of this Boolean function would require the use of an AND and OR gate (parallel collection of series connected relays).

It is usually difficult to derive and simplify the resulting I/O Boolean function from the truth table with more than two inputs and one output by using Boolean algebra and theorems. A reduction technique tool has been proposed based on K-maps, consisting of a graphical representation of the truth table where columns and rows correspond to each logic input variable. Here, adjacent row and column designations differ by only one bit. This is called a gray code sequence. For example, with three inputs, the gray code sequence format as a change of one binary between sequences would be $000 - 001 - 011 - 010 - 110 - 111 - 101 - 100$. In the case of n-input variables in the truth table, these would correspond to 2^n boxes in K-maps. Hence, developing a reduced I/O Boolean function would consist of:

1) Constructing the truth table.
2) Selecting an appropriate K-map with 2^n boxes for n inputs from the truth table.
3) Applying the gray code sequence for the row and column designations of the K-maps.
4) Copying the 1s and 0s from the output locations in the truth table to the corresponding boxes in the K-maps. If the variable is undetermined, an output value X should be assigned in that box.
5) Encircling all adjacent 1s in the same column and in the same row of the K-maps.
6) Writing the Boolean function and the product term for each circle (group).
7) Simplifying the resulting I/O functions by retaining only the common variables. Undetermined logic output variables X can be discarded in the minimization process. In this case, the resulting set of product terms would be combined using an OR logic function to derive the minimized Boolean function.
8) Writing SOPs for all previous Boolean expressions to derive an I/O Boolean function (sum of the common variables from each group).

K-maps can be used for up to eight input variables. Above that, the simplification can be readily programmed by a computer-aided design tool such as Quine-McCluskey's technique,

Table 6.3 Truth table for three inputs and one output.

A	B	C	Y
0	0	0	0
0	0	1	0
0	1	0	1
0	1	1	0
1	0	0	1
1	0	1	0
1	1	0	1
1	1	1	1

Table 6.4 Corresponding K-map for three inputs.

in order to determine a minimum SOP expression as a sum of minterms, or Petrick's technique, in order to determine all possible minimum SOP solutions. Tables 6.3 and 6.4 provide examples of a three-input-and-one-output truth table and its corresponding K-map.

Using the grouping method for a simplified SOP, the Boolean equation for each group is:

Group (in green) $=\overline{A}(B\overline{C}) + A(B\overline{C})$ with $(B\overline{C})$ as common variables
Group (in blue) $=(A)B(\overline{C}) + (A)\overline{B}(\overline{C})$ with $(A\overline{C})$ as common variables
Group (in red) $=(AB)(\overline{C}) + (AB)(C)$ with (AB) as common variables

The simplified SOP Boolean expression is:

$$\overline{Y} = B\overline{C} + A\overline{C} + AB$$

6.2.2.2 Sequential Modeling Tools: Sequence Table, Switching Theory, and State Diagram

For all discrete events within each operating cycle, the operating sequence can be summarized into a sequence table containing, for each sequence, information related to the input transition conditions and the energized or de-energized system output values. I/O Boolean functions characterizing the sequence table can be determined as follows:

1) List in a table all input and output devices involved in process operations, with their respective logic variables. If possible, add information regarding the types of operating modes (level or push) and their initial conditions.
2) From the process functional analysis, construct the sequence table for each operating cycle such that the operating sequences are listed in chronological order in the first column. In the following columns, list their corresponding activated system inputs and the logic value for each system output, respectively.

3) From the analysis of the sequence table, search for any change in a set of system output values that has the same combination of input values but results in different system output values. Those identical sequences (i.e. where the same previous sequence output values along with the same system inputs cause different system outputs) lead to a confusing situation where input variables identical to previous sequence output values can set out more than one set of system output values.

4) In order to differentiate the identical sequences from all other sequences in the operating cycle, as many virtual output variables as necessary can be created.

5) For each system (virtual and real) output variable, derive the I/O Boolean function by identifying the input transition conditions and the output and virtual variables that need to be activated or deactivated (i.e. turn those variables ON and OFF) such that:

$$Output_i[1] = \sum all\ input\ transition\ conditions\ activating\ Output_i \tag{6.3}$$

While:

$$Output_i[0] = \prod all\ input\ transition\ conditions\ deactivating\ Output_i \tag{6.4}$$

Then:

$$Output_i = (Output_i + Output_i[1]) \bullet (\overline{Output_i[0]}) \tag{6.5}$$

Example 6.1 Consider a system with two logic input variables (push buttons START and STOP) that are used to activate and deactivate the system output variable MOTOR_STARTER. This defines the process state of the motor ON or OFF, as defined in Table 6.6.

In the case of the motor starter, $MOTOR_{STARTER}[1] = START$ and $MOTOR_{STARTER}[0] = STOP$, the I/O Boolean function will result in:

$$MOTOR_{STARTER} = (MOTOR_{STARTER} + MOTOR_{STARTER}[1]) \bullet \overline{MOTOR_{STARTER}[0]}$$
$$= (MOTOR_{STARTER} + START) \bullet \overline{STOP}$$

This logic-based method of process modeling allows a system output to be turned or maintained at zero (OFF), especially when there is a conflict between input transition conditions (e.g. pushing start and stop in the same time).

Another method to capture the behavior of discrete event systems is the state diagram. A state diagram is a graphical technique that depicts the behavior of discrete event systems through an evaluation of the state output change under a specific activation of input transition conditions. Two types of diagrams may be used: (i) the Moore type, where the state output value depends on the present state; and (ii) the Mealy type, where the state output value depends on the present state as well as the state input transition conditions. The drawing of a state diagram requires the following assumptions: (i) state transitions are only activated by the system input status change; (ii) a system is always in one state and only one state; (iii) states are described as the combination of state output values; (iv) states are equivalent if the same system inputs produce an identical combination of state outputs; (v) with n system inputs, there must be at least 2^n outgoing arrows per circle (state); and (vi) system input variables are classified into either (a) the level-type operating mode, where the input transition condition (value) is maintained as active until the state output value changes (e.g. a light switch with a spring to hold position) or (b) the pulse-type operating mode, where the input transition condition actuates the state output and returns to its initial condition immediately after (e.g. emergency stop of machine tools). The Moore and Mealy state diagrams allow the derivation of I/O Boolean functions based on the type of solid-state device chosen (e.g. flip-flops and latches). A simplified version of

Table 6.5 State table based on a Mealy machine.

Present state	System input	Next state	System output

Table 6.6 Truth table of a starter motor.

START	STOP	MOTOR_STARTER
0	0	MOTOR_STARTER
0	1	0
1	0	1
1	1	0

a state diagram may also be used for discrete event systems, as it can easily be converted into programming languages using the transition equation method. All of these state diagrams are presented in this section. First, a general procedure to sketch state diagrams consists of the following steps:

1) Identifying all system inputs that cause process transitions (e.g. sensing devices, control panel push buttons, etc.), as well as all system outputs (actuating devices, control panel light indicators, etc.).
2) Establishing a list of all system operating sequences as a sequence table or all possible inputs combination that activate system outputs as a truth table.
3) Defining a state as a combination of system outputs that are activated by a set of process inputs.
4) Constructing a state (excitation) transition table with at least three columns: present state (the value for each possible state at time t); system input (all possible input transition conditions); and next state (the value for each possible state at time $t + 1$). Based on step 3, a column capturing system output may be added, as illustrated in Table 6.5. The next state is defined using solid-state devices.
5) Converting the present state variable values, the system input, and the next state into binary values using the gray code sequence format.

Example 6.2 Revisiting Example 6.1, the relationship between those variables is given through the truth table operating conditions as presented in Table 6.6. By choosing a D-flip-flop, the corresponding state transition table can be derived, as depicted in Table 6.7.

Hence, the resulting K-maps can be obtained as illustrated in Table 6.8; an example of the logic circuit equivalent is display in Figure 6.2.

Applying the SOP method, the I/O Boolean yields:

$$MOTOR_{STARTER} = START \bullet \overline{STOP} + MOTOR_{STARTER} \bullet \overline{STOP}$$
$$= (MOTOR_{STARTER} + START) \bullet \overline{STOP}$$

Using the D-flip-flop, considering N as the number of system states (system output combination) and n as the number of flip-flops, the design rule is given by:

$$2^n \geq N$$

Table 6.7 State transition table of a starter motor.

MOTOR_STARTER Present state (Q[t])	Process inputs START	STOP	MOTOR_STARTER Next state (Q[t + 1])	D
MOTOR_STARTER	0	0	*MOTOR_STARTER*	0
0	0	1	0	0
0	1	0	1	1
0	1	1	0	0
1	0	0	1	1
1	0	1	0	0
1	1	0	1	1
1	1	1	0	0

Table 6.8 K-maps of starter motor.

		MOTOR_STARTER STOP 00	01	11	10
START	0	0	0	0	1
	1	1	0	0	1

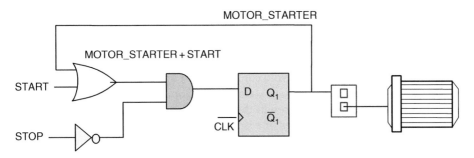

Figure 6.2 Logic circuit of motor starter with a D-flip-flop.

Here, with $n = 1$, the D-flip-flop gate output can be summarized as:

$$Q_1(t + 1) = START \bullet \overline{STOP} + Q_1(t) \bullet \overline{STOP} = (Q_1(t) + START) \bullet \overline{STOP}$$

As such:

$$D = (Q_1(t) + START) \bullet \overline{STOP} = (MOTOR_{STARTER}(t) + START) \bullet \overline{STOP}$$

6) From the finite-machine type chosen (Mealy or Moore), constructing either: (a) a state table based on the Mealy machine with four columns: (1) present state, q; (2) process inputs $(x_i y_i)$; (3) next state, for example, illustrated by $(q_0, 0)$; and (4) system output (value of the system output); or (b) a state table based on the Moore machine, with the particularities of q_{xx} and Z being, respectively, the state (present or next) values and the system output values. This is illustrated in Tables 6.9 and 6.10.

Table 6.9 State table based on the Mealy machine.

q	00	01	11	10
		$x_i y_i$		
q_0	$q_0, 0$	$q_0, 1$	$q_1, 0$	$q_0, 1$
q_1	$q_0, 1$	$q_1, 0$	$q_1, 1$	$q_1, 0$

Table 6.10 State table based on the Moore machine.

q	01	01	11	10	z
			$x_i y_i$		
q_{00}	q_{00}	q_{01}	q_{10}	q_{01}	0
q_{01}	q_{00}	q_{01}	q_{10}	q_{01}	1
q_{10}	q_{01}	q_{10}	q_{11}	q_{10}	0
q_{11}	q_{01}	q_{10}	q_{11}	q_{10}	1

7) Sketching the state diagram by representing the system states with circles, each arrow corresponding to each possible input transition condition. An example of a Mealy state diagram is illustrated in Figure 6.3, where the process state changes from state 1 to state 2 due to the activation of the input transition condition, causing the process output to be set at 0, as illustrated in Figure 6.3. In the case of a Moore state diagram, the process outputs are indicated within the state, as they are independent of the input transition condition.

8) Applying the binary coding on system input transition conditions and state outputs. They are usually written above each arrow and separated by a slash (/), as depicted in Figure 6.3(b) and (d)

Using the sequence table, and in order to avoid confusion as in the case of a system operating with multiple alternative sequences, the system state change from input transition conditions and energized outputs could be reduced through the sketching of a simplified state diagram. Figure 6.4 depicts the generic sketching rules of a simplified state diagram. Here, through the construction of arrows leaving and arriving, each state, i (S_i), can be sketched as such: (i) arrows arriving to S_i correspond to those input transition conditions that turn ON state S_i and keep it ON while deactivating a preceding state of S_i; (ii) all arrows leaving S_i correspond to input transition conditions that turn OFF S_i and activate a following state of S_i.

Hence, the sketched state diagram described in Figure 3.5 can be converted into Boolean functions using:

$$STATE_i = \left(STATE_i + \sum_{j=1}^{n}(T_{i,j} \bullet STATE_j) \right) \bullet \prod_{k=1}^{m} \overline{(T_{l,k} \bullet STATE_{l+1,k})} \tag{6.6}$$

$$O_m = \sum S_{state\_O_m\_energized} \tag{6.7}$$

with $\prod_{k=1}^{m} \left(\overline{T_{l,k} \bullet STATE_{l+1,k}} \right)$ being all next states and involved input transition conditions deactivating the $STATE_i$, while $\sum_{j=1}^{n}(T_{i,j} \bullet STATE_j)$ is all previous states and involved input transition conditions that activate the $STATE_i$. Recall that $STATE_i$ represents the state of the system at

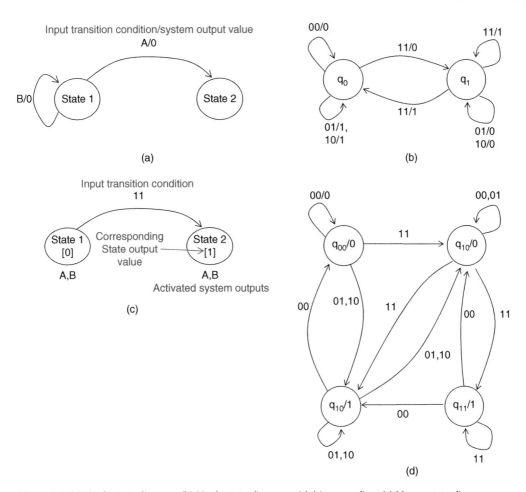

Figure 6.3 (a) Mealy state diagram. (b) Mealy state diagram with binary coding. (c) Moore state diagram. (d) Moore state diagram with binary coding.

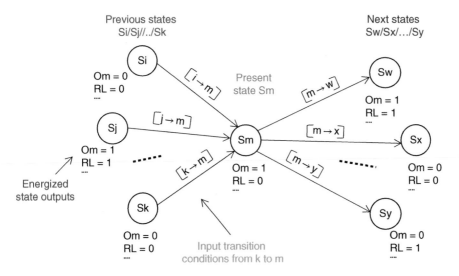

Figure 6.4 Generic state diagram sketching.

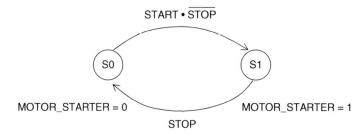

Figure 6.5 A two-state motor starter state diagram.

the ith operating sequence characterized by all the output logic values, n is the number of input transitions leading to the ith state, m is the number of input transitions out of the ith state, $T_{i,j}$ is the logical input conditions of a transition from the jth state to the ith state, and $T_{i,k}$ is the logical input conditions of a transition from the ith state to the kth state. If it is not possible to reach a state, a virtual intermediary state with identical output values to the preceding state should be created and inserted between the two.

Example 6.3   Recall Example 6.1 on the motor starter system. The corresponding state diagram is illustrated in Figure 6.5, such that $Motor_{starter} = 0$ is in state 0 (S0) and $Motor_{starter} = 1$ in state 1 (S1). A system state change occurs only when the start push button is activated, while simultaneously the stop push button is not activated ($START \bullet \overline{STOP}$). The system returns to its initial system state 0 only when $STOP = 1$.

Here, applying Equation (3.9) for two states, S0 and S1, leads to Boolean equations describing S0 and S1 such as:

$$S0 = (S0 + S1 \bullet STOP) \bullet \left(\overline{S1 \bullet START \bullet \overline{STOP}} \right)$$
$$S1 = \left(S1 + S0 \bullet START \bullet \overline{STOP} \right) \bullet \left(\overline{S0 \bullet STOP} \right)$$

In order to solve these Boolean equations, it will be necessary to have the initial conditions of each state and the order of execution. Equation (3.5) can only be applied if such information is not available, otherwise, once the state I/O Boolean functions are described, the system outputs can be defined as functions of the states. Using Equation (3.10), it can be found that the system output ($MOTOR_{STARTER}$) ($MOTOR\_STARTER$) is ON in state 1 such that:

$$MOTOR_{STARTER} = S1$$

Substituting $MOTOR_{STARTER}$ into S1 (S0 being discarded, as there is no output variable activated), this equation yields:

$$MOTOR_{STARTER} = (MOTOR_{STARTER} + START \bullet \overline{STOP}) \bullet \overline{STOP}$$

which can be simplified into:

$$MOTOR_{STARTER} = (MOTOR_{STARTER} + START) \bullet \overline{STOP}$$

With this method, the I/O Boolean functions are directly obtained from the state diagram with little effort. Instead of hardwired solid-state device implementation as in the previous example, this I/O Boolean function can be implemented using a software-based logic programmable device. However, some safety measures should be inserted to prevent overlapped sequence execution. This could be done by ensuring sufficient elapsed time of transition between multiple new and old states during the activation and deactivation of output devices.

In the following sections, illustrative design steps of logic controllers for several industrial processes are presented, from system functional analysis to formal modeling, based on the following methodology list:

1) process schematics;
2) I/O involved equipment listing;
3) table of sequence or truth table;
4) switching theory or state diagram;
5) state transition table and K-maps;
6) I/O Boolean function;
7) timing diagram;
8) logic control circuitry.

6.3 Production Systems

6.3.1 Portico Scratcher

Consider a four-stage pozzolana removal process as depicted in Figure 6.6. First, by pushing StartPB, the scratcher driven by motors M5 and M6 positions itself transversally just above the pozzolana stockpile. Then, while the secondary and primary arms, driven by the activation of motors (M2A and M3A), are moving down over the pozzolana stockpile, the motors (M1 and M4) scratch the pozzolana from one side to another, up to the conveyor, for 139 s. After a short stop, the motor contactors M2A and M3A reposition for 2 s farther down the pozzolana stockpile. Then, for another 139 s, M1 and M4 scratch the pozzolana. This scratching stage lasts until the primary arm reaches the lower-level limit switch LS1. Here, both motor contactors M2B and M3B are activated to retract simultaneously until the secondary arm reaches the upper position given by LS2. At this position, motors M5 and M6 move to a new transversal position above the pozzolana and the scratching stage of the process can restart up to the lower-level limit switch LS1. The conveyor belt collects the pozzolana scratched continuously toward the next cement-drying station.

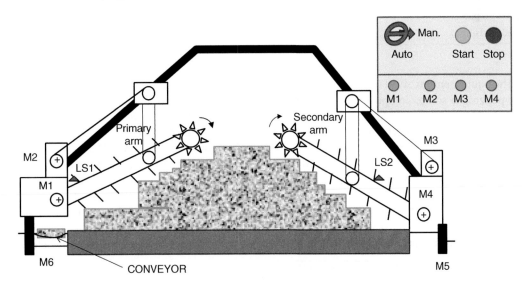

Figure 6.6 Schematic of the cement pozzolana scratching process.

Table 6.11 Equipment involved in the cement pozzolana scratching process.

Input equipment	Symbol	Output equipment	Symbol
Start push button	*Start*	Transversal motion motors	$M5, M6$
Stop push button	*Stop*	Control panel Stop red LED	$L1$
Portico transversal position switch	*PD*	Control panel Start green LED	$L2$
Primary arm lower-position switch	*LS1*	Clockwise motion (up) of primary arm (M2)	$M2B$
Secondary arm upper-position switch	*LS2*	Anticlockwise motion (down) of primary arm (M2)	$M2A$
Timer delay 1 (139s)	*T[139]*	Clockwise motion (down) of primary arm (M3)	$M3A$
Timer delay 2 (2s)	*T[2]*	Anticlockwise motion (up) of primary arm (M3)	$M3B$
		Scratching motors (pozzo removal)	$M4, M1$
		Timer1	$T1$
		Timer2	$T2$

Table 6.12 Sequence table for the cement pozzolana scratching process.

	Inputs							Outputs							
Sequences	**START**	**STOP**	**LS1**	**LS2**	**PD**	**T[139]**	**T[2]**	**M1**	**M2**	**M5**	**Dir**	**T1**	**T2**	**M2A**	**M2B**
Transversal motion	1	0	0	1	0	0	0	0	0	1	0	0	0	0	0
Moving down	0	0	0	0	1	0	1	0	1	0	0	1	0	1	0
Scratching and removal	0	0	0	0	1	1	0	1	0	0	0	0	1	0	0
Moving down	0	0	0	0	1	0	1	0	1	0	0	1	0	1	0
End scratching and removal pass	0	0	1	0	0	0	0	0	1	0	1	0	0	0	1
End of rising	0	0	0	1	0	0	0	0	0	0	0	0	0	0	0

The equipment involved is listed in Table 6.11. From the functional analysis, the sequences of the pozzolana scratching process are presented in Table 6.12. It should be noticed that motors (M3 and M2), (M1 and M4), or (M5 and M6) run simultaneously, usually in the same direction, except for M2 and M3, which run in the opposite direction. Only M2 and M3 are bidirectional motors; thus, M2 has two motor contactors (M2A, M2B; for down motion, M2A is ON when DIR is set to 0, while for up motion, M2B is ON when DIR is set to 1) and M3 has two motor contactors (M3A and M3B). Therefore, the number of outputs can be reduced from eight to six. The transversal positioning is discarded in the sequence table (Table 6.12). The activation of the photodetector (PD=1) indicates that the scratcher is over the pozzolana and needs transversal repositioning.

From an analysis of the sequence table with switching theory, the I/O Boolean equation yields:

$$M1 = (M1 + T[139] \bullet \overline{T[2]}) \bullet \overline{STOP}$$

$$M1 = (M1 + T[139] \bullet \overline{T[2]}) \bullet \overline{T[2]} \bullet T[139]$$

$$M2 = (M2 + T[2] \bullet PD \bullet \overline{LS2} \bullet T[2] \bullet \overline{T[139]} + \overline{LS1})$$

$$\bullet \overline{T[139] \bullet T[2]} + \overline{T[2] \bullet PD} \bullet LS1 + LS2$$

$$M5 = (M5 + LS2) \bullet \overline{T[2] \bullet PD \bullet \overline{LS2}}$$

$$R1 = (R1 + \overline{T[2] \bullet PD} \bullet LS1) \bullet LS2$$

The state diagram derived from the sequence table (Table 6.12) is depicted in Figure 6.7(a), while the resulting timing diagram is shown in Figure 6.7(b). Equivalent I/O Boolean functions can also be obtained from this state diagram.

6.4 Biomedical Systems

6.4.1 Robot-Assisted Surgery

A robot-assisted laparoscope-based surgery for an incisional biopsy operation consists of removing a fragment of tissue for pathological diagnosis. After patient positioning, the preparatory phase consists of anesthesiological injection into the patient tissue area. Then, a robot arm with a manipulator tool performs the skin incision. The four robotic arms (two manipulators: one for the cutting operation and one for laparoscope handling) are inserted into the patient in the docking phase. Once docking around the specific organ is complete, the surgeon completes the biopsy from the remote console operation. If necessary, the drainage tools are also inserted. After biopsy operation (cutting of tissue or anatomical specimen), an undocking phase allows the retrieval of anatomical specimens by returning the robotic arm to its initial position. Finally, the suture operation is performed. Once complete, the patient can be safely transferred to the recovery room in the post-operating phase. The apparatus used is depicted in Figure 6.8(a).

Activation of the start button causes the robot arm motor extension (MT) to be energized by the contactor (KME) such that the piston rod of robot arm 1 extends it for a 2 s time delay ($T1/TD[10\,sec]$). When the time delay occurs, the robot arm is extended from level 1 ($EX0$) to reach level ($EX1$) as an overhaul docking approach. Then, the motor-driven cutter (biopsy) MT2 contactor (KMB) is activated while the piston rod of the robot arm simultaneously advances halfway up ($EX2$). At this position, the motor ($MT2/KMB$) is de-energized while activation of the contactor (KMR) forces the rod to retract up to its initial position ($EX0$), where the contactor (KMR) is deactivated. The system remains at rest awaiting reactivation of the Start push button. It should be noticed that this biopsy operation is done between the incision operation and the suturing or skin closure operation. The system depicted in Figure 6.8(b) has one cycle made up of six (06) sequences, as presented in Table 6.13.

Recall that the major steps in the formal modeling process consist of: (1) the identification of sequences causing binary change (between levels 0 and 1); and (2) the identification of problematic sequences, either (i) those having a similar input combination causing different system outputs while having an identical previous output combination (this is called the input differentiation problem) or (ii) those duplicated sequences having a similar output combination resulting from identical input conditions and previous output combinations (this is called the output differentiation problem). If there is any such case, virtual variables should be added to differentiate between these events. This analysis is summarized in Table 6.14. Furthermore, it is convenient to include the Stop push button in the resulting Boolean function as a mandatory interlocking operation.

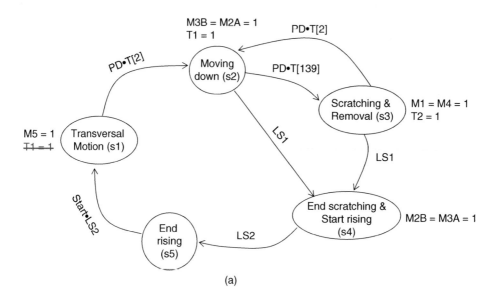

(a)

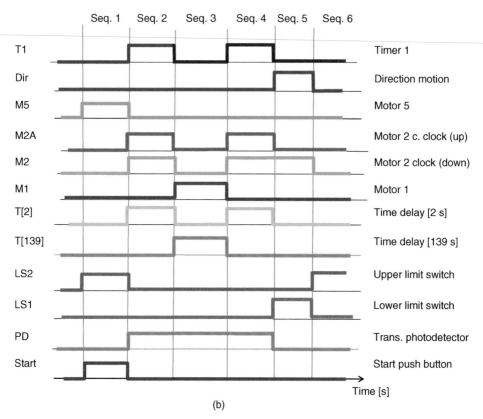

(b)

Figure 6.7 (a) State diagram of the cement pozzolana scratching process. (b) Timing diagram of the cement pozzolana scratching process.

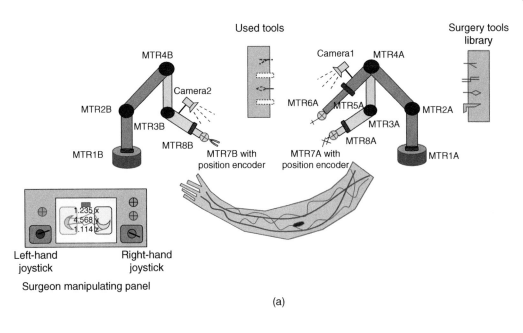

(a)

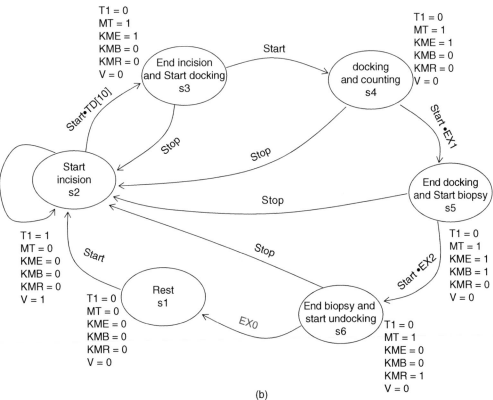

(b)

Figure 6.8 (a) Robot-assisted biopsy surgery. (b) State diagram of the robot-assisted biopsy operation. (c) Timing diagram for the robot-assisted biopsy process.

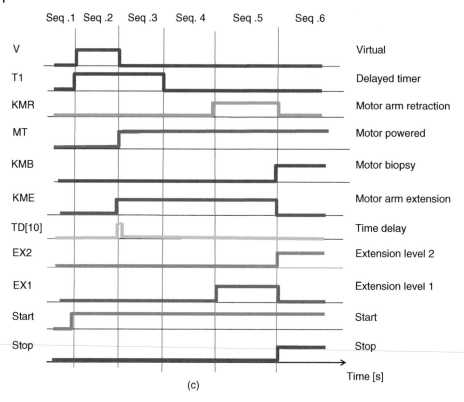

Figure 6.8 (Continued)

Table 6.13 Equipment involved in the biopsy operation process.

Input equipment	Symbol	Output equipment	Symbol
Start push button	*Start*	Downward robot arm motor 1 (MT1)	*KME*
Stop push button	*Stop*	Upward robot arm motor 1 (MT1)	*KMR*
No extention	*EX0*	Biopsy motor 2 (MT2)	*KMB*
Extention level 1	*EX1*	Counter on/off	*CT*
Extention level 2	*EX2*	Timer on/off	*T1*
Counter up	*CM1*	Position motor-powered	*MT*
Timer delay	*TD[10]*	Biopsy motor-powered	*MT2*

Using the state diagram method, after completing the functional analysis of the drilling machine process, the process can be described through six system states, as depicted in Figure 6.8(b). Hence, the I/O Boolean equations will result in:

$$KME = (KME + TD[10]) \bullet (\overline{S1} + S2) \bullet \overline{STOP}$$
$$MT = (MT + TD[10]) \bullet (START + S2) \bullet \overline{STOP}$$
$$KMR = (KMR + S1) \bullet (\overline{S2} + S1) \bullet \overline{STOP}$$
$$T1 = (T1 + V \bullet START) \bullet (V + TD[10]) \bullet \overline{STOP}$$
$$V = (V + START) \bullet TD[10]$$
$$TD[10] \equiv Timercpt$$

Table 6.14 Sequence table analysis for the drilling machine process using the switching theory (check correctness).

Sequences	Input devices					Output devices					Changing					Virtual
	Start	EX0	EX1	EX2	TD[10]	KME	KMB	MT	KMR	T1	OP	OP	M	D	T1	V
Rest	0	0	0	0	0	0	0	0	0	0						0
Start incision	1	0	0	0	0	0	0	0	0	1	X^1				X^1	1
End incision and start docking	1	0	0	0	1	1	0	1	0	1		X^0	X^1			0
Stop docking	1	0	0	0	0	1	0	1	0	0	X^1		X^0		X^1	0
Start biopsy	1	0	1	0	0	1	0	1	1	0	X^0	X^1			X^0	0
End biopsy and undocking	1	1	0	1	0	0	1	1	0	0						0
Rest	0	0	0	0	0	0	0	0	0	0						0

Usually, I/O Boolean functions derived from the sequence table approach have fewer Boolean functions than those of the state diagram. Hence, the sequence table method is suitable for logic control programs with a smaller size or shorter execution time. However, the state diagram offers much more (straightforward) Boolean equations that are easily used to validate the process operation modeling. The corresponding timing diagram is sketched in Figure 6.8(c).

6.4.2 Laser Surgery Devices

Carbon dioxide laser-based surgery is a medical operation where a laser cuts tissue by emitting an intense beam of light, irradiating it through a photochemical effect. Such an operation follows the sequence of: (i) positioning the laser over the area to be treated using a motorized reflector (mirror) to incise the patient's skin; (ii) tissue dissection by heating, which starts by varying the wavelength, pulse frequency, and light power; (iii) heating, which continues until the tissue is vaporized when the temperature of the treated area reaches a set value (as detected by thermosensor); and (iv) coagulation of the tissue, which is produced by varying the wavelength up to level 2 (WAVEL2). Consider a laser-based tumor tissue treatment as illustrated in Figure 6.9(a), whose devices are listed in Table 6.15.

Typical surgical routine procedures are, sequentially, incision and drainage, central line placement, biopsy or tissue evaporation, and, finally, wound closure or laceration repair by coagulation, to stop bleeding. Initially, during tumor tissue-based treatment, the laser beam is activated ($CR1 = 1$) once mirror positioning is complete, as indicated by the low photodetector ($PD = 1$). Then, the laser beam wavelength is set to level 2 ($CR2 = 1$) when the treated area temperature rises to the temperature given by thermodetector level 2 ($TD1 = 1$). This corresponds to the coagulation and temperature reduction processes, which are summarized in Table 6.16. Here, the control relay for the inflow of the laser beam is always activated once the Start button is pushed. Detectors (PD) and ($TD1$) operate on level mode. A memory function may be required to maintain the laser beam to level 2 until the burning temperature given by ($TD1$) is reached.

From the sequence table analysis, only ($CR2$) changes its logic output value (activated/deactivated). A timer is used so that the coagulation phase is processed once a thermodetector ($TD1$) detects the upper tissue evaporation temperature limit. In order to ensure proper treatment time, the laser beam is alternately activated and deactivated every 15 s for a total

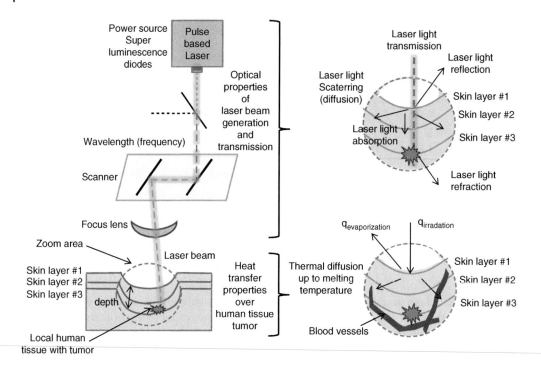

Figure 6.9 Laser surgery operating schematic.

Table 6.15 Equipment involved in a toxic liquid treatment tank.

Input equipment	Symbol	Output equipment	Symbol
Start push button	*START_PB*	Control relay 1 heating laser beam activated	*CR*1
Mirror position photodetector	*PD*	Control relay 2 coagulating laser beam wavelength level 2	*CR*2
Thermodetector	*TD*1	Wavelength level 2 LED indicator	*WAVEL*2

Table 6.16 Sequence table analysis for a toxic liquid treatment tank.

Sequences	Input devices			Output devices		Analysis	
	START_PB	*TD*1	*PD*	*CR*1	*CR*2	Changing	Virtual
1) Laser positioning and skin incision	1	0	1	1	0	X	
2) Tissue dissection by heating	1	0	0	1	0		
3) Tissue vaporized	1	1	0	1	0	X	
4) Tissue coagulated	1	0	0	1	1		

period of 200 s. The resulting I/O Boolean equations are given by:

$$CR1 = (CR1 + START\_PB) \bullet \overline{TD1}$$

$$WAVEL2 = CR1$$

From sequence table analysis, only $CR2$ changes its logic output value (activated/deactivated), so that $TD1 = 1$ is required for $CR2 = 1$, resulting in $CR2[1] = TD1$, while $TIMER[15s] = 1$ is required for $CR2 = 0$, causing $CR2[0] = TIMER[15s]$. Hence, using the equipment listed before, the I/O Boolean function for $CR2$ yields:

$$CR2 = (CR2 + CR2[1]) \bullet \overline{CR2[0]}$$

Thus:

$$CR2 = (CR2 + TD1) \bullet \overline{TIMER[15s]}$$

$$TIMER_{200\,sec}$$

$$LASER\_ON = TIMER[200\ s]$$

$$TIMER_{15SEC} = TIMER[200\ s]$$

In the state diagram format, with $S2 = CR2$ and $S1 = \overline{CR2}$, using Equation (3.9), the state Boolean function would lead to:

$$S2 = (CR2 + TD1 \bullet S1 \bullet TD1 \bullet \overline{PD} \bullet S2) \bullet (\overline{PD \bullet S1})$$

which is equivalent to:

$$CR2 = (CR2 + TD1 \bullet CR2 \bullet TD1 \bullet \overline{PD} \bullet CR2) \bullet (\overline{PD \bullet \overline{CR2}}) = CR2 + TD1 \bullet PD$$

The equivalent Mealy and Moore state diagrams are depicted in Figures 6.10 and 6.11. The transition table could be derived as summarized in Table 6.17.

Figure 6.10 Mealy state diagram of the laser surgery process.

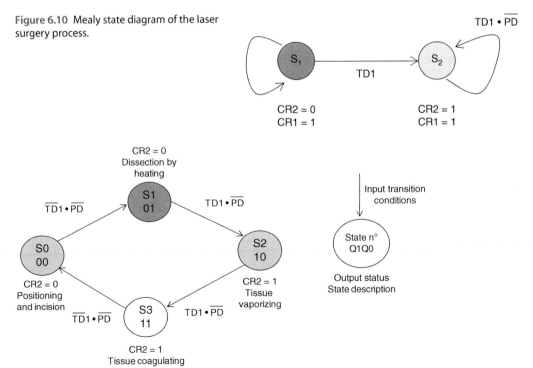

Figure 6.11 Moore state diagram of the laser surgery process.

Using the current state ($Q1$, $Q0$), $TD1$, and PD from the state transition table (Table 6.17), it is possible to fill the corresponding value of the next state (CR2) in the excitation table. This is equivalent to the Karnaugh table summarized in Table 6.18.

The corresponding state transition table for the laser surgery processes whose operating sequences are summarized in Table 6.17 is given in Table 6.19.

Using the sequence table, the corresponding K-map is summarized in Table 6.20.

Using the D-flip-flop, the state Boolean function is given by:

$$CR2(t+1) = TD1 + \overline{CR2(t)} \bullet PD = \overline{CR2(t)} \bullet \overline{PD} \bullet \overline{TD1} + CR2 \bullet TD1 \bullet \overline{PD}$$

Table 6.17 Laser surgery process state transition table.

State (Q1, Q0)	TD1	PD	Next state (Q1, Q2)	Corresponding output (CR2)
00	0	0	01	0
01	1	0	10	0
10	0	0	11	1
11	0	1	00	1

Table 6.18 Corresponding K-maps of the laser surgery process.

		TD1 PD			
		00	01	11	10
(Q1, Q0)	00	1	X	X	X
	01	X	X	X	0
	11	X	1	X	X
	10	1	X	X	X

Table 6.19 Laser surgery process state transition table.

CR2(t)	TD1	PD	CR2(t + 1)
0	0	0	1
1	1	0	1
1	0	0	0
0	0	1	0

Table 6.20 Corresponding K-maps.

		TD1 PD			
		00	01	11	10
CR2	0	1	0	X	X
	1	0	X	X	1

6.5 Transportation Systems

6.5.1 Elevator Motion Systems

Consider a three-floor elevator system as depicted in Figure 6.12(a). The elevator starts an up or down motion when a direction (Up or Dn) is selected. When the elevator reaches the requesting floor, the cabin gate opens (OG) for 30 s and then closes (CG). Then, based on the selection of the ith floor by the *Pbi* (*Pb*1, *Pb*2, or *Pb*3) selected, the motor contactor for the selection direction (*M*1 or *M*2) turns on until the corresponding floor limit switch *SWi* (*SW*1, *SW*2, or *SW*3) is activated. Then, the motor stops (*M*1 or *M*2 deactivated) and the cabin gate opens for another 30 s, after which it is closed (CG). The gate detector (GPD) is used to activate the gate opening contactor by detecting person presence over the gate ramp. Also, when the elevator is waiting at a specific floor level, pressing the same floor level push button will keep the door open. Otherwise, it will request that the elevator moves to this specific level. In order to model such an elevator system, I/O Boolean functions can be derived from the combinational logic based on either a truth table covering all possible floor selections or advanced floor-counting strategies and computing devices.

For space reasons, only a two-floor elevator motion cycle (floor 1–floor 2 and floor 2–floor 1) is considered and analyzed here, using a sequence table and a state diagram. In order to reduce the number of variables, it is considered that the floor selection combines two variables: the direction and the selected floor level. For example, pushing button *Pb*1 activates *Pb*1 and Dn, while pushing button *Pb*2 activates *Pb*2 and Up. Table 6.21 lists the equipment involved.

I/O Boolean equations derived from the sequence table established in Table 6.22 are:

$$OG = (OG + Pb1 + SW2 + Pb2 \bullet T[30s] + SW1 \bullet D1) \bullet \overline{AU} \bullet \overline{T[30s]}$$

$$CG = (CG + T[30s]) \bullet \overline{AU} \bullet \overline{T[30s]}$$

The I/O equations with respect to the time diagram are:

$$MTR_{UP} = M1 = (M1 + SW1 + \overline{T[30s]} + V3) \bullet \overline{AU} \bullet \overline{SW2}$$

$$MTR_{DOWN} = M2 = (M2 + \overline{SW2} + \overline{T[30s]} + V4) \bullet \overline{AU} \bullet SW1$$

$$T1 = (T1 + Pb1 + SW2 + Pb2 \bullet T[30s] + SW1 \bullet V5) \bullet \overline{T[30s]}$$

Using the following virtual output variables, their Boolean equations yield:

$$V1 = (V1 + SW1) \bullet \overline{T[30]}$$

$$V3 = (V3 + SW1 \bullet T[30s]) \bullet (SW2 + \overline{T[30s]})$$

$$V4 = (V4 + SW1 \bullet \overline{T[30s]}) \bullet (SW2 + \overline{T[30s]})$$

$$V5 = (V5 + SW1) \bullet \overline{T[30s]}$$

An equivalent state diagram is sketched in Figure 6.12(b). Notice that the state diagram offers wider and easier modeling options as more than one operating cycle can be represented. The corresponding system-state Boolean equations yield:

$$W1 = (W1 + D \bullet SW1 + C1 \bullet (Pb1 + GPD)) \bullet \overline{SW1} \bullet T[30s]$$

$$C1 = (C1 + SW1 \bullet SW1 \bullet T[30]) \bullet \overline{Pb2} \bullet \overline{SW1} \bullet (Pb1 + GPD)$$

$$U = (U + C1 \bullet Pb2) \bullet \overline{SW2}$$

$$W2 = (W2 + U \bullet SW2 + C2 \bullet SW1 \bullet (Pb1 + GPD)) \bullet \overline{SW2} \bullet T[30s]$$

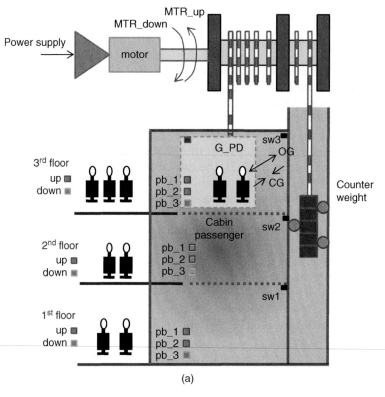

(a)

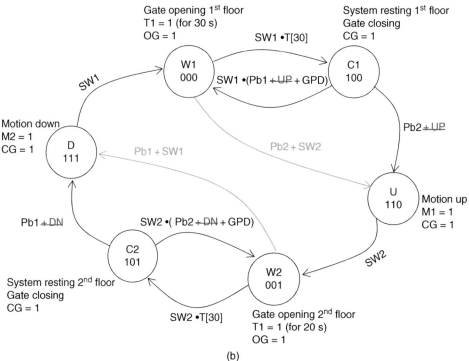

(b)

Figure 6.12 (a) Three-floor elevator system. (b) State diagram of a two-floor elevator system (UP/DN is optional). (c) Elevator timing diagram. (d) Circuit logic of a two-floor elevator system. (e) Connection diagram of the crane motion logic control system.

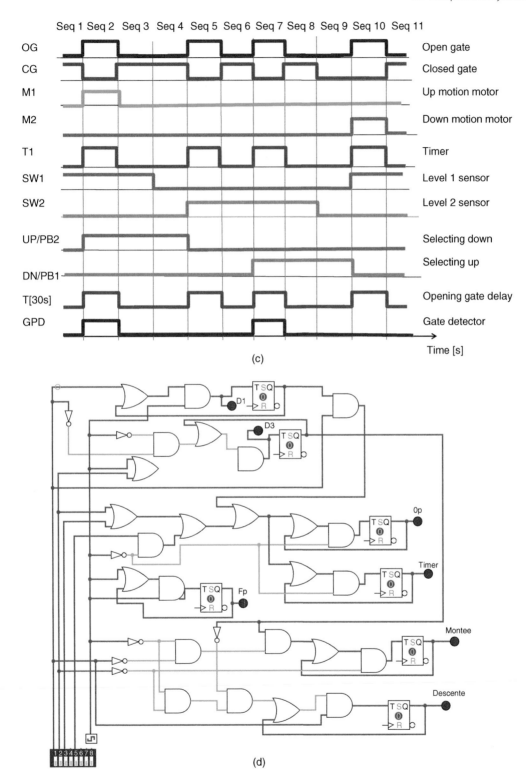

Figure 6.12 (Continued)

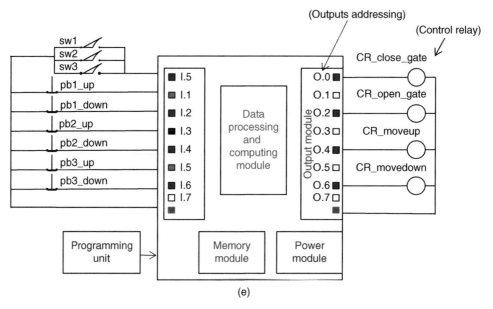

Figure 6.12 (*Continued*)

Table 6.21 Equipment involved in a three-floor elevator system.

Input equipment	Symbol	Output equipment	Symbol
Selecting up motion	Up	Bidirectional open cabin motor	OG
Selection down motion	Dn	Bidirectional closed cabin motor	CG
Level 1 sensor	$SW1$	Bidirectional up motor	$M1$
Level 2 sensor	$SW2$	Bidirectional down motor	$M2$
Level 3 sensor	$SW3$	Timer [ON/OFF]	$T1$
Calling for floor 1	$Pb1$		
Calling for floor 2	$Pb2$		
Calling for floor 3	$Pb3$		
Timer delay 1	$T[30]$		
Emergency stop	AU		
Cabine gate photodetector	GPD		

$$C2 = (C2 + W2 \bullet SW2 \bullet T[30]) \bullet \overline{Pb1} \bullet \overline{SW2 \bullet (Pb2 + GPD)}$$

$$D = (D + C2 \bullet Pb1) \bullet \overline{SW1}$$

with the system output Boolean equations given by:

$$CG = C2 + D + C1 + U$$

$$OG = W1 + W2$$

$$M1 = U$$

$$M2 = D$$

$$T1 = W1 + W2$$

Table 6.22 Sequence table analysis of a two-floor elevator system using switching theory.

Sequence	Input devices						Output devices					Changing					Virtual			
	GPD	SW1	SW2	Pb1	Pb2	T[30]	OG	CG	M1	M2	T1	OG	CG	M1	M2	T1	V1	V3	V4	V5
Rest	0	1	0	0	0	0	0	1	0	0	0						0	0	0	0
Selecting and door open	1	1	0	0	1	0	1	0	0	0	1	X^1				X^1	0	0	0	0
Door closed	1	0	0	0	0/1	1	0	1	0	0	0	X^0	X^1			X^0	0	0	0	0
Move up $1 \to 2$	0	0	0	0	0/1	0	0	1	1	0	0		X^0	X^1			0	0	0	0
Door open	0	0	1	0	0	0	1	0	0	0	1	X^1		X^0		X^1	0	1	1	0
Door closed	0	0	1	0	0	1	0	1	0	0	0	X^0	X^1			X^0	0	1	1	0
Selecting and door o	1	0	1	1	0	0	1	0	0	0	1	X^1	X^0			X^1	0	1	1	0
Door closed	0	0	1	0/1	0	1	0	1	0	0	0	X^0	X^1			X^0	0	1	1	0
Move down $2 \to 1$	0	0	0	0/1	0	0	0	0	0	1	0		X^0		X^1		0	0	0	0
Door open	0	1	0	0	0	0	1	0	0	0	1	X^1			X^0	X^1	1	0	0	1
Door closed	0	1	0	0	0	1	0	1	0	0	0	X^0	X^1			X^0	1	0	0	1
Rest	0	1	0	0	0	0	0	1	0	0	0		X^0				0	0	0	0

Table 6.23 Listing of input and output devices involved in an automated fruit-picker system.

Input equipment	Symbol	Output equipment	Symbol
Middle range sensor	MR	Product output solenoid	S
Low range sensor	LR		
Re Scan	R		

These logic functions should be equivalent to the I/O Boolean functions derived from the sequence table analysis using switching theory. The resulting timing diagram and logic controller circuit are depicted in Figures 6.12(c) and (d), while the wiring diagram when controlled by a programmable logic controller is shown in Figure 6.12(e).

$$D = (D + C2 \bullet Pb1) \bullet \overline{SW1}$$

6.5.2 Fruit-Picker Arm

Consider an automatic robot-assisted fruit-picker system with input and output devices as summarized in Table 6.23. Depending on the position of the fruit with respect to the robot arm, either the low or the middle range is activated, as depicted in Figure 6.13(a).

Table 6.23 can be converted into the process formal model using a Moore state diagram, as shown in Figure 6.13(b), or an equivalent Mealy state diagram, as shown in Figure 6.13(c).

Using the gray sequence coding of system state outputs and input transition conditions as summarized in Table 6.24, it is possible to describe the process using a Moore state diagram with binary coding, as shown in Figure 6.14.

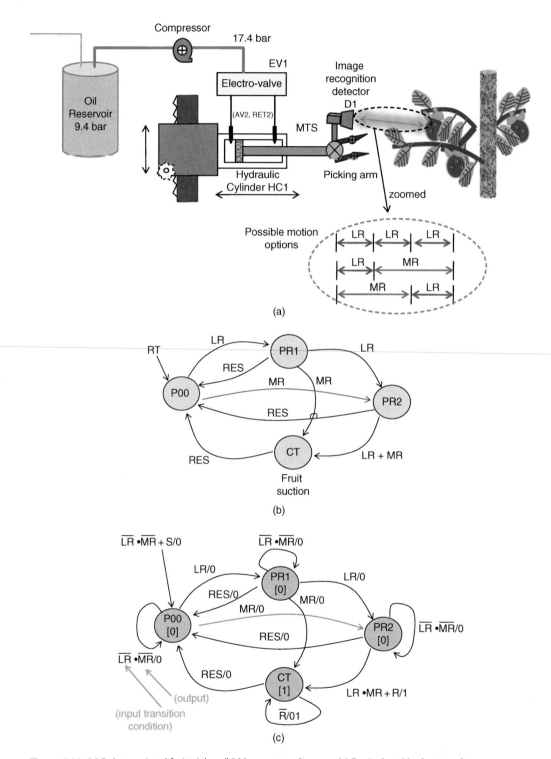

Figure 6.13 (a) Robot-assisted fruit picker. (b) Moore state diagram. (c) Equivalent Mealy state diagram.

Table 6.24 Binary coding of input transition conditions and fruit-picker system states.

Fruit position	>200 cm	200 cm	100 cm	0 cm
State	P00	PR1	PR2	CT
Binary coding	00	01	10	11
Pulse motion				
CASE1	00	LR	LR	LR
CASE2		MR	MR	LR
CASE3		LR	MR	MR
Binary coding [DN]	00	01	10	11

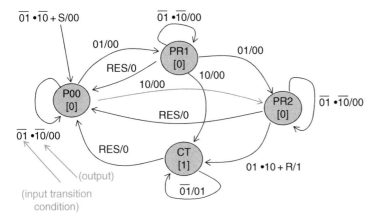

Figure 6.14 Equivalent Mealy state diagram with binary coding.

Alternatively, from the system I/O binary coding, it is possible to describe the process using the state transition table and D-flip-flop depicted in Table 6.28. Therefore, K-maps can be used to derive the corresponding I/O Boolean functions, as shown in Tables 6.23–6.28.

6.5.3 Driverless Car

A typical driverless vehicle contains the following embedded mechatronic systems: (i) antilock brake system (ABS), allowing the wheels to smoothly stop their rotation when a mechanical brake is activated; (ii) traction control system (TCS); (iii) vehicle dynamics control (VDC); (iv) electric ignition for fuel air combustion; (v) engine management and transmission control; (vi) airbag activation; (vii) air-conditioning system; (viii) seat-belt control; (ix) mirror

Table 6.25 D_1 Karnaugh table.

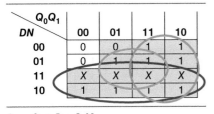

DN \ Q_0Q_1	00	01	11	10
00	0	0	1	1
01	0	1	1	1
11	X	X	X	X
10	1	1	1	1

$D_1 = Q_1 + D + Q_0 N.$

Table 6.26 D_0 Karnaugh table.

Q_0Q_1 DN	00	01	11	10
00	0	1	1	0
01	1	0	1	1
11	X	X	X	X
01	0	1	1	1

$$D_0 = \overline{Q_0}N + Q_0\overline{N} + Q_1N + D.$$

Table 6.27 *Open* output Karnaugh table.

Q_0Q_1 DN	00	01	11	10
00	0	0	1	0
01	0	0	1	0
11	X	X	X	X
10	0	0	1	0

$$Open = Q_1Q_0.$$

Table 6.28 State transition table with a D-flip-flop gate.

Present state	Present state Q1	Q0	Inputs MR	LR	Next state	Next state D1	D2	Output Open
P00	0	0	0	0	P00	0	0	0
			0	1	PR1	0	1	0
			1	0	PR2	1	0	0
			1	1	CT	X	X	X
PR1	0	1	0	0	P00	0	1	0
			0	1	PR1	1	0	0
			1	0	PR2	1	1	0
			1	1	CT	X	X	X
PR2	1	0	0	0	P00	1	0	0
			0	1	PR1	1	1	0
			1	0	PR2	1	1	0
			1	1	CT	X	X	X
CT	1	1	0	0	P00	1	1	1
			0	1	PR1	1	1	1
			1	0	PR2	1	1	1
			1	1	CT	X	X	X

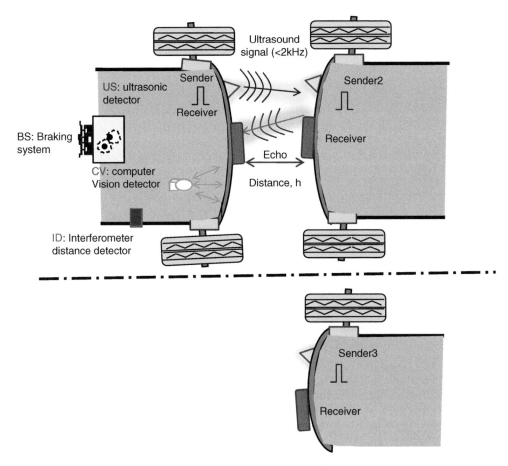

Figure 6.15 Unmanned vehicle with embedded navigation free collision system.

control; (x) instrument cluster; (xi) climate and vehicle front light activation control; (xii) park distance control system; (xiii) self-parallel parking, averaging fuel tank level; (xiv) alcohol test and engine activation; (xv) windows lift systems; and (xvi) obstacle-free and anti-collision navigation system. In that regard, regulatory and safety policies require that vehicles are reliably collision-free and that aerial and terrestrial unmanned mobile devices have reliable and robust-proven obstacle detection and anti-collision systems to ensure free navigation. Hence, this embedded free navigation system should relay on three sensors to provide information on the distance and number of nearest objects detected along the navigation path. The object distance is assessed using three different distance measurement technologies: a computer vision detector (CV); an ultrasonic detector (US); and an interferometer distance detector (ID), as depicted in Figure 6.15. The brake system (BS) of the unmanned vehicle is energized if any of the detectors is actuated. All detectors are normally closed-contact switches connected in series.

Hence, they maintain the BS output energized status even in the event of a detector electric circuit wiring failure (open switch contacts, broken wire connections, open relay coils, burned fuses, etc.). In short, the BS should be open if and only if two of the three detectors (CV, ID, US) acknowledge an object within the unauthorized distance range. This is considered to be a three-inputs and one-output combinational system, as described by the eight combinations of the truth table in Table 6.29.

Table 6.29 Truth table when two out of three detectors are checked.

	ID	US	CV	Output (BS)
	0	0	0	0
	0	1	0	0
	0	0	1	0
	1	0	0	0
$\mathrm{Prod}1 = ID \bullet US \bullet \overline{CV}$	1	1	0	1
$\mathrm{Prod}2 = ID \bullet \overline{US} \bullet CV$	1	0	1	1
$\mathrm{Prod}3 = \overline{ID} \bullet US \bullet CV$	0	1	1	1
$\mathrm{Prod}4 = ID \bullet US \bullet CV$	1	1	1	1
				$SOP = \mathrm{Prod}1 + \mathrm{Prod}2 + \mathrm{Prod}3 + \mathrm{Prod}4$

Here the I/O relationship is given by:

$$Output(BS) = ID \bullet US \bullet \overline{CV} + ID \bullet \overline{US} \bullet CV + \overline{ID} \bullet US \bullet CV + ID \bullet US \bullet CV$$

In case there are more output conditions than inputs, it is suitable to use the POS method to derive an I/O Boolean function. Here, for each ith row of the truth table, the logic input combination activating any output is given by:

$$Sum(i) = \sum inputs \tag{6.8}$$

Hence, the POS method captures the Boolean relationship between all input combinations activating a jth output by:

$$Output(j) = POS = \prod_j sum(i) \tag{6.9}$$

This could be implemented using an OR gate feeding into an AND gate (series collection of parallel connected relays). If it is desired to have an unmanned vehicle with a free navigation system and outputs such as object presence detector failure detection in addition to BS activation, the POS method could be as summarized in Table 6.30. Recall that if two of the three

Table 6.30 Truth table when using a virtual disagreement detector.

	ID	US	CV	Output (BS)	Detector disagreement
	0	0	0	0	0
$Sum1 = \overline{ID} \bullet US \bullet \overline{CV}$	0	1	0	0	1
$Sum2 = \overline{ID} \bullet \overline{US} \bullet CV$	0	0	1	0	1
$Sum3 = ID \bullet \overline{US} \bullet \overline{CV}$	1	0	0	0	1
$Sum4 = \overline{ID} \bullet TD \bullet CV$	0	1	1	1	1
$Sum5 = ID \bullet TD \bullet \overline{CV}$	1	1	0	1	1
$Sum6 = ID \bullet \overline{TD} \bullet CV$	1	0	1	1	1
	1	1	1	1	0
					$POS = Sum1 \bullet Sum2 \bullet Sum3 \bullet Sum4 \bullet Sum5 \bullet Sum6$

detectors (ID, US, CV) acknowledge a true object presence, the BS is open and the failure of one of three detectors is assumed as a detector disagreement. When detector outputs are all in logic level 0 or 1, all detectors agree. From the truth table in Table 6.30, there are two outputs (but six output activation combinations) and three inputs. The implementation of this detector disagreement is expected to offer a safer logic execution of process control.

6.6 Fail-Safe Design and Interlock Issues

The *fail-safe* design principle ensures a continuous execution of system operation in the event of predefined automation system component failure. This makes its control system as tolerant as possible to likely wiring or component failures. Applications of fail-safe design are critical in the generation and distribution of electric power to hospital facilities, telecommunications systems, water treatment systems, inner-city traffic lights, highway automated gates, and other important social infrastructures. Here, large circuit breakers are opened and closed by electrical control signals from protective relays. In the event of excessive current, the relay should be chosen such that its switch contact could be open to interrupt a signal while initiating a redundant alternative power source almost at the same time. Hence, this would be done with no disruption of service supply. The *interlock* design principle consists of setting a preventive action and technology to ensure no harm comes to the shop-floor operator and there is no destruction of involved system equipment. This is done through a set of detectors/sensors that activate system interruption in the case of predicted failure. Thus, the interlock design consists of implementing some interlocking switches at key locations in the system plan layout with defined activation rules.

Usually, some system operating safety measures are implemented through a three-hierarchical SFC level that could be described as follows: (i) a safety SFC (GS) managing process safety modes such as emergency start, general process faults, process initialization, after process defaults start, and so on; (ii) traded SFC (GT) defining the type of operational modes such as overall process start and selection of types of modes (semi-automatic, manual, automatic); and (iii) SFC operational modes such as the manual operation SFC (GM) or the production SFC (GPNA).

6.6.1 Logic Control Validation (Commissioning)

General specifications of automation system commissioning can be summarized through a set of test protocols regarding: (i) each hardware and software automation system component; (ii) an integrated software and hardware automation system (especially for data processing) for normal production operation conditions and for compliance between control and command execution of automated system actions before changing operating conditions, such as the speed variation control of a motor; (iii) assessment of operating system conditions (e.g. check detector activation condition); (iv) interlock assessment ahead of system failure, such as operator security and detector activation; (v) whole automation system commissioning and compliance tests of electrical wiring; and (vi) final electrical wiring diagrams and power flow diagram commissioning. In the case of hardwired circuit design, the process simulation for the designed logic controller could be validated through computer tools such as Verilog, ABEC, VHDL, and PALASM.

Exercises and Problems

1) Derive a logic circuitry using an R-S logic gate from the following Boolean function:

$$Light = (Light + PD) \bullet \overline{ENGINE_{IGNITION}}$$

2) Consider the anti-braking system given in Figure 6.16, where, the accumulator oil level detector is designed to activate the pump (M_PUMP) at high level ($L1$) and deactivate it at low level ($L2$) when the speed wheel is not at rest (WS). Design a logic control algorithm for this anti-braking system.

3) Derive a logic circuitry using an R-S logic gate from the following Boolean equation:

$$M = (M + Td[5s] \bullet SW1) \bullet \overline{SW1} \bullet SW2 \bullet STOP$$

4) Derive the I/O Boolean function corresponding to the logic circuitry for the control of a delayed motor pump (M_PUMP), as illustrated in Figure 6.17.

5) It is desired to control battery overcharging and discharging from a solar cell in order to keep the battery charged at peak level without overcharging, as depicted in Figure 6.18. It is recommended that you never discharge a deep-cycle battery below 50% of its capacity, in order to significantly extend its life. Based on detectors for low and high ($L50, L95$) battery charging levels, a cut-off of the battery from load voltage (BLS: a logic variable for the system related to a battery low-voltage switch) is activated. Also, in solar normal insulation conditions, solar cells generate a voltage high enough to charge batteries. However, if the solar cell voltage is lower than the required battery charging voltage, the charging current should be zero, as it is blocked by the diode. Among connected loads, there are lighting and alarm systems. Consider process input variables ($L50, L95$) interfacing with the logic controller and process output variables controlling the relay of power cut-off to load (BLS) and battery charging-only light indicators (LBC: a logic variable for the light indicator of a battery). For any change in input combination, the values are captured by the truth table in Table 6.31.

   a) Derive the simplified I/O Boolean functions from the truth table and K-maps, if necessary.

   b) Build the equivalent logic circuitry.

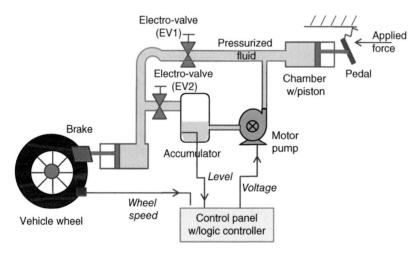

Figure 6.16 Vehicle anti-braking system.

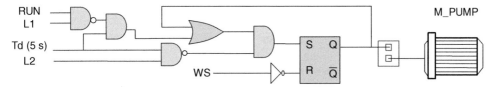

Figure 6.17 Logic controller circuit for a delayed pump activation of an anti-braking system.

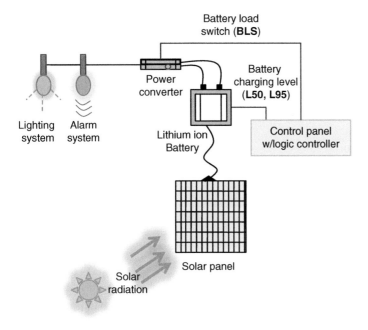

Figure 6.18 Battery charging process.

Table 6.31 Truth table of process outputs.

L50	L95	BLS	LBC
0	0	1	1
0	1	0	0
1	0	0	1
1	1	0	0

    c) How could this logic control system be modified to consider the variation of solar insinuation with respect to a consistent battery charging process? (Hint: ensure continuous battery charging process).

   6) Considering the Karnaugh table of a battery lighting system (CR_LOAD) with two inputs ($L50, L100$) given in Table 6.32, derive the resulting I/O Boolean function.

6.1 Consider an automatic-fastening seat belt, ignition interlock, and seat-belt warning system consisting of a sound alarm and a seat belt control relay, which are activated if an occupant is detected by an occupant position sensor (oS) (i.e. the driver seat only;

Table 6.32 K-maps of a motor starter system.

| | | L100 L50 | | | |
		00	01	11	10
CR_LOAD	0	0	0	X	1
	1	1	0	0	1

Table 6.33 Truth table of a belt fastening process.

Ignition key sensor (iS)	Occupancy sensor (oS)	Fastening belt switch (AS)	Car horn (CH)	Belt fastening motor (BFM)	Comment
0	0	0	0	0	No alarm
0	0	1	0	0	No alarm
0	1	0	0	0	No alarm
1	0	0	0	0	No alarm
1	1	0	0	0	No alarm
1	0	1	1	1	Intrusion
0	1	1	1	1	Intrusion
0	0	1	0	0	Alarm active
1	1	1	1	1	Intrusion

passenger is not included at this time) upon activation of the ignition key sensor (iS) when the key is turned. A furthermore, the fastening active switch (AS) is used to activate/deactivate the alarm once the safe system conditions are satisfied. The truth table in Table 6.33 summarizes the three inputs (an occupancy seat-belt switch, ignition key sensor, and fastening active switch) and two car alarm outputs (fastening motor and combined ignition interlock and car horn).

a) By examining the truth table, derive the I/O Boolean functions.
b) Derive the equivalent logic circuit for the fastening seat belt, ignition control, and alarm warning system. For further safety compliance, it is desired to expand the system for all three passengers. Update the truth table and the I/O Boolean equations accordingly.
c) Consider that such a system would tighten the seat belt in the case of crash occurrence. Then, the seat belt goes from normal (NS) to severe (SF) fastening level. This is presumably activated either by a sudden brake (SB) or by the front car collision sensor (CS). Adapt the truth table accordingly, and design the logic control circuitry. Combine it with the logic control system found in (a) and (b).

6.2 Consider the automatic activation of a vehicle's adaptive front lighting system based on environmental and climate conditions, as depicted in Figure 6.19. Here, lights are turned on and kept ON when any of certain surrounding weather conditions are met. The control method for the lamp should include the distance of the oncoming vehicle, illumination in the vicinity, and the decision on whether to use a high or a low beam. It is also considered that there are two possible light intensity levels: low medium (LL)

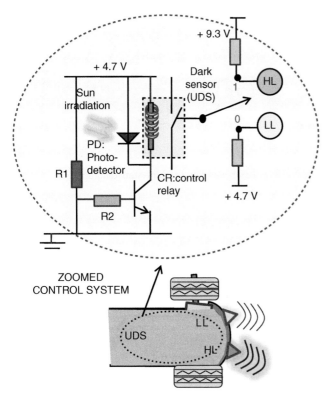

Figure 6.19 Automatic vehicle adaptive front lighting system.

Figure 6.20 Equivalent state diagram for automatic lightning system.

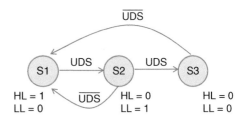

and high-intensity (HL), based on a dark sensor (UDS: a logic variable for an ultraviolet light dark sensor). Considering the state diagram in Figure 6.20, migrate the logic control circuitry accordingly.

It is desired to adjust the light beam from a high to a low intensity level if an obstacle appears at a distance from 5 to 50 m. Also, the light activation should consider the distance of the oncoming vehicle and illumination in the vicinity to adjust the use of a high or a low light beam. Describe how this logic control could be adapted.

6.3 Consider the two state diagrams depicted in Figure 6.21(a) and (b). Write the feasible I/O Boolean states and the output equations describing them.

6.4 A logic circuit for the activation and deactivation of a remotely controlled electric motor-driven pump is illustrated in Figure 6.22. Here, the motor is energized by three digital sensors: an infrared remote control source facing a photodiode (PD), a switch turned ON, and a limit switch LS set always to 1. These signals are connected to a

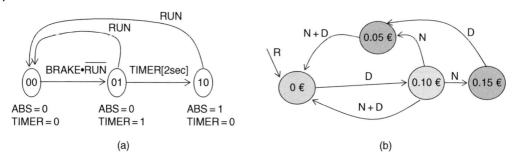

Figure 6.21 (a) Anti-braking system state diagram. (b) Four-state vending machine state diagram.

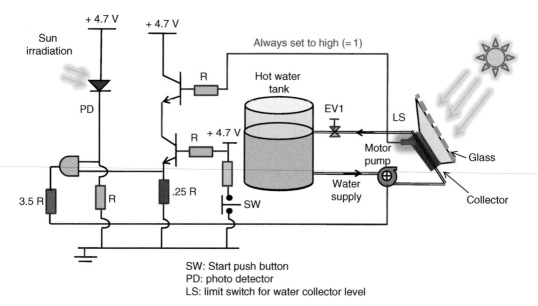

SW: Start push button
PD: photo detector
LS: limit switch for water collector level

Figure 6.22 Motor-driven pump control circuit.

two-input AND gate, all corresponding to an I/O logic Boolean function that can be summarized by:

$$MOTOR = Switch \bullet PD$$

It is desired to upgrade the circuit of logic control for motor pump switching operations, according to the truth table in Table 6.34, including deactivation operations (e.g. PD being OFF and switch set to 1).

a) Derive the upgraded I/O Boolean function of this system corresponding to the truth table in Table 6.35.
b) Build the circuit to implement the logic controller derived in (a).

6.5 Consider a pedestrian road crossing with individual lamp adjustment of three unknown variables *LAMP*1 (green), *LAMP*2 (amber), and *LAMP*3 (yellow). The controller logic operates based on possible values of input-variable push buttons PB1 and PB2, as shown in Table 6.34. For example, when *LAMP*1 is set to 1, neither button is pushed. When *LAMP*3 is set to 1, either button is pushed. When *LAMP*2 is set to 1, both buttons are pushed. Using the truth table in Table 6.34, derive the I/O multiple Boolean equations.

Table 6.34 Truth table of outputs *LAMP*1, *LAMP*2, and *LAMP*3.

PB1	PB2	LAMP1	LAMP2	LAMP3
0	0	1	0	0
1	0	0	0	1
0	1	0	0	1
1	1	0	1	0

Table 6.35 K-maps of a starter motor system.

		PD SWITCH			
		00	01	11	10
MOTOR	**0**	0	0	X	X
	1	X	X	1	1

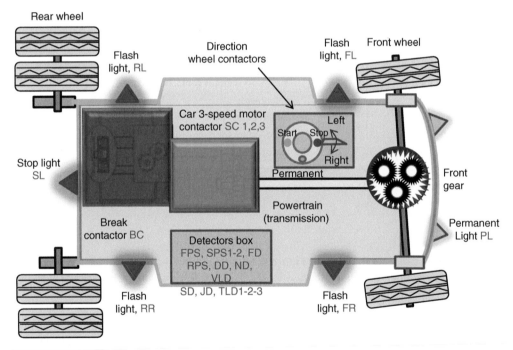

Figure 6.23 Driverless car system.

6.6 Consider a driverless car with an automatic speed direction and crash avoidance system, and the generic navigation state diagram depicted in Figures 6.23 and 6.24. The input and output variables used to model this system are depicted in Table 6.36. Key vehicle maintenance fault detectors (for engine, flat tire, etc.) have been discarded for simplification. It is assumed that the car trajectory is known in advance.

a) Based on the GPS-given trajectory, derive a sequence table.

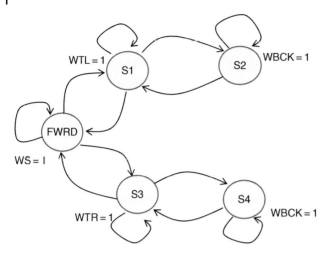

Figure 6.24 State diagram for an anti-collision system in a driverless car.

Table 6.36 Equipment involved in an electric driverless car motion control system.

Input devices	Symbol	Output devices	Symbol
Start push button	Start	Left flash light indicators (rear and front)	RL, FL
Stop push button	Stop	Right flash light indicators (rear and front)	RR, FR
Front position sensors	FPS	Car direction motor contactor	MC
Side position sensors 1, 2	$SPS1, SPS2$	Car three-speed motor contactor	$SC1, 2, 3$
Relative position sensor	RPS	Break contactor	BC
Front car presence detector	FD	Permanent light	PL
Side car presence detector	SD	Car direction motor contactor for backward motion	WBCK
Traffic light detector (R,Y,G)	$TLD1, 2, 3$	Car motor contactor for motion to the right	WTR
Junction detector	JD	Car motor contactor for motion along a straight line	WS
Day and night detectors	DD, ND	Car motor contactor for motion to the left	WTL
Visibility length detector	VLD		

    b) Based on the state diagram in Figure 6.24, complete with I/O variables for such car crash avoidance processes (Hint: s1, s2, s3, s4 are left forward, left backward, right forward, right backward, respectively).

    c) Derive an equivalent I/O Boolean function.

6.7 Consider the thermal control laser-based cutting machine depicted in Figure 6.25. Here, once the Start push button is activated, the cutting process consists of positioning the motorized laser mirror at a predetermined angle θ by moving from position PD1 toward PD2 using a motor contactor (KM1), while the focus lens position is maintained at PD0. Once PD2 is activated, the focus lens (MTL) moves from PD0 to PD3 and the laser beam is generated from power-supplied (KMP) activation of timer T1 for 2 s (T2 s). Then, the mirror is positioned to PD1 by the motor contactor (KMR). Based on the sequence table in Table 6.37, derive the equivalent I/O Boolean functions. A specific

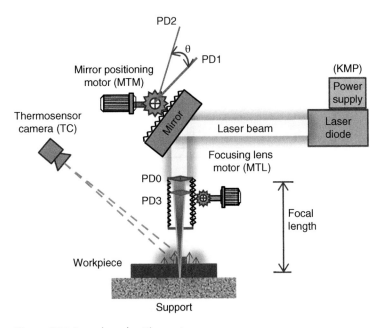

Figure 6.25 Laser-based cutting system.

Table 6.37 Sequence table of the laser-based cutting process.

Sequences	Input devices							Output devices					
	Start	Stop	T[2s]	PD1	PD2	PD0	PD3	KMP	KM1	MTL	KMR	T1	TD
1) Rest	0	0	0	1	0	1	0	0	0	0	0	0	0
2) Start	1	0	0	1	0	1	0	0	1	0	0	0	0
3) Advance to PD2	0	0	0	0	0	1	0	0	1	0	0	0	0
4) PD2 activated	0	0	0	0	1	1	0	0	1	0	0	1	0
5) Focus lens advance PD0 to PD3	0	0	0	0	0	1	0	0	0	1	0	0	0
6) PD3 activated	0	0	0	1	0	0	1	0	0	1	0	0	0
7) Laser cutting for 2 s (Timer ON)	0	0	0	0	1	1	1	1	0	1		1	0
8) Timer delay off	0	0	1	1	0	1	1	1	0	1	0	1	1
9) Retract to PD0	0	0	0	1	0	0	0	0	0	1	1	0	0
10) Retract to PD1	0	0	0	1	0	0	0	0	0	0	1	0	0

melting temperature detected by the thermal sensing camera could also interrupt the cutting process. Operating sequences are summarized in Table 6.37.

6.8 a) Write an assembly language program that computes the following control equation:

$$A1 \leftarrow (A2 + B \bullet C) \bullet \overline{D + E}$$

with data memory locations of $A1, A2, B, C, D,$ and E (0001, 0010, 0011, 0100, 0101, and 0110, respectively).

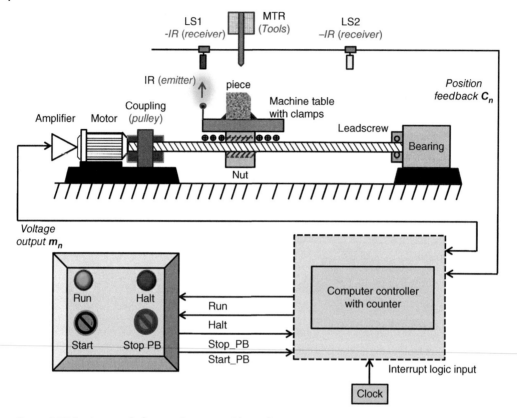

Figure 6.26 Logic control of system for screw table motion system.

b) Write an assembly language program that computes $A \times B$ using addition operations within a register.

c) Consider the screw table motion system illustrated in Figure 6.26. When the Start button is pushed, the Halt lamp is turned OFF while the RUN lamp is turned ON and the motor contactor (KM) is energized to provide an output voltage $m(n)$ to the stepper motor. The motor contactor (KM) is deactivated when either limit switch LS1, limit switch LS2, or the Stop button is activated. This also sets the Run lamp OFF, while the Halt lamp turns ON. In order to design a logic controller, write two assembly language programs: (i) using the asynchronous mode, where the system is synchronized with external event occurrence using an interrupt program; and (ii) for the synchronous mode where the system uses a clock signal to ensure synchronization with an event occurrence.

6.9 Recall the system where a motor rotates by one angular increment in position whenever a pulse is received at its Step_motion() input signal. The Dir_motion() input signal determines the rotational direction (logic level 0 for counterclockwise motion, logic level 1 otherwise).

a) For a motor resolution of 250 steps per revolution, design a logic controller to perform a three-quarter revolution in a clockwise direction and two revolutions in another direction. Edit an interrupt-based logic controller program for the system shown in Figure 3.52 to output signals STEP and DIR (a logic variable for the motor direction of motion) to the motor. A 1 GHz clock is used to construct interrupts setting STEP pulse output rate at 50 000 Hz.

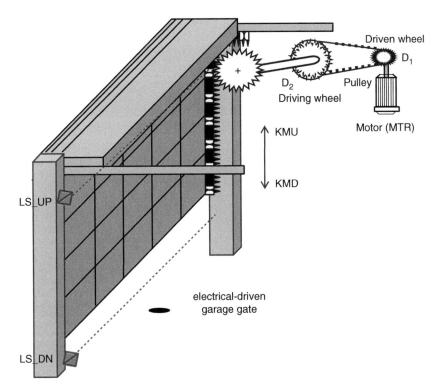

Driven wheel

D_1

D_2

Pulley

Driving wheel

KMU

KMD

Motor (MTR)

LS_UP

LS_DN

electrical-driven
garage gate

Figure 6.27 Automatic garage gate.

b) Consider a 12 cm motion (*LS*1 to *LS*2) with a pitch of 6 rev cm^{-1}. Design a logic controller program where LS1 and LS2 are used to reverse the rotational direction of the motor-driven table.

c) By defining input and output variables and setting their logic addresses, design an assembly-language program to continually move the table from one limit switch to another at a constant velocity of 1.8 cm s^{-1}.

6.10 Consider the garage door opener with up and down limit switches (LS_U) and (LS_D) indicating the bidirectional (with contactors KMU and KMD) motorized garage door position depicted in Figure 6.27.

a) Sketch a four-state state diagram corresponding to the operation of this garage door (Hint: states are door open, door opening, door closed, and door closing).

b) Design a logic circuit to perform the required logic function.

c) Design a computer interface develop a corresponding logic control program to perform this logic function.

6.11 Pneumatic rubber actuators are used to design a snake-like robot capable of a smooth motion for the exploration of rugged terrain or of a smooth-walled pipe, as depicted in Figure 6.28. When the bidirectional motor moves in one direction, this increases pressure in adhesive segment *i* (chamber), resulting in its stretching along the joint axial direction, while extending (reducing pressure) the adhesive segment *i* + 1 in the opposite position. The instant direction of the joint actuation motor is given by a pulse width modulation (PWM)-generated signal. These motorized actuators change the position of the adhesive segments (chambers), which in turn change their internal pressure in proportion to two

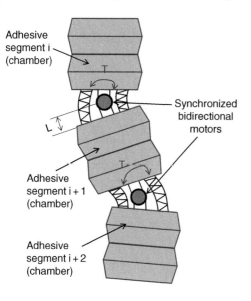

Adhesive segment i (chamber)

Synchronized bidirectional motors

Adhesive segment i + 1 (chamber)

Adhesive segment i + 2 (chamber)

Figure 6.28 Snake robot.

bidirectional motors capable of moving left, right, forward, and backward through the activation of contactors (KML, KMR, KMF, KMR, respectively).

Anti-collision is based on four obstacle detectors (OBS_L, OBS_R, OBS_F, OBS_B) locating the obstruction object on the left, right, front, or backside. Such a system can be used for rescue operations in hazardous environments, and is effective for searches in underground spaces and for disaster victim search inspections over rubble in quake-devastated zones, both remotely and where limited direct human intervention is possible. If necessary, upgrade the state diagram in Figure 6.29 to ensure a coherent anti-collision snake robot motion and derive the I/O Boolean functions, subsequently.

6.12 For compliance with driving safety and security regulations, a car engine (AIRBAG) must validate some checks before it can start. The I/O Boolean function for ENGINE could be described as:

$$AIRBAG = (AIRBAG + FCONTACT\_SW \bullet SCONTACT\_SW) \bullet \overline{BRAKE}$$

a) Develop an high-level language (C, Phyton) or assembly language programs (using AND? OR, LOAD, etc.) to develop this logic using the I/O interface addresses given in Table 6.38.

6.13 Consider a solar-powered electric car, as depicted in Figure 6.30. In order to maximize the battery charging process, it is suitable to ensure maximum solar irradiation exposure despite any possible obstruction (only over xy plan). This is done by establishing and maintaining an approximately 90° orientation of the attached solar panel with solar irradiation. Hence, intermittent panel orientation can be done through two bidirectional motorized supports, whose direction is activated by the respective contactors (KM1A, KM1B, KM2A, KM2B, KM3A, KM3B, KM4A, KM4B). These supports allow the two panel pairs to be positioned along the x- and y-direction based on sunlight detection. Assume that each 2 s period of activation of the step motor of each panel pair produces a force capable of moving the panel orientation by 2.5° in x or y direction.

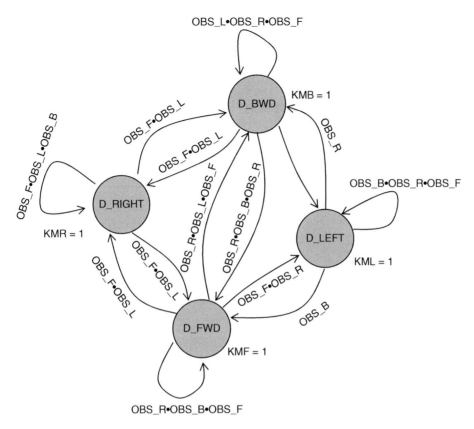

Figure 6.29 State diagram for snake robot motion.

Table 6.38 I/O interface addresses for a car engine safety program.

Process variables	Variable type: address	bit
AIRBAG	Output: 0 (O/0.0)	0
FCONTACT_SW	Output: 0 (O/0.1)	1
BRAKE	Input: 1 (I/1.0)	0
SCONTACT_SW	Input: 1 (I/1.1)	1
AIRBAG	Input:1 (I/1.2)	2

It is assumed that the conversion from 2D geodetic coordinate system into geographic and Cartesian coordinate system is given by:

$$H = sun\ to\ vehicle\ terrestrial\ distance$$

$$X(t) \approx H \cos(lattitude\ reading) \cos(longitude\ reading)$$

$$Y(t) \approx H \cos(lattitude\ reading) \sin(longitude\ reading)$$

Please note that the latitude and longitude readings are given in degrees.
It is desired to derive a logic control algorithm panel positioning process by performing time-based ON/OFF step motor operations.

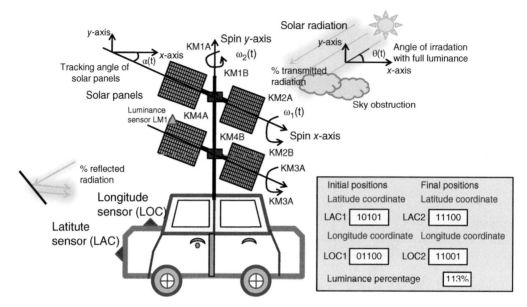

Figure 6.30 Satellite with spin-control thrusters.

a) List all involved input and output devices, including those required for the tracking system of the panel orientation trajectory (Hint: list any timers and counters involved).
b) For a xy planar trajectory, develop the sequence table describing the panel positioning process and the equivalent state diagram.
c) Write states and process-output Boolean expressions from the state diagram.
d) According to (c), write an assembly language program to implement the logic control algorithm.
e) Write the equivalent ladder diagram.
f) Design a logic circuit to perform the required I/O Boolean functions derived previously.
g) Sketch the corresponding timing diagram and discuss your logic control algorithm.

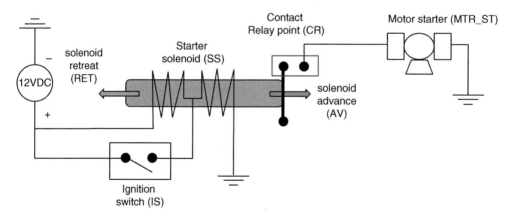

Figure 6.31 Relay-based control logic of a car engine starter.

Table 6.39 Sequence state table for a car engine starter.

Sequences	System inputs		System outputs		
	IS	CR	AV	RET	MTR_ST
1) Solenoid advance	1	0	1	0	0
2) Motor running	1	1	1	0	1
3) Solenoid retreat	0	1	0	0	1
4) Motor driving end	0	0	0	1	0

6.14 The relay logic circuit for a vehicle engine starter is drawn in Figure 6.31, while the operating sequence is summarized in the sequence table depicted in Table 6.39. It is desired
to migrate the control system of this electromechanical relay-based motor starter into a
logic control algorithm. Here, the activation and deactivation of the start and stop push
buttons are considered as inputs in order to energize and de-energize a 12 V DC starter
motor system.
   a) Based on the system schematic and table of sequence, derive I/O Boolean equations
   capturing the starter motor operating sequences.
   b) Build the equivalent logic control circuit.
   c) Write the assembly language program to implement this logic control algorithm.

6.15 Recall the elevator for a three-story building illustrated in Figure 1.8. It is desired to
design a logic controller using: (i) as inputs, three-floor selection push buttons (Pb1, Pb2,
Pb3) located outside of the elevator at each floor (instead of Up/Down push buttons) and
three-floor limit switches located on the floor strip (sw1, sw2, sw3), and (ii) as the output,
the elevator bidirectional motor (M1, M2).
   a) Derive a state diagram for this elevator, ensuring a suitable floor selection.
   b) Derive the corresponding I/O Boolean equations.
   c) Develop an equivalent ladder logic program that implements the resulting I/O
   Boolean equations.

6.16 It is desired to design a logic controller for a medical tablet press machine, capable of
producing a tablet batch size of 100. A powder processing granulator substation feeds
this machine with granules of a uniform size. Then, the machine presses granules in
20 recipients for 3 s. Afterward, it applies a sweetened paste around the bitter-tasting
tablets in batches of 100, before drying them for 5 min. Eventually, the dried tablets are
collected manually and sent to the quality control station, while the machine is fed with
a new batch of granules. When the start button is pressed, a light green indicator turns
ON. The loading of a new batch of granules is launched on the recipients encapsulated
by a motor-driven wheel, and the motor is actuated until a switch (Switch 1) is closed.
Consequently, for 10 s, the tablet press machine is activated. Then, the motor moves further until another switch (Switch 2) is closed, launching the release of the resulting tablet
from the recipient. When the motor displaces the wheel for more than 12 s without
activating the switch (Switch 2), a machine failure occurs, causing the motor to stop and
a red light indicator to turn ON. The remedy operations consist of switching the machine
to its manual operating mode, then reverting it to its initial state and pushing the start

Table 6.40 Some input and output variables involved in an automated drug conditioning process.

Process inputs	Process outputs
I:1 – start push button	0 : 1 – green light indicator
I:2 – stop push button	0 : 2 – machine ON
I:3 – switch 1 – batch of granules out of the recipient	0 : 3 – red light indicator
I:6 – switch 2 – batch of granules out of the recipient	0 : 4 – amber light indicator
	0 : 5 – motor ON

button twice. At the end of each successful granule batch process, a counter should update the count of the drug condition and an amber light indicator should be set ON.

a) Update Table 6.40 of listing of the variables needed to indicate when each state is ON, as well as any timers and counters used.
b) Draw a state diagram for this drug conditioning process from the injection of the granules batch into the recipients up to the pressing operation occurring before the sweet coating operation.
c) Write an I/O Boolean expression for each transition in the state diagram.
d) Sketch a wiring diagram for the programmable logic controller.
e) Write the corresponding ladder logic diagram for the state.

6.17 Consider an electric vehicle (MOTOR) that is energized when a simultaneously recorded fingerprint (FGP) presses the start push button (START), the driver's safety belt detector (BELT) is closed, the in-board engine failure detector (ENGINE) is at level 0, and the

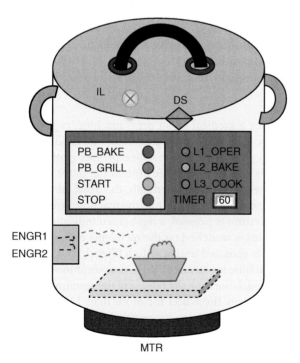

Figure 6.32 Microwave oven system.

in-vehicle driver alcohol-level breath detector (ATL) has finished and is OK. The electric vehicle will turn OFF immediately if the stop push button is activated or an engine problem is detected.

a) Develop an I/O Boolean equation for this system.

b) Build the corresponding logic circuit for this system, using the D-flip-flop device.

c) Construct the equivalent ladder diagram.

6.18 Figure 6.32 depicts a microwave oven with two mutually exclusive operating cycles: cooking and defrost operating cycles. It is considered that, when the cook cycle selection (PB_cook) and (START) push buttons are pressed, the microwave emitter (EM), the light indicator (L_cook), and the motor (MTR)-driven rotating table are energized at 100% power for 90 s. Similarly, if the defrost cycle selection (PB_defrost) and START push buttons are pressed, the microwave emitter (EM), the light indicator (L_defrost), and the motor (MTR) driving the rotating table are energized at 60% power for 30 s, then de-energized for 30 s and re-energized for another 30 s. It should be indicated that: (i) the cycle selection must be completed before the START push button can be activated; (ii) the activation of either the PB_cook or the PB_defrost push button during the heating process will have no effect on the cycle execution; (iii) the last operating cycle selected is perpetuated from cycle to cycle, except if another selection occurs; (iv) no cycle may start if the microwave is not properly closed, meaning the door is unlatched (DS = 0); and (v) when the door is unlatched (DS = 0), an inside microwave light (IL) is energized. Any activation of the STOP push button or opening of the door (DS) causes an immediate termination of the operating cycle. L_power and L_prog are light indicators for the microwave power supply being ON and the heating cycle being in progress, respectively.

a) List in a table all I/O devices involved for both cycles.

b) Establish the sequence table for both cycles.

c) Derive the state diagram and subsequent I/O Boolean functions.

d) Derive the corresponding logic controller program in a ladder diagram.

e) Build the circuitry for the logic controller using a D-flip-flop gate.

Figure 6.33 Mixing tank system.

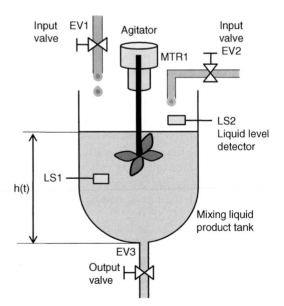

6.19 Figure 6.33 illustrates an automatic mixing station. Consider the following operating sequences:

a) Open valve 1 (*EV*1) until level 1 (*LS*1) is reached for the first liquid.
b) Then close valve 1 (*EV*1).
c) Open valve 2 (*EV*2) until the level of limit switch 2 (*LS*2) is reached for the second liquid.
d) Then close valve 2 (*EV*2).
e) Start the motor (*MTR*1) and agitate the two liquids for up to 20 min.
f) Then stop the motor (*MTR*1).
g) Open valve 3 (*EV*3) for up to 5 min.
h) Then close valve 3 (*EV*3).
i) Repeat the mixing process.
   i) List all I/O devices involved in the mixing process sketch the state diagram derive the I/O boolean functions corresponding to the operating sequences just described.
   ii) Draw the wiring diagram and the corresponding ladder logic program for this mixing process.

6.20 Consider an automated fruit picker, as shown in Figure 1.21.
The states of the fruit-picking process consist of: (i) State 1, relating to the positioning of the conveyor and the advance of the robot arm at a feed rate; (ii) State 2, relating to the pressure-based clamping; (iii) State 3, relating to the robot arm cutting off the fruit; and (iv) State 4, relating to the unclamping of the fruit from the robot arm. The cycle then begins again at State 1.

a) Using the I/O devices shown in Figure 1.21, draw the state diagram for this fruit-picking process.
b) Derive the equivalent sequence table.
c) Derive the corresponding I/O Boolean functions.

6.21 Figure 6.34 illustrates the 3D printing process on a customized food printing. It is required to design a logic control for the discrete selection of ingredients in this 3D customized food printing process. It is recommended to use an ON/OFF control for the motorized pumps in charge of ensuring the injection of food paste and the heaters. To position the plate, a table is driven by bidirectional motor (Motor 4A, Motor 4B). x-axis limit swicthes indicate the table positioning. While bi-directional motor 3 (Motor 3A and Motor 3B) along with bidirectional motor extruder (Motor EL, Motor ER) ensure positoning over z and x axes respectively. The ingredient flow rate is maintained constant by the pressure valve.

a) List all I/O process devices involved, with their assigned addresses, and then draw the wiring diagram schematic.
b) Using the operating conditions present here and the schematic in Figure 6.34, draw the state diagram and the equivalent sequence table.
c) Sketch the corresponding SFC for this logic, based on the corresponding mnemonics of process devices.

6.22 Consider an artificial solar-powered climate system, as depicted Figure 6.35. A chamber-based indoor climate is defined by a suitable warm temperature and humidity (i.e. the balance of heat over the oxygen level), promoting the growth of vegetables. It is suitable to logically control (ON/OFF control) the pump humidifier and heater to

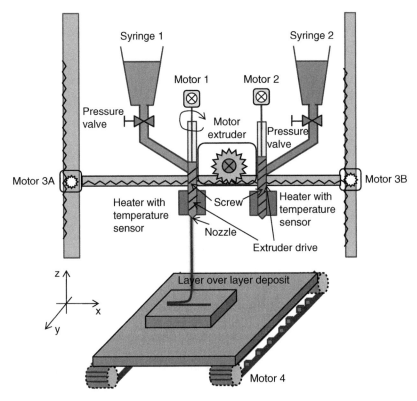

Figure 6.34 Three-dimensional printing process.

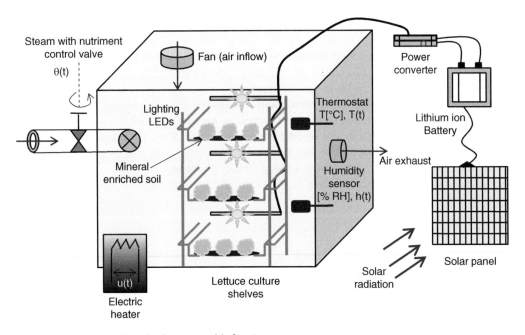

Figure 6.35 Chamber-based indoor vegetable farming.

supply water steam to maintain a specific humidity, while the solar-powered lamp is used to heat the chamber. Inside chamber temperature and humidity are sensed using detectors.

a) List all I/O devices involved in the automatic farming process.
b) Determine the sequence table and fill the corresponding state transition table for this automatic farming system.
c) Draw the equivalent state diagram.
d) Derive the subsequent state Boolean functions.

Bibliography

1 Alciatore, D.A. and Histand, M.B. (2012). *Introduction to Mechatronics and Measurement Systems*, 4e. McGraw-Hill.
2 Ashar, P., Devadas, S., and Newton, A.R. (2012). *Sequential Logic Synthesis*. Springer.
3 Bemporad, A. and Morari, M. (1999). Control of systems integrating logic, dynamics, and constraints. *Automatica* 35 (3): 407–427.
4 Bolton, W. (2016). *Mechatronics: Electronic Control Systems in Mechanical and Electrical Engineering*. Pearson.
5 Bollinger, J.D. and Duffie, N.A. (1989). *Computer Control Machines and Processes*. Addison-Wesley.
6 deSilva, W. (2010). *Mechatronics: A Foundation Course*. CRC Press.
7 Isermann, R. (2003). *Mechatronics System Fundamentals*. Springer.
8 Jack, H. (2005). *Automatic Manufacturing Systems with PLCs*, available at: https://www.pacontrol.com/download/plcbook5_0.pdf
9 Katz, R.H. and Borriello, G. (2004). *Contemporary Logic Design*, 2e. Prentice-Hall.
10 Kuphaldt, T.R. (2007). *Lessons in Electric Circuits*, 4e, vol. IV. Digital.
11 Lampérière-Couffin, S. and Lesage, J.J. (2000). Formal verification of the sequential part of PLC programs. In: *Discrete Event Systems* (ed. R. Boel and G. Stremersch). Springer.
12 Mano, M.M. and Ciletti, M.D. (2006). *Digital Design*, 4e. Prentice-Hall.
13 Nof, S.Y. (2009). *Handbook of Automation*. Springer.
14 Olsson, G. (2003). *Industrial Automation*. Lund University.
15 Petruzella, F.D. (2010). *Programming Logic Controllers*, 4e. McGraw-Hill.
16 Thayse, A. (1981). Boolean calculus of differences. In: *Lecture Notes in Computer Science*, vol. 101 (ed. G. Goos and J. Hartmanis). Berlin: Springer.
17 Roth, C.H. (2002). *Digital System Design Using VHDL*. Thomson.
18 Vahid, F. (2007). *Digital Design*. Wiley.

7

Hybrid Controller Design

7.1 Introduction

Electrical-driven system operations have alternated continuous and discrete characteristic behavior. These classical operating conditions require the development of hybrid control strategies in order to maintain a smooth and predictable performance. Due to their digital nature, the design of a hybrid control algorithm can be associated with a monitoring tool into a set of decision-making tools ensuring that the process runs safely and reliable decisions are made in the presence of faults or undesired disturbance. It is expected that such a system will provide commands based on distributed control-oriented decisions for large-scale plant operations.

However, the design of monitoring and hybrid control systems is quite complex due to constraints related to: (i) the combined continuous and discrete event nature of process operations, which makes adequate modeling of operating system dynamics difficult; and (ii) the dependency of system data on the sensing and detection methods and technologies used, which can lead to incomplete and uncertain data as well as a variety of data formats.

In this chapter, an integrated methodology for the design and implementation of monitoring and hybrid control systems for electrical-driven systems and process operations is presented. First, requirements for the design of the hybrid control system, along with the gathering and the analysis of the system data structure, are defined. Then, formal system models are developed to integrate discrete and continuous process operations. This lays the foundation for the design of the monitoring and hybrid control system. Finally, some case studies on hybrid control of electrical-driven systems are presented.

7.2 Requirements for Monitoring and Control of Hybrid Systems

Most electrical-driven systems and processes display compound behavior with discrete and continuous behavioristic characteristics. For each system, an associated controller has to be designed according to its operating characteristics and performance with respect to productivity, optimization, reliability, safety, continuous operations, and even stability objectives. Consequently, control of a hybrid system has to adopt integrated logic and continuous control algorithms as well as supervision programs. These control design objectives are translated into software, hardware, and mechanical design specifications to ensure integrability and interoperability in terms of diagnostic, monitoring, control, maintenance, predictability, and safety applications. This chapter will focus on monitoring and hybrid control software integration.

Control Of Mechatronic Systems: Model-Driven Design And Implementation Guidelines,
First Edition. Patrick O.J. Kaltjob.
© 2021 Patrick O.J. Kaltjob. Published 2021 by John Wiley & Sons Ltd.

The monitoring and hybrid control software specifications are defined based on a combination of operational and technical requirements, so that the resulting system can:

1) ensure the real-time safety and predictability of system performance in terms of modeling the pattern of normal operations;
2) ease system scalability, configurability, and maintainability by the structural design and technological classification of sensing and data acquisition technology, as well as system data structure and formatting.

7.2.1 Requirements for Hybrid Control System Design

Mechatronic systems can be modeled as continuous-state systems or discrete-event systems, leading, respectively, to I/O differential equations and I/O Boolean functions. Consequently, control of hybrid system requires: (i) the software integration of logic and continuous control algorithms, as well as supervision programs, into hybrid control strategies; and (ii) data integration by the design of built-in interoperability and intelligent information-processing systems combining multiple functions for control and monitoring decisions. Challenges in the development of a practical design approach for hybrid control of mechatronic systems and electrical-driven processes are: (i) to develop a suitable mathematical model capturing their continuous and discrete event behavioristic characteristics; and (ii) to integrate the control systems with respect to technological constraints and operational characterization (discrete and continuous) (e.g. time delays, signal-to-noise ratios, etc.).

7.2.2 Requirements for Operations Monitoring System Design

System operations monitoring is a function that consists of recognizing anomalies in the behavior of a dynamic system and its underlying faults. Hence, system operation monitoring for abnormal behavior poses a challenging problem of process operations observability and diagnosis, which can be approached through full-scale sensor-based process data-gathering and model-based process data analysis. Commonly encountered system operating monitoring objectives are:

1) Monitoring the execution of process operations in order to take appropriate counteraction in the event of abnormal process behavior.
2) Detecting faults from measurable process variables with regard to tolerances, and generating alarms for the process operator.
3) Diagnosing fault features through symptoms of process anomalies.
4) Selecting remote processing maintenance operations and counteractions to fault occurrences.

Such hybrid controls for system operations should be able to maintain the process values close to targeted process values automatically. Subsequently, information related to monitoring and hybrid control systems acts as variables that can be classified as follows:

1) Reporting variables related to system data analysis, such as automation field device preventive maintenance (e.g. variables indicating detection and diagnosis of a tool failure).
2) Configuration variables related to system operating configurations, which are parameters of the operating conditions (e.g. feeding rate, spindle speed, cutting tool geometry, depth of cut) and controller settings. In addition, some variables and features characterizing the process performance or product quality (e.g. vibration main frequency or surface roughness) may be included.

3) Operations monitoring variables related to the execution status of process actuating devices in real time, without discontinuity.
4) Fault monitoring variables related to safety requirements and fault management, which include: (i) safety variables to warn against personnel and equipment harm; (ii) performance variables to monitor process performance degradation; and (iii) fault (anomalies) variables for the detection and diagnosis of fault occurrence.
5) Control variables related to process command and control operations, which are manipulated input variables of actuating devices.
6) Intermediated variables related to data exchange between the database and the computing unit (e.g. mean variance in process data analysis routines).

Hence, the control and monitoring software can be divided among the following program modules:

1) The process configuration module, which contains algorithms related to the definition of process operating conditions (logic and feedback), the controller configuration and the timing or sequential execution of manipulated variables values for process actuators.
2) The process monitoring module, which contains algorithms related to the reporting in real time of process devices operating status. Some process variables could be estimated from process data. This should allow remote supervision of execution operations from top management to field operator.
3) The command and control module, which contains hybrid control algorithms (logic and feedback).
4) The data availability module, which contains algorithms related to ensure availability and reliability of gathered process data from the data acquisition, transmission and sensing devices up to the archive in the historical database. Some redundancy strategies can be used.

Table 7.1 lists typical variables for a monitoring system.

7.2.3 Process Interlock Design Requirements

Based on identified operating and failure modes, process stop modes are designed as protective measures to maintain the process equipment under normal physical and safe operating conditions. These protective measures can be performed by devices or systems providing the maximum protection with the minimum hindrance to the normal execution of process operations. As an illustrative example, this can be implemented by placing some interlocking switches at key locations in the process plan layout with defined activation rules, achieved either by (i) preventing access during dangerous motion or by (ii) preventing dangerous motion during access.

Preventing access during dangerous motion is achieved by using (i) fixed enclosing guards along the system motion path, which does not require access and (ii) movable guards with interlocking switches, in case access is required. This is expected to be interlocked with the power source in order to ensure the hazard power routed through this switch can be switched OFF.

Table 7.1 Commonly used variables for monitoring software design.

Interlock variables	Threshold values	Intermediate variables	Manipulate variables	Configure variables	Function variables	Sensors equipment	Actuator equipment

Some interlocking switches activate devices to lock the security guard door until the system is in a safe state. Typical industrial applications use combined solutions with movable guards and interlocking switches. There are also other solutions, such as two-hand control, which can prevent access while a system is under dangerous operating conditions (e.g. two start buttons must be operated at the same time to run a painting robot).

Preventing dangerous motion during access is used when frequent access is required along the system motion path. Here, an interlocking system or device is used to prevent dangerous motion while allowing unrestricted access. An interlock switch isolates the power source once any intrusion or external presence is detected. This is done by using detective motion devices, such as: (i) *photoelectric light curtains emission*, which encloses the perimeter of the system motion path; (ii) *pressure-sensitive safety mats*, which guard a floor area around the system motion path; (iii) *emergency stop devices* manually operated at the control panel level, which, when they are actuated, must be latched in before a stop command can be activated; and (iv) *pressure-sensitive edges*, which are mounted to the edge of the system motion path. It is important for the controller to have failsafe measures and to be able to stop the system quickly after switching off the power source. Any activation of a process interlock should not cause a hazardous situation, but should move the process into the initial state or an intermediary one.

7.3 Design Methodology for Monitoring and Control Systems

Due to the dual nature of mechatronic systems operation, a hybrid control system is usually required. A hybrid control system consists of combining sequential logic control and feedback control algorithms. In this section, a monitoring and hybrid control system design methodology is presented.

Step 1. According to the design requirements identified, a generic design methodology for a monitoring and control system consists of:

1) Performing a technical audit of the process components in order to establish performance indexes (e.g. stability, precision, speed of response, robustness, and safety), along with productivity indexes and even energy consumption indexes. Then, comparing these indexes with typical standards or benchmarks:
2) Listing the devices and equipment involved in the process operations and their characterizing information, such as:
   a) the operating conditions (e.g. input voltage, operating range, command profile, sequence execution, technical specifications, etc.) of the actuating equipment (analog/digital, e.g. a motor or solenoid);
   b) the operating conditions (e.g. measurand, operating range, resolution, etc.) of the process-sensing devices and detectors;
3) Sketching the piping and instrumentation diagram (P&I diagram) and the process flow diagram (P&F diagram). In addition to the identification of process components and subsystems, P&I and P&F diagrams display some key piping and instrument details for process control, activation, and shutdown. Some symbols used to draw several diagrams (P&F, P&I, electrical wiring diagrams) can be found in the literature (see Figure 7.1(a) and (b)).
4) Analyzing the process operating conditions in order to list the:
   a) process operating cycles;
   b) process safety constraints.

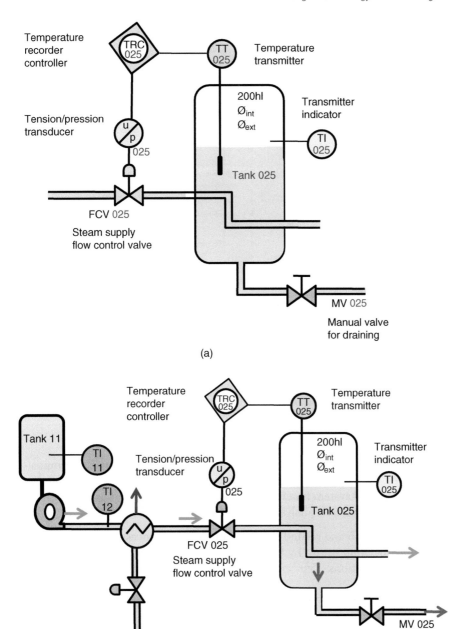

Figure 7.1 (a) Example of a P&I diagram process temperature control system. (b) Example of a P&F diagram for a process temperature control system.

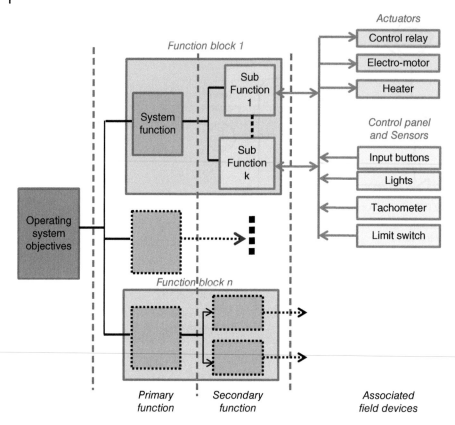

Figure 7.2 FAST decomposition method.

Step 2. Prior to monitoring and hybrid control design, a functional analysis and decomposition should be done using the functional analysis system technique (FAST), as illustrated in Figure 7.2. This results in several system functional blocks.

Proceeding with system behavioristic analysis using the FAST method consists of:

1) Identifying the global function and related primary functions for process control objective.
2) Performing functional decomposition by breaking down each primary function into secondary functions or subfunctions. Typically encountered blocks from FAST-decomposed processes are listed in Table 7.2.
3) Listing all constraints for each key primary function (technical experience is required) according to the following groups:
   a) operational capacity constraints (max. level, max. travel distance, max. temperature, etc.);
   b) operational time constraints.

Table 7.2 Listing of typical FAST expected results.

Primary function	Constraints	Secondary function	Associated operational tasks	operational tasks for equipment protection	Operational tasks for operator security	Operational tasks for setting up (preparation)

4) Listing all operational tasks associated with each secondary function, such as:
   a) active devices related to each process operating task;
   b) conditions of activation;
   c) conditions of termination.
5) Adding all necessary tasks involved in setting up activities such as preparatory tasks: power supply activation checking of the start-up operational condition; loading of work pieces for processing or even initialization; and referencing of associated tools.
6) Establishing a descriptive list of actions and events sequences under the various process operational "modes," including "normal production," "emergency," "start-up," and "shut down" modes, as well as special modes that may not occur very frequently. These process actions and events must be related to each specific mode (e.g. the "Auto" push button [PB] turns the AC motor on).

Step 3. The system operating specifications and subsystem functional blocks should be identified. This might be done in order to integrate all functional blocks related to the operating modes using the structural analysis and design technique (SADT) (dependency chart). Subsequently, this is achieved by:

1) Establishing a coordination between the process operating tasks through:
   a) The classification between manual, semi-automatic, and automatic operatio;
   b) The definition of the activation/termination condition of each task.

Step 4. The control scheme should be selected (e.g. open, closed-loop control feedback, predictive control, adaptive control, or logic control), as depicted in Figure 7.3.

Step 5. For each system functional block, control algorithms should be designed, as discussed in previous chapters. This can be done through:

1) Control logic design using techniques such as flow charts, state diagrams, and petri nets.
2) Continuous controller design using techniques such as model predictive control (MPC) or proportional-integral-derivative (PID) controller design.

This is based on the modeling of the execution of process operations through:

1) Sequential logic process formal modeling using Boolean algebra or a finite-state machine
2) Difference-based equations for a formal model of discrete even-system behavior or differential equations for a physics-based model of continuous system behavior.

The hybrid control devices should then be sized and subsequent control algorithms developed by:

1) Defining the computing hardware of the control unit and sizing the detection equipment.
2) Implementing logic and continuous control algorithms through programming languages such as SFC, ladder and function block, timers and counters, Boolean mnemonics, structured text (ST), instruction list (IL), or more common programming language environments (C++, Java, etc.).

Step 6. The resulting control algorithms should be combined within a hybrid control system. Usually, a logic control unit activates and deactivates system functional block start and stop operations while a continuous control unit maintains the tracking of the continuous command signal. This results in a hybrid system modeling for subprocess sequencing, as depicted in Figure 7.4.

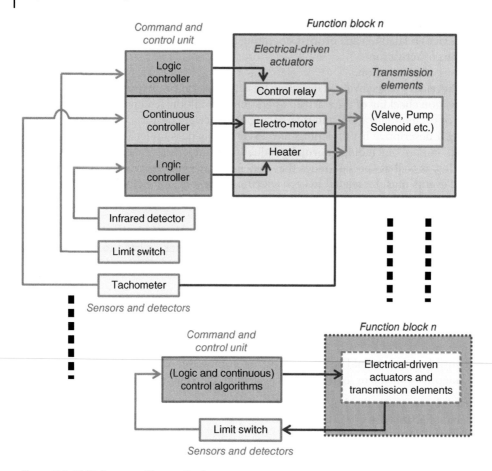

Figure 7.3 FAST decomposition method.

Step 7. For each system operating sequence, an interlock system design should be performed in order to ensure safe operation. This can be done by:

1) Defining safety inputs, output variables, and interlocks based on safety and regulatory requirements. All of these variables are expected to be complete after the functional analysis.
2) Establishing a list of effects from potential failures within the sensors, the actuators, or even the whole process, such as drops in hydraulic pump pressure and failures in top and bottom limit switches.
3) Adding tasks associated with the protection of the device for each secondary function, especially interlocks for the process stop condition.
4) Adding tasks associated with operations and operator security for each secondary function, especially an emergency stop in case of accident.
5) Designing system interlocks and forcing action by designing logic programming that fulfills safety requirements by:
   a) checking that cycles of operation and logic meets safety specifications;
   b) designing algorithms for verification and validation.

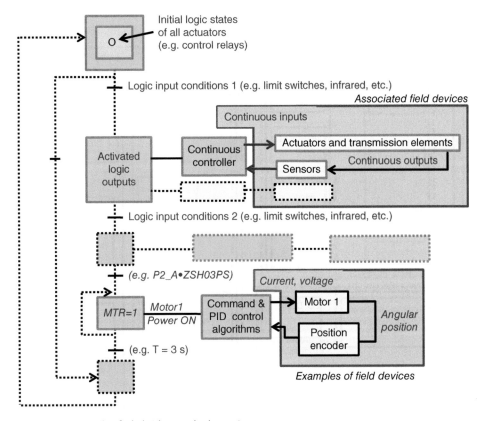

Figure 7.4 Example of a hybrid control schematic.

Step 8. Normal process operating conditions should be maintained by designing software and hardware components so that the mechatronic systems are based on specific requirements regarding how data should be gathered, analyzed, and reported. This involves integrating the process operating modes, control parameters, and interlocks using the SADT method (dependency chart) and the start and stop process operational mode graphical analysis (similar to GEMMA [in French, *Gestion des modes de marches et d'arrets*], see Figure 7.5). Among variables involved there are:

1) Process control input variables representing sensing devices and detectors, as well as visualizing units (man-machine interface, MMI).
2) The equivalent binary (specify pulse or level type) or continuous process variables assigned to each input and output from each process sensing device and detector, along with the actuating equipment.
3) Manipulated variables, representing outputs of the controller unit.
4) Input and output signals from the MMI (e.g. ResetPushButton, StopPushButton, LimitSwitch, etc.), with their technical specifications (e.g. pulse or level type, etc.).
5) The equivalent binary or continuous process visualizing variables assigned to each input and output signal.

Consequently, a bottle-washing process operating mode using process functional blocks with process failure recovery functions is structured using sequential function chart (SFC), also

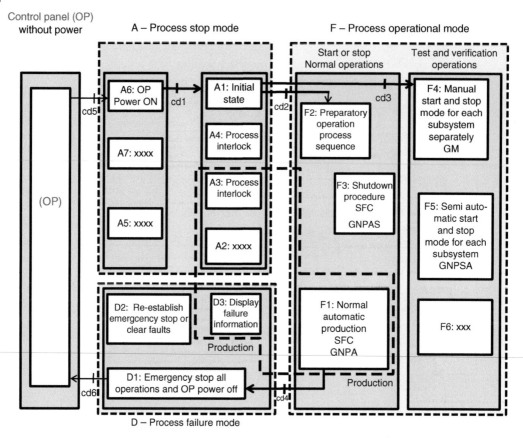

Figure 7.5 Process start- and stop-mode graphical analysis.

known as "Grafcet" in French. This is a graphical language for the implementation and specification of sequential algorithms. SFC hierarchy can be decomposed to three levels:

1) Safety SFC (GS), which structures modes related to process operating safety (emergency start, general process faults, process initialization, after-process defaults start, etc.).
2) Traded SFC (GT), which defines process operating modes (semi-automatic, manual, automatic) as well as the initial process start mode.
3) Execution of any of the available operating modes, such as:
   a) the manual operating mode SFC (GM), which structures manually activated start and stop process operating sequences separately;
   b) the normal start automatic mode SFC (GNPA), which automatically executes the start mode of process operating sequences;
   c) the normal stop automatic mode SFC (GNPAS), which automatically executes the stop/shutdown mode of process operating sequences;
   d) the normal semi-automatic mode SFC (GNPSA), which semi-automatically executes start and stop/shutdown modes of process operating sequences.

Step 9. The process automation project documentation should be edited using connection diagrams, such as for:

1) the electrical power supply diagram (power supply of current/voltage required) and command diagram;
2) the electronic circuit (if used);

3) the I/O wiring connection diagram (in/out of controller device);
4) the detailed system configuration and I/O wiring connection diagram (I/O address assignments);
5) the internal storage address assignments and storage register assignments.

7.4 Examples of Hybrid Control and Case Studies

In order to illustrate the process monitoring and control design methodology, three industrial case studies are developed in this section. In addition, features related to the implementation of the detection and diagnosis tools for induction motors are investigated. Some internal monitoring devices such as a watchdog can be used to supervise the execution of the controller program operations, to report irregularities and difficulties, and even to stop the machine or process in the case of an emergency not otherwise covered.

7.4.1 Elevator Motion System

An example of a control system for a crane-based vertical motion process is illustrated in Figure 7.6, while its feedback block diagram and the logic control connections are shown in Figure 7.7. Figure 7.8 displays the FAST decomposition of the elevator motion process, and the corresponding GEMMA is given in Figure 7.9.

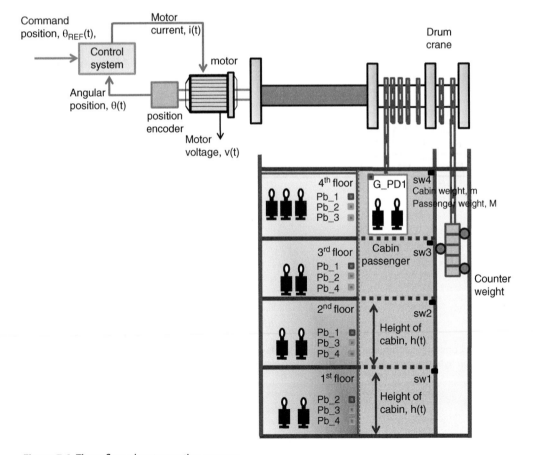

Figure 7.6 Three-floor elevator motion process.

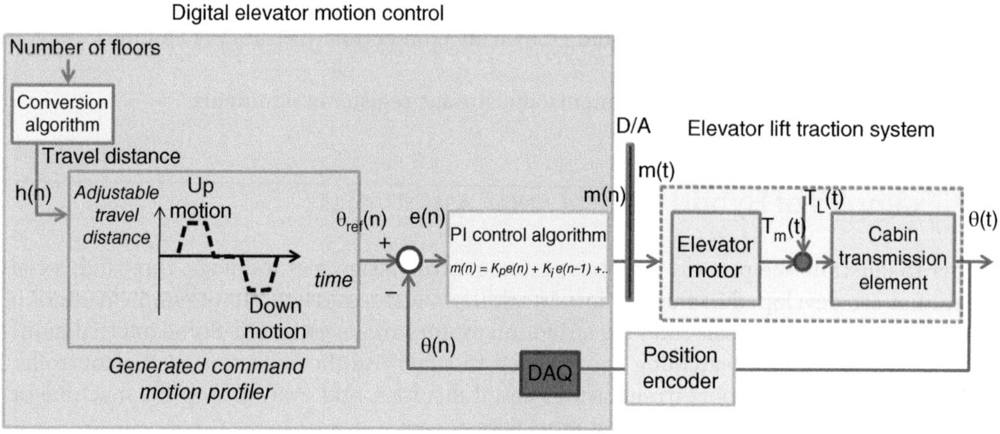

Figure 7.7 (a) Continuous command and control of an elevator lift traction motion process. (b) SFC of an elevator motion hybrid control. (c) Block diagram of an elevator motion feedback control system.

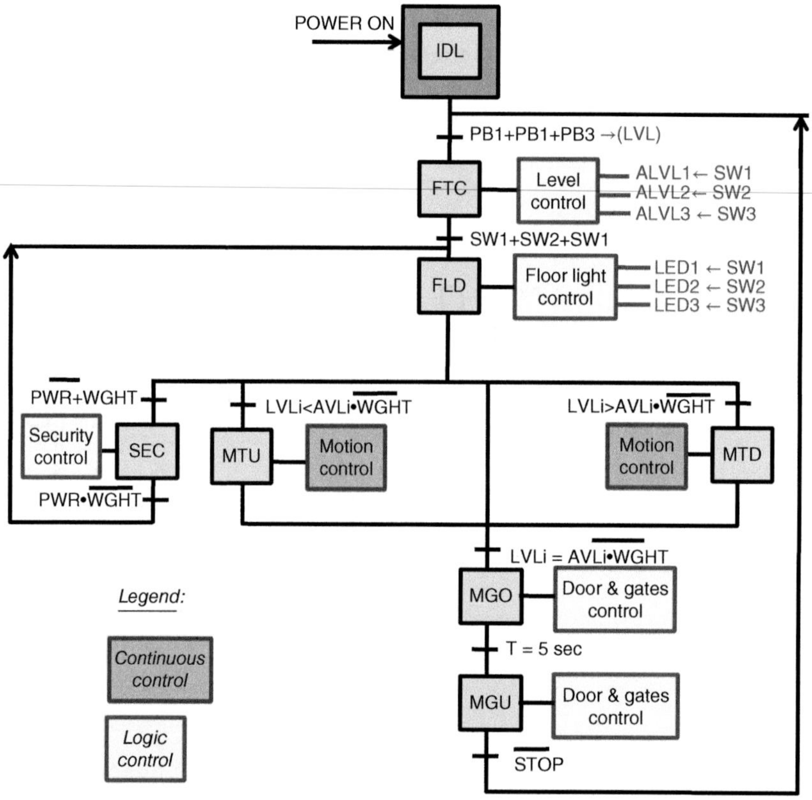

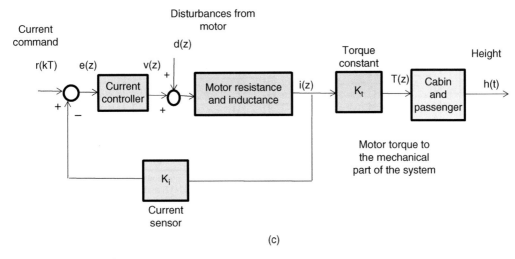

(c)

Figure 7.7 (*Continued*)

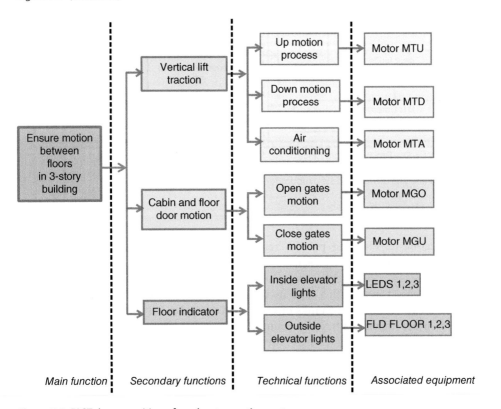

Figure 7.8 FAST decomposition of an elevator motion system.

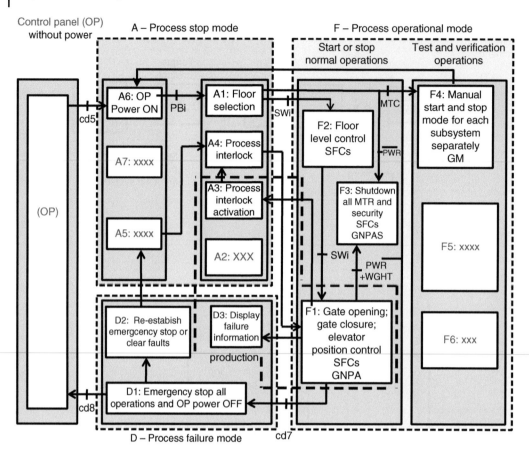

Figure 7.9 Graphical analysis of start and stop modes for an elevator motion process.

7.4.2 Bottle-Cleaning Process

A thermochemical-based cleaning process removes contaminants and organisms within empty bottles in a brewery plant. This is expected to result in safe-to-drink products. At the beginning of this process, dirty bottles are uploaded into a motorized handling system. This handling system has 750 trains, with a capacity of 50 bottles each. This system passes through three successive washing tanks (two tanks of water, maintained at 30 and 55°C, respectively, and one tank of water mixed with soda, maintained at 80°C). At the end of this washing process, cleaned bottles are off-loaded from the handling system for storage, as illustrated in Figure 7.10. Key devices and equipment used in the bottle washing process are summarized in Table 7.3. All motors involved in bottle handling, loading, and off-loading systems that should have synchronized motions are on/off-controlled. The two electrovalves EV4 and EV5, connected to motor pump MTR1 at the outlet of the hot water/steam tank, are used to control water temperature levels. The injected steam is mixed with cold water to increase tank water levels. This is done through thermostat-based temperature measurement (temperature

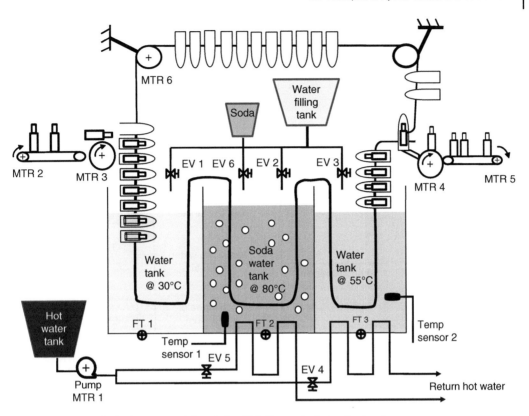

Figure 7.10 Simplified schematic and description (P&I diagram) of a bottle-cleaning process.

sensors 1 and 2). Similarly, soda and water fill the tanks through electrovalves (EV1, EV2, EV3, and EV6).

During the operating of such a bottle-washing process, it should be noted that: (i) the water-and-soda tank should not be overfilled, due to a defective anti-over drop limit switch (such a situation has a high risk of burning the skin of shopfloor operators); (ii) the high water tank temperature level and the poor synchronization of the bottle-handling system should prevent bottle breakdown; (iii) the command and control of process operations should be monitored to ensure real-time adjustments; (iv) on/off-controlled electrovalves regulate the supply of steam to maintain the water tank around 55–80°C; and (v) periodically, tanks are emptied. Using the FAST method, a decomposition of the bottle-washing process would yield: (a) filling and emptying of tanks; (b) heating of soda and water tanks; (c) control of tank temperatures; and (d) uploading and off-loading of bottles, as shown in Figure 7.11. Subsequently, as shown in Figure 7.4(a), the hierarchy of SFC can be drawn as the graphical analysis of process operating sequences for start, stop, and failure modes, with their activation conditions as depicted in Figure 7.12. Then, it is possible to logically control the soda tank filling subprocess, as described through an SFC in Figure 7.13(b), while the block diagram in Figure 7.14 summarizes the closed-loop control of the coupling between the tank temperature

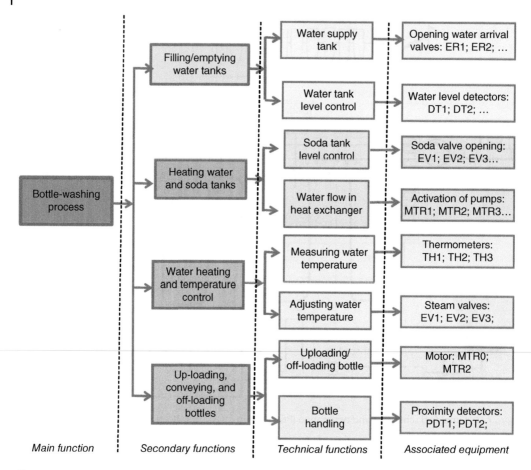

Figure 7.11 Functional analysis using the FAST method.

and the conveyor speed. The corresponding control and command wiring diagram is shown in Figure 7.13(c).

7.4.3 Cement-Drying Process

In a cement production plant, pozzolana with around 15% humidity is supplied from a quarry. A 3% humidity is required for the subsequent grinding process. Hence, a mandatory drying process is performed. This consists of injecting heavy fuel and compressed air to be burned into a flame chamber. Then, the generated hot air (between 700 and 810°C) is drawn into the kiln dryer tube to dry the fed humid pozzolana. The hot air leaves the tube with a temperature around 100°C, while the rotation of the kiln propels the dried pozzolana along the tube. Here, the dried pozzolana is extracted, while dust is removed by a separate filter, as illustrated in Figure 7.15(a). Table 7.4 summarizes the pozzolana drying-process devices and equipment, the corresponding process variables, and the process operating sequences.

Table 7.3 Process equipment listing.

Equipment	Technical characteristics	Operating conditions	Associated process operations and control strategy
AC pump MTR1	$P = 18.5$ kW	Oil-filled $n = 1500$ rev min^{-1} $p = 80$ bar	Star-based start-up of asynchronous motor for water tank injection at 80, 55°C. On/off controller
AC motor MTR6	50 Hz, 1.1 kW	1430 rev min^{-1}	Direct-based start-up of asynchronous motor for bottle-handling chain in 30, 80, 55°C tanks. On/off controller
Conveyor MTR5	$P = 7.5$ kW $Q = 250$ m^3 h^{-1}	$n = 1500$ rev min^{-1} $p = 80$ bar	Asynchronous motor, four limit switches, conveyor speed = 7 05·10^{-2} m s^{-1}
On-loading/ off-loading bottle MTR2, MTR3, MTR4, MTR5	50 Hz, 1.1 kW	1430 rev min^{-1}	Direct-based start-up of asynchronous motor for bottle on/off-loading handling system. On/off controller
Electrovalves (EV4, EV5)	Total volume of heated liquid: 93l Total volume of heating liquid: 21l	Heating liquid: Pmax = 6 bar, Tmax = 100°C. Heated liquid: Pmax = 7 bar, Tmax = 170°C.	Water heating stage of 80 and 55°C
Thermostat	Range 0–5 bar and 0–150°C	0–10 bar 0–150°C	Transmitter 4–20 mA Maintain water temperature at 80 and 55°C
Electrovalves (EV1, EV2)	DN = 38.1 mm, 0–170°C; 0–10 bar;	p = 2–4 bar 121–155 °C	ASI-based pneumatic (ON/OFF regulation) Steam through heat exchanger for water tanks at 52 and 62.5°C and command by pressurized air-supply electrovalves
Manual valve for steam arrival	DN = 125 mm 0–40°C; 0–10 bar	p = 3–4 bar	Valve to supply steam for thermal heat exchanger

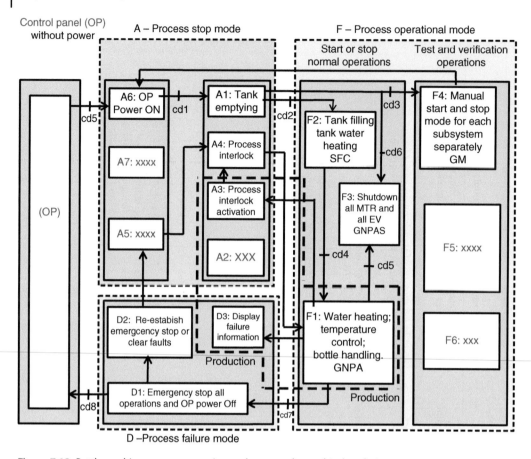

Figure 7.12 Bottle-washing process operating and stop mode graphical analysis.

In order to generate a flame in the combustion chamber, two asynchronous motor-driven pumps (MTR1, MTR4) inject fuel and compressed air toward a lighter that has ignited the flame. As an alternative to fuel, butane gas can be driven as a replacement combustible. Simultaneously, a motorized conveyor (MTR6) carries the humid pozzolana into a motorized rotary dryer tube (MTR5) connected to the flame chamber. In order to ensure complete combustion, the air-to-fuel ratio and the pressure in the flame chamber should be maintained constant by varying the opening of the dilution and combustion fans (MTR3, MTR2). Electrovalve EV1 feeds the pozzolana from the silo, while electrovalve EV2 extracts it from the rotary kiln. Table 7.5 summarizes the pozzolana dryer operating sequences. From the process audit, a high temperature variation within the flame chamber produces a frequent shutdown of the dryer tube. The temperature can be reduced by controlling the fuel inflow rate. Also, fluctuation of the air-to-fuel ratio due to the combustion and dilution fans could cause an incomplete combustion due to the zero pressure in the dryer tube. The performance of the ON/OFF control scheme of the pressure and temperature within the flame chamber is assessed and compared to industrial standards, as shown in Table 7.4. This can be used to set the design objectives of control loops from Figure 7.15(c). Here, it is required to migrate from an ON/OFF control scheme

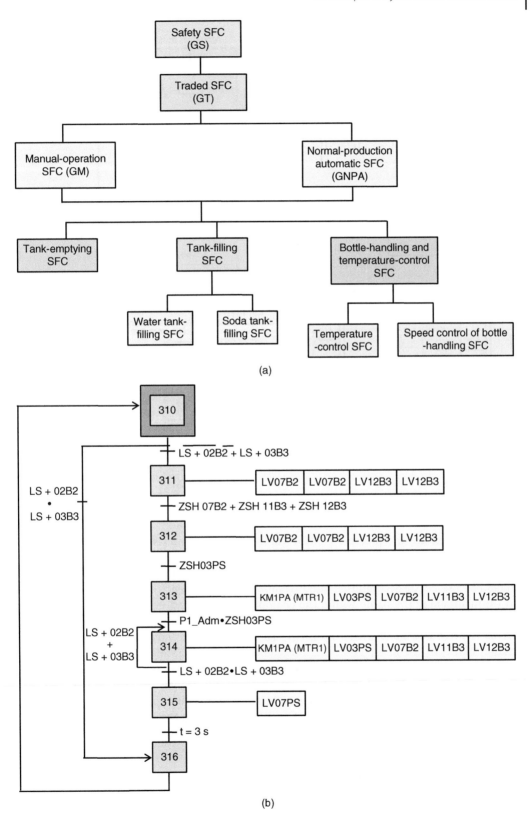

(a)

(b)

Figure 7.13 (a) SFC-based hierarchy of a bottle-washing process. (b) SFC for soda tank filling for a bottle-washing process. (c) Control and command schematic for a bottle-washing process.

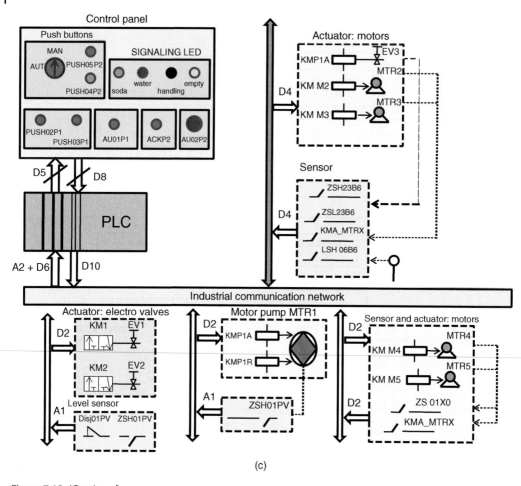

(c)

Figure 7.13 (*Continued*)

to a PID temperature and pressure control with settling time and rise time less than 450 and 150 s, respectively.

Using the FAST method and the process variables defined in Figure 7.15(b), the dryer cement process can be decomposed into two subprocesses: (i) injecting pozzolana subprocess; and (ii) maintaining the flame within the chamber, as illustrated in detail in Figure 7.15(b). The corresponding graphical analysis of the start and stop operating modes of the dryer process is presented in Figure 7.15(d). An example of a logic control program using the function block for the dryer process is presented in Figure 7.15(e). Figure 7.16 depicts the overall control and command wiring diagram of the pozzolana drying process. It might be noticed, for example, that there are eight digital output signals from the programmable logic controller toward the eight field contactors activating the motor HP pump and electrovalves (e.g. for MTR4, XVS AA09, and XSV BAC07) through the Modbus-based industrial network.

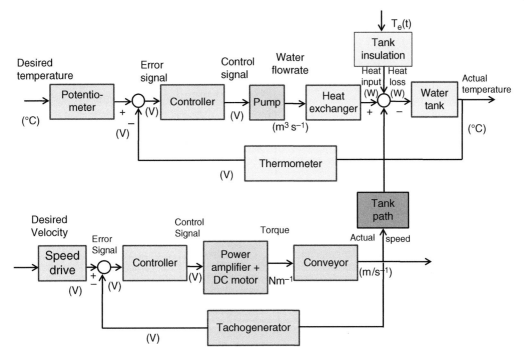

Figure 7.14 Subprocess function block diagrams.

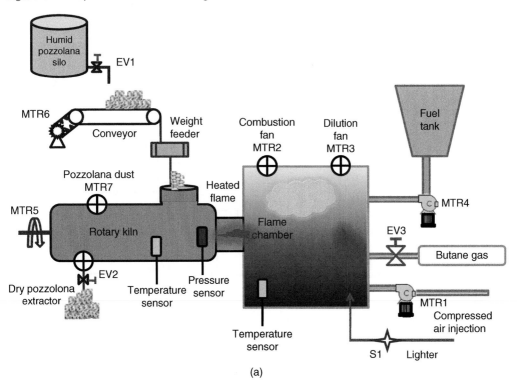

(a)

Figure 7.15 (a): Simplified process schematic and description of a cement-drying process. (b) Function analysis of the cement-drying process using a FAST method. (c) Feedback control block diagram of the cement-drying process. (d) Stop and start graphical analysis of the cement-drying process. (e) Example function block for pozzolana dryer process shutdown.

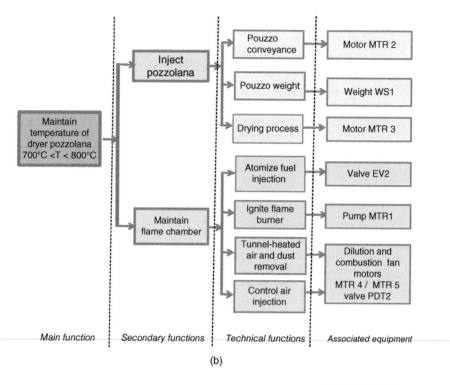

(b)

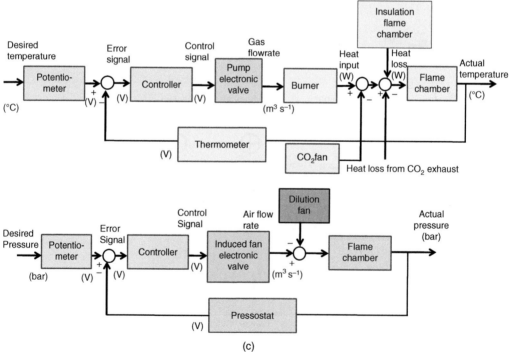

(c)

Figure 7.15 (Continued)

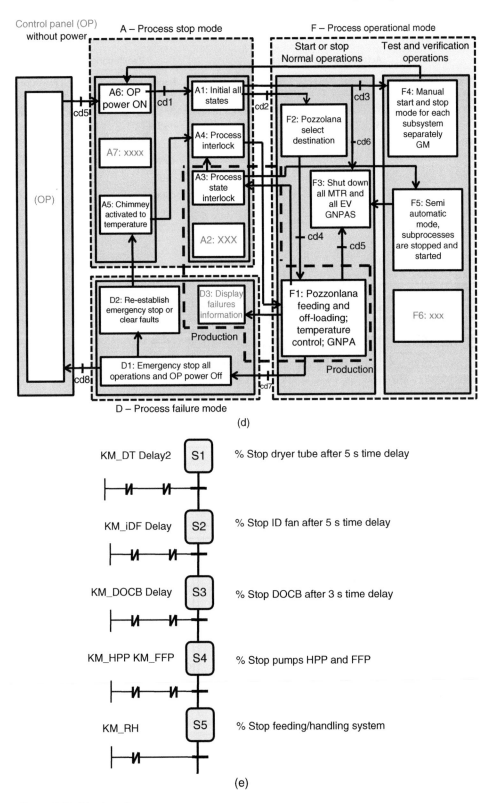

(d)

(e)

Figure 7.15 (*Continued*)

Table 7.4 System performance audit and specifications of the cement-drying process.

Characteristics	Actual system performance		Benchmarked values	
	Temperature	Pressure	Temperature	Pressure
Set-point values	760°C	−2 mbar	760°C	−2 mbar
Precision	±20°C	±2 mbar	<±2°C	±2 mbar
Steady state	0.578	0.19	1	1
Overshoot (%)	0	0	<2%	<2%
Peak time (s)	680	768	<400	<500
Settling time (s)	425	576	<300	<400
Rise time (s)	197	211	<100	<150

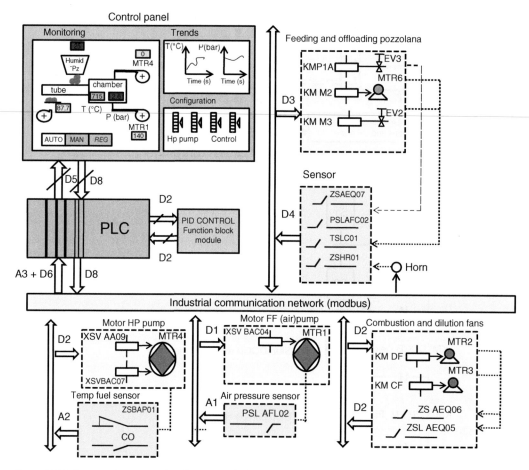

Figure 7.16 Cement dryer control and command schematic.

Table 7.5 Start and stop operating mode of cement-drying process.

Operating mode	Operating sequences	Variables	Variables types	Description	Interlocks
Startmode	Select pozzolana destination	Direct grinding, Silo2	Bool, Bool	Select grinding machine or Pozzolana silo	AU
	Start pozzolana dust extractor	Start_Aux, Start_PDE	Bool, Bool		
	Start dryer tube motor, set horn on	Start_D, KMK, Elev, PB_Conv, KM_DT	Bool, Bool, Bool	Activate crampon and direction conveyor belt	
	Fuel feeding and high-pressure fuel pump	Start_RH, S1 EV3, MTR1	Bool, Integer, Integer	Activate heater; low fuel temperature $90°C < T < 120°C$; low compressed air pressure	Range: $5\,bar < P < 7\,bar$ Fuel thermostat; shut down flame chamber and fuel pumps
	Light flame-chamber burner, start flame chamber; Start dilution fan and combustion fan	Start_FC, KM_DE, KM_CF	Bool	Fault, such as excess flame chamber output temperature, absence of flame, no compressed air	Flame fault from detector; flame chamber output temperature $700°C < T < 810°C$ Shut down fuel pumps and flame chamber
	Start humid pozzolana extractor	Start_PE	Bool		
Stop mode	Stop pozzolana extractor (weight feeder)	Stop_PE	Bool	Stop Pozzolana extractor	
	Shut down flame chamber, close fuel electronic valves, stop dilution fan and combustion fan	Stop_FC, KM_DE, KM_PE, KM_CF	All Bool		
	Shut down dryer, dryer output crampon and direction conveyor belt, ID fan	Stop_DT, KM_IDE, KM_DCB, KM_Elev	All Bool		
	Shut down pozzolana dust extractor	Stop_PDE, KM_PDE	Bool, Bool		
	Shut down auxiliaries	Stop_Aux, KM_Aux	Bool, Bool		
	Shut down re-heater	Stop_RH, KM_RH	Bool, Bool		
	Emergency stop	AU	Bool		Power supply to EVs, fuel and air supply system, and combustion fan

Exercises and Problems

7.1   Commonly used in the pharmaceutical industry, a granule-forming process (granulator) schematic is shown in Figure 7.17. A feed stream of powder is sprayed between an outer stationary cylinder and an inner rotating one. As the motor-driven cylinder rotates, the fluidizing air drains out the particles of granules formed throughout the granulator. The rotational speed of the cylinder defines the final shape and size of the granules. The valve is used to adjust the injection flow rate of the powder. Increasing the valve on the edge by 1% causes an increase in the diameter of the granules of 2 mm. A 2.5% increase in the agitating motor speed will decrease the thickness in the center position by 1 mm. A 2% increase in the heating temperature will increase the thickness in the center by 3 mm and reduce the edge thickness by 0.5 mm.

A sensor measures the thickness of the formed granules. Once the particle powder is deposited into the chamber, the granulator process consists of: (i) injecting air and fluid; (ii) heating the air; (iii) stirring the wet particles; and (iv) forming granules from the particle grouping.

a)   List all sensing devices and actuating equipment involved in the granule forming process and their I/O variables.

b)   Perform a process decomposition using the FAST method.

c)   After defining controlled and process variables, draw closed-loop block diagrams for the control of granule feed, and size.

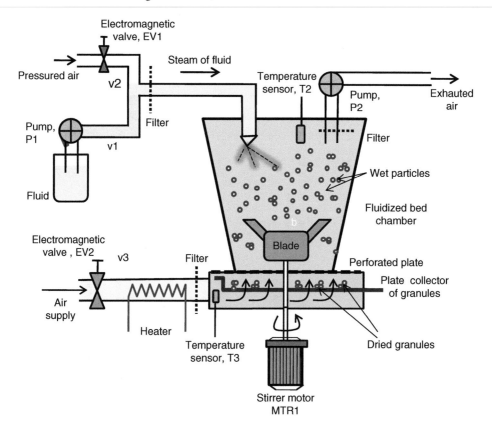

Figure 7.17 Granulator process schematic.

d) Update the list of process binary variables with those from the control panel enabling the stop and start activation of the process and the indication of one of the two granule sizes.

e) Using functional analysis and SFC, list the operating modes and operating cycles, along with their operating sequences.

f) Develop a generic logic and feedback control for this process for various defined granule sizes.

7.2 A distillation column is used for the separation of liquids, such as crude oil at different boiling points, that are widely used in the petrochemical industries. Figure 7.18 gives an overview of a typical crude oil distillation column and the equipment involved in it. To separate gasoline from asphalt, the reflux rate (at the top) is adjusted to control the distillate composition, while the reboiler steam rate (at the bottom) is varied to control the bottom composition. Any change in the feed rate to the column acts as a disturbance to the process. Among the equipment involved are a condenser, a vapor-liquid separator, a compressor, and a stripper.

When the composition of gasoline in the top distillate stream is below a set point (at the top), the flow rate of the cold liquid refluxes into the column and the reflux flow increases the gasoline purity of the distillate stream. However, the additional cold liquid works its way down the column, beginning to cool the bottom. This allows an increase in the gasoline flow out of the bottom stream. As the composition moves off the set point, the flow of steam into the reboiler increases, heating the bottom of the column and causing an increase of hot vapors traveling up it. This eventually causes the top of the column

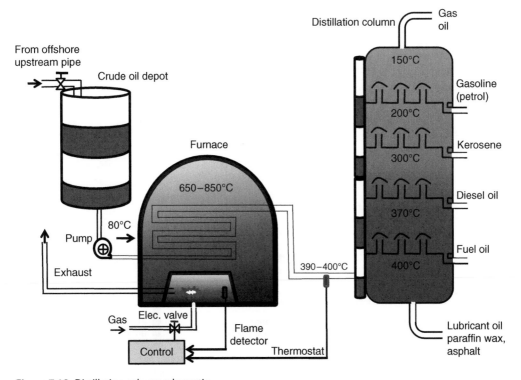

Figure 7.18 Distillation column schematic.

to heat up. Similarly, as the top of the column heats, the gasoline purity in the distillate stream becomes too low. In response, at the top, there is an increase in the flow of cold reflux.

a) List all devices and equipment involved in the distillation column control and their corresponding I/O variables.

b) After defining controlled variables and process variables, draw and label closed-loop block diagrams for feed control, boiler temperature control, and reflux control.

c) Update the list of process binary variables with those from the control panel enabling the start and stop activation of the process when temperature threshold values are reached per stage.

d) From the functional analysis, list operating modes and operating cycles, along with their operating sequences.

e) List three types of safety measures and the corresponding interlocks that need to be implemented.

f) Develop an SFC normal production program integrating the logic control with the feedback control.

g) Sketch the control and command schematic.

7.3 Consider the milk bottle-filling process illustrated in Figure 7.19. Develop a monitoring and control system for this process. A sensor detects the position via a limit switch, waits 0.6 s, and then fills the bottle until a photosensor detects a filled condition. After the bottle is filled, outfeed motor M2 is energized and moves the filled bottle; after 0.5 s, the

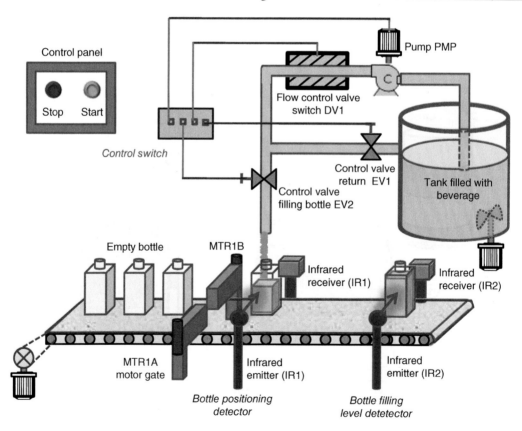

Figure 7.19 Milk bottle-filling process.

fed motor M1 is energized and conveys the next bottle. This control system includes start and stop push buttons for the process.

a) List all actuating equipment and sensing devices or detectors involved in this process, as well as the corresponding I/O variables.
b) Using the FAST method, derive the operating sequences.
c) Update the process safety measures, along with start and shutdown operating modes, to derive interlocks.
d) Design the expected process logic control and feedback control loops.

7.4 It is desired to design a control and monitoring system for activation/deactivation of either bi-directional motor or motor-driven pump using AC motors AC motor with variable-speed drives. This allows migration from a hardwired control to a programmable logic control and monitoring system. As such, the operator panel station performs potentiometer speed control (speed regulator), forward and reverse direction selection, and selection of the manual or automatic operation running mode of the variable-speed drive, and even start and stop push buttons, through run/jog switches. Figure 7.20 shows an operator station used to manually control a variable-speed drive. Note that the start, stop, run/jog, potentiometer, and forward/reverse field devices are connected with the same names that are used in the control program.

a) Using all sensing and actuating devices in Figure 7.20, design the control panel and complete the wiring diagram with connection links in/out of the control unit according to an operating strategy defined.
b) To complete the migration, design the logic controller of AC motors and speed feedback controller (when required).

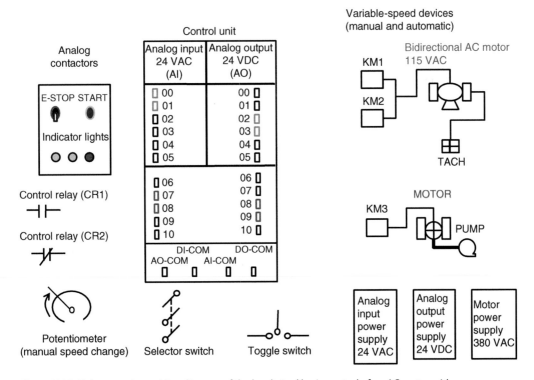

Figure 7.20 To be complete wiring diagram of the hardwired logic control of an AC motor with a variable-speed drive.

7.5 Recall the thermal plant and connected electrical supply network illustrated in Figure 7.21. It is desired to powerflow and monitoring system. As such:
a) List the voltage and power control equipment, operating sequences, and associated constraints.
b) Decompose the process operations using the FAST method.
c) Based on operating conditions, establish the SADT or GEMMA graph of the start and failure modes, as well as interlocks.
d) Define the control panel with monitoring variables for this power plant.
e) Propose an integrated sequential logic and feedback control system block diagram.
f) Draw the command, power, and I/O wiring diagrams of the hybrid system.

7.6 It is desired to control the flow of water at a hydroelectric dam in order to maintain generated electric power close to a forecasted daily load. Sketch a control panel (with monitoring variables) and corresponding logic control system as illustrated in Figures 7.22 and 7.23. The process equipment is listed in Table 7.6.

7.7 In a cement-milling station, a portico scratcher with a capacity of 200 tons per hour allows raw cement material, such as gypsum and dry and wet pozzolana, to be conveyed from a storage hall to the grinding or drying cement subprocesses. First, high-speed transversal positioning of the portico scratcher above the raw cement material deposit is completed. Based on the two motorized arms (primary and secondary), the scratcher then moves from right to left across the deposit, to the conveyor-based handling system, and on to other cement subprocesses, as illustrated in Figure 7.24. Tables 7.7 and 7.8 list the equipment involved and operating sequences, respectively.

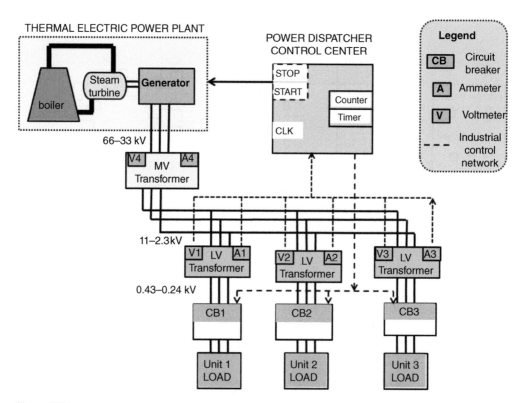

Figure 7.21

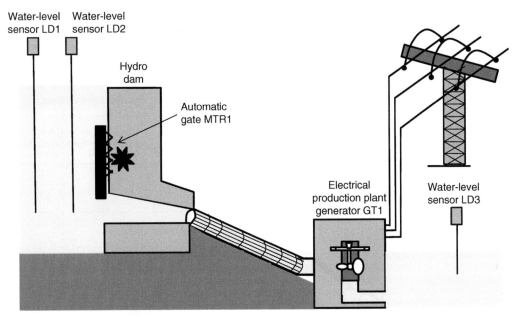

Figure 7.22 Simplified process schematic of a water control system in a hydroelectric dam.

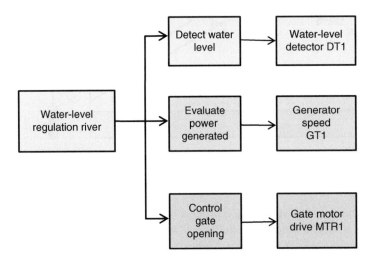

Figure 7.23 FAST analysis of a water control system in a hydroelectric dam.

Based on the information provided:
a) Create an SFC using the start and stop modes based on the graphical analysis approach.
b) For each actuating equipment, define the control variables and types as well as type(s) activation conditions required for SFC.
c) Develop all required SFC (security, production, etc.) (Hint: interlocks formal modeling).
d) Based on a defined control panel, derive the generic logic control algorithm for the scratching process and the resulting control and command schematic.

Table 7.6 Process equipment listing.

Equipment	Technical characteristics	Operating conditions	Associated process operations and control strategy
Generator GT1	Power 11 kW	Max. load arm: 35 tons Speed: 0.4 m s^{-1}	Asynchronous motor with squirrel cage for power generation
Gate motor drive MTR1	Power 6 kW	Max. load arm: 35 tons Speed: 0.4 m s^{-1}	Asynchronous motor with squirrel cage for water gate motion
Water-level sensor LD1, LD2, LD3	Rittmeyer pressure-based sensor	Length of carrier: 80 m Speeds: low 2.6 m mm^{-1}; high 16 m mm^{-1}	

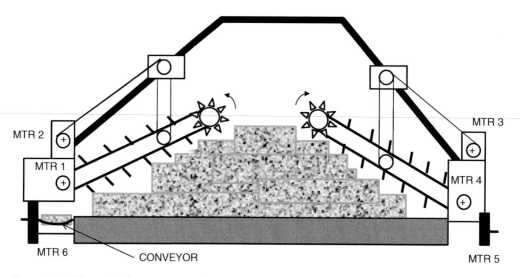

Figure 7.24 Schematic of a cement pozzolana scratcher process.

7.8 In the brewery industry, the tunnel pasteurization process consists of the thermal treatment of beer in bottles in several temperature-controlled water-based media, reducing bacterial growth. This extends the beer's shelf-life. As illustrated in Figure 7.25, the filled bottles are conveyed at a constant speed of 7 mm s^{-1} through several water showers at temperatures of 30, 52, 62, and 30°C, respectively. Some Motorized pumps are used to maintain the water temperature, at a flow rate of 250 m^3 s^{-1} on a closed loop between the water tank and the heat exchanger (170°C). Table 7.9 summarizes the pasteurization operating sequences.

The process function blocks for the temperature control can be decomposed into: (i) control air flow rate; (ii) flame chamber and control of flame intensity; (iii) flame chamber and fuel pressure regulation; (iv) burner and ignite flame block. Using the information provided:

a) Develop an interlocking logic (if necessary) for this process.

b) Create an SFC using the start and stop operating modes based on the graphical analysis approach.

Table 7.7 Process equipment listing.

Equipment	Technical characteristics	Operating conditions	Associated process operations and control strategy
Primary robot arm scratcher MTR1	Power 11 kW	Max. load arm: 35 tons Speed: 0.4 m s^{-1}	Asynchronous motor with squirrel cage for primary robot arm motion
Secondary robot arm scratcher MTR4	Power 6 kW	Max. load arm: 35 tons Speed: 0.4 m s^{-1}	Asynchronous motor with squirrel cage for secondary robot arm motion
Vertical motion of secondary robot arm scratcher MTR3	Force 2000 kg	Length of vertical trip: 80 m; speeds: low 2.6 m mm^{-1}; high 16 m mm^{-1}	
Vertical motion of primary robot arm scratcher MTR2	Force 1600 kg Power at low speed 0.83 kW; at high speed 5 kW	Length of vertical trip: 40 m; speeds: low 2.6 m mm^{-1}; high 16 m mm^{-1}	
Cross-motion portal MTR5 and MTR6	Power 5 kW	Cross-motion scratching speed: 0.9 m mm^{-1}; high speed 9.4 m mm^{-1}; brake torque max. 2 (10 mdaN)	

    c) Develop all required SFC (security, production, etc.) (Hint: interlocks formal modeling).

    d) Based on a defined control panel, derive the generic logic control algorithm for the pasteurization process and resulting control and command schematic.

7.9 Beer fermentation is a three-stage process. First, in the primary stage, the yeast is added to the malt seed in water-filled fermentation tanks. This mixture is kept closed for 8–12 days. During this stage, the temperature must be maintained at 12°C and the pressure at 1 or 2 bar while the CO_2 resulting from alcohol metabolism is evacuated, as illustrated in Figure 7.26. At the end of CO_2 production, the temperature of the generated beer in the fermentation tank is reduced to 5°C, while the pressure is maintained at 1 or 2 bar, to ease the solidification of the remaining yeast in the tank. Then, the temperature of the beer is brought down to −1°C and the pressure kept at 1 or 2 bar, to ease the collection of solid-based yeast deposit at the bottom of the tank. Finally, the yeast deposit is removed and the filtered beer is output through a dedicated electrovalve. Figure 7.27 illustrates the functional decomposition of this fermentation process and Table 7.10 presents the fermentation process operating sequences. Based on the function subprocess block diagrams for temperature and pressure control in the fermentation tank:

    a) Develop an interlocking logic for this process.

    b) Create an SFC using the start and stop operating mode-based graphical analysis approach.

    c) Based on a defined control panel, derive the generic logic control algorithm for the pasteurization process and resulting power supply control and command schematic.

Table 7.8 Process operating modes and sequences.

Operating mode	Operating sequences	Variables	Variable types	Description	Interlocks
Start mode	Start transversal positioning motion	MTR6, MTR5	Bool, Bool, Bool	High-speed transversal displacement and positioning scratcher portal	Fault due to poor synchronization of MTR5 and MTR6 during transversal motion (anti-twisting)
	Positioning of primary scratcher robot arm	MTR1	Bool	Low-speed position of primary scratcher robot arm above pozzolana (72°)	
	Positioning of secondary scratcher robot arm	MTR4	Integer, FBD	Low-speed position of secondary scratcher robot arm above pozzolana (68°)	Detect rotation of secondary scratcher robot arm
	Pozzolana scratching by winch of secondary scratcher robot arm (139 s)	MTR3	Repeat 1 successively	Low-speed pozzolana scratching by winch of the secondary robot arm for 139 s	If primary scratcher robot arm > 23 A, stop MTR3
	Pozzolana removal by winch of primary scratcher robot arm (139 s)	MTR2	Repeat 2 successively	Low-speed pozzolana removal by winch of primary robot arm for 139 s	If primary scratcher robot arm > 73 A, stop MTR2
	Repositioning of primary scratcher robot arm	MTR1	Repeat 3 successively until (72° − delta = 17°)	Low-speed position of primary scratcher robot arm above pozzolana (72° − delta)	If collision between secondary and primary arms, stop MTR2 and MTR3
	Repositioning of secondary scratcher robot arm	MTR4	Repeat 4 successively until (68° − delta = 13°)	Low-speed position of secondary scratcher robot arm above pozzolana (68° − delta)	

(Continued)

Table 7.8 (Continued)

Operating mode	Operating sequences	Variables	Variable types	Description	Interlocks
	Position referencing before transversal motion	MTR1, MTR2, MTR3, MTR4		Both primary and secondary arms to 72° and 68°, respectively	
	Start transversal positioning motion by 50 cm	MTR6, MTR5		Low-speed displacement of portal by 50 cm	
	Repositioning scratching etc.	All motors involved		Repeat previous step until scratcher portal reaches pozzolana length of 60 m	
Stop mode	Stop secondary scratcher robot arm	MTR3	Bool		
	Stop secondary scratcher robot arm	MTR4			
	Stop primary scratcher robot arm	MTR2	Bool		
	Stop primary scratcher robot arm	MTR1	Bool		
	Stop transversal positioning motion	MTR6, MTR5			

Table 7.9 Pasteurization process operating sequence.

Operating mode	Operating sequences	Variables	Variable types	Description	Interlocks
Start mode	Fill water	PB_START, all electrovalves (EVx)	Bool, Bool, Bool, Bool, etc.	Activation valves for inflow tanks	
	Turn on heat exchangers	EHX1, EHX2		2 bars < P < 4 bars 117°C < T < 131°C	
	Open steam electronic valves to supply water at specified temperature to the tanks	EV1, EV2, EV3			
	Start pumps to heat water tanks to 52 and 62°C in closed loop	MTR1, MTR2, MTR3 Sensor_temp, PID_temp_control	Integer, FBD		Pumps MTR2 and MTR3 start if tanks' water temperatures are reached
	Conveying bottle	LS1, LS2, LS3, LS4, MTR4	Bool, Bool, Bool, Bool	Move bottle filled with beer into pasteurization tunnel Constant conveyor speed	MTR4 fails if temperature out of reference Four limit switches detect conveyor travel distance (stop and start)
	Temperature control	MTR2, EHX1/2, EHX2, EV1, EV2		Temperature control via steam flow through heat exchangers using electrovalve and thermostatic valve	Tank pressure (P < 1.85 bars) temp_faults out of order (Temp_sensor) Stop conveyor
Stop mode	Stop filling water	LS_level	Bool		
	Stop conveyor	MTR4			
	Close steam valve	EV1, EV, EV3			
	Stop pump	MTR1, MTR2			
	Remove yeast	Timer, EVx	Bool		
		Timer, EVx	Bool		

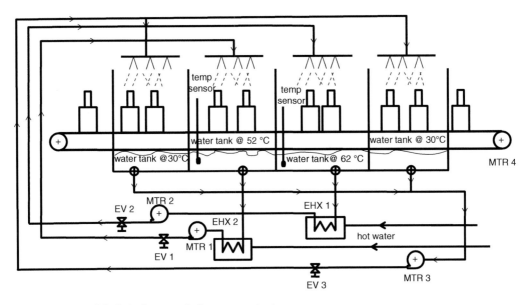

Figure 7.25 Simplified P&I diagram of a beer pasteurization process.

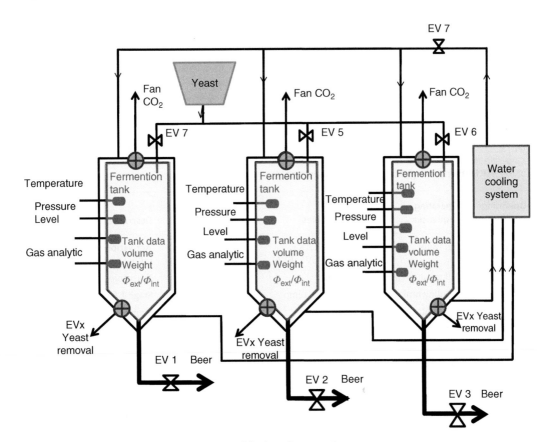

Figure 7.26 Simplified schematic description of the beer fermentation process.

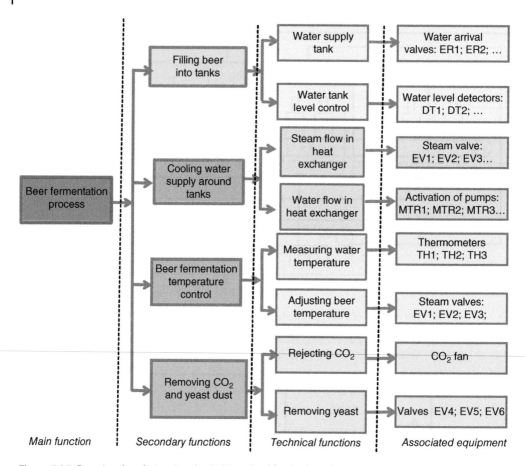

Figure 7.27 Functional analysis using the FAST method for the beer fermentation process.

Table 7.10 Fermentation process operating sequence.

Operating mode	Operating sequences	Variables	Variables types	Description
Start mode	Fill water	PB_START, EV4, EV5, EV6	Bool, Bool, Bool, Bool, etc.	Activation valves for inflow tanks
	Fill yeast (20 min)	Timer	Bool	Yeast injection
	Cooling tank at 12°C	Sensor_temp	Integer	Stage 1 fermentation
	Fermentation (5 d)	Timer	Bool	Stage 2 fermentation
	Cooling tank at 5°C	Sensor_temp	Integer	Stage 2 fermentation
	Cooling tank at −1°C	Sensor_temp	Integer	Stage 3 fermentation
	Beer filtration	EVx	Bool	Removal of beer
Stop mode	Stop filling water	LS_level	Bool	Fermentation preparation
	Stop fermentation	PB_STOP		End fermentation
	Remove yeast	Timer, EVx	Bool	Yeast collection
	Start cleaning (2 h)	Timer, EVx	Bool	Cleaning tank

7.10 Consider the challenges related to consensual decision-making in distributed problem solving by a cooperative scheme involving multiple mobile agents, such as autonomous underwater vehicles, unmanned aerial vehicles, or automated highway systems. In order to achieve such consensus, each vehicle must communicate with its neighbors in a coordinated fashion to avoid collisions and adjust its trajectory with respect to the targeted sequence points. Hence, the key characteristic of such a cooperative scheme is that data is exchanged with nearby agents (i.e. localized communication), due to limitations in the communication bandwidth.

  a) Using a schematic, sketch the cooperating agent structure, the type of data exchanged, and the cooperative agent components.
  b) Considering that each vehicle has a circular protection zone in a radius around it, how could the time delay of the consensus convergence be affected under the limited information exchanged and the dynamic interaction topology at a traffic junction. (Hint: traffic volume, junction configuration, vehicles speed among parameters to be considered)?

Bibliography

1 Arbel, A. and Seidmann, A. (1984). *Selecting a Microcomputer for Process Control and Data Acquisition*. IIE Transactions, Taylor & Francis.

2 Bailey, D. and Wright, E. (2003). *Practical Scada for Industry*. Newnes.

3 Bosh, J. (2003). *Design and Use of Industrial Software Architectures*. Newnes.

4 Bower, J.M. (1997). *Control of Sensory Data Acquisition*. Elsevier.

5 Boyer, S.A. (2009). *SCADA: Supervisory Control and Data Acquisition*, 4e. ISA.

6 Chiang, L.H., Russell, E.L., and Braatz, R.D. (2001). *Fault Detection and Diagnosis in Industrial Systems*. Springer.

7 Ding, S.X. (2008). *Model-Based Fault Diagnosis Techniques: Design Schemes, Algorithms, and Tools*. Springer-Verlag.

8 Erickson, K. and Hedrick, J. (1999). *Plant Wide Process Control*. Wiley.

9 Barlett, T.L.M. (2010). *Industrial Automated Systems: Instrumentation and Lotion Control*. Cengage Learning.

10 Luyben, W. (1990). *Process Modeling, Simulation and Control for Chemical Engineers*. McGraw-Hill.

11 Ogunnaike, B. and Ray, W. (1994). *Process Dynamics, Modeling and Control*. Oxford University Press.

12 Marlin, T. (1995). *Process Control: Designing Processes and Control Systems for Dynamic Performance*. McGraw Hill.

13 Trigeassou, J.-C. (2011). *Electrical Machines Diagnosis*. Wiley-ISTE.

14 Seborg, D.E., Edgar, T.F., Mellichamp, E.A., and Doyle, F.J. (2011). *Process Dynamics and Control*, 3e. Wiley.

15 Smith, C. and Corripio, A. (1997). *Principles and Practice of Automatic Process Control*. Wiley.

16 Bullo, F., Cortes, J., and Martinez, S. (2009). *Distributed Control of Robotic Networks*. Princeton University Press.

17 Mesbahi, M. and Egerstedt, M. (2010). *Graph Theoretic Methods in Multiagent Networks*. Princeton University Press.

18 Francis B. (2014). *A Course on Distributed Robotics*, Notes for CDC Bode Lecture.

19 Wei, R. and Cao, Y. (2011). *Distributed Coordination of Multi-Agent Networks*. Springer.

20 Kaltjob, P. and Duffie, N.A. (2005). Real time, cooperative, predictable decision-making in Heterarchical manufacturing systems design: requirements, generic methodology and applications. *Journal of Manufacturing Systems* 34 (2): 153–160.

21 Bai, H., Arcak, M., and Wen, J. (2011). *Cooperative Control Design*. Springer.

22 Bertsekas, D.P. and Tsitsiklis, J.N. (1997). *Parallel and Distributed Computation: Numerical Methods*. Athena Scientific.

23 Komenda, J., Masopust, T., and Schuppen, J. (2013). Supervisory control of distributed discrete-event systems. In: *Control of Discrete-event Systems*, v. 433 Lecture Notes in Control and Information Sciences, vol. 10 (ed. C. Seatzu, M. Silva and J.H. van Schuppen), 107–126. Springer.

24 Jadbabaie, A., Lin, J., and Morse, A.S. (2003). Coordination of groups of mobile autonomous agents using nearest neighbor rules. *IEEE Transactions on Automatic Control* 48 (6): 998–1001.

25 Lynch, N. (1999). *Distributed Algorithms*. Morgan Kaufmann.

8

Mechatronics Instrumentation: Actuators and Sensors

8.1 Introduction

Sensors and actuators are key devices involved in information, energy, or material processing within mechatronic systems. Most actuators are electric-driven and convert the electric energy supplied into mechanical energy or work. They include devices such as motors, control valves, heaters, solenoids, relay switches, and electromagnets. Usually, they are associated with mechanical, fluidic, or thermal transmission elements, which drive actions related to the dynamics of solid, liquid, and gas substances or to chemical reactions. Consequently, mechatronic system dynamics can be captured through binary logic or continuous-time variables. Depending on their time-varying characteristics, these variables are gathered by the detection or measurement instrumentation of process variables through transducers or sensors, producing signals correlated to the variations of the physical input conditions. Transducers transforming one form of energy into another are either: (i) active transducers, when they are able to generate their own energy (e.g. thermocouples); or (ii) passive transducers, when they require additional energy (e.g. strain gauge). There are various types of signals, such as: (i) analog voltage signals; (ii) amplitude- or frequency-modulated signals; (iii) pulse-width-modulated signals; and (iv) square-wave signals, representing binary values of 1 or 0. Usually, analog process-level inputs (voltage) are transformed into digital output ranges (bits) so that they can be processed by a computer. Figure 8.1 displays a generic system interface. Digital sensors produce logic-level outputs that can be directly interfaced to a control computer, while analog sensors produce voltage outputs that require an analog-to-digital converter in order to be interfaced to a computer. Furthermore, these sensing devices can be either contact (e.g. tachometers) or contact-free (e.g. encoders and interferometer types) devices.

However, the sensing challenge is to obtain a good estimate of a physical quantity. Indeed, converting physical phenomena into measurable voltages or current signals often requires signal conditioning, such as amplifiers that boost current and voltage. Therefore, understanding the differences and limitations of different sensing devices allows suitable ones to be selected for the design of a computer controller system.

In this chapter, classical models of electrical-driven actuating systems, along with the dynamics of their associated transmission element dynamics, are presented. Some process models involving various transmission elements are described, in order to illustrate their real-life applications. In addition to electrical-driven actuator characterization, sensor types and measuring principles are covered. Dynamic characterizations are presented for various physical system variables such as linear and rotational position, speed, acceleration, force, torque, chemical content, distance, flow, temperature, proximity, level, and pressure. In addition, signal conditioning principles and categorization (zero-, first-, second-order sensors, etc.), sensor and detector design specifications, and device characteristics such as

Control Of Mechatronic Systems: Model-Driven Design And Implementation Guidelines,
First Edition. Patrick O.J. Kaltjob.
© 2021 Patrick O.J. Kaltjob. Published 2021 by John Wiley & Sons Ltd.

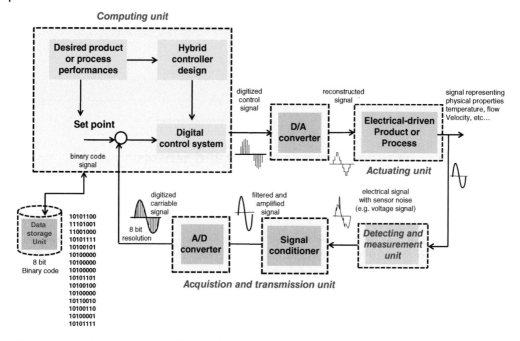

Figure 8.1 Sensing and data-acquisition chain.

linearity, sensitivity, limitations, calibration, and saturation are detailed. Finally, the sensing and operating principles of dynamic sensors such as the dynamics model, time domain, and frequency response characteristics are developed.

8.2 Actuators in Mechatronics

Electric actuators can generate either a binary output signal (i.e. contact relay) or continuous signal (stepper motor). Such electrical-driven actuating systems can be classified based on their electrostatic or electromagnetic – and even electrothermal – design principles. Their coupled transmission elements generate force to motion, pressure to flow, heat, and so on. Recent advances in nanotechnologies have led to the development of low-scale actuators (nano or micro).

Electromechanical actuators usually operate based on the electrostatic (electrostatic actuators) or the electromagnetic (electromagnetic actuators) principle. Electrostatic actuators use charged particle motion within the induced field to generate adequate force, vibration, pressure, and temperature variation displacement of a membrane or beam. They include piezo-electric actuators, which deliver strain-based actuation from piezo-materials by generating a voltage proportional to the applied mechanical deformation. Such piezo-materials use either ceramics (i.e. lead-zirconate-titanate) or polymers (i.e. polyvinylidene fluoride). Surrounded by electrodes or attached to them they allow strain-based actuation of up to a few kilovolts per millimeter. Among micro-device applications are: (i) position systems, such as micromanipulation (microrobot with piezo-legs), vibration oscillation suppression systems, valves (micro-pumps in artificial hearts), and ultrasonic motors; and (ii) handling systems, such as microgrippers and laser printers (which use charged particles from the toner, which move based on the electrophotographic principle).

Electromagnetic actuators are based either on electromagnetic force (Lorentz) or on electromagnetic induction (Faraday) principles, so that induced magnetic fields around a ferrous stator cause an object to move within the air gap separating them. Hence, they convert electrical energy into rotational or linear mechanical energy. They include linear solenoids, which work by energizing a solenoid (coil) that generates a magnetic field along which a ferromagnetic object can move, causing an increase in flux linkage. Their typical applications include actuations of valves, switches, and electromechanical relays. There are also rotary solenoids, where a ball moves along inclined bearing raceways. Other electromagnetic actuators include electromechanical motors such as motors, AC motors, and stepper motors. As summarized in Table 8.1, there are various types of electric motor, including series-excited DC motors, DC shunt motors, DC separately excited motors, AC asynchronous motors, AC synchronous motors, brushless AC motors, stepper motors, and linear motion motors. All have the following components: stator, rotor, field coil, commutating device (to control DC motor current flow), armature rotor winding, brush, and slip rings. Due to their wide industrial applications, the modeling and selection methodologies for various types of motors are presented in the following sections. Table 8.1 summarizes commonly encountered electrical-driven actuating systems.

8.3 Electromechanical Actuating Systems

Electromechanical actuators are mainly driven by electric motors or solenoids and operate on electromagnetic and electrostatic principles, as described in the following subsections.

8.3.1 Solenoids

Based on the electromagnetic (ferromagnetic) principle, solenoids provide linear or rotary motion. They operate by moving an iron core inside a wire coil, as depicted in Figure 8.2. Initially, the iron core is maintained outside the coil by a spring. When a voltage is applied across the coil, the resulting current generates a magnetic field around the coil to produce a magnetic force capable of moving the core. Solenoids can create voltage spikes. Typical electromechanical switches, such as relays and contactors, operate on the basis of this principle; so too do pneumatic valves, car door openers, and so on. Due to their low voltage and current requirements, such devices can be connected to a logic control unit (e.g. programmable logic controller).

A typical example is when a current is going through the coil and inducing a magnetic force that can balance the force of gravity and cause a train (which is made of a magnetic material) to be suspended. This could be equivalent to a RL circuit $V(t)$ connected in series to a coil whose flowing current $i(t)$ creates a force capable of balancing upward the train of mass m at a mid level $h(t)$. The system model around the equilibrium (the train is suspended) can then be given by:

$$\begin{cases} m\dfrac{d^2h(t)}{dt^2} = mg - \dfrac{Ki(t)^2}{h(t)} \\ V(t) = L\dfrac{di(t)}{dt} + i(t)R \end{cases}$$

Based on the AC inductive (magnetic) principle, a ferromagnetic core moves axially within the hollow cylinder, inducing the variation of the magnetic flux linking the primary winding to each of the secondary windings, as depicted in Figure 8.3. Any deviation is related to the position of the ferromagnetic tube. From Faraday's law, the electromagnetic induction is given by:

$$V = -N\frac{d\phi_B}{dt} = -N\frac{d(BA)}{dt} \tag{8.1}$$

where V represents the induced voltage, N the number of turns of the coil, ϕ_B the magnetic flux, B the magnetic field, and A the cross-sectional area of the coil. A current variation produces

Table 8.1 Typical electrical-driven actuating systems.

Electric actuators	Actuator types	Description
Binary actuators (switching circuits)	Thyristor, bipolar, solid-state relay, diode, MOFSET, transistor	Electronic switching devices offering high-frequency response
Electromagnetic actuators (electromechanical)	Direct current (DC) motor	Wounded field (separately excited) where speed varies with armature voltage or field current
		Wounded field (shunt) with constant-speed application and low starting torque
		Wounded field (series) with high starting torque and where speed varies with low load
		Wounded field (compound) with low starting torque
		Permanent magnet (variable speed but fixed magnetic field)
		Brushless (electronic commutation) fast response
	Alternating current (AC) motor	Induction motor
		Synchronous motor
		Universal motor (AC/DC power supply)
	Stepper motor	Hybrid (changes pulse signal into rotational motion)
		Variable reluctance (precision motion but requires switching devices)
Electromagnetic actuators (non-electromechanical)	Relay contact, electromagnet, solenoid	Applications requiring large force
Electrohydraulic actuators	Electromagnetic cylinder	Linear motion cylinder
	Electrovalves	Directional valves, pressure valves
	Electropump, fan, blower	Hydraulic motor (gear, vane, piston, rotary, reciprocating)
Electrostatic actuators	Piezo-electric motor	Voltage proportional to applied stress (laser printer)
Electrothermal actuators	Heater	
	Furnace	
Low-scale material devices (micro/nano)	Shape memory alloys	Temperature-sensitive
	Dielectric elastomer actuators	Spring-sensitive
	Ionic polymers	
	Electrostrictive	
	Piezo-electric material	Micropumps, micromotor, microvalve, microrobot (micromanipulation, cave exploration)
	Electroactive polymers	
	Electromagnetics	Door locks, HVAC control

Figure 8.2 Typical energized and unenergized solenoids.

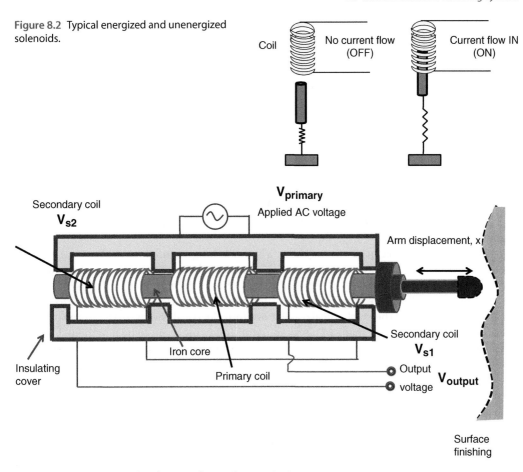

Figure 8.3 Solenoid-based surface-roughness characterization.

a magnetic field and consequently changes the orientation of the source. For small motions, the position of the core is proportional to the level of the output signal voltage. Otherwise, the relationship is nonlinear. The windings are designed to allow a linear relationship between the core position and the output voltage, such that the primary winding voltage is given by:

$$V_{primary} = V \sin \omega t \tag{8.2}$$

The voltage difference as a function of core linear position x results in:

$$V_{sec\,ondary} = kVx \sin \omega t \tag{8.3}$$

where k is a constant and x is the core linear position. By its value and sign, the level of the output voltage indicates the magnitude of the induced winding voltage difference, as well as its in or out phase status in relation to the primary coil applied signal. It is given by:

$$V_{output} = kVx \tag{8.4}$$

8.3.2 Digital Binary Actuators

Digital binary actuators are used to enable or disable event occurrences or systems operations by flipping between two discrete states. Binary actuators can be either: (i) bistable actuating systems, where they are designed around material with bistable properties, such as electromechanical relays and solid state devices; or (ii) discrete actuated systems, where they associate switching power electronic devices with actuators such as electric motors and hydraulic-powered cylinders. Hence, they are light-weight and are embedded in a load structure such that they can change the dynamic behavior autonomously (e.g. vibration attenuation and noise cancelation). Furthermore, they do not need power to maintain each stable state. Among bistable actuators, there are:

- *Dielectric elastomer actuators* (DEAs), which have a ratcheting transmission that acts as a power spring. High-speed switching DEAs are suitable for robotic systems.
- *Shape memory alloys* (SMAs), which display metal alloy phase contractions in the face of thermal variations. They are very sensitive to environmental conditions (e.g. temperature, humidity, dust).
- *Ionic polymers*, which can vary their volume once when some of their ions are absorbed by their polymer microstructure.
- *Electroactive polymer actuators*, which generate a decreasing force as their deformation increases.
- *Piezo-actuators*, which can have a stacked or a laminar design configuration. Laminar-design actuators consist of piezo-electric strips with electrodes bonded onto them, while stacked actuators consist of thin wafers of piezo-active material between metallic electrodes in parallel connection. Both are suitable for: (i) *suppressing oscillations*, thanks to piezo-active materials that convert the mechanical oscillations into electrical energy; (ii) *microrobot* applications, where the legs are made of piezo-actuators, allowing them to be lengthened, shortened, or bent in response to an applied voltage at the electrodes; (iii) *micropump* applications, where their diaphragms are actuated by piezoactuators, allowing I/O check valves to be opened for fluid pumping; (iv) *micromanipulators*, which convert contraction into gripping operations and are suitable for positioning applications; (v) *microdosage* devices, which allow a high-precision dosage of liquids within a range of nanoliters; and (iv) *piezo-motors*, which convert oscillations into a continuous motion, resulting in an elliptical motion over the contact area. Various oscillations offer the ability to develop different kinds of piezo-motors: longitudinal, transversal, shear, and torsional.

There are also micro and nano devices (less than 15 mm in size) with embedded electronic circuitry, such as electrostatic motors, which use the electrostatic principle to generate actuating forces.

Example 8.1 Consider the formal modeling of a logic function for the motor pump activation of a solar heating system illustrated in Figure 8.4. Here, the motor pump is activated by a pulse-type start push button and two digital sensors: irradiation from a solar source toward a photodiode (PD) and a limit switch turned ON (closed) when the collector is not full; all of this is connected to a two-input AND gate, corresponding to the logic Boolean function given by:

$$Pump = SW \cdot PD \cdot LS$$

An alternative solution could be to implement it using program-based logic devices (PLDs) that are made of integrated circuits (several hundred logic gate structures). These include: (i) a programmable logic array (PLA), which has an AND layer in the middle and an OR layer level at output; and (ii) a programmable array logic (PAL), which has a programmable AND layer and a fixed OR layer.

8.3.3 DC Motors

Motors are the most commonly used actuators in mechatronics. In terms of electrical-driven actuators used for precision vertical motion, direct current (DC) motors are preferred. DC motors have two distinct circuits: an outside set of coils, called a field stator circuit, and an

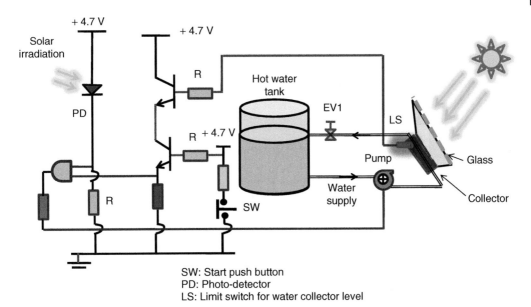

SW: Start push button
PD: Photo-detector
LS: Limit switch for water collector level

Figure 8.4 Solid state circuit for control of a solar thermal heating system.

inside set, called an armature rotor circuit. When a DC voltage, $V_f(t)$ is applied as excitation, the induced force within the coil causes the rotation of the rotor armature (i.e. due to generating torque). Depending on the armature and field circuit configuration, the application of Kirchhoff's voltage law into the field and armature circuits respectively yields:

$$V_f(t) = R_f i_f(t) + L_f \frac{di_f(t)}{dt} \tag{8.5}$$

and:

$$V_a(t) = R_a i_a(t) + L_a \frac{di_a(t)}{dt} + E_b(t) \tag{8.6}$$

where R_f and L_f represent the resistance and the inductance of the field winding, respectively. The current $i_f(t)$ produced in the winding generates the magnetic field necessary for rotor rotation in the armature circuit, such that the voltage applied across the armature terminals is $V_a(t)$ and the current flowing in the armature circuit is $i_a(t)$. R_a and L_a are the resistance and the inductance of the armature winding, respectively, and E_b is the total voltage induced in the armature. Then, applying the induction effect within the motor armature winding, the torque for a multi-turn coil armature conductor is given by:

$$T_m(t) = 2NB_v l i_a(t)r = K\phi i_a(t) = K_t i_a(t) \tag{8.7}$$

where l is the conductor length, r is the radius of the armature conductor about the center of rotation, L is the axial length of the conductor, B_v is the average flux density under a pole, A is the area of the coil, and ϕ is the flux/pole in the Weber and torque constant K_t, depending on the coil geometry. The direction of the current flow determines the rotation direction of the motor shaft. Thus, using Faraday's law, the generated EMF voltage induced in the motor by several coils wound on the rotor can be obtained as:

$$E_b(t) = 2NB_v lr(\omega_m(t)) \tag{8.8}$$

Hence, during the transient period, the applied armature voltage $V_a(t)$ varies with the current (producing torque proportional to emf $E_b(t)$) until it equals the applied voltage at steady state such that:

$$E_b(t) = V_a(t) - R_a i_a(t) \tag{8.9}$$

Here, if any current flows in the coil, there is no further acceleration, meaning the rotor turns at a constant speed. Hence, the electrical power delivered is equivalent to the mechanical power, so that:

$$E_b(t)i_a(t) = \omega(t)T_m(t) \tag{8.10}$$

Electrical connections of armature and field windings are used to classify DC motors. As such, the field windings can be either self-excited or separately excited, meaning that the terminals of the winding can be connected across the input-voltage terminals or fed from a separate voltage source. Furthermore, in self-excited motors, the field winding can be connected either in series or in parallel.

For example, in the case of the circuit series configuration (i.e. a self-excited series DC motor), the flux is directly proportional to the armature current. These motors are called series wound DC motors or universal motors, as they can run the same whether using an AC or a DC voltage source. Hence, the induced voltage results in:

$$V_a(t) = E_b(t) + (R_f + R_a)i_a(t) + (L_f + L_a)\frac{di_a(t)}{dt} \tag{8.11}$$

As long as the flux is directly proportional to the field current, the torque developed in the rotor can be expressed as:

$$T_m(t) = (K_f i_a(t))K i_a(t) = K_f K i_a(t)^2 \tag{8.12}$$

where K_f is a constant that depends on the number of turns in the field winding and the geometry of the magnetic circuit. Under steady-state operating conditions, the armature current results in:

$$i_a(t) = \frac{V_a(t)}{R_a + R_f + K_f K \omega(t)} \tag{8.13}$$

The rotor torque developed yields:

$$T_m(t) = K_f K \frac{V_a(t)^2}{(R_a + R_f + K_f K \omega(t))^2} \tag{8.14}$$

As illustrated in Figure 8.5, consider a DC motor with a permanent magnet configuration having a linear torque-speed operating range. Table 8.2 summarizes the process variables and parameters involved. The DC motor shaft is connected to a gearbox with a given ratio K_g:

$$K_g = \frac{n_i}{n_m} = \frac{\theta_i}{\theta_m}$$

where n_i and n_m are the number of teeth on the load-side and the motor-side gears, respectively, when a load with inertia J_L is attached to the output shaft of the gearbox. Therefore, the load inertia reflected at the motor shaft is given by:

$$J_{L \to M} = \frac{1}{K_g^2} J_L$$

$$N_r = \frac{N_{PL}}{N_{PM}} = \frac{z_{PL}}{z_{PM}}$$

$$\omega_m(t) = N_r \omega_L(t)$$

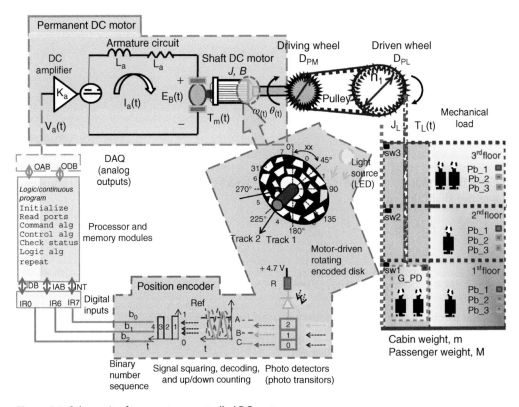

Figure 8.5 Schematic of an armature-controlled DC motor.

Table 8.2 Control variables and system parameters of a DC motor.

Parameters	Description and values	Variables	Description and values
K_a	Amplifier gain (V V^{-1})	$V_a(t)$	Apply DC input voltage (V)
L_a	Armature inductance (Henry)	$T_m(t)$	Motor torque (N.m)
R_a	Armature resistance (Ohms)	$i_a(t)$	Armature current (A)
K_b	Back emf constant (V s^{-1} rad^{-1})	$E_b(t)$	Induced armature EMF (V)
B	Motor damping (N.m.s^{-1} rad^{-1})	$\omega(t)$	Motor velocity (rad s^{-1})
J	Motor inertia (N.m.s^{-2} rad^{-1})	$\theta(t)$	Motor position (rad)
J_L	Load inertia (N.m.s^{-2} rad^{-1})	$T_L(t)$	Load torque (N.m)
K_t	Motor torque constant (N.m^{-1} A^{-1})	$i_f(t)$	Field current (A)
K_m	Motor constant (V s^{-1} rad^{-1})		
K_e	Tachometer gain (V s^{-1} rad^{-1})		
K_g	Gearbox ratio (%)		
L_f	Field inductance (Henry)		
R_f	Field resistance (Ohms)		

Table 8.3 Coil location per number of stator phases.

Number of phase m	1	2	3
ϕ_{Oj} of phase 1	0	0	0
ϕ_{Oj} of phase 2	90	60	45
ϕ_{Oj} of phase 3	—	120	90
ϕ_{Oj} of phase 4	—	—	135

The specific formula for the load inertia can be derived by applying Newton's balance equation to the motor shaft, this yields:

$$J\frac{d\omega(t)}{dt} + B\omega(t) = T_m(t) - T_{L\rightarrow M}(t)$$

with:

$$T_{L\rightarrow M}(t) = J_{L\rightarrow M}\frac{d\omega(t)}{dt} - \frac{1}{K_g^2}J_L\frac{d\omega(t)}{dt}$$

Thus, the mechanical part of the motor and load can be rewritten as:

$$J_{eq}\frac{d\omega(t)}{dt} + B\omega(t) = T_m(t)$$

$$J_{eq} = J + \frac{1}{K_g^2}J_L$$

From its configuration, a linear relationship is assumed between the motor torque and the induced current $i(t)$, such that:

$$T_m(t) = K_m i_a(t)$$

In general, K_m and K_e are considered equal when neglecting the losses due to the electromechanical power conversion. Therefore, it is also assumed that:

$$K_e = K_m = K_b$$

After algebraic manipulations to eliminate the variables $T_e(t)$, $E_b(t)$, and $i_a(t)$, the relationship between the applied armature voltage $V_a(t)$ and the load position $\theta_L(t)$ or the output DC motor rotor position $\theta(t)$ (assuming a synchronized motion such as $(\theta_L(t) = \theta(t))$) can be derived as:

$$K_m i_a(t) = J_{eq}\frac{d\omega(t)}{dt} + B\omega(t)$$

which is equivalent to:

$$K_m\frac{di_a(t)}{dt} = J_{eq}\frac{d^2\omega(t)}{dt^2} + B\frac{d\omega(t)}{dt}$$

Also:

$$E_b(t) = K_b\omega(t)$$

$$V_b(t) = R_a i_a(t) + La\frac{di_a(t)}{dt} + E_b(t) = \frac{R_a}{K_m}\left(J_{eq}\frac{d\omega(t)}{dt} + B\omega(t)\right) + \frac{La}{K_m}\frac{di_a(t)}{dt} + K_b\omega(t)$$

Thus, the model of the DC motor velocity system can be obtained such that:

$$\frac{d^2\omega(t)}{dt^2} + \frac{R_aJ_{eq} + L_aB}{L_aJ_{eq}}\frac{d\omega(t)}{dt} + \frac{R_aB + K_bK_m}{L_aJ_{eq}}\omega(t) = \frac{K_m}{L_aJ_{eq}}V_a(t)$$

by rearrangement, this yields:

$$\omega(t) + \frac{R_aJ_{eq} + L_aB}{L_aJ_{eq}}\frac{d\omega(t)}{dt} + \frac{L_aJ_{eq}}{R_aB + K_bK_m}\frac{d^2\omega(t)}{dt^2} + = \frac{K_m}{R_aB + K_bK_m}\omega(t)$$

Thus, the speed response $\omega(t)$ is dominated by the mechanical subsystem. This model is valid for DC motors with negligible friction and constant thermal operating characteristics. This is assimilated to a third-order model if the position rather than the velocity is controlled.

By neglecting the effect of the armature inductance L_a, the model becomes a first-order model given by:

$$\frac{R_aJ_{eq}}{K_m}\frac{d\omega(t)}{dt} + \frac{R_aB + K_bK_m}{K_m}\omega(t) = V_a(t)$$

which can be rewritten as:

$$\frac{R_aJ_{eq}}{R_aB + K_bK_m}\frac{d\omega(t)}{dt} + \omega(t) = \frac{K_m}{R_aB + K_bK_m}V_a(t)$$

The transfer function in Laplace domain is given by:

$$G(s) = \frac{\Omega(s)}{V_a(s)} = \frac{K}{1 + \tau_{em}s}$$

Alternatively, the DC motor modeling of the internal current control can be given by:

$$G_I(s) = \frac{I_a(s)}{V_a(s)} = \frac{J_{eq}s + B}{L_aJ_{eq}s^2 + (R_aJ_{eq} + L_aB)s + R_aB + K_bK_m}$$

If the short characteristic time dynamics were neglected, a position-based model of DC motor could be:

$$\frac{1}{\omega_n^2}\frac{d\theta^3(t)}{dt^3} + \frac{2\xi}{\omega_n}\frac{d\theta^2(t)}{dt^2} + \frac{d\theta(t)}{dt} = KV_a(t)$$

Here, when the mechanical part is a dominant component of the DC motor model, it can be considered from operating conditions as an integration process. Thus, the model can be approximated such that:

$$\frac{d\theta(t)}{dt} = KV_a(t)$$

8.3.4 AC Motors

Despite DC motors being the most commonly encountered, alternating current (AC) motors such as synchronous, asynchronous, and induction (single or three-phase) motors are suitable for applications requiring high torque. In the case of three-phase induction motor models, voltage, current, and magnetic flux quantities are expressed in terms of complex state space vectors. The resulting state space models are valid for any instantaneous variation of voltage and current, so they describe their performance under both steady-state and transient operations. Because an induction motor displays time-variant electrical and mechanical system dynamics, a commonly used modeling strategy is to approximate it by a DC-like motor model after successive transformations. The resulting induction motor model returns a constant input voltage and frequency. Hence, using Faraday laws, for the three stator phases, a global expression in the matrix form yields:

$$\begin{bmatrix} V_{sa}(t) \\ V_{sb}(t) \\ V_{sc}(t) \end{bmatrix} = R_s \begin{bmatrix} i_{sa}(t) \\ i_{sb}(t) \\ i_{sc}(t) \end{bmatrix} + \frac{d}{dt} \begin{bmatrix} \psi_{sa}(t) \\ \psi_{sb}(t) \\ \psi_{sc}(t) \end{bmatrix} \tag{8.15}$$

Similarly, the three rotor phases can be written as:

$$\begin{bmatrix} V_{ra}(t) \\ V_{rb}(t) \\ V_{rc}(t) \end{bmatrix} = R_s \begin{bmatrix} i_{ra}(t) \\ i_{rb}(t) \\ i_{rc}(t) \end{bmatrix} + \frac{d}{dt} \begin{bmatrix} \psi_{ra}(t) \\ \psi_{rb}(t) \\ \psi_{rc}(t) \end{bmatrix} \tag{8.16}$$

where R_s, R_r are stator and rotor winding equivalent resistances, while $V_{sa,b,c}(t), V_{ra,b,c}(t),$ $\psi_{sa,b,c}(t), \psi_{ra,b,c}(t), i_{sa,b,c}(t), i_{ra,b,c}(t)$ are stator and rotor three-phase motor a, b, c voltage, flux, and current, respectively. Consider a 2-from-3 reference frame transformation given by $T_{23} = \frac{\sqrt{2}}{3} \begin{bmatrix} 1 & \frac{-1}{2} & \frac{-1}{2} \\ 0 & \frac{\sqrt{3}}{2} & -\frac{\sqrt{3}}{2} \end{bmatrix}$. Thus, using Equation (8.16), the dynamics equations become:

$$T_{23}[V_{abcs}(t)] = T_{23} \left\{ R_s[i_{abcs}(t)] + \frac{d[\psi_{abcs}(t)]}{dt} \right\} \tag{8.17}$$

$$[V_{\alpha\beta s}(t)] = R_s T_{23}[i_{abcs}(t)] + \frac{d[T_{23}\psi_{abcs}(t)]}{dt} \tag{8.18}$$

This can be rewritten as:

$$\begin{bmatrix} \psi_{\alpha\beta s}(t) \\ \psi_{\alpha\beta r}(t) \end{bmatrix} = \begin{bmatrix} \psi_{s\alpha} \\ \psi_{s\beta} \\ \psi_{r\alpha} \\ \psi_{r\beta} \end{bmatrix} = \begin{bmatrix} L_s & 0 & M \cdot P(\theta) \\ 0 & L_s & \\ M \cdot P(-\theta) & L_r & 0 \\ & 0 & L_r \end{bmatrix} \begin{bmatrix} i_{s\alpha} \\ i_{s\beta} \\ i_{r\alpha} \\ i_{r\beta} \end{bmatrix} = \begin{bmatrix} L_s & 0 & M \cdot P(\theta) \\ 0 & L_s & \\ M \cdot P(-\theta) & L_r & 0 \\ & 0 & L_r \end{bmatrix} \begin{bmatrix} i_{\alpha\beta s}(t) \\ i_{\alpha\beta r}(t) \end{bmatrix} \tag{8.19}$$

where $P(\theta) = \begin{bmatrix} \cos\theta & -\sin\theta \\ \sin\theta & \cos\theta \end{bmatrix}$ and $L_s = l_s - m_s; L_r = l_r - m_r; M = \frac{3}{2}m_{sr}$. It is desired to transform induction motor need to be subsequently behaving as time-invariant DC motors at a steady state. These $i_{s\alpha}$ and $i_{s\beta}$ from the stator reference frame (α, β) into the d-q reference frame, rotating at the same speed as the angular frequency of the phase currents. The resulting i_{sd} and i_{sq} components will then be independent of time and speed. Considering the d-axis aligned with the rotor flux, the transformation is illustrated in Figure 8.6, where θ_{field} is the rotor flux position.

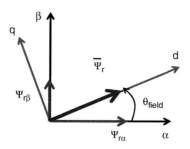

Figure 8.6 Park transformation.

Consider the case of a squirrel-cage induction motor with stator and rotor winding as depicted in the complex d-q equivalent circuit of an induction motor (neglecting rotor leakage inductance). Here, the voltage feeds the squirrel-cage induction motor model in a d-q synchronously rotating frame and ensures that i_{qs} delivers the desired electromagnetic torque such that:

$$\theta_s = \theta + \theta_r \tag{8.20}$$

Then, the stator and rotor components become:

$$[x_{\alpha\beta s}] = P(\theta_s)[x_{dqs}] \tag{8.21}$$

$$[x_{\alpha\beta r}] = P(\theta_s)[x_{dqr}] \tag{8.22}$$

After transformation of the motor quantities from a reference frame to the general reference frame, which is the frame attached to the rotor flux linkage space vector with direct axis (d) and quadrature axis (q) (d-q coordinates), the motor model is as follows:

$$\begin{cases} V_{ds}(t) = R_s i_{ds}(t) + \dfrac{\psi_{ds}(t)}{dt} - \omega_s \psi_{qs}(t) \\ V_{qs}(t) = R_s i_{qs}(t) + \dfrac{\psi_{qs}(t)}{dt} - \omega_s \psi_{ds}(t) \end{cases} \tag{8.23}$$

$$V_{dr}(t) = 0 = R_r i_{dr}(t) + \frac{\psi_{dr}(t)}{dt} - (\omega_s - \omega)\psi_{qr}(t)$$

$$V_{qr}(t) = 0 = R_r i_{qr}(t) + \frac{\psi_{qr}(t)}{dt} - (\omega_s - \omega)\psi_{dr}(t) \tag{8.24}$$

The electromagnetic torque can be expressed by using space vector quantities as:

$$T_m(t) = \frac{2}{3} n_p (\psi_{s\alpha}(t) i_{s\beta}(t) - \psi_{s\beta}(t) i_{s\alpha}(t)) \tag{8.25}$$

Thus, the mechanical model of the induction motor is given by:

$$J \frac{d\omega(t)}{dt} + B\omega(t) = \frac{n_p}{2} (T_m(t) - T_L(t)) \tag{8.26}$$

where n_p is the number of motor pole pairs (H) and $T_m(t)$ is the electromagnetic torque (N.m). R_s, R_r, L_s, L_r, M, J, and B are, respectively, the stator phase resistance (Ω), the rotor phase resistance (Ω), the stator phase inductance (H), the rotor phase inductance (H), the mutual (stator to rotor) inductance (H), the moment of inertia of the rotor and the load (kg m^{-2}), and the damping coefficient of the bearing (kg m^{-2} s^{-1}). This motor rotates at a speed of $\omega(t) = d\theta(t)/dt$ (rad s^{-1}) (i.e. with direct and quadrature axes x, y), where $\theta(t)$ is the angle between the direct axis of the stationary reference frame (α) attached to the stator and the real x-axis of the general reference frame.

8.3.5 Stepping Motors

A stepping or stepper motor is an incremental electric actuator consisting of wound fields and a permanent magnet. Such incremental motors are either permanent magnet, hybrid, or variable reluctance motors. Each received signal pulse of current by one of the stator windings generates rotational motion of a specific number of degrees (fixed angular step). Therefore, each pulse from the command generation system corresponds to an angular positioning and defines the step mechanical increment. Thus, the pulse frequency determines the motor angular velocity. Incremental motors can be either unipolar with n rotor poles for $360/n$ steps or bipolar. Their rotors could have as many as 200 poles producing up to 1.8° per step, that is, a stepper motor can move by 90°, 45°, 18°, or even a fraction of a degree per pulse. In the case of variable reluctance magnet stepper motors, they have toothed rotors and stator winding. Hence, rotation is induced by the reduction of magnetic reluctance between the stator poles and rotor poles. The motor step angle is determined by the number of teeth on the rotor, while the excitation type determines the stepping rates. The hybrid stepper motor has two multi-toothed armatures covering a cylindrical permanent magnet and defining the motor step angle between 0.9° and 5°. A permanent magnet stepper motor has a cylindrical permanent magnet. The alignment of the stator magnetic field with the permanent magnet ensures its rotation by a step angle between 45° and 120°.

These motors have a low torque capability and use on/off switch outputs for position control. Usually, the stepper motor control is an open loop through to the sequencing pulse-generation system executed within a personal integrated controller (PIC). Stepper motors can be found in machine tools, typewriters, printers, watches, pointing mechanisms for antennas, mirrors for space applications, telescopes, and so on.

The number of steps per revolution of the rotor, S, can be calculated as:

$$S = 2mN_r \tag{8.27}$$

where N_r is the number of rotor pole pairs and m is the number of stator phases. The angular step or number of degrees a rotor should turn per step is:

$$\Delta\phi = \frac{360}{S} \tag{8.28}$$

If a sinusoidal characteristic is assumed for the magnetic field in the air gap, the contribution of each phase j to the motor torque $T_{mj}(t)$ yields:

$$T_{mj}(t) = K_m \sin(n\phi(t) - \phi_{oj}(t))i_j(t) \tag{8.29}$$

where K_m is the motor constant, $\phi(t)$ is the actual rotor position, $\phi_{oj}(t)$ is the location of the coil j in the stator, $i_j(t)$ is the current in the coil as a function of time, and n is half the number of rotor teeth. The relationship between the supplied voltage $U_j(t)$ and the coil properties is given by:

$$U_j(t) = E_{bj}(t) + Ri_j(t) + L\frac{di_j(t)}{dt} \tag{8.30}$$

$E_{bj}(t)$ is the electromotive force induced in phase j in each coil. It is given by:

$$E_{bj}(t) = K_m \sin(n\phi(t) - \phi_{oj}(t))\omega(t) \tag{8.31}$$

where $\omega(t)$ is the angular velocity of the rotor, R is the resistance of the coils, and L is the inductance of the coil. The total torque produced by the stepper motor is:

$$T(t) = \sum_{j=1}^{k} T_{mj}(t) = J\frac{d\omega(t)}{dt} + B\omega(t) + T_f(t) \tag{8.32}$$

The arrangement of the coils has to be considered for more than one phase. Table 8.3 summarizes the angular position of the coils in the stator for various phases.

8.3.6 Transmission Mechanical Variables

Mechanical transmission elements (e.g. gearbox, pulley, screw) connected to an electrical-driven motor allow the transfer of their rotational speed to the attached load. Based on the

Table 8.4 Dynamical equations for the motion and torque of some mechanical transmission systems.

Mechanical transmission types	Equations	
	Motion	**Inertia and torque**
Belt pulley (timing belt) Electric motor — Driving wheel D_{PM} — Driven wheel D_{PL} — Load — Pulley	$N_r = \dfrac{N_{PL}}{N_{PM}} = \dfrac{D_{PL}}{D_{PM}}$ $\omega_m(t) = N_r \omega_L(t)$	$J_{total} = J_{PL \to M} + J_{L \to M} + J_M + J_{PM}$ derived from: $J_{PL \to M} = \left(\dfrac{1}{N_r}\right)^2 \dfrac{J_{PL}}{\eta}$ $J_{PL} = \dfrac{m_{PL} D_{DL}^2}{8}$ $J_{L \to M} = \left(\dfrac{1}{N_r}\right)^2 \dfrac{J_L}{\eta}$ $J_{PM} = \dfrac{m_{PM} D_{DM}^2}{8}$
Screw-wheel drive Electric motor — D_{wheel} Screw wheel — Load — Screw	$N_r = \dfrac{N_L}{N_M} = \dfrac{z_{wheel}}{z_{screw}}$ $\omega_m(t) = N_r \omega_L(t)$	$J_{total} = J_{L \to M} + J_{wheel \to M} + J_M + J_{screw}$ derived from: $J_{L \to M} = \left(\dfrac{1}{N_r}\right)^2 \dfrac{J_L}{\eta}$ $J_{wheel} = \dfrac{m_{wheel} D_{wheel}^2}{8}$ $J_{wheel \to M} = \left(\dfrac{1}{N_r}\right)^2 \dfrac{J_{wheel}}{\eta}$ $J_{wheel} = \dfrac{m_{screw} l_{screw}^2}{12}$
Crankshaft Electric motor — D_{PL} — l L_{MS} — Load	$N_r = \dfrac{N_L}{N_M}$ $\omega_m(t) = N_r \omega_L(t)$	$J_{total} = J_{L \to M} + J_{b \to M} + J_b + J_{ma}$ derived from: $J_{b \to M} = \left(\dfrac{1}{N_r}\right)^2 \dfrac{J_b}{\eta}$ $J_{wheel} = \dfrac{m_b l_b^2}{12}$ $J_{ma} = \dfrac{m_{ma} l_{ma}^2}{8/12}$
Cylindric gear Electric motor — D_{GM} D_{GL} — Load — Driving screw wheel — Driven screw wheel	$N_r = \dfrac{N_{GL}}{N_{GM}} = \dfrac{D_{GL}}{D_{GM}}$ $\omega_m(t) = N_r \omega_L(t)$	$J_{total} = J_{L \to M} + J_{GL \to M} + J_M + J_{GM}$ derived from: $J_{GL \to M} = \left(\dfrac{1}{N_r}\right)^2 \dfrac{J_{GL}}{\eta}$ $J_{GL} = \dfrac{m_{GL} D_{GL}^2}{8}$ $J_{GM} = \dfrac{m_{GM} D_{GM}^2}{8}$ $J_{L \to M} = \left(\dfrac{1}{N_r}\right)^2 \dfrac{J_L}{\eta}$

Table 8.5 Parameters and variables used.

Symbol	Definition	Units
J_M	Motor inertia	kg.m^2
J_L	Load inertia	
J_{PL}	Passive pulley inertia	
J_{PM}	Active pulley inertia	
$J_{PL \rightarrow M}$	Passive pulley to motor inertia	
$J_{L \rightarrow M}$	Load to motor inertia	
J_{GL}	Passive gear wheel inertia	
J_{GM}	Active gear wheel inertia	
J_{ma}	Crankshaft follower inertia	
J_b	Crankshaft cam inertia	
$J_{ma \rightarrow M}$	Crankshaft to motor inertia	
J_{wheel}	Wheel inertia	
$J_{wheel \rightarrow M}$	Wheel to motor inertia	
J_{screw}	Screw inertia	
J_{total}	Total inertia	
m_b	Mass of crankshaft	kg
m_{screw}	Mass of screw	
m_{PL}	Mass of passive pulley	
m_{PM}	Mass of active pulley	
m_{wheel}	Mass of wheel	
m_{ma}	Mass of crankshaft follower	
η	Efficiency	No unit
N_r	Power transmission ratio	
N_M	Motor rotational speed	rev.min^{-1}
N_L	Load rotational speed	
N_{PL}	Passive pulley rotational speed	
N_{PM}	Active pulley rotational speed	
N_{screw}	Screw rotational speed	
N_{wheel}	Wheel rotational speed	
$\theta_m(t)$	Motor shaft angular position	rad
$\theta_L(t)$	Load angular position	
$\omega_m(t)$	Motor shaft angular velocity	rad.s^{-1}
$\omega_L(t)$	Load angular velocity	
D_{PL}	Diameter of passive wheel	Meters
D_{PM}	Diameter of active wheel	
$T_{L \rightarrow M}(t)$	Load to motor torque	N.m
$T_L(t), T_m(t)$	Load torque, motor torque	
z_{screw}	Number of filets of screw	No unit
z_{wheel}	Number of wheel teeth	

load torque model and their inertia at the level of the motor shaft, the relationship between the load torque and the required motor shaft speed yields:

$$T_{req}(t) = J_{total}\frac{d\omega_m(t)}{dt} + B\omega_m(t) \tag{8.33}$$

$$T_{L \to M}(t) = \frac{T_L(t)}{\eta N_r} \tag{8.34}$$

Then, the total torque required is given by:

$$T_{m\_max}(t) = T_{req}(t) + T_{L \to M}(t) \tag{8.35}$$

$$\omega_m(t) = N_r\omega_L(t) \tag{8.36}$$

A safety torque, including particular starting and braking torques, should be added. Tables 8.4 and 8.5 present a generic torque equation, with the symbols used for some typical mechanical transmission systems.

8.4 Electro-Fluidic Actuating Systems

Electrical-driven fluidic actuating systems, such as electric motors associated with pumps, fans, blowers, and compressors, raise the mechanical energy of a fluid, causing an increase in and control of the flow rate and pressure, or a motion or elevation of the fluid. Fluidic actuating systems usually have volumetric flow rate as their input and pressure difference as their output. Electro-fluidic actuating systems are classified based on the type of fluid, the flow rate, and the pressure. While typical fans are used to move gas-based fluids because they require a high-volume and low-pressure differential, blowers are suitable for a similar volume but moderate pressure differential. In the case of fluids requiring a large pressure differential, compressors are used. Usually, fluidic actuating systems offer a higher power-to-weight ratio, better speed and acceleration responses, and easier directional change compared to mechanical elements for material handling and transport applications (automotive, aerospace, etc.). Fluids can be characterized by density (mass per unit of volume), viscosity (resistance to deformation), and bulk modulus (compressibility of fluid).

8.4.1 Electric Motorized Pumps

Electric pumps transform electrical energy into potential hydraulic energy by raising the fluid pressure through the volume space reduction during inlet-to-outlet pump-ports travel. They take the form of either fixed-flow gears or variable-flow pumps (where the flow can be variable or not). A pump can be associated to an electric valve for directional flow control, as well as to increase velocity of the fluid. The volume of fluid that is moved during one revolution is called the pump displacement D. This pump displacement and the maximum rotational speed ω_{max} determine the pump capacity Q so that:

$$Q = D\omega_{max} \tag{8.37}$$

Typical pumps are either positive-displacement pumps or centrifugal pumps. *Positive-displacement pumps* are of either rotary or reciprocating type. They generate fluid motion by mechanically displacing segmented fluid through a discharged nozzle. Reciprocating pumps move fluid by varying the fluid pressure using a diaphragm, while rotary pumps add kinetic energy to the fluid by raising the flow rate, which in turn raises the pressure of the fluid as it

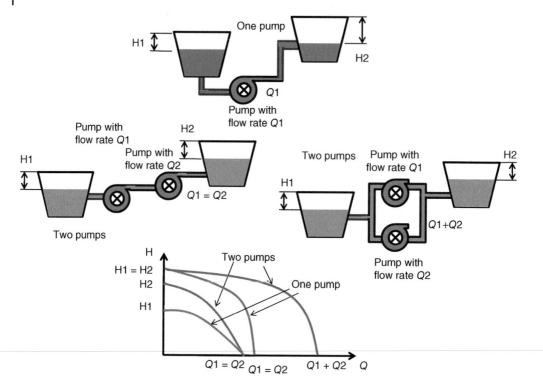

Figure 8.7 Serial and parallel pump connections.

exits the discharged nozzle. Rotary pumps can use other devices to perform fluid compression, including gears, lobes, and screws. Other displacement pumps include volumetric piston pumps, which have either an axial or a radial configuration, corresponding to one or more cylinders with a piston sliding into each. Here, the piston displacement determines the fluid flow. *Centrifugal pumps* use a rotating impeller to increase the fluid pressure and velocity. The center of the rotating impeller receives an inflow of fluid, which in turn is mechanically accelerated consequent to its rotation before it leaves through the side of the pump.

The operational characteristics of a pump can be obtained by plotting the curves of the head, H, the power, P, and the efficiency of the pump, according to the flow rate, Q, for a number of constant velocities, N, as illustrated in Figure 8.7. It should be noted that the efficiency takes a maximum value and then suffers a reduction. The ideal condition of the operation is achieved when the total head, H, and the flow rate, Q, have the same values as the maximum efficiency. This point is considered the "operating point."

The choice between pumps depends on the fluid type to be pumped, as well as the expected head and flow rate. The pumps can be set in series, such that the overall head, H, is the sum of each pump flow rate. When connected in parallel, the flow rate, Q, is increased by adding the total head, H, of each pump working by itself. In this case, the curve H versus Q can be obtained by adding the flow rates of each pump operating by itself with the same total head. The affinity laws or fan laws are as follows.

The flowrate is a function of the impeller speed (as diameter) and is given by:

$$Q_2 = Q_1 \left(\frac{N_2}{N_1} \right) = Q_1 \left(\frac{D_2}{D_1} \right) \tag{8.38a}$$

The head as a function of the impeller speed or diameter is given by:

$$H_2 = H_1 \left(\frac{N_2}{N_1} \right)^2 = H_1 \left(\frac{D_2}{D_1} \right)^2 \tag{8.38b}$$

The power as a function of the impeller speed or diameter is given by:

$$BHP_2 = BHP_1 \left(\frac{N_2}{N_1} \right)^3 = BHP_1 \left(\frac{D_2}{D_1} \right)^3 \tag{8.38c}$$

The power requirement of a pump can be derived from the overall head developed, as well as the mass of fluid to be pumped per unit time. This is given by the product of the shaft velocity ω and the mass flow rate $Q(t)$. Hence, pumps are defined by the relation between the fluid pressure P, the flow $Q(t)$, the shaft torque $T(t)$, and the velocity $\omega(t)$, given by:

$$Power = \Delta P Q(t) = T(t)\omega(t) \tag{8.39a}$$

If D_v is the volumetric displacement of the pump (m rad^{-1}) then:

$$\begin{cases} T(t) = D_v \Delta P \\ Q(t) = D_v \omega(t) \end{cases} \tag{8.39b}$$

For a fluidic system such as a pump, the load torque of the fluid in the tank can be estimated by:

$$T_L(t) = \frac{gQ(t)h(t)}{\omega(t)} \tag{8.39c}$$

with g being the gravitational acceleration, $Q(t)$ the flow rate, $h(t)$ the tank height, and $\omega(t)$ the pump rotational speed.

Other electrical-driven fluid movers include *centrifugal fans*, where the speed of air stream as fluid increases due to a rotating impeller, being highest when the fluid reaches the end of the blades. The speed is then converted into pressure by the blade shapes. *Axial fans* move an air stream along the axis of the fan. Typical axial flow fans are propeller, tube-axial, and vane-axial types. All fans are driven by electric motors, such as pumps. There are also *blowers*, capable of producing negative pressures for vacuum systems. These include centrifugal blowers and positive-displacement blowers, both of which are driven by electric motors. Centrifugal blowers are similar to centrifugal pumps. The gear-driven impeller can rotate at speeds up to 15 000 rpm. Positive-displacement blowers have rotors that trap air and push it through the housing in order to provide a constant volume, even if the system pressure varies.

8.4.2 Electric-Driven Cylinders

Electric cylinders transform the fluid (pneumatic or hydraulic) flow rate and pressure into a mechanical force and velocity via piston positioning actuated by the electromagnetic principle. Figure 8.8 illustrates a typical cylinder structure, which could be unidirectional or bidirectional (single- or bidirectional and double-acting), with a rod on one or both sides of the piston (single or double rod). With electric cylinders, a fluid is pumped into one side of the cylinder under pressure, causing that side to expand and creating a linear force advancing the piston. This force is proportional to the cross-sectional area of the cylinder. For a double-acting and double-rod cylinder, the velocity is given by:

$$v(t) = \frac{4q(t)}{\pi(D^2 - d^2)} \tag{8.40a}$$

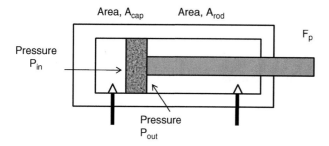

Figure 8.8 Single-acting, pressured air-powered cylinder.

During retraction, the force is given by:

$$F_p = P_{in}\frac{\pi(D^2 - d^2)}{4} - P_{out}\frac{\pi(D^2)}{4}$$

(8.40b)

During extension, the velocity and force are given by:

$$v(t) = \frac{4q(t)}{\pi D^2}$$

(8.40c)

$$F_p = \frac{\pi D^2}{4}(P_{in} - P_{out}) + \frac{\pi d^2}{4}P_{out}$$

(8.40d)

with P_{in}, P_{out}, A_{in}, and A_{out}, being the head-end chamber pressure, the rod-end chamber pressure, the head-end piston area with a diameter D, and the rod-end piston area with a rod diameter d, respectively. Based on the orifice equation, the flow rate is given by:

$$q(t) = CA\sqrt{\frac{2}{\rho}(P_p - P_{in})} = kx\sqrt{\frac{2}{\rho}(P_p - P_{in})} = A_{in}v(t) + \frac{V_1}{\beta}\frac{dP_{in}}{dt}$$

(8.40e)

8.4.3 Electrovalves

Electrovalves dictate: (i) whether or not the fluid is flowing (On/Off valves) or its direction of flow (bidirectional sliding valves); (ii) the desired pressure valve outlet, independent of inlet flow pressure variations (pressure regulator valves); (iii) flow rate regulation, using an orifice with a variable area (flow rate regulator valves); and (iv) the controlled displacement, speed, and force of the solenoid through flow rate or pressure difference, in order to ensure actuating system positioning (flow-proportional valves, pressure control-proportional valves, or servo valves). These valves can be electric-powered as well as hydraulic or pneumatic, depending on the type of driven fluid (liquid or gas). They can be fully or partially closed or opened, in a position dictated by signals transmitted from the controlling unit. Basic valve types are: (i) ball valves; (ii) butterfly valves; (iii) diaphragm valves; (iv) gate valves; (v) plug valves; (vi) globe valves; and (vii) eccentric valves.

In terms of configuration, each valve has a number of connections for port inlets and outlets; this is known as the number of the *n*-way position. Normally, open/closed terminology defines the valve status when the power is OFF. Commonly encountered valves are designed as two-way, three-way, and even four-way position valves, as shown in Figure 8.9(a). They are used for ON/OFF fluid flow control, unidirectional and single-acting cylinders, and bidirectional and double-acting cylinder operating functions. Three-way control valves have inlet, outlet, and exhaust ports, which can be closed or open, as illustrated in Figure 8.9(a). When there is no energy supplied, the outlet and exhaust ports are interconnected. When the solenoid is

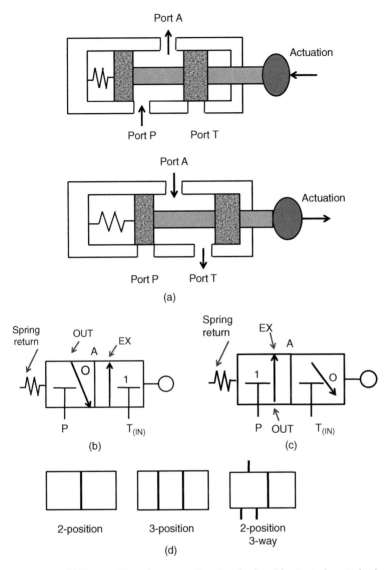

Figure 8.9 (a) Two-position, three-way directional solenoid actuated control valves (open and closed). (b) Equivalent open symbolic two-position, three-way directional control valves. (c) Equivalent closed symbolic two-position, three-way directional control valves. (d) Some symbolic directional control valves.

energized, the inlet and outlet ports are interconnected. There are also two-way control valves (one inlet port and one outlet port), such that the valve is considered open when the energy is OFF, and three-way control valves (three ports, such as one inlet and two outlets), as illustrated in Figure 8.9(b) and (c).

Considering that the flow pressure P is produced in the fluid flow, the valve behavior is described by the following nonlinear equation relating the flow and the valve trim:

$$q(\theta) = Cf(\theta)\sqrt{\frac{\Delta P}{g}} \tag{8.41}$$

where $\theta(0 \leq \theta \leq 1)$ is the valve position, g is the fluid density (Kg/L), C is the valve coefficient decided by the valve size, and $f(\theta)$ is the valve trim type for different plugs, given by:

$$f(\theta) = \begin{cases} \theta \ \ for \ linear \\ \sqrt{\theta} \ \ for \ \ quick...opening \\ R^{\theta-1} \ \ equal \ \ percentage \end{cases} \tag{8.42}$$

where R is the rangeability or ratio between the lowest and highest fluid flow rates, such that a larger value of R will indicate better accuracy of the valve position, which is given by:

$$R = \frac{Q_{max}}{Q_{min}} \tag{8.43}$$

An equal percentage is suitable for a large pressure drop, while a non-equal one is used where a fairly constant pressure drop is expected. Quick opening is used for intermittent service, requiring a large and instant flow rate. The opening angle θ depends on the opening sectional area A and the flow pressure β (including a constant factor given by fluid viscosity characteristics), such that:

$$\theta = \frac{A}{\beta} P \tag{8.44}$$

Thus, the valve coefficient is given by:

$$C = Q_{max}\sqrt{\frac{G_t}{\Delta P_v}} \tag{8.45}$$

with Q_{max} being the maximum flow through the valve ($1 \ min^{-1}$), $\Delta P = P_{in} - P_{out}$ the up- and downstream pressure drop across the valve (bar), and G_t the liquid-specific gravity. The maximum flow through the valve Q_{max} can be estimated from:

$$Q_{max} = Av \tag{8.46}$$

The airflow rate (for a pneumatic valve) is given by:

$$q(t) = K_q\theta(t) \tag{8.47}$$

Considering that the overall flow pressure in a system ΔP_{total} is delivered by a pump or a compressor, the variation in opening angle of the valve causes a flow change, which in turn produces a variation in pressure difference ΔP_v across the valve. The pumping-required overall pressure difference ΔP_{total} can be estimated from ΔP at the maximum flow such that:

$$\Delta P_{total} = \Delta P_v + \Delta P_s = \left(\frac{q_{max}}{C_v}\right)^2 g_s + kq^2_{max} \tag{8.48}$$

where:

$$\Delta P_v = \Delta P_{total} - \Delta P_s = \left(\frac{q_{max}}{C_v}\right)^2 g_s + kq^2_{max} - kq^2 \tag{8.49}$$

As a rule of thumb, ΔP_{total} should be around $1/3$–$1/4 \ \Delta P_{total}$ at a nominal flow rate. In order to control a motorized valve, it is required to control its electric motor in revolutions per minute (rpm), position, acceleration, and torque.

8.5 Electrothermal Actuating Systems

Electrothermal actuating systems, such as heaters, heat pumps and furnaces, can convert electrical energy into heat. Based on the Joule principle, the current passing through a

Table 8.6 Dynamics equations governing heat transfer.

Conduction	$q''_{cond} = -K\dfrac{dT}{dx}$	ε_{cond} is the heat flux in (W m^{-2})
		K is the thermal conductivity
Convection	$q''_{cond} = h(T_s - T_\infty)$	h is the heat transfer coefficient (W K^{-1} m^{-2})
		T_S is the surface temperature (K)
		T_∞ is the fluid temperature (K)
Radiation	$E = \varepsilon\sigma T_R^4$	E is the emissive power (W m^{-2})
		T_R is the surface temperature (K)
		σ is the Stefan–Boltzmann constant $\left(5.67 \times \dfrac{10^{-8}\,W}{m^2 K^4}\right)$
		ε is the radiative emissivity $(0 < \varepsilon < 1)$
Conductive thermal resistance R_{th} (the ability of an object to transfer heat between two points)	$q_{cond} = q''_{cond}\,A = \dfrac{-KA\Delta T}{L}$ $R_{th} = \left\|\dfrac{\Delta T}{q_{cond}}\right\| = \dfrac{AL}{kA}$ $= \rho_{th}\dfrac{L}{A}$ in (K w^{-1})	ΔT is the temperature change (K) q is the heat flow (m^3 s^{-1}) L is the length of the longitudinal rod (m) A is the cross-sectional area (m^2) q_{cond} is the heat conduction (W s^{-1})
Heat capacity and constant time for heating and cooling	$\tau = R_{th}C_{th}$ $Q = sh \cdot m \cdot \Delta T = C_{th}\Delta T$	Q is the stored thermal energy (W) C_{th} is the heat capacity (J kg^{-1}) sh is the specific heat (J K^{-1} kg^{-1}) m is the mass (kg) τ is the time constant (s)

resistor converts the electrical energy into heat energy. For example, heat pumps extract the heat energy from the ambient air or from the ground and raise the temperature via heat exchanged with a fluid boiling at low temperature. Then, the resulting vapor is compressed and condensed to a liquid form in a condenser. This can be used, for example, to heat the space inside a building; commonly encountered thermal actuating systems include furnaces (induction, electric, muffle). Among the advantages of electric heating methods over other forms (e.g. gas-based) are: (i) a better distribution of heat energy compared to chemical combustion; (ii) a refined temperature control; (iii) near real-time heating from a cold start up; and (iv) immediate and instantaneous shutdown. Typical heat transfers encountered are summarized in Table 8.6, along with the formulas governing their dynamics behavior.

Thermal actuating systems can be controlled using either contactor or thyristor control. Contactor control consists of a contactor device used to switch the current to the thermal actuator ON/OFF. This switching method is slow to respond to the signal and is more suitable for fluids with high thermal mass, such as water. Thyristor devices switch as often as several hundred times a second, providing a much finer response to the desired heat level. This method allows precise temperature control. It is suitable for fluids with a low thermal mass, such as fuel gas, or that can burn at higher temperatures, such as triethylene glycol (TEG).

Physical phenomenon

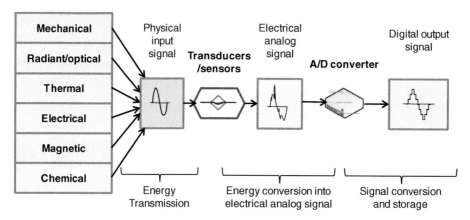

Figure 8.10 Typical electrical transducers and their computer interfaces.

8.6 Sensors in Mechatronics

An electrical sensor, also called a *transducer*, is a device that receives energy from one system and transmits it to another in a different form (preferably electrical over mechanical or acoustical). The electrical output of a transducer may be analog, digital, or frequency-modulated. This section covers only analog output transducers, which can be classified into two major categories:

1) *Active transducers*, which generate an electrical signal directly in response to physical parameters and do not require external power to operate, such as piezo-electric sensors and photocells.
2) *Passive transducers*, which require external power to operate, such as strain gauges and thermistors.

Figure 8.10 summarizes the commonly encountered electrical transducers.
Transducers can be classified according to one of the following criteria:

1) *Power supply* requirements, which define if a sensor is active (i.e. it generates an electrical signal in response to an external stimulus, such as a piezo-electric sensor) or passive (i.e. without an external power supply, such as a thermocouple).
2) *Nature of the sensor output signal*, which can be digital (binary), such as a shaft encoder, or analog (continuous in magnitude and in temporal content), such as object temperature or displacement.
3) *Measurement operational mode*, which is the proportionality of the difference between the initial condition of the instrument and the measurand of interest. With the null mode, the influence and measurand of interest are balanced (equal) but opposite in value.
4) *Input/output signal dynamic relationships* (zero, first, second order, etc.) of the sensor.
5) Type of *measurand* (mechanical, thermal, magnetic, radiant, chemical, etc.).
6) Type of *measured variable* (resistance, inductance, capacitance, etc.).

Table 8.7 summarizes typical sensing methods for various types of variable.
Parameters used to characterize a transducer's ability to convert a physical phenomenon into electrical signals are:

Table 8.7 Typical variables and associated sensing methods.

Measurand	Sensing methods
Displacement/position	Resistive, capacitive, opto-electric, Hall effect, variable reluctance
Distance	Triangulation, measuring wheel, radar, echelon, capacitive/inductive proximity
Temperature	Thermistor (NPC, PTC), infrared radiation, thermocouple
Pressure	Piezo-resistive, capacitive, piezo-electric, strain gauge
Velocity	Hall effect, opto-electronic, variable reluctance
Luminance	Photoresistor, photodiode, phototransistor

- *The input range*, corresponding to the range between the highest and the lowest sensed values measured for a specific variable of the physical phenomenon.
- *The threshold*, defined as the lowest change in input that is able to induce an output variation.
- *The transfer function*, illustrated by calibration curves expressing the relationships between input and output signals.
- *The sensitivity*, corresponding to the slope of the calibration curve. It can be determined by assessing the ratio between the output signal and output signal low variations. The thermometer sensitivity is estimated by:

$$S = \frac{\Delta V}{\Delta T} \text{in V}^\circ\text{C}^{-1} \tag{8.50}$$

- *The accuracy* of a sensor, estimated from the standard deviation to express the difference between the value given by the sensor and the real physical value.
- *Hysteresis*, which estimates the variation of output signals induced when increasing and decreasing input signals are applied several times.
- *Resolution*, which characterizes the ability of the sensor to detect small process variable changes (i.e. the lowest process variable value detectable).
- *The precision* of a measurement device, expressed through accuracy and resolution. It is estimated by measuring the same process variable several times under identical conditions.
- *Error characterization*, integrating the accuracy, resolution, and precision.
- *Repeatability*, corresponding to the ability of the sensor to produce the same value when subjected to the same conditions. It is determined by the variation of values observed after several measurements under the same conditions.
- *Nonlinearity* (often called *linearity*), given by the highest deviation of the calibration curve from a straight line.
- *The decay time*, defined as the time required by a sensor signal to return to its initial value after a step change.
- *The bandwidth* of a sensor, corresponding to the sensor frequency range.
- *The dynamic response*, given by the frequency range obtained for usual sensor operations.
- *The slew rate*, expressing the output accuracy induced by an input change.
- *The environmental limiting parameters* (e.g. humidity, temperature, dust, dirt/oil, corrosives, and pressures).
- *Sensor calibration*, providing the expected relationship between the input and output signals.
- *Zero point/offset*, defined as the resulting output when the initial input value is zero; *saturation* is reached when input variations are no longer able to induce significant output change.
- *The measurement linearity* or *calibration curve*, corresponding to the rate of deviation from the ideal linear calibration curve. Subsequently, the maximum linearity error is the maximum deviation obtained.

- *The impedance of the measurement equipment,* estimated from the ratio between the input and the output voltages of the crossing current. For example, low-power input signals would require a measurement instrument with a high input impedance, while the signal output from a signal conditioning unit would require a low-output impedance instrument.
- *The operating temperature,* corresponding to the optimal range of temperature of the sensor.
- *The signal-to-noise ratio,* expressing the relationship between the signal values and the output noise.

Typical sensors used for measurement and detection are summarized in Table 8.8.

8.6.1 Measurement Instruments

Measuring instruments can be classified based on their measuring method, applied as their energy conversion or transducing method.

8.6.1.1 Relative Position (Distance)

Instruments used to derive the presence or position of an object are called relative position or distance sensors. They may operate through direct contact or without any contact and may be classified based on the various measurement methods used:

1) *Optical distance sensors or detectors* involve a photoresistor or photocell, which reacts to the reception of an infrared light signal by: (i) varying its resistance; (ii) inducing the current flow in a circuit (photodiode); or (iii) modifying the magnitude of the current flow (phototransistor). The spatial configuration of such photoelectric sensors may be: (i) an opposite positioning of the light-emitting component and the light-receiving component, such that the presence of an object disrupts the signal; (ii) a same-side positioning (retroreflective) of the light-emitting and light-receiving components that requires a mirror to send back the emitted signal; or (iii) a diffuse design close to the retroreflective design, which ensures that the signal is sent back to a broad surface instead of a sole point. Devices such as photoresistors operate as interferometer linear-type sensors and require a commutator to separate the emitted infrared light from ambient light. The commutator transforms the emitted signal into a high-frequency signal, which can be decrypted by the receiver component. Such sensors have the advantage of providing a high resolution (μm) and a broad range (up to a few meters). Photodiodes and phototransistors may, respectively, be applied to binary applications with signal disruption and distance (linear position) measurement. Examples of such applications include access door security.
2) *Resistive-based position sensors* include rotary potentiometers and rheostats. They operate via a wiper rotating between two terminal resistant elements related by a conductive element. The observed voltage depends on the contact position of the wiper, which determines the overall resistance of the circuitry.
3) *With magnetic position sensors and detectors,* the current crosses a magnetic field through a conductor component and the voltage induced is proportional to the object position such that the distance from the sensor is given by:

$$V_h = \frac{R_h I B}{t} \tag{8.51}$$

with V_h, $R_h I$, t, and B being the Hall voltage, the material-dependent Hall coefficient, the current, the element thickness, and the magnetic flux density, respectively. Hall-effect

Table 8.8 Some commonly encountered sensors.

Measured	Sensing methods and description
Linear and angular position and velocity sensors	Optical encoder (absolute and incremental) angular position measurement
	Electrical tachometer (generator or magnetic-pickup velocity measurement)
	Hall effect sensor (position measurement with high resolution but low range)
	Interferometer (laser-based position measurement)
	Capacitive transducer (high-frequency dynamics position measurement)
	Magnetic pickup
	Gyroscope (angular position measurement)
	Vision (presence/absence; check position)
	Ultrasonic (flow/obstruction; counting; fluid or gas level; object shape)
Acceleration sensors	Seismic accelerometer
	Piezo-electric load cells
Flow sensors	Rotameter
	Ultrasonic type
	Electromagnetic flow meter
	Pitot tube
Force, pressure, torque sensors	Dynamometers
	Ultrasonic stress sensor
	Tactile sensor
	Strain gauge
Temperature noncontact sensors and detectors	Thermal imagers
	Radiation thermometers
	Fiber optic
Temperature contact sensors and detectors	Phase-change devices
	Bimetallic thermometers
	Filled-system thermometers
	Cryogenic temperature sensors
	Thermocouple (for operating range 200°C to over 1100°C)
	Thermistors (for operating ranges over 110°C)
	Resistance temperature detector (RTD)
	Thermidiode, thermo transistors (nano devices like chips)
	Infrared type (wavelength resolution)
	Infrared thermography (temperature distribution measurement)
Smart sensors	Optical fiber (use as a strain sensor), temperature sensor (high resolution and range), high resolution and range (force sensor, level sensor)
	Magnetoresistive (use as force sensor, torque sensor)
	Piezo-electric (accelerometer, force sensor, strain sensor)
Proximity sensor	Photoelectric
	Capacitance
	Hall effect
Light sensor	Photodiodes
	Photoresistors
	Phototransistor
	Photo conductors
Binary noncontact detector	Inductive proximity to detect metal object
	Capacitive proximity to detect dielectric object
	Optical presence to detect an object breaking a light beam or reflecting light (use for counting; check passing/obstruction of objects; monitor the presence, absence, or position of an object)
Binary contact detector	Mechanical contact (switch) to detect an object

magnetic sensors are affected by vibration. However, such devices operate with a high resolution for a short distance range (between 0.2 and 5 mm). Their applications include security systems and multi-axis robot arms (where they control the position and speed). Some Hall effect detectors may only be activated or deactivated when a specific value of the magnetic field intensity is reached. As such, they only apply to an object presence or absence detection, or to a particular position or reference position detection. In the latter case, an incremental encoder is required.

4) *Inductive position-based sensors* use the variation of the inductance proportional to the gap between a targeted metallic object and the magnetic field, which results in an oscillator voltage variation. When the targeted metal object moves within the proximity of the coil, it induces an eddy current in the metal object surface that causes a secondary magnetic field which interferes with the probe magnetic field. Such a device is used for the detection of metal objects over short distances.

5) *Optical, ultrasonic, or microwave distance-measuring sensors* are based on the principle of wave propagation through a fluid medium. As such, these devices can be: (i) radio-based or radar (using time-of-flight [TOF], frequency-modulation, or phase-shift measurement techniques); (ii) laser light-based (using TOF); or (iii) sound-based. Such measurement devices All are sensitive to temperature and target obstruction. Distance formulas using TOF or frequency techniques based on the propagation time of the various signal types (ultrasonic, radio frequency, optical energy) from the source to the targeted object are presented in Table 8.9.

Among other reflective-based measurement techniques, *interferometry* is based on the assessment of the intensity of the returning signal. Initially, the generated laser light is split into two different paths (a path directed toward the detector with a fixed length and a path directed toward the moving object with an attached retro reflector). The distance to the moving target can be estimated as the relative travel wavelength of both signal paths. Based on vibration type, object shape and size, laser light speed, reflectivity, and direction, the resulting measurement distance can be calculated. Such sensors are commonly encountered in devices such as navigation guidance systems, 3D robot positioning systems, object detection (collision avoidance) systems, and security systems. Microwave *position sensors* use a microwave frequency signal reflected onto the targeted object and returning back to the sensor to indicate object presence or absence within the radio frequency field. With *ultrasonic position sensors*, the emitted signal is ultrasonic and it is possible to measure distance from 1 cm up to few meters, as depicted in Figure 8.11. With the phase-shift measurement technique, the frequency-modulation component at the receiver side is used to derive the time shift with the emitted signal. Hence, the distance to the object is given by:

$$distance = \frac{f(t) - f(t-T) \times light\_speed}{4 f_{modulated} f_{deviation}} \tag{8.52}$$

Table 8.9 Distance formula using a TOF technique.

Measurement set up	Distance formula
Emitter and receiver with same location (same sensor)	$distance = \frac{TOF}{2} signal - speed$
Emitter and receiver with different locations (receiver attached to targeted object)	$distance = TOF\, signal - speed$

Figure 8.11 Ultrasound-based distance measurement operation.

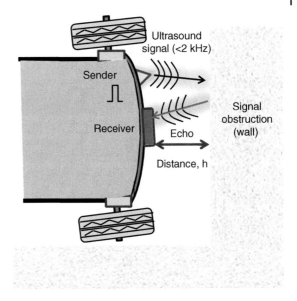

where $f_{modulated}$ is the signal modulated frequency, $f_{deviation}$ is the total modulated frequency deviation, and $f(t)$ and $f(t - T)$ are, respectively, the emitted and the received signal frequency. Usually, an ultrasonic sensor emits a sound above the normal hearing threshold of 16 kHz. The time that is required for the sound to travel to the target and reflect back is proportional to the distance to the target and is given by:

$$h = \frac{Ct}{2} \tag{8.53}$$

with C and t being, respectively, ultrasound speed on the media and time taken for the signal to travel from sender to receiver. The ultrasound signal is sensitive to the ambient temperature and humidity. These contactless sensors are suitable for applications such as the measurement of contaminated fluid levels in tanks and the measurement of distances up to 10 m, regardless of target shape and material.

6) When the measurement is based on a switch's position, *electromechanical switches* (e.g. limit switches) provide binary information about the switch status, so: closed or open, actuated or not actuated, allowing current to flow or not. Each switch has at least one pole with two points of contact.

7) *Capacitance distance sensors* use gap variation between two metallic plates to derive the distance proportional to their electrical capacitance. This allows a linear position and proximity measurement. As illustrated in Figure 8.12, the capacitance variation of passive and contact-free sensors is given by:

$$C = \frac{q}{V} = k\varepsilon_0 \frac{A}{d} \tag{8.54}$$

with C being the capacitance, q the charge (in Coulomb), V the potential difference between the two plates (V), ε_0 the dielectric constant, d the permeability constant, A the effective plate area (in m^2), and d the plate separation (in m). This delivers high-resolution position measurement. Such devices are highly sensitive to temperature and humidity variation. They are usually applied to detect the presence or absence of metallic objects or of a liquid or powder with a dielectric material constant greater than that of air.

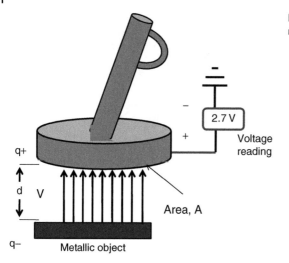

Figure 8.12 Capacitor-based distance measurement principle.

8) *Distance measurement sensors with embedded computer vision systems* collect and process a targeted object photograph to derive its physical characteristics, such as: (i) dimension; (ii) position; (iii) appearance (surface finish, color, gloss, presence of defects); and (iv) markings (logos, characters, etc.). Figure 8.13 depicts the process of distance measurement using a computer vision system. Depending on the light positioning, the distance is given by:

$$D = \frac{fH}{h} \tag{8.55}$$

9) *Position measurement sensors using the force of gravity* involve the use of mercury conductivity measurement to close or open a switch. Such sensors are sensitive to temperature.

10) *Position measurement sensors using magnetostrictive time* consist of a moving magnet (target) and a magnetostrictive wire. A current pulse passes through the magnetostrictive wire over the acting magnet's field and generates an ultrasonic pulse in the wire. The position of the magnet is related to the time delay between the current pulse and the detected ultrasonic pulse.

11) *Position measurement sensors using a potentiometer* consist of a resistive element sliding along a wiper, as depicted in Figure 8.14. When a voltage is applied across this resistive element, the resulting output AC voltage magnitude is found to be linearly proportional to the position of the brush over the resistive element. Though such sensors have the same operating principle as the linear variable differential transformer (LVDT), potentiometers do not require signal conditioners: they measure the absolute rotational position and can rotate without limits (multiple 360°), making the wiper jump from one end of the resistor to the other.

For linear displacement measurement, x, the output voltage can be derived from:

$$V_0 = kVx \tag{8.56a}$$

For rotary displacement measurement, θ, the output voltage will be:

$$V_0 = kV\theta \tag{8.56b}$$

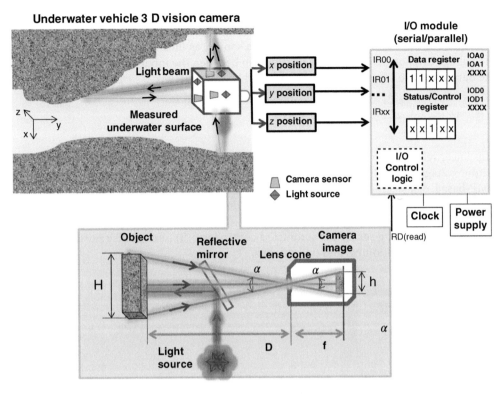

Figure 8.13 Measurement principle using a vision system.

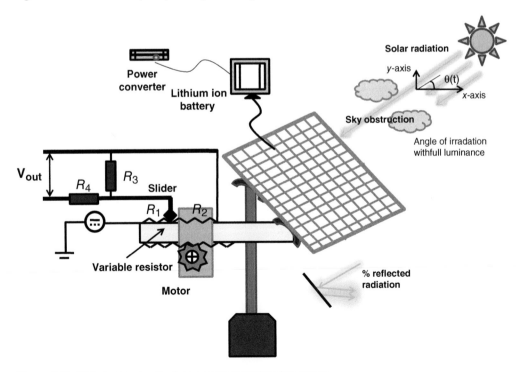

Figure 8.14 LMP electric circuitry for an antenna positioning system.

A linear motion potentiometer (LMP) consists of: (i) a wiper attached to a mobile object whose position is monitored; (ii) a potentiometer whose resistance is related to the displacement of the wiper; and (iii) a DC voltage source, as illustrated in Figure 8.15. Using the voltage division properties, the relationship between the input voltage signal, V_{cc}, and the output voltage signal, V_{out}, can be captured by the equation:

$$V_{out} = \frac{R_0 + \frac{R_2}{R_L}}{R_1 + \frac{R_2}{R_L}} V_{cc} \tag{8.56c}$$

This linear relationship expresses that for every position there is a given value of electrical output voltage. Table 8.10 summarizes some characteristics of a typical LMP.

LMP-based displacement measurements are sensitive to the friction during wiper displacement and the speed of the moving object due to the contact mechanism, as well as potentiometer resistance temperature variations. Typical applications of LMPs include: (i) a solenoid positioning sensor for automated controllers; (ii) satellite dish positioning; (iii) automatic valve positioning; and (iv) robotics.

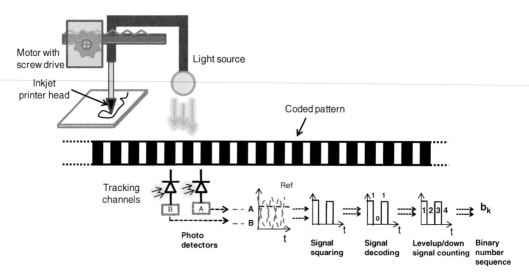

Figure 8.15 Principle behind an incremental encoder with two tracks.

Table 8.10 Example LMP sensor characteristics.

Power supply	5–40 VDC
Dynamic range	0–6.15 in.
Repeatability	0.5–1 mm error
Accuracy	0.02–15%
Sensitivity	4 mm V^{-1}
Resolution	Infinite (<0.01 mm)
Hysteresis	0.02%
Operating temperature	−65–105°

8.6.1.2 Angular Position Measurement Using an Encoder and a Resolver

Resolvers and optical encoders are devices used to achieve an angular position measurement. Based on the photo-interrupter principle, optical encoders convert motion into an electrical pulse train, and the resulting electrical pulses are decrypted by the circuitry. Hence, the components of a rotary optical encoder consist of: a signal (light)-emitting diode (LED), a light guidance and shaping device (optical mirror), a photosensitive receiver, and a coded rotating wheel (disk) directly in contact with the motor controller through its shaft. The series of transparent (binary level 1) or opaque (binary level 0) angular or linear slots on the surface of the disk are designed according to Gray code binary number patterns. The light emitted by the LEDs through the transparent zones of the disk is detected by photodetectors (photodiodes). The disk rotation induces a succession of square wave signals corresponding to the encoder output signal. This output signal is normally a binary or a Gray code number. The code is then converted into the displacement measurement within the decoder through a pulse-counting device. The number of pulses per revolution is directly related to the number of slots on the disk, or a multiple thereof.

Optical encoders can be classified in two categories/families: (i) incremental encoders, where the angular position of a motor is derived from the incrementation or decrementation of generated pulses; and (ii) absolute-position encoders, where the shaft angle position is estimated with precision. These two families include variants, such as (i) absolute multiturn encoders and (ii) tacho-encoders for data processing. Overall, the average velocity can be derived using the time between pulses.

With *incremental optical position encoders*, an initial angular position is arbitrarily set to zero and the direction of the rotating object attached to the coded disk is derived using two 90°-out-of-phase sinusoidal shape-like signals (channels A and B) generated from photodetectors. The two channels, which can be divided into several intervals with equal angles, are alternately opaque and transparent. Here, the number of divisions corresponds to the resolution or number of periods. The light beam passing through a transparent slot activates two out-of-phase photodetectors, which subsequently generate a pulse-like signal that will be squared as follows: if A signal and B signal are above the reference signal, the square circuit signal is set to 1; otherwise, it is set to 0. The signal decoding is done through a pulse-counter device. Here, the reference point can be lost when the power supply is temporally interrupted. The motion direction is given by the phase relationship between the A and B pulse trains; that is, the indication of which signal leads the other. The number of slots and the disk circumference define the encoder resolution. The total count (pulse) defines the total distance.

The motion direction of the motor shaft can be derived from the relative 90° ($\Delta\theta/4$) phase shift between signals A and B. Figures 8.15 and 8.16(a) illustrate the opaque area arrangements on the disk of the incremental optical linear and angle encoders for an inkjet printer motor positioning system and rotational motor position, respectively. This has the advantage of simplifying the encoder, because only two photodetectors are required. However, external counter hardware has to be used in order to develop an absolute position measurement from the output of the incremental encoder. The initial or reference position is given through a Z-signal in comparison with signals A and B. It is designed by a transparent slot in the gray coded disk, and after each revolution it allows the motor position to be reset, as depicted in Figure 8.16(b). The angular spacing of the opaque areas is $\Delta\theta$.

Angular position measurement can also be performed with counter hardware. At each transition in A and B, a pulse is generated and sent to the counter. The binary output of the counter can be as many bits as needed to represent the required range of motion. Each pulse received during

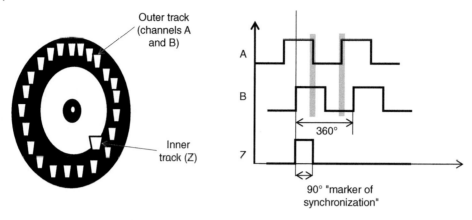

Figure 8.16 (a) Principle of an incremental encoder with two tracks. (b) Detection principle for the direction of rotation and synchronization.

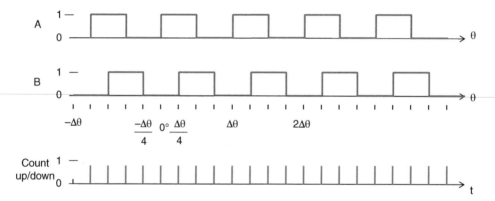

Figure 8.17 Binary signals generated by an incremental encoder.

a positive rotation causes the binary counter output to increase by one (count up). Provided that the counter is properly initialized such that its output is zero at the desired initial angular position, it gives an absolute reading of the angular position. Similarly, a separate set of pulses can be generated for a negative rotation and sent to the counter. In this case, each pulse causes the counter output to decrease by one (count down) along the bidirectional encoder angular position. The output of the counter changes by four for an angular rotation of $\Delta\theta$, so the effective resolution is $\Delta\theta/4$. Incremental optical angle encoders are available with 100 000 or more opaque areas, or "lines," on the disk. However, the dynamic response of the photodetectors limits the rotational velocity of the sensor. If the maximum frequency at which photodetectors can operate is known, the maximum rotational velocity of the encoder can be found from:

$$\omega_{\max} = \frac{f_{\max}\Delta\theta}{6} \tag{8.57}$$

where $f_{\max}$ is the photodetector cut-off frequency in Hz and $\Delta\theta$ is the incremental angle. Figure 8.17 illustrates some binary signals generated by an incremental encoder.

The accuracy of such an angular measurement device depends on the latch-based pulse-counting circuit, which is very sensitive to unpredictable noise interference. Hence, in order to detect possible measurement errors, it is important to continuously compare generated signals A, B, and Z. Such a technique also allows the reconstruction of the initial

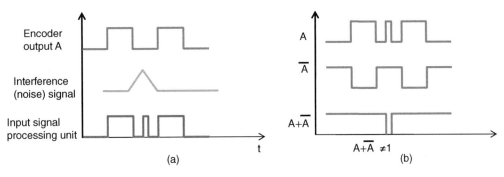

Figure 8.18 (a) Counting error on encoder channel A. (b) Non-counting interference signal.

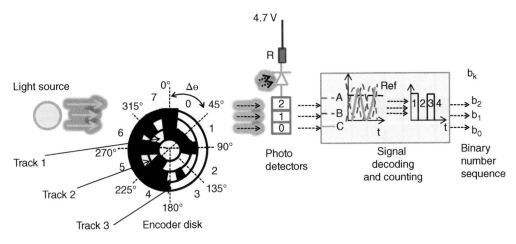

Figure 8.19 Principle of an absolute encoder.

signal in case of counting error. Figure 8.18(a) and (b) summarizes the mitigation of encoder signal interference.

Absolute rotary encoders are similar to incremental sensors except that the disk is designed around three rather than two tracks, which are used and arranged such that direction of rotation is embedded with the angular position signal, corresponding to a unique binary word (a binary Gray code) that increments or decrements accordingly. For example, an increment from seven to eight in numerals corresponds to a change from 0100 to 1100 in Gray code. Hence, such encoders can produce the actual position image of a moving part by producing a set of logic-level signals representing a binary word proportional to its angular position. Figure 8.19 illustrates an absolute optical angle encoder with three bits of binary position output. The angular position encoder does not require digital-to-analog conversion (DAC).

Each bit is generated by one photodetector according to the arrangement of opaque areas on the disk. As shown in Figure 8.19, the half-opaque and half-transparent track gives information on the direction of motion, while the other tracks provide information on the angular position of a quarter of the disk motor shaft. The higher the number of windows per ring, the higher the encoder resolution, which varies from two to thousands of windows per ring. Collectively, these bits generate the binary code for any angular position, as shown in Table 8.11.

At the boundaries between some regions on the disk, more than one bit changes simultaneously in the binary output, in order to avoid large errors in position measurement. For example,

Table 8.11 Example of binary codes generated by absolute rotary encoder.

n	θ(deg)	b_1(deg)	b_2(deg)	b_3(deg)
0	0–45	0	0	0
1	45–90	0	0	1
2	90–135	0	1	0
3	135–180	0	1	1
4	180–225	1	0	0
5	225–270	1	0	1
6	270–315	1	1	0
7	315–360	1	1	1

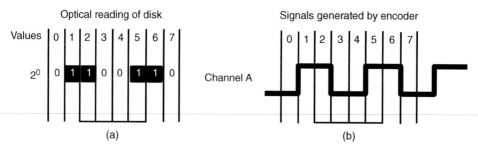

Figure 8.20 (a) Bit-based optical reading of a disk. (b) Binary signal-generated encoder.

when the disk rotates from three to four, the binary output changes from 011_2 to 100_2 through intermediate combinations such as 111_2, if they are not synchronized.

Figure 8.20 depicts a signal-encoding example. It could be that the number of parallel outputs corresponds to the number of bits or tracks on the disk. Also, the code generated by an absolute encoder can be either natural binary (pure binary) or reflected binary (Gray binary code). Overall, with an absolute encoder, the derived code is repeated after one revolution and several geared encoders can be used to increase the resolution or the range of angular position sensing. However, absolute-angle encoders are useless when a large number of output bits is needed to meet range and resolution requirements. Hence, usually, incremental optical angle encoders are used for position-sensing applications, while absolute encoders are applied to movement and position monitoring.

Linear optical encoders are similar to optical angular encoders, with the coded wheel (disk) replaced by a coded bar and a slider that carries the optical and electronic components. The distance measurement is operated along the bar, allowing a resolution from a few micrometers to about 3 m at an operating speed of about 600 mm s^{-1}. There are also *angular resolvers*, which are shaft-oriented angular sensors based on an AC magnetic principle. Here, two coils (primary and secondary) provide sinusoidal-type outputs, which are compared.

8.6.1.3 Velocity Measurement
The direct measurement of velocity can be performed using analog transducers, as:

1) *Magnet-and-coil velocity*-sensing devices, which are based on the Faraday law of induction, and consist of a solenoidal coil with several turns of wire. Here, the voltage induced into the

coil is proportional to the velocity of the magnet. These sensors are suitable for displacements less than 0.5 m and greater than 10 mm.

2) *Tachometer AC (or DC) generators*, which are attached to a rotating shaft such that its rotation induces a voltage directly proportional to its angular velocity. If ω is the armature speed and k_t is the tachometer gain, the DC commutating output voltage yields:

$$V_o = k_t \omega \qquad (8.58)$$

It should be noted that this generator voltage device and its hooked armature produce a noise over the DC output of the tachometer with a fundamental frequency directly proportional to the angular velocity. Such tachometer generators are suitable for shaft speeds above 5000 rpm. This technique introduces some drag into the system and is not appropriate for motion-tracking applications.

3) *Counter-type velocity sensors*, which involve counting electrical pulses over a fixed interval of time and converting the result into velocity.

4) *Linear velocity transducers* (LVTs), which are inductive devices that consist of a rod, called the core (a permanent moving magnet), and two electrical coils connected in series opposition, as illustrated in Figure 8.21. Based on the Lorentz law, the core moving through the coils generates a current flow through the wire, which is proportional to the magnet velocity transducer. The voltage generated by the transducer is given by:

$$V_o = Blv \qquad (8.59)$$

with B being the component of the flux density normal to the velocity, l the length of the conductor, and v the magnet linear velocity. The magnet can be directly mounted on a moving mechanical element without electrical connections. The output voltage V_o must be converted from analog to digital. Using any position sensor, the velocity measurement can be estimated through *digital differential position algorithms* such as the Euler backward approximation:

$$\omega_i = \frac{\theta_i - \theta_{i-1}}{T} \qquad (8.60)$$

where ω_i is the average velocity (rather that instantaneous angular velocity), θ_i is the current position, θ_{i-1} is the previously sampled position, and T is the sampling interval. It is suitable to use a trapezoidal approximation for refined velocity estimation. It should be noted that the velocity resolution is highly dependent on the sampling period T and the position resolution. The sensor operating intervals are around 50 mm.

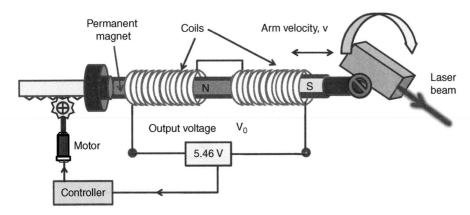

Figure 8.21 Linear velocity transducer (LVT) device for a laser-based cutting process.

8.6.1.4 Acceleration Measurement
Acceleration measurement can be derived through:

1) Digital differential velocity algorithms, in the case of angular motion.
2) Accelerometers based on Newton's principle, in the case of linear motion. This is achieved in one of several ways. (i) Through the displacement measurement of a seismic mass moving freely within a cage surrounded by a spring-damper support structure, its relative motion being recorded and converted into voltage levels. Such devices are called electromechanical accelerometers. Here, each motion generates a magnetic force opposed and proportional to the displacement from the zero position. This force depends on the length of the conductor within the field, the density of the magnetic field, and the value of the angle between the conductor and the magnetic field. (ii) From the force-balance principle applied on the core of an electromagnet, such as in the high-speed train accelerometer illustrated in Figure 8.22. (iii) By using one electrode of a piezo-electric crystal as a vertical vibration sensor (accelerometer), as illustrated in Figure 8.23. Such devices are called piezo-electric accelerometers due to the charge generated when force is applied on the crystal induced from an acceleration. (iv) By using the plate of a differential capacitor as the seismic element. For seismic accelerometers based on resistive, capacitive, inductive, or piezo-electric material, the acceleration motion of a specific mass is captured through its resulting force, $F(t)$, given by:

$$F(t) = m\frac{d^2x(t)}{dt^2} + b\frac{dx(t)}{dt} + kx(t)$$ (8.61a)

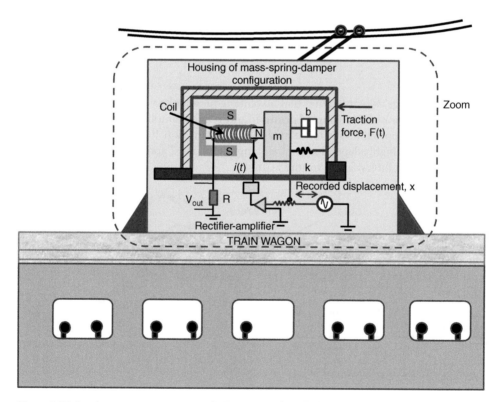

Figure 8.22 Accelerometer measurement device mounted in a high-speed train.

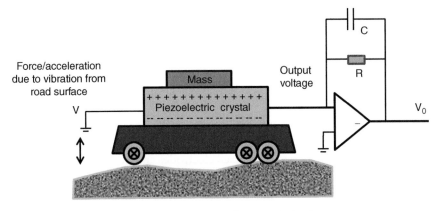

Figure 8.23 Piezo-electric accelerometer circuit for vertical vibration measurement.

where m is the mass, b the damping coefficient, and k the spring constant of the mass-damper-
spring apparatus. Hence, the cage apparatus dynamics equation relating the mass motion $x(t)$, the accelerometer motion $y(t)$, and the angle between gravity and the motion axis α results in:

$$m\frac{d^2(y(t) - x(t))}{dt^2} + b\frac{d(y(t) - x(t))}{dt} + k(y(t) - x(t)) = mg\cos\alpha - m\frac{d^2x(t)}{dt^2} \quad (8.61b)$$

It should be noted that the natural frequency of the fixed mass bounds the frequency application interval. In addition, such a sensory system is highly sensitive to environmental conditions (temperature, dust, magnetic field effects, humidity, etc.). Typical applications of accelerometers are to reduce system vibration and noise by adjusting speeds or to actively eliminate chatter.

For a piezo-electric accelerometer, the piezo-electric material is compressed or shear-stressed over the cross-sectional area A of the material crystal during any motion (vibration) of the apparatus. Its motion is proportional to the electric charge given by:

$$q(t) = k_i m\frac{d^2x(t)}{dt^2} \quad (8.62a)$$

with k_i being the piezo-electric coefficient of piezo-electric materials (quarts, ceramic crystals, etc.). It is possible to estimate the deformation in the crystal and the resulting voltage $V(t)$ across it as:

$$V(t) = k\frac{F(t)}{A} \quad (8.62b)$$

where k is the piezo-electric voltage constant of the crystal. Applying Newton law's, the acceleration of the mass, $a(t)$, is a function of the voltage produced across the crystal, such that:

$$a(t) = V(t)\frac{A}{kM} \quad (8.63a)$$

The voltage produced by the piezo-electric accelerometer is detected by a charge amplifier that is proportional to acceleration. Such accelerometers are sensitive to temperature variations, causing nonlinearity in measurement, and to angular acceleration, causing rotation-induced errors. Typical applications include vibration detection, navigation systems and gyroscopes, and vehicle acceleration monitoring.

Other accelerometers include strain-gauge piezo-electric accelerometers, where the piezo-electric crystal material is connected to the mass apparatus. Here, during any motion of the apparatus, a change in the length of the strain gauge causes a variation of the conductors' resistivity ρ, such that:

$$\frac{dR/R}{dL/L} = 1 + 2\delta + \frac{d\rho/\rho}{dL/L} \qquad (8.63b)$$

where $\frac{d\rho/\rho}{dL/L}$ indicates the resistance change due to piezo-resistivity. There are also electrostatic accelerometers, which consist of an electrode pivoting around fixed electrodes. Hence, based on Coulomb's law, any variation of the gap between the fixed electrodes and the pivoting electrode produces an electrical signal.

8.6.1.5 Force Measurement

A force is any action on an object (or a subject). It is considered dynamic when it results in a deformation or a displacement and is considered static when it does not result in an accelerating motion. Among the methods of measuring a force, there are those using:

1) *Levers* to compare the applied weight force with the displacement of a known mass attached over a spring as counterweight. This displacement is measured using a gyroscope, a position sensor, or a velocity sensor, sometimes within a simple spring-balance scale.
2) *Accelerometers*, where the unknown force is balanced by the gravitational force of a known mass such that its acceleration yields:

$$a_c(t) = \frac{F(t)}{m} \qquad (8.64a)$$

The unknown force can also be balanced by an electromagnetically developed force such that $F(t) = F_{em}(t)$, with $F_{em}(t)$ being the electromagnetical force at equilibrium. Hence, converting the force to a fluid pressure and measuring that pressure, it yields:

$$P(t) = \frac{F(t)}{A} \qquad (8.64b)$$

where $P(t)$ is the pressure to be measured and $F(t)$ is the acting force on object surface A.
3) *Correspondence*, (defined by Hooke's law) between the deflection of an elastic object (*strain gauge, piezo-electric crystal, force-sensing resistor*) and the applied static force. Hence, the strain gauge length increases as its area of cross section decreases, causing an increase in its resistance, as depicted in Figure 8.24. The resulting strain gauge length and its cross-sectional area are used to find a new resistance, which is given by:

$$R = \frac{V}{I} = \rho\frac{L}{A} = \rho\frac{L}{wt} \qquad (8.65a)$$

with $R, l, V, A, t, w,$ and ρ being the resistance of the wire, the length, the voltage, the cross-sectional area of the conductor, and the thickness, width, and resistivity of the material, respectively. Commonly encountered strain gauges are (i) wire, (ii) foil, and (iii) semiconductor. Usually, these sensors transform any change in resistance into a small voltage scale through a Wheatstone bridge circuit. In the case of wire strain gauges, a fine wire element is typically cemented into a thin sheet of paper, Bakelite, or Teflon. The sensitivity of a material to the strain, called the gauge factor (GF), is given by:

$$GF = K = \frac{dR/R}{dl/l} \qquad (8.65b)$$

with ΔR being the change in the initial resistance in ohms, R the initial resistance in ohms (without strain), and Δl the change in the length in meters, and l the initial length in meters

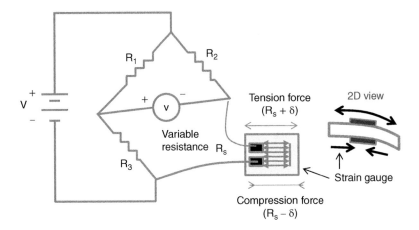

Figure 8.24 Quarter bridge strain gauge circuit.

(without strain). The strain gauge is highly sensitive to temperature. Piezo-electric materials relate elastic crystal deformation to electrical potential. Hence, among the piezo-electric materials used to measure the applied force are quartz, Rochelle salt, lithium sulfate, and barium titanate ceramic.

4) *Capacitive force transducers*, where the membrane small elastic deflection from applied force induces a variation of capacitance, which is converted into a voltage through an electronic circuit. Here, with two plates separated by an air gap, the capacitance of the sensor is given by:

$$C = \varepsilon_0 \varepsilon_r \frac{A}{h} \tag{8.66}$$

with C being the capacitance, ε_0 the dielectric constant of free space, ε_r the relative dielectric constant of the insulator, A the overlapping area for two plates, and h the thickness of the gap between the two plates. Typical applications of force sensors include multiple-axis robotic handling, and manufacturing assembly operations, and impact shock force in a car crash.

5) *Magnetoresistive force sensors*, where temperature-sensitive sensors change their resistivity within a magnetic field (such as Bismuth).

6) *Tactile force sensors*, where the pressure distribution across the closed spaced array of a robot arm fingertip is measured. Typical industrial applications of such sensors include manufacturing, robot handling, and the surface identification of product shapes.

8.6.1.6 Torque Measurement

A torque is an action causing a momentum around a rotational axis. Thus, in a torque sensor using a shaft made of ferromagnetic material, the torque can be derived from the variation in magnetic properties caused by the strain stress. Such sensors generate voltage output signals that are magnetoelastically or magnetoresistively dependent on either solid–solid or liquid–solid interactions. Among those sensors are torquemeters (noncontact strain gauges) and torkducers (magnetoelastic transducers).

8.6.1.7 Flow Measurement

Gas or liquid, or a mixture of gas and liquid, can be sensed in terms of mass or volumetric flow rate and flow direction (turbulent or laminar, compressible or not, free of particles or nonhomogeneous and time-varying). A commonly encountered device for fluid flow velocity measurement is called a flowmeter. These are designed using various measurement approaches, such as: (i) the fixed-point fluid flow velocity at a pipe cross-section; (ii) the average fluid flow velocity at a pipe cross-section; and (iii) the mass flow rate or total volume of fluid flowing through a pipe. The flow velocity profile is related to the fluid instrumentation design (pipes, valves, etc.) and the fluid properties (laminar, turbulent, etc.), which are characterized by the Reynolds number. Hence, the selection of a flowmeter requires knowledge of the fluid type (gas, solid, liquid), the fluid properties (viscosity, density, corrosivity, etc.), the fluid piping and other installation conditions (pipe diameter, pipe length, Reynolds number, valve type and character-istics, etc.), the expected dynamic characteristics (dynamic response, accuracy, etc.), and the operating conditions (temperature, pressure, etc.). Typically encountered flowmeters can be classified into:

1) *Contact-based sensors,* such as:
   a) *Differential pressure flowmeters,* which involve deriving a differential measurement of the flow on at least two points along the fluid flow direction. Sensing device configurations include venturi, orifice, flow nozzle, elbow, and Pitot tube static. Such flowmeters are suitable for velocity flow measurement of gas and low-viscosity liquids.
   b) *Positive displacement flowmeters,* which involve deriving the flow rate proportionally to a shaft rotation. Such a sensing device configuration includes a rotameter (variable-area in-line flowmeter) or a turbine meter, a cylinder or plug, and a variable aperture.
   c) *Turbine flowmeters,* which involve measuring the fluid flow through rotor speed mea-surement. Hence, the magnetic-based sensor converts the average rotor speed into volt-age levels proportional to the fluid average velocity or volumetric rate in a pipe. The sensing device configuration includes an impeller, an axial turbine, and a propeller.
   d) *Mass flowmeters,* which involve deriving the flow rate from an induced voltage propor-tional to the current required to maintain the temperature constant in a heating element across the pipe. The sensing device configuration includes a thermal anemometer. Such flowmeters are suitable in slow-motion gas flows through large pipes, such as in building heating, ventilation, and air conditioning (HVAC) systems.
2) *Contact-free sensors,* such as:
   a) *Electromagnetic flowmeters,* where a fluid behaving as a conductor moves along a pipe containing electrodes. This produces a magnetic field across its section wall. The fluid is ionized such that its motion induces a voltage proportional to the fluid mass flow rate, which can be measured by the electrodes based on Faraday's law. Such a fluid, used as an electromagnetic flowmeter (pulsed DC magnetic, AC magnetic, etc.), is suitable for low-speed flows of liquids, sludges, or slurries.
   b) *Ultrasonic flowmeters,* where high-frequency sound-wave signals are sent through a fluid toward receivers. On the principle that the sound transmission speed is directly propor-tional to the flow rate of the fluid, any distortion of the receiver signal is proportional to the flow velocity as given by the Doppler effect. Here, the transmission delay is sen-sitive to the velocity profile and the temperature. Such flowmeters are suitable for clean (particle-free), subsonic fluids, as small particles could disturb the signal.
   c) *Displacement pressure flowmeters,* which involve measuring a fluid flow rate through the pressure difference across an orifice, as illustrated in Figure 8.25, such that the volumetric flow rate is given by:

$$Q = \frac{C}{\sqrt{1 - \beta^4}} \varepsilon \frac{\pi}{4} d^2 \sqrt{\frac{2(p_{in} - p_{out})}{\rho}} \tag{8.67}$$

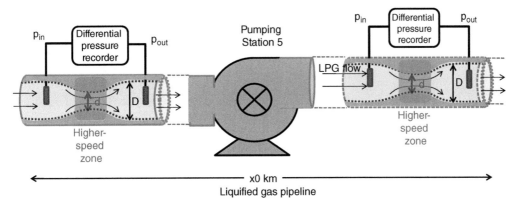

Figure 8.25 Displacement pressure flowmeter with LPG flow rate within a pipeline.

with p_{in} and p_{out} being the downstream and upstream pressure of the orifice plate, respectively, ρ the density of the fluid upstream of the orifice plate, d the downstream hole diameter of the orifice plate, and β the ratio d/D, with D being the upstream internal pipe diameter.

8.6.1.8 Pressure Measurement

Considering pressure as the force acting on a surface, it is possible to measure it as force per unit area. Pressure measurement is usually achieved within a chamber and can give: (i) an absolute pressure, where the chamber containing the fluid is sealed at 0 Pascal; (ii) a gauge pressure, where the chamber is vented to the atmosphere; (iii) a differential pressure, where another pressure is acting on the chamber; or (iv) a sealed pressure, where the chamber is sealed at a given pressure (different to 0 Pascal). Usually, fluid pressure transducers consist of elastic diaphragms dividing two chambers (the working chamber and the reference chamber). The measurement is given by the difference in pressure between the two. Pressure transducers transform a device motion or deformation proportionally to an applied pressure.

Commonly used transducers are: (i) the Bourdon tube, where an elliptical metal tube deforms when there is a differential between inside and outside pressure; (ii) the Bellow tube, which is a metal tube that is flexible under applied pressure; and (iii) the diaphragm, which is a circular plate that is elastic under pressure. As summarized in Figure 8.26(a), some of these sensors can use a resistance-based strain gauge or their own outer walls as differential capacitors and so derive the corresponding electrical signal output. As illustrated in Figure 8.26(b), baffle-based pressure-measurement devices are embedded into unmanned underwater vehicles.

Other types of pressure sensors are:

1) Measurement devices using a variable capacitance pressure that consists of changing a physical property of one of the variables (e.g. distance d). This varies the capacitance C, as depicted in Figure 8.27 for the case of a water dam. The capacitance is given by:

$$C = \frac{KS(N-1)}{d} \tag{8.68}$$

with $K, S, N,$ and d being the dielectric constant of the material between the plates, the area of one side of one plate, the number of plates, and the distance between two adjacent plates, respectively.

2) Variable-resistance pressure measurement devices, which are capable of detecting large resistance variations. They usually operate in potentiometer circuits or bridge circuits (strain gauge transducers).

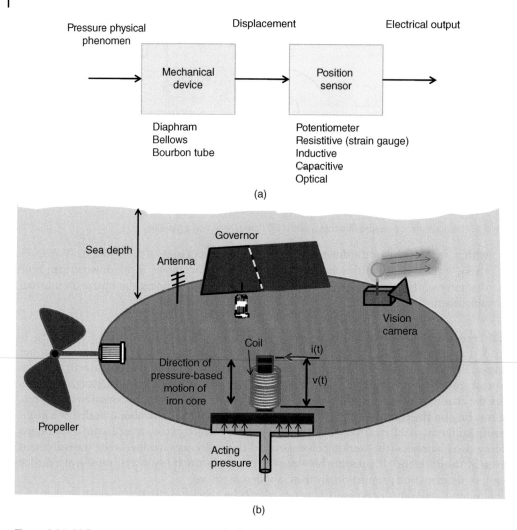

Figure 8.26 (a) Pressure measurement principle. (b) Baffle-based pressure measurement schematic.

3) Measurement devices using variable inductance pressure, which operate on the principle that the voltage drop across a coil generates a magnetic field proportionally to the rate of time-based current variation.

4) Measurement devices using piezo-electric pressure, which are based on the material properties involved in generating an electrical potential when subject to a mechanical strain. Among the materials used are quartz, Rochelle salt, ammonium dihydrogen phosphate, and even ordinary sugar.

8.6.1.9 Liquid-Level Measurement
Liquid-level sensors can involve contact or can be contact-free, and ensure either continuous or threshold-value measurements. Typical liquid-level sensors include:

1) Classical *float-type liquid-level sensors*, which measure either continuous or discrete points of liquid level.

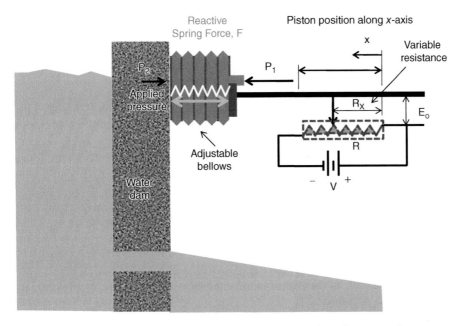

Figure 8.27 Capacitance pressure sensor based on a Bellow transducer for a water dam.

2) *Hydrostatic pressure liquid-level sensors*, where the weight of the liquid in the tank is proportional to the differential pressure between the surface and the bottom.
3) *Electrical capacitance liquid-level sensors*, where the capacitor is a metal bar within a concentric tank that is open at both ends, which determines the level of liquid with respect to the liquid level in the tank walls. Such a sensor is usually combined with signal conditioning devices for use in such applications as average fuel-level indicators.

Commonly used contact-free liquid-level sensors include:

1) *Capacitive proximity switches*, which carry out point measurements of liquid level.
2) *Ultrasound sensors*, where an attenuated ultrasound energy signal sent from a transmitter above the liquid is collected by the receiving unit and compared to the reference signal when the gap is filled with air. This provides a liquid-level point measurement. A signal conditioning circuit is usually required. Typical applications of such sensors include sea water depth measurements and fish localizers on industrial fishing boats, using echo-ranging transducers, and powder and grain level measurements in brewery silos.
3) *Radio frequency sensors*.
4) *Electro-optical-based sensors*.

Example 8.2 It is desired to evaluate the flow rate of a corrosive fluid in a pipeline. For safety compliance reasons, contactless measurement devices are preferred. Typical measurement methods are summarized in Figure 8.28.
The appropriate sensing method should thus be chosen from the following candidates:

1) differential pressure in two pipeline points;
2) rotor speed of attached turbine-based measurement;
3) electromagnetic-based measurement;
4) ultrasound Doppler effect-based measurement.

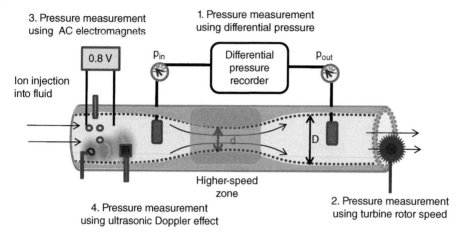

Figure 8.28 Four types of pipe-based fluid flow rate measurement.

For this pipeline, the ultrasonic method appears to be the most suitable and cost effective, as it uses a contact-free device, as required.

8.6.1.10 Radio Frequency-Based Level Measurement

This method, also named RF capacitance or just RF, can be applied to the detection of the level of liquids, slurries, or granules contained in a vessel. Various designs are available for the measurement of a process level at a specific point, at multiple points, or continuously over the entire vessel height. Radio frequencies used vary from 30 kHz to 1 MHz. For the RF capacitance measurement principle, the device consists of an electrical capacitance located between two conductors positioned a specific distance apart. The first is a vessel wall and the second is a measurement probe or an electrode. The two conductor areas are normally positioned. The medium between the two conductors is insulated, meaning that the level measurement does not involve a conducting material. The capacitance is given by:

$$C = E\frac{KA}{d} \tag{8.69}$$

with C, E, K, A, and d being the capacitance in picofarads (pF), free space permittivity, the insulating material dielectric constant, the conductor's effective area, and the distance between the conductors, respectively. The value of K is directly proportional to the system charge storage capacity.

8.6.1.11 Smart and Nano Sensors

Smart and nano sensors are usually low-scale devices based on electrostatic, piezo-resistive, piezo-electric, or electromagnetic sensing principles. They can be made of different materials, such as silicon and polymer. More specifically, they include:

1) *Optic fiber-based smart sensors* (magnetorestrictive, piezo-electric, optic fibers, etc.), which use the proportionality between the intensity or spectrum of their material (glass and silica) physical characteristics and the variations in the sensed system strain, force, temperature, liquid level, vibration, and so on.
2) *Electrostatic or capacitive-based smart sensors*, which involve transducing a force, vibration, pressure, or temperature variation into the displacement of a membrane or beam. This displacement then induces variations of membrane capacitance. These electrostatic sensors

are used as accelerometers, sound sensors, gyroscopes, pressure sensors, and even tactile sensors.

3) *Piezo-resistive-based smart sensors*, which transduce an applied mechanical strain or deformation into a magnitude electrical field through variations of the resistive material dimension. Typical materials used include alloy and silicon. Some application examples are piezo-resistive and capacitive-based micropressure sensors, where the diaphragm deformation is transduced into pressure differences, and the use of polymer-based material to monitor the high temperature of an internal combustion engine. Another example is accelerometers used in vibration measurements or in vehicle anti-collision systems.

4) *Thermal-based smart sensing devices*, which involve producing heat from a current flowing through a resistive element over an ambient fluid. When a fluid with a lower temperature passes through the device, it reduces the temperature of the heated element by heat convection. The temperature variation of the element provides the flowmeter (anemometer) information on the fluid direction and its flow rate. The mechanical momentum transfer principle is another flow-sensing technique.

5) *Light-based smart sensors*, which use phototransistors, photodiodes, or photoresistors such that their exposure to light intensity is proportional to their resistance variation, which is then converted into voltage change. These sensors can also be used as photodetectors, which are activated by a thermal light source such as a diode, quartz halogen, or fluorescent light, or daylight. Such light sources are capable of producing various electromagnetic signal wavelengths: 30 nm for ultraviolet light, between 400 and 700 nm for visible light, and up to 0.3 mm for infrared light. Here, the absorbed radiation capacity of the detector is not proportional to the emitted signal wavelength (i.e. pyroelectric). Typical applications of such lights sensors include fire detectors, infrared-based intruder alarms, and object shape capture devices.

6) *Image sensors*, which consist of an image-recording component combined with image data-processing software. First, image generation and encoding is achieved through light irradiance of the image source. Then, the image is projected over a plane, on which the charge of the image spatial distribution is presented. The charge of the image is accumulated, stored, and converted into a voltage signal. Here, a charge sample is called a picture element or pixel. Color sensors are achieved by placing color filters over the individual pixels. Typical applications include machine vision detecting systems in manufacturing operations or public event surveillance, as well as robotic guidance.

7) *Radiant-based detectors*, in which the absorbed photons capacity is proportional to the induced mobility of electrons (i.e. electrical charge), causing changes in their resistance values (i.e. photoresistors) or current or voltage values (i.e. photodiodes or phototransistors). Using the semiconductor functioning principle, photodetection is the conversion of the thermal excitation of electrons into current. The signal produced is proportional to light irradiance and is sensitive to temperature. Typical applications of phototransistors are coin detectors in vending machines, level sensors, and proximity sensors.

Other customized control measurement devices are: (i) microelectromechanical-micro-mechatronic systems (MEMS); and (ii) nanoelectromechanical-nanomechatronic systems (NEMS).

8.6.2 Detection Instruments

Binary instruments can only detect a state that is either true or false. This can be done either by direct contact or by proximity between the detector and the object. Examples of methods used to detect physical phenomena are: (i) the inductive proximity to detect if a metal object is nearby; (ii) the capacitive proximity to detect if a dielectric object is nearby; (iii) the optical presence to detect if an object passes through a light beam; and (iv) the mechanical contact applied to the detection of a contact force between an object and a switch. Typical binary instruments are:

- Sinking and sourcing switches, which switch the current flow ON or OFF. When the current flows out, these are known as sourcing switches.
- Plain switches, which switch the voltage ON or OFF.
- Solid-state relays, which switch AC outputs.
- Transistor–transistor logic (TTL), which uses voltage values 0 and 5 V to indicate logic levels.

Typical applications of such binary devices include material handling operations, the detection of vehicle presence in car parking, the detection of gas or liquid levels, the detection of object shape and position, and the tracking of a tagged object using radio frequency identification (RFID).

8.6.2.1 Electromechanical Limit Switches

The electromechanical limit switch is a detection device made of an actuator and electrical contact components. The detection is performed through its physical contact with a moving object or subject. The resulting electrical signal is sent to a signal processing system. Such limit switches are commonly encountered in automated systems.

8.6.2.2 Photoelectric Sensors

Photoelectric sensors detect various opaque to nearly transparent objects or subjects. Here, an LED generates infrared light pulses (wavelenghts between 850–950 nm). The generated pulses can be received by a photodiode or phototransistor, which amplifies a photoelectric current. Comparison of the generated signal to a reference threshold results in a discrete signal indicating whether object/subject presence is detected. There are several factors affecting the performance of these detection systems, such as: (i) the distance between the sensor and the targeted object or subject; (ii) the characteristics of the targeted object or subject, especially in terms of material reflectivity, transparence, color, and size; and (iii) environmental conditions, such as ambient light or the background. Such devices are used in automatic doors and barriers for security access management.

8.6.2.3 RFID-Based Tracking and Detection

RFID is a set of automatic identification devices using protocol-based radio wave signal exchange between a targeted subject or an object embedded within an electronic tag and a remote reader (transponder) component. Initially, an electromagnetic signal is sent from the transponder toward the tag antenna, where the flowing current charges the integrated capacitor. When the capacitor threshold voltage value is reached, it releases a time-varying amplitude radio-based signal over the tag coils, corresponding to the electronic tag information encoded back to the transponder. The transponder demodulates and stores this encoded signal. Figure 8.29 summarizes the RFID operating principle.

There are three types of tags: (i) passive tags, which produce surface acoustic waves on reception of a request signal from the transponder; (ii) semi-passive tags with embedded capacitive diodes enabling the production of a small current; and (iii) active tags with an integrated circuitry capable of data processing. There are two types of RFID communication protocol: tag talk first (TTF) and interrogator talk first (ITF). In either case, a subroutine should be included to avoid tag data collision when simultaneous tag code readings are required. The tag code standard is defined by the global identifier electronic product code (PEC) (96 bits). Among the advantages of RFID identification are: (i) its ability to modify data contained in the tag; (ii) its use of non metallic materials for read/write access; (iii) its insensitivity to environmental factors such as dust and so on; (iv) its high recording capacity (one thousand characters for one

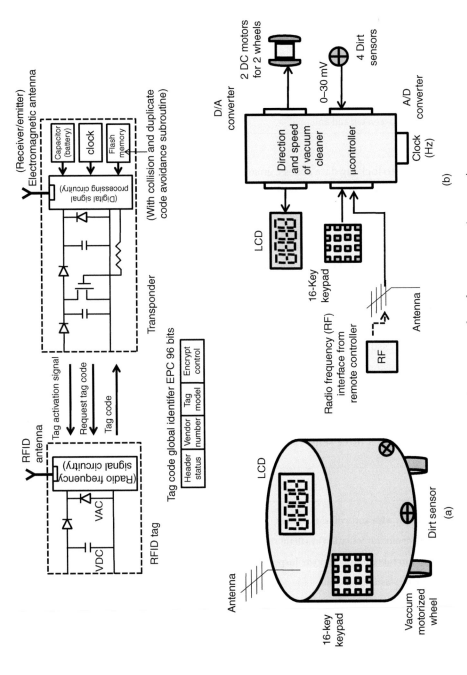

Figure 8.29 RFID detection principle. (a) Vacuum cleaner system. (b) Computer interfaces for vacuum motion control.

Table 8.12 RFID signal properties and characteristics.

Frequency	Advantage	Application
125–134 kHz	Immunity to environment (water, metal)	Animal identification
13.5 MHz	Standard antenna/tag communication protocols	Tracking books in libraries Access control Payment systems
850–950 MHz	Long communication range	Product management in distribution
2.45 GHz	High transfer speed between tag and antenna	Vehicle tracking (freeway tolls)

tag); and (v) its ability to lock tag data access, ensuring data confidentiality. The low cost of RFID tags offers the prospective of replacing traditional barcodes with product tracking in several sectors, including supply chain and transportation, along with new applications including security and payment. Some radio wave frequencies define the distance coverage and thus determine possible RFID applications, as summarized in Table 8.12.

8.6.2.4 Binary Devices: Pressure Switches and Vacuum Switches

Pressure or vacuum switches within a hydraulic or pneumatic circuit are used to monitor and measure pressure switches. They convert a pressure variation into a discrete electrical signal through the pressure-driven movement of a diaphragm, a piston, or a bellow, which activates actuating electrical contacts mechanically. The resulting resistance variation produces a signal proportional to the pressure.

Exercises and Problems

8.1 Find the suitable electric motors among the applications in the following table (justify your choice) and list three sensors that could be associated with them.

Applications	Electric motors
1) Wheelchair with anti-collision system	1) Permanent magnet DC motor
2) Turbulent motion actuation in a flight simulator	2) Brushless DC motor
3) Snake-arm robot for nuclear site inspection	3) Synchronous AC motor
4) Automatic washing machine	4) Stepper motor
5) Pick-and-place robot in a truck assembly line	5) Asynchronous AC motor

8.2 a) Indicate three different methods for operating a continuous temperature metering system installed inside tanks.

b) Sketch and explain the principle of resistance thermometers.
c) For each of the measurement devices listed in the following table, choose the suitable process applications.

Measurement devices	Process applications
1) Resistance thermometer	1) Temperature measurement in high-rate flow
2) Oscillation piston flowmeter	2) Flow metering of water
3) Orifice plate	3) Flow metering in corrosive fluid
4) Thermocouple	4) Mass metering
5) Ball valve	5) Temperature measurement inside a tank
6) Inductive-based flow meter	6) Flow metering in a large-diameter pipe

8.3 a) For the following systems, list the types of electric actuating systems encountered:
  i) automatic soda dispenser vending machine;
  ii) artificial hearth;
  iii) automatic adjustable sound system;
  iv) bottle milk filling system;
  v) automatic car air bag;
  vi) automatic car parking system;
  vii) dialysis blood treatment;
  viii) robot arm for endoscopic surgery.
 b) Identify at least 2 typical examples of sensing devices encountered for:
  i) concussion patient recovery through a stability test;
  ii) automatic window adjustment in a vehicle;
  iii) automatic surveillance camera with facial recognition;
  iv) situation awareness and obstacle avoidance in a mobility system.

8.4 a) Derive the resolution of an absolute seven-binary-track rotary optical encoder.
 b) What is the voltage level output of a linear operating range (K-type) thermocouple at 195°C, if it is capable of generating a voltage of 30 mV at 155°C and of 195 mV at 175°C?
 c) A potentiometer measures a robot angle such that the power supply connected across it is 3.5 V and the total wiper travel is 270°. The wiper arm is directly connected to the rotational joint. Derive the corresponding potentiometer output voltage for 43° if there is a voltage offset of 4.3 mV.
 d) A gearbox with a 7:1 ratio is associated to a motor. For a position encoder attached to this motor, derive the position of a geared downshaft required to be positioned to 0.1°. Derive its minimum resolution.
 e) Use the equations for a permanent magnet DC motor to explain how it can be used as a tachometer.
 f) What is the key difference between position measurements obtained through encoders and those obtained through potentiometers?

8.5 An accelerometer instrument is used to access the relative seismic mass variation via a miniature LVDT. The undamped natural frequency of the accelerometer instrument is 10 kHz and the effective stiffness is $K = 2000\,\text{N m}^{-1}$.
   a) Determine the mass m and damping coefficient c required, if a damping ratio of 0.707 is targeted.
   b) Determine the sensitivity of the accelerometer in volts per g at a frequency of 1500 Hz, given that the sensitivity of the LVDT is $250\,\text{mV mm}^{-1}$.

8.6 A 270 Ω potentiometer measures the linear displacement of piston-based robot arm, while the associated voltmeter with a resistance of 2.7 kΩ converts this position measurement into a voltage unit. What is the reading error when the potentiometer slider is at a 38% position based on a voltage scale of 10 V?

8.7 For each of the following statements, select the correct option:
   a) An accelerometer provides accurate dynamic measurements of seismic phenomena (i) at frequencies well below its natural frequency; (ii) at frequencies above its natural frequency; (iii) at any frequency; or (iv) none of these.
   b) A tachometer is used to assess the real motor velocity of $20\,\text{rad s}^{-1}$. The measurements are done five times, leading to the following results:

Reading	1	2	3	4	5
Velocity (rad s^{-1})	19.7	20.1	20	19.8	19.9

   i) The bias of the instrument is (difference between mean and true value): (i) $19.90\,\text{rad s}^{-1}$; (ii) $0.10\,\text{rad s}^{-1}$; (iii) $0.05\,\text{rad s}^{-1}$; or (iv) none of these.
   ii) The accuracy of the tachometer is: (i) 1.5%; (ii) 2.0%; (iii) 0.5%; or (iv) none of these.
   iii) The linearity of an instrument defines: (i) the maximum deviation of the output from the real value; (ii) the minimum change in input required to reflect a variation in the output; (iii) the maximum deviation of the output from the best-fit line; or (iv) none of these.
   iv) The precision of an instrument expresses: (i) the closeness of the data to the mean value (precision); (ii) the closeness of the data to the true value (accuracy); (iii) the closeness of the data to the median value; or (iv) none of these.
   v) Derive an increment for a three-bit ADC with a 0–10 V range.
   vi) What is the maximum linearity error?

8.8 It is desired to design a small autonomous vacuum cleaner with a 16-key multiplexed keypad and liquid-crystal display (LCD) as the user interface for the motion control of the front and rear wheels separately, as depicted in Figure 8.29(a) and (b). It should have an antenna to communicate with the base station (remote control). It should detect dirt concentration and location by estimating dirt particles (DP) per cm². In addition, it should have two DC motor-driven wheels. Also it should use a DP sensor to adjust each motor speed based on amount of dirt detected with a ADC conversion time of 50 microsec) while a 5-bit ADC converter should be used to sample each motor speed and direction. The microcontroller provides four additional control bits to determine the motor directions (00, both motors stopped; 01, left motor goes in reverse while right motor goes forward; 10, left motor goes forward while right motor goes in reverse; 11, both motors move forward). Finally, the vacuum should have an obstacle avoidance system. Considering this vacuum system, answer the following questions:

a) What are the possible actuating and sensing devices involved?

b) How many bits are required to collect data from the dirt sensor, if it has a range of 0–6000 DP cm^{-2} and a resolution of 100 DP cm^{-2}?

c) If this sensor has a maximum slope of 1000 DP s^{-1} and is capable of generating from 0 to 30 mV, what is its accuracy?

d) The motor controllers accept a voltage between -12 and $+12$ V. A positive voltage causes them to spin the attached wheel in a forward direction. A negative voltage causes them to spin it in a reverse direction. What is the resolution of the DACs used to control the motors?

e) Draw a block diagram for the vacuum cleaner, including as much detail as possible, such as ports of I/O devices sensors, actuators, DACs, logic, and so on. Add any components, as long as their functions are described.

8.9
a) A two-fluid manometer uses a combination of kerosene $r = 756$kgm^{-3} and alcohol-diluted water $r = 866$kgm^{-3}. The diameters of the reservoir and the connecting U-tube are 50 and 6 mm, respectively. (i) What is the sensitivity (change in output/change in input) of the two-fluid manometer? (ii) If the U-tube were inclined, at what angle would the sensitivity be doubled?

b) A data acquisition card that samples at a rate of 100 000 samples per second is used to measure a signal given by $v = 2\sin(2\pi 15700t)$, where t is in seconds. Determine whether there is an aliasing problem.

c) A type T thermocouple is used to measure the temperature of hot water. The reference junction is exposed to ambient air $T = 25°C$, while the measuring junction is placed in boiling water. Determine the measured voltage.

d) Consider a thermocouple with a resolution of $0.01°C$ and a sensitivity of 5 V °C^{-1}. What is the smallest voltage variation it can display?

8.10
A five-bit ADC with a 0–4.77 V range is used to monitor an oven temperature operating in the 190–$350°C$ range. The resistance of the thermistor as a function of temperature is given as:

$$R = 100 \times e^{800/(T+273)}$$

where R is in ohms and T is in degrees Celsius. The input voltage and ballast resistor are 5 V and 700 Ω, respectively, as illustrated in Figure 8.30.
Complete the following table:

Temperature (thermistor) (°C)	Resistance (thermistor) (Ω)	Output voltage (thermistor) (V)	Binary output of A/D converter	Recorded temperature (°C)
190	3900	3.11	0111	39.6
225	2867	2.33		
260				
310				

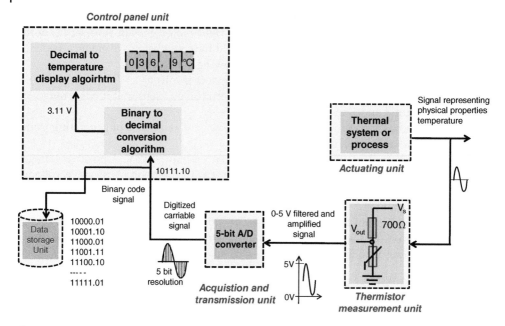

Figure 8.30 RTD sensor with an A/D data acquisition interface.

8.11 a) List four applications of DC motors in a vehicle and specify the types of DC motors encountered.

b) How can the stepper motor speed and its motion resolution be derived?

c) List three embedded intelligent and wearable sensing devices for human health monitoring.

d) List three piezo-electric actuating systems in medical monitoring technology.

e) List the types of motors used in: (i) a low-inertia car 2D painting robot; (ii) a moving conveyor of six-packs of 10–50 cl water bottles; and (iii) automatic car windows.

f) In the case of high-radial dynamics load conditions, among DC motorsthe DC, which key design parameters would affect motor selection? Explain why.

g) Among sensor characteristics such as resolution, slew rate, linearity, accuracy, and precision, which one captures any output variation after the smallest input change?

8.12 Consider two position-measurement devices $[p_1, p_2]$ connected in series, with 3.5 mV mm^{-1} being the sensitivity of p_1. If, for a 2.5 mm linear displacement, an overall measurement voltage signal of 8.5 mV is recorded, what will be the measurement sensitivity of p_2 considering an offset voltage at p_1 of -0.2 mV? What is the measurement sensitivity of p_2 when there is a zero-offset voltage of p_1?

8.13 A position resolver is directly attached to the screw of a lathe machine. The screw has a per-rev pitch of 2.5 mm. The phase shift between the resolver's excitation and the feedback signals is incremented by 1 every time a clock pulse is generated, as derived by a counter. The resolver excitation frequency is 900 Hz and a 2.5 µm position resolution is required to ensure an accurate control of machine motion. What is an appropriate clock frequency?

8.14 Two electrical-driven mining carts loaded with raw material for a total weight of 5 tons are moving upward along an 8% hill, as illustrated in Figure 8.31. The driven motor is attached to the drum (0.8 m diameter) and advances at a constant speed of 60 km h^{-1}.

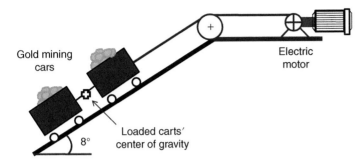

Figure 8.31 Electric-driven carts moving uphill.

a) Assuming that the friction coefficient between the car wheels and the rail surface is 0.60 (kg s^{-1}), what is the minimum motor torque required? How much energy will be consumed (in kWh) after a travel distance of 120 m? Assume the weight of both carts acts at their center of gravity, which is at the middle of the junction cable.

b) For the case of carts moving downwards at a 90 km h^{-1} constant speed, what is the power delivered by the motor?

c) What constant speed profile and power rates are required to move the carts uphill along 90 m of travel distance within a 3 min time interval?

8.15 Revisiting the system depicted in Figure 8.31, the motion power profile is given such that there is an acceleration phase from zero to a maximum power of 2000 kW in 25 s, then a 45 s period of constant velocity, requiring a power of 1500 kW, and finally a deceleration or braking period, where power is reduced from 500 W to zero uniformly across 10 s. For such a motor, what is the suitable power (kW) rating for root mean square (RMS) power?

8.16 Detection of toxic chemicals in the environment is carried out using electrochemical sensing devices based on a chromatographic technique through sample solution separation. Here, in case of diffusion of charged species in the test environment, measurements are collected through amperometric, thermal conductimetric, or potentiometric techniques. The resulting signals are amplified and processed through key feature-identification signal processing. What could be the key elements for chemical weapon detection using two detection techniques for compliance checks?

8.17 Electrical-driven actuators are used in the design of a snake-like robot capable of a smooth and collision-free motion over rugged terrain (or through a smooth-walled pipe). Each snake-like robot segment weighted 5.5 Kg for total 23.5 Kg. Its maximum linear speed of 80 mm/sec is achieved and each associated motor should have a rated torque less than 30 N.m. The collission-free motorized actuators change the position of adhesive segments (chambers), which alter their internal pressure proportionally to the motor rotational angle $\theta(t)$, as illustrated in Figure 8.32.
Select type of electric motors to drive robot joints with 2D wheels which are suitable for such intermitent operations, and describe the sensing devices involved.

Figure 8.32 Pneumatic-actuated snake-like robot.

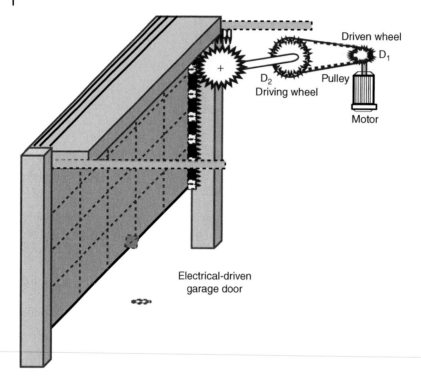

Figure 8.33 Screw-based gear wheel motion for a garage door.

8.18 A bidirectional motor moves at a constant speed of 180 rpm to drive a screw-wheel connected to a garage door, as illustrated in Figure 8.33. It is a 24-tooth driven wheel motor, while the driving wheel has 32 teeth. The motor moves through a 16-tooth pinion along a 96-tooth rack for a 9 cm pitch.

a) Select a DC motor suitable for such a garage door motion.

b) Describe a transmission element (e.g. screw, gear, wheel) configuration capable of quadrupling the motion speed along a travel distance of 50 cm for the selected motor?

c) When choosing a stepper motor with a 1.8° per step, what number of steps is required to cover the same distance?

8.19 A port gantry system moving a container as depicted in Figure 8.34 has a 2D motion as summarized in Figure 8.35.

a) Determine the continuous power rating of such a motor, with a duty cycle as summarized in the following table.

b) What type of electric-driven motor for 2D container handling should it be?

Gantry operations	Closing of the gripper	Hoisting	Opening of the gripper	Lowering of the gripper	Rest
Duration (s)	8	15	5	15	20
Power required (W)	45	95	35	55	0

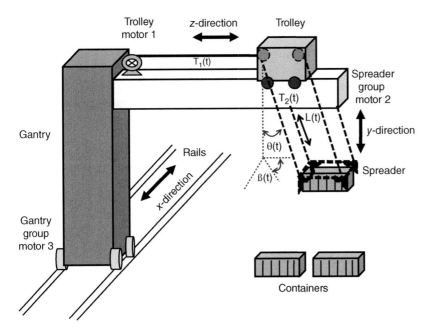

Figure 8.34 Gantry crane schematics.

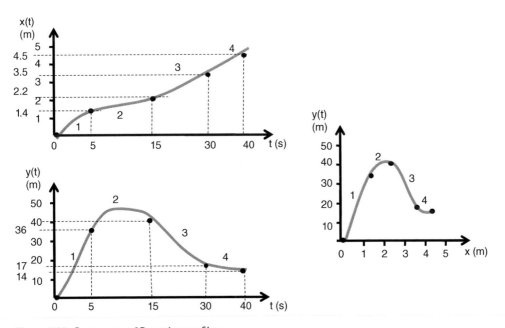

Figure 8.35 Gantry crane 2D motion profiles.

8.20 a) For fluid motion along a pipeline with minimum/maximum characteristics of flow rate: 3.05, 12.1 l s^{-1}, pressure: 3.5, 33.1 psi, temperature: 45, 75°C, select a suitable control valve from among the following, based on their valve coefficient/diameter size parameters: 650/4, 1050/8, 2105/16 cm.

b) What is the discharge velocity of a 3.48 specific gravity fluid that supposedly fills a tank at 12 m height? How would it be possible to double its discharge velocity? How would the discharge differ if the fluid were a gas rather than a liquid?

8.21 a) After a chemical reaction, the resulting solution toxic level T_C measurement is obtained indirectly through several instruments (T_Y, T_V, T_X), such that it can be obtained through:

$$T_C = (T_V^2 + 3.2T_Y)T_X$$

Thus, using the following measurement results, derive the resolution measurement of the toxic level for the resulting solution:

$$T_Y = 0.2\,\text{ppb} \pm 0.01$$

$$T_V = 5.6\,\text{ppb} \pm 0.3$$

$$T_X = 12.8\,\text{ppb} \pm 1.1$$

b) What is the most suitable control valve (valve coefficient/diameter size: 710/6, 1110/9, 2460/18 cm) for a fluid with the following characteristic minimums and maximums: flow rate 3.05, 12.1 l s^{-1}; pressure 3.5, 33.1 psi; temperature 35, 55°C?
c) What is the discharge velocity for a liquid with specific gravity 3.48 used to fill a tank at 18 m height? How is it possible to double this discharge velocity?

Bibliography

1 Alciatore, D.G. and Histand, M.B. (2003). *Introduction to Mechatronics and Measurement Systems*, 2e. McGraw-Hill International Editions.
2 Baker, B. (2005). *Real Analog Solutions for Digital Designers*. Elsevier.
3 Bishop, R.H. (ed.) (2002). *The Mechatronics Handbook*. ISA, CRC Press.
4 Cetinkunt, S. (2006). *Mechatronics*. Wiley.
5 de Silva, C.W. (2010). *Mechatronics: A Foundation Course*. CRC Press,.
6 Fraden, J. (2004). *Handbook of Modern Sensors*, 3e. Springer.
7 Gottlieb, I. (1994). *Electric Motors and Control Techniques*, 2e. McGraw-Hill.
8 Herman, S. (2013). *Industrial Motor*, 7e. Cengage Learning.
9 Isermann, R. (2003). *Mechatronic Systems: Fundamentals*. Springer.
10 Hughes, A. (2005). *Electric Motors and Drives: Fundamentals, Types, and Applications*, 3e. Newnes.
11 Johnson, J.L. (1992). *Basic Electronics for Hydraulic Motion Control*. Penton IPC.
12 Jury, E. and Tsypkin, Y.Z. (1971). On the theory of discrete systems. *Automatica* 7: 89–107.
13 Karnopp, D.C., Margolis, D.L., and Rosenberg, R.C. (2012). *System Dynamics: Modeling, Simulation, and Control of Mechatronic Systems*, 5e. Wiley.
14 Kazmierkowski, M.P., Krishnan, R., Blaabjerg, F., and Irwin, J.D. (2002). *Control in Power Electronics: Selected Problems*. Academic Press.
15 Krishnan, R. (2001). *Electric Motor Drives: Modeling, Analysis and Control*. Prentice-Hall.
16 Ljung, L. (1999). *System Identification: Theory for the User*, 2e. Prentice-Hall.
17 Lyshevski, S.E. (2000). *Nano- and Micro-Electromechanical Systems: Fundamental of Micro- and Nano-Engineering*. CRC Press.

18 Manring, N. (2013). *Fluid Power Pumps and Motors: Analysis, Design and Control.* McGraw-Hill.

19 Manring, N. (2005). *Hydraulic Control System.* Wiley.

20 Moran, M.J., Shapiro, H.N., Munson, B.R., and Dewitt, D.P. (2002). *Introduction to Thermal Systems Engineering: Thermodynamics, Fluid Mechanics and Heat Transfer.* Wiley.

21 Onwubolu, G.C. (2005). *Mechatronics Principles and Applications.* Elsevier.

22 Petruzella, F. (2009). *Electric Motors and Control.* McGraw-Hill.

23 Pinsky, M. and Karlin, S. (2010). *An Introduction to Stochastic Modeling.* Academic Press.

24 Sul, S.-K. (2011). *Control of Electric Machine Drive Systems.* Wiley-IEEE Press.

25 Ripka, P. and Tipek, A. (2007). *Modern Sensors Handbook.* ISTE Ltd.

26 Sashida, T. and Kenjo, T. (1993). *An Introduction to Ultrasonic Motors.* Oxford University Press.

27 Singh, R.P. and Heldman, D.R. (2001). *Introduction to Food Engineering*, 3e. Academic Press.

28 Shumway-Cook, A. and Woollacott, M.H. (1995). *Motor Control: Theory and Practical Applications.* Williams & Wilkins.

29 Soloman, S. (1998). *Sensors Handbook.* McGraw-Hill Handbooks.

30 Shumway, R.H. and Stoffer, D.S. (2010). *Time Series Analysis and Its Applications: With R Examples.* Springer.

31 Herman, S.L. (2010). *Industrial Motor Control*, 6e. Cengage Learning.

32 Suzuki, Y., Tani, K., and Sakuhara, T. (2000). Development of new type piezo-electric micro-motor. *Journal of Sensors & Actuators* 83: 244.

33 Uchino, K. (1997). *Piezo-electric Actuators and Ultrasonic Motors*, 349. Kluwer.

34 Yeaple, F. (1996). *Fluid Power Design Handbook*, 3e. Dekker.

A

Stochastic Modeling

Operating data-based mathematical characterization of system behavior can be derived into either non parametric or parametric models. A generic parametric model identification and coefficient estimation procedure consists of:

1) Preprocessing the dataset, which entails: (a) spitting it into estimation and validation data for model identification and validation, respectively; and then (b) examining the data via plots to check for: (i) the missing data or visible errors in data; (ii) any aliasing effects (due to poor choice of sampling time); (iii) any trends or offsets; and (iv) high- or low-frequency disturbances.
2) Selecting the structure and the order of the system model. This iterative procedure consists of fitting various model structures and orders.
3) Estimating the "best" model within the selected model structure by deriving the model coefficients that "best fit" with the dataset.
4) Determining the correctness of the model by analyzing the residual errors obtained in the estimation process and determining the confidence regions for the estimated coefficients.

A.1 Discrete Process Model State-Space Form

The state-space model describing the relationship between the discrete system input, the sampled system output, and the system noise signal is given by State Equation (A.1), representing: (i) the state variable $x(k+1)$ at time instant $k+1$ in terms of the previous state variable $x(k)$ at time k; and (ii) the observation with output $y(k)$, such that:

$$x(k+1) = A(q)x(k) + B(q)u(k) + w(k) \tag{A.1}$$

$$y(k) = C(\theta)x(k) + v(k) \tag{A.2}$$

with $A(q) = 1 + a_1 q^{-1} + \ldots + a_{na}q^{-na}$; $B(q) = b_1 + b_2 q^{-1} + \ldots + b_{nb}q^{-nb+1}$; $w(k)$ being the white noise representing the uncertainties on the process model or input disturbances, $v(k)$ being the white-noise sequence representing noise measurement, and $u(k)$ being the exogenous input.

The stochastic state-space system model is given by:

$$\hat{x}(k+1) = A(\theta)\hat{x}(k) + B(\theta)u(k) + K(\theta)e(k) \tag{A.3}$$

$$y(k) = C(\theta)\hat{x}(k) + e(k) \tag{A.4}$$

where $\hat{x}(k)$ represents the estimated state at instant k, $e(k)$ is the error or white noise, and $K(\theta)$ is the Kalman gain matrix; $A(\theta)$ $B(\theta)$ $C(\theta)$ are the process and measurement, respectively. For a SISO model, the generic transfer function representation yields:

$$y(k) = G(q, \theta)u(k) + H(q, \theta)e(k) \tag{A.5}$$

Control Of Mechatronic Systems: Model-Driven Design And Implementation Guidelines,
First Edition. Patrick O.J. Kaltjob.
© 2021 Patrick O.J. Kaltjob. Published 2021 by John Wiley & Sons Ltd.

with:

$$G(q, \theta) = C(\theta) \frac{B(\theta)}{(q^{-1}I_{n*n} - A(\theta))^{-1}} + I_{m*m} \tag{A.6}$$

$$H(q, \theta) = \frac{C(\theta)K(\theta)}{(q^{-1}I_{n*n} - A(\theta))^{-1}} + I_{m*m} \tag{A.7}$$

A.2   Auto-Regressive Model with an eXogenous Input: ARX Model Structure

In this case, the discrete process model is represented by the difference equation:

$$y(k) + a_1 y(k-1) + \ldots + a_{na} y(k-na) = b_1 u(k-nk) + b_2 u(k-nk-2)$$
$$+ \ldots + b_{nb} u(k-nk-nb+1) + e(k) \tag{A.8}$$

with the sequence $\{y(k), y(k-1), \ldots, y(k-na)\}$ representing the current and past process output measurements with order na, the sequence $\{u(k-nk), u(k-nk-2), \ldots, u(k-nk-nb+1)\}$ representing the past process input records with order nb and input delay nk, $e(k)$ being the error signal (process and measurement noise) describing independent and identically distributed zero-mean white noise with variance λ, and $a_i's$ with $i = 1, 2, 3, \ldots, na$ and $b_j's$ with $j = 1, 2, 3, \ldots, n$ being the coefficients of the difference equation or the model parameters.

This SISO Auto-Regressive with an eXogenous input (ARX) model structure consists of estimation of output value at instant k, $y(k)$ using an auto-regression of na past values: $\{y(k-1), \ldots, y(k-na)\}$, the past recorded input(s), and a model error estimate.

A.3   The Auto-Regressive Model – AR Model Structure

In this case, the current output value of the discrete process model is given as a function of na past output values and the current unknown error estimate, such that:

$$y(k) = a_1 y(k-1) - a_2 y(k-2) - \ldots - a_{na} y(k-na) + e(k) \tag{A.9}$$

A.4   The Moving Average Model – MA Model Structure

Here, the current output variable is represented as a weighted sum or moving average of the past and present nc error estimates, such that:

$$y(k) = e(k) + c_1 e(k-1) + c_2 e(k-2) + \ldots + c_{nc} e(k-nc) \tag{A.10}$$

This is compactly expressed as:

$$y(k) = C(q)e(k) \tag{A.11}$$

with:

$$C(q) = 1 + c_1 q^{-1} + \ldots + c_{nc} q^{-nc} \tag{A.12}$$

Equation (A.12) illustrates the filtering of the white-noise signal $e(k)$ (usually considered with zero mean and variance of σ^2) through a filter $C(q)$. Knowledge of $C(q)$ allows one to determine, for instance, the energy distribution of the random disturbance of the process.

A.5 The Auto-Regressive Moving Average Model – ARMA Model Structure

In this case, the model representation is a combination of the AR and MA models. It describes the current output value as a regression of its past values and the error model term as a moving average, such that:

$$(1 + a_1 q^{-1} + \ldots + a_{na} q^{-na}) y(k) = (1 + c_1 q^{-1} + \ldots + c_{nc} q^{-nc}) e(k) \tag{A.13}$$

The ARMA model allows the current output $y(k)$ to be derived by passing white noise through a filter $C(q)/A(q)$, such that:

$$\frac{C(q)}{A(q)} = \frac{1 + a_1 q^{-1} + \ldots + a_{na} q^{-na}}{1 + c_1 q^{-1} + \ldots + c_{nc} q^{-nc}} \tag{A.14}$$

A.6 The Auto-Regressive Moving Average with eXogenous Input Model – ARMAX Model Structure

Adding the external input to the ARMA model results in an ARMAX model. Here, the current output depends regressively on its *na* previous values and the past *nb* input values, as well as *nc* error values, such that:

$$y(k) = -a_1 y(k - 1) - \ldots - a_{na} y(k - na) + b_1 u(k - nk) + \ldots + b_{nb} u(k - nk - nb + 1)$$
$$+ e(k) + c_1 e(k - 1) + \ldots + c_{nc} e(k - nc) \tag{A.15}$$

In a more compact notation, this gives:

$$A(q)y(k) = B(q)u(k) + C(q)e(k) \tag{A.16}$$

These SISO parametric models are readily extended to MISO models with an additional input variable.

A.7 Selection of Model Order and Delay

The model orders are integers characterizing the complexity of a model or its number of poles and zeros; that is, for ARX structure (n_a, n_b) or ARMAX structure (n_a, n_b, n_c) with input delay n_k. This requires prior knowledge of physical model properties including the observed input delay. These model orders are checked for correctness against routines, for example by (see Table A.1):

1) Analyzing peaks obtained from the spectral of sequenced data and the analysis on correlation from frequency responses. Usually, for the ARX and ARMAX models, the order of $A(q)$ is given by:

$$OrderA(q) = 2 \times number\_peaks$$

2) Performing model-order reduction analyses by observing potential cancelation in pole-zero pairs in a *p-z* map.

Table A.1 Typical model structure encountered for a system model.

Model type	Applications	Limitations
AR: $A(q)y(k) = e(k)$	Used to predict behavior of the time series based only on past output values	
MA: $y(k) = C(q)e(k)$	Used to predict noise correlation	
ARMA: $A(q)y(k) = C(q)e(k)$	Used to predict behavior of the time series from past error values only	
ARX: $A(q)y(k) = B(q)u(k - nk) + e(k)$	Noise model is $1/A(q)$ Models requiring both control (input) and disturbance effects Used to obtain simple models with good signal-to-noise ratio	Cannot model noise and dynamics independently (noise model coupled to dynamics)
ARMAX: $A(q)y(k) = B(q)u(k - nk) + C(q)e(k)$	Noise model is $C(q)/A(q)$ Used when dominating disturbance entry at the input	Assumes the random sequence $e(k)$ is identically distributed and passes via C models only serially correlated random effects of the system
ARIMAX: $A(q)y(k) = B(q)u(k - nk) + \dfrac{C(q)}{\Delta}e(k)$ $\Delta = 1 - z^{-1}$	Better models the noise (not just as serially correlated or moving average $e(\mathbf{k})$), but adds an integral term Δ, allowing it to model the process noise with drifting behavior	
Box Jenkins: $A(q)y(k) = \dfrac{B(q)}{F(q)}u(k - nk) + \dfrac{C(q)}{D(q)}e(k)$	Provides additional flexibility when modeling noise Used when noise is not the input, but rather measurement disturbance	May be computationally complex due to the need to find many parameters
Output Error: $y(k) = \dfrac{B(q)}{F(q)}u(k - nk) + e(k)$	Noise model assumed to be 1. Used to parameterize only system dynamics and not the noise model	

A.8 Parameter Estimation Methods

Once the model structure is selected and the model order is set, the next step in the stochastic-based system modeling procedure is to estimate the model's parameters (a_i, b_i, etc.). Among the possible techniques, the linear least-squares (LSs) estimation is the most straightforward. This consists of deriving the estimated model's coefficient that "best fits" a set of empirical data through a convex optimization over a cost function as the sum of squared prediction errors from the model structures represented in a linear or pseudo-linear regression. For example, consider a model structure given by the difference equation for input/output measurements. The system model yields:

$$y(k) = -a_1 y(k-1) - \ldots - a_{na} y(k-na) + b_1 u(k-nk) + \ldots + b_{nb} u(k-nk-nb+1) + e(k)$$

$$y(k+1) = -a_1 y(k) - \ldots - \ldots - a_{na} y(k-na-1) + b_1 u(k-nk) + b_2 u(k-nk-2)$$

$$+ \ldots + b_{nb} u(k-nk-nb+1) + e(k+1)$$

$$\vdots$$

$$y(k+N-1) = -a_1 y(k+N-2) - \ldots - a_{na} y(k-na-(N-1)) + b_1 u(k-nk)$$

$$+ \ldots + b_{nb} u(k-nk-nb+1-(N-1)) + e(k+N-1) \quad \text{(A.17)}$$

In matrix form, for $k = 0, 1$, the set of N equations here yields:

$$\begin{bmatrix} y(k) \\ y(k+1) \\ \vdots \\ y(k+N-1) \end{bmatrix}$$

$$= \begin{bmatrix} -y(k-1) & -y(k-2) & \cdots & -y(k-1) & u(k-1) & \cdots & u(k-nb) \\ -y(k) & -y(k-1) & \cdots & -y(k-na-1) & u(k) & \cdots & u(k-nb-1) \\ \vdots & \vdots & \vdots & \vdots & \vdots & \vdots & \vdots \\ -y(k+N-2) & -y(k+N-3) & \cdots & -y(k+N-(na+1)) & u(k+N-2) & \cdots & -y(k+N-(nb+1)) \end{bmatrix}$$

$$\times \begin{bmatrix} a_1 \\ a_2 \\ \vdots \\ a_{na} \\ b_1 \\ b_2 \\ \vdots \\ b_{nb} \end{bmatrix} + \begin{bmatrix} e(k) \\ e(k+1) \\ \vdots \\ e(k+N-1) \end{bmatrix} \quad \text{(A.18)}$$

This is equivalent to:

$$Y = X\theta + \varepsilon$$

with $Y = \begin{bmatrix} y(k) \\ y(k+1) \\ \vdots \\ y(k+N-1) \end{bmatrix}$ being the vector of N output observations, $\varepsilon = \begin{bmatrix} e(k) \\ e(k+1) \\ \vdots \\ e(k+N-1) \end{bmatrix}$ being

the vector of N random errors (all assumed white noise), $\theta = [a_1, a_1, \ldots, a_{na}, b_1, b_2, \ldots b_{nb}]^T$ being the vector of parameter values to be determined,

$$X = \begin{bmatrix} -y(k-1) & -y(k-2) & \cdots & -y(k-na) & u(k-1) & \cdots & u(k-nb) \\ -y(k) & -y(k-1) & \cdots & -y(k-na-1) & u(k) & \cdots & u(k-nb-1) \\ \vdots & \vdots & \vdots & \vdots & \vdots & \vdots & \vdots \\ y(k+N-2) & -y(k+N-3) & \cdots & -y(k+N-(na+1)) & u(k+N-2) & \cdots & u(k+N-(nb+1)) \end{bmatrix}$$

being the matrix of past observations of output $y(k)$ and control input $u(k)$ for N measurements, **X** denoting the measurement vector of known values of na previous $y(k)$ observations and nb previous $u(k)$ observations, and with $\varphi(k) = [-y(k-1)\dots - y(k-na)..u(k-1)\dots u(k-nb)]$. Thus, the matrix **X** is equally denoted as:

$$X = \begin{bmatrix} \varphi^T(k) \\ \varphi^T(k+1) \\ \vdots \\ \varphi^T(k+N-1) \end{bmatrix}$$

In this way, Equation (A.18) readily yields:

$$\begin{bmatrix} y(k) \\ y(k+1) \\ \vdots \\ y(k+N-1) \end{bmatrix} = \begin{bmatrix} \varphi^T(k) \\ \varphi^T(k+1) \\ \vdots \\ \varphi^T(k+N-1) \end{bmatrix} \begin{bmatrix} a_1 \\ a_2 \\ \vdots \\ a_{na} \\ b_1 \\ b_2 \\ \vdots \\ b_{nb} \end{bmatrix} + \begin{bmatrix} e(k) \\ e(k+1) \\ \vdots \\ e(k+N-1) \end{bmatrix} \tag{A.19}$$

Every row of the equation can therefore be written as:

$$\begin{bmatrix} e(k) \\ e(k+1) \\ \vdots \\ e(k+N-1) \end{bmatrix}$$

$$y(k) = \varphi^T(k)\theta + e(k) \tag{A.20}$$

Equation (A.20) is a linear regression equation from the ARX model and is used in the formulation for the LS estimation. Vector θ represents a *set of parameters* (model coefficients) to be determined, while $\varphi(k)$ is a measurement vector of the past observed I/O signals, which are used to estimate the next value of the system output signal $y(k)$. A similar formulation can be determined for the other model structure, yielding the pseudo-linear regression. Among commonly used parameter estimation techniques are the LS and recursive least-square (RLS) methods.

A.9   LS Estimation Methods

The LS approach is an off-line parameter (model coefficient) identification technique based on N measurements. The technique consists of using an initial guess of the model parameters, $\hat{\theta}(0)$, to predict the system output at instant k, given by $y(k) = \varphi^T(k)(\hat{\theta}\ k-1)$. Then, the difference between the predicted output, $\hat{y}(k)$, and the measured system output, $y(k)$, is calculated. Hence, the prediction error, $e(k)$ can be defined as:

$$e(k) = y(k) - \hat{y}(k) \tag{A.21}$$

Consequently, a cost function $J(k)$, defined as the sum of squared errors, measures this prediction error. Based on the entire set of N *observed data*, the cost function $J(k)$ is defined for

$k = 1 \dots N$ as:

$$\min J(k) = \frac{1}{N} \sum_{k=1}^{N} e(k)^2 = \frac{1}{N} \sum_{k=1}^{N} (y(k) - \hat{y}(k \mid \theta))^2 \tag{A.22}$$

This cost function is solved as an optimization problem to find a global minimum with respect to the parameters. Finding this minimum of the prediction error is often termed the LS estimation method. With the ARX model, finding a unique solution (global minimum) to the optimization problem using the LS method is quite straightforward, involving solving for θ such that:

$$\frac{\partial J(k)}{\partial \theta} = 0 \tag{A.23}$$

A.10 RLS Estimation Methods

The online identification mode parameters are constantly tuned with real-time new observed data. The algorithm used to estimate these new system parameters is given by:

$$\hat{\theta}_{k+1} = \hat{\theta}_k + K_k \{y_{k+1} - \varphi_k^T \, \hat{\theta}_k\} \tag{A.24}$$

with values defined as:

$$y_k = \varphi_k^T \, \hat{\theta}_k + e_k$$

$$\varphi_k^T \, \hat{\theta}_k = \hat{y}_k$$

$$K_k = \frac{P_k \varphi_k}{\lambda + \varphi_k^T P_k \varphi_k}$$

$$P_k = \frac{1}{\lambda} \left[\frac{P_{k-1} \varphi_k \varphi_k^T P_{k-1}}{\lambda + \varphi_k^T P_{k-1} \varphi_k} \right]$$

this yields:

$$\hat{\theta}_{k+1} = \hat{\theta}_k + P_k \varphi_k \left[\frac{e_k}{\lambda + \varphi_k^T P_k \varphi_k} \right] \tag{A.25}$$

where y_k is the output measurement at the kth time step, φ_k describes the measurement vector of past observation of input and output at the kth time step, $\hat{\theta}_k$ corresponds to vector of the model's parameter estimate, K_k is the smoothing factor, which tunes the effect of the new measurement on parameter estimate, P_k is the estimated error of variance, and λ is the forgetting factor (with $(0 < \lambda < 1)$), ensuring that older measurements carry less weight in the new parameter estimate.

A.11 Model Validation

The model validation check determines the correctness of the model's orders and the parameter estimates. Among these statistical goodness-of-fit metrics are: (i) Akaike's information criterion (AIC); (ii) final prediction error (FPE); (iii) minimum description length (MDL); and (iv) best fit criteria, as described in Table A.2.

Table A.2 Commonly used statistical criteria for model order validation.

Method: statistical criteria	Definition	Comment
Best fit	$FIT = \left(1 - \dfrac{\|y - \hat{y}\|}{\|y - \bar{y}\|}\right) \times 100$ where: y = measured output $\hat{y}$ = simulated output $\bar{y}$ = mean value of output	Best model minimizes FIT 100% fit corresponds to a perfect fit 0% fit indicates the fit guesses the output to be a constant ($\hat{y} = y$)
AI Criterion	$AIC = \log V + \dfrac{2d}{N}$ where: V = loss function (see Equation (2.39)) d = number of estimated parameters N = Number of estimate values in dataset.	Best model minimizes AIC Investigates model quality when model is simulated on different dataset $AIC \approx FPE$ for $d \ll N$
FPE criterion	$FPE = \left(V + \dfrac{2d}{N}\right)$	Most accurate model minimizes FPE Estimates model fitting error when model is used to predict new outputs $FPE \approx AIC$ for $d \ll N$
MDL criterion	$MDL = V\left(1 + \dfrac{d \ln N}{N}\right)$	Most accurate model minimizes MDL Based on V and a penalty for the number of terms used

A.12 Prediction Error Analysis Methods

Once results are derived from parameter estimation techniques (such as LSs) and chosen as the "best fit" based on one of the goodness-of-fit statistics criteria, the resulting residuals (or prediction errors), $e(k)$, are carefully examined as they describe what part of the data the model could not describe. The error residuals should be carefully examined for obvious trends. The plot of the prediction errors (residuals) versus N measurement data may appear to follow a pattern (trend). When the plot has no trend, and nor do the number of sign changes, the prediction errors are random. This makes the correctness of the LS-based model impossible. Random error plot results can be checked by counting the sign changes in the prediction errors. When there is pattern, this may indicate that the model is not correct. Hence, any pattern, such as any gradual rise or data curve shaping in the plot of prediction errors, invalidates the coefficients' estimates. Among the residual analysis techniques for residuals, randomness, and independence are: (i) auto-correlation (via a whiteness test); (ii) normality check; and (iii) cross-correlation of the residuals and inputs.

A.13 Estimation of Confidence Intervals for Parameters

The confidence interval is the region within which an observation is expected to fall with a given probability. For residuals, therefore, the confidence interval corresponds to the range of residual values with a specific probability that they are statistically insignificant for the system. This means that, if the residuals are normally distributed, the confidence interval of the parameters can be derived. Estimation of the confidence information is done by deriving the individual confidence limits of the coefficients based on an assumption of independence of their estimation. As such example, the confidence interval is given by $100(1-\alpha)\%$. This is used if estimates: (i) are independent; (ii) if it contains relatively small elements; and (iii) if it can relax requirements in practice.

A.14   Checking for I/O Consistency for Different Models

A model validation method involves comparing its output results to those obtained from measured output data. This can be done by dividing the system measurement dataset into an estimation dataset (to derive the model coefficient) and a validation data set. The model is then simulated with the validation system input data, and its output results compared with the measured system output data.

B

Step Response Modeling

The step response test is a graphical modeling approach to model systems based on their dynamical responses to step input. From the analysis of the system step response, the general form of system parameters (gain, delay, time constants) can be estimated. Some model simplifications can be done by representing higher-order dynamics properties in a dead time term in the resulting model. Generic models from the step change and resulting response are summarized in Table B.1. Systems with differentiation cause the step input response to have an unbound amplitude response. Hence, their system parameters cannot be modeled using this approach. Thus, the magnitude of the step input signal to be used in the system must consider avoiding an unbound system response. Other limits of this modeling approach are that: (i) a higher than second-order dynamics model is covered; (ii) it is more suitable for systems with integration-dominating dynamics; and (iii) it is valid for small step changes around operation conditions.

Table B.1 Generic step response model types and corresponding transfer functions.

Step response type	Continuous process model	Discrete time process model
Gain plus delay process	$c(t) = Km(t - D)$	$c(n) = Km(n - d)$ $K = \dfrac{C_{ss}}{M}$ $d = \dfrac{D}{T}$ Here, d must be an integer; therefore, any fractional part is usually rounded up to the next integer
Integration plus delay process	$\dfrac{dc(t)}{dt} = Km(t - D)$	$c(n) = c(n - 1) + Km(n - d - 1)$ $K = \dfrac{\left.\dfrac{dc(t)}{dt}\right|_{ss}}{M} \approx \dfrac{C_{ss}}{M}$ $d = \dfrac{D}{T}$

Control Of Mechatronic Systems: Model-Driven Design And Implementation Guidelines,
First Edition. Patrick O.J. Kaltjob.
© 2021 Patrick O.J. Kaltjob. Published 2021 by John Wiley & Sons Ltd.

Table B.1 (Continued)

Step response type	Continuous process model	Discrete time process model
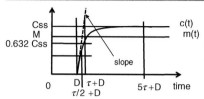 First-order plus delay process Output reaches a steady state value of css after a time interval of about 5τ.	$\tau\dfrac{dc(t)}{dt} + c(t) = Km(t - D)$ with $\tau = 2(t_{0.63} - t_{0.39})$	Overdamped second- and higher-order systems can be approximated as a first-order system with dead time. DT model: $c(n) = e^{-T/\tau}c(n-1)$ $\qquad + K(1 - e^{-T/\tau})m(n - d - 1)$ $K \approx \dfrac{C_{ss}}{M}$ $d = \dfrac{D}{T}$ and $D = t_{0.632} - \tau$
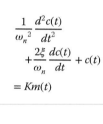 Underdamped second-order process	$\dfrac{1}{\omega_n^2}\dfrac{d^2c(t)}{dt^2}$ $\quad + \dfrac{2\xi}{\omega_n}\dfrac{dc(t)}{dt} + c(t)$ $\quad = Km(t)$	$c(n) = a_1 c(n-1) + a_2 c(n-2)$ $\qquad + Kb_1 m(n-1) + Kb_1 m(n-2)$ with: $a_1 = 2e^{-\xi\omega_n T}\cos\omega_d T$ $a_2 = 2e^{-2\xi\omega_n T}$ $b_1 = 1 - \dfrac{\xi\omega_n}{\omega_d}e^{-\xi\omega_n T}\sin\omega_d T$ $\qquad - e^{-\xi\omega_n T}\cos\omega_d T$
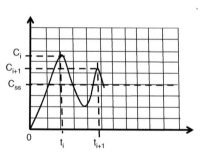 Log decrement method		$b_2 = e^{-\xi\omega_n T}\left(e^{-\xi\omega_n T} + \dfrac{\xi\omega_n}{\omega_d}\sin\omega_d T\right.$ $\qquad\left. - \cos\omega_d T\right)$ with: $\omega_d = \omega_n\sqrt{1 - \xi}$ $\xi = \dfrac{1}{2\pi(j-1)}\ln\left(\dfrac{C_i - C_{ss}}{C_{i+1} - C_{ss}}\right)$ For a larger damping ratio: $\xi = \dfrac{\ln\left(\dfrac{C_i - C_{ss}}{C_{i+1} - C_{ss}}\right)}{\sqrt{4\pi^2 + \ln\left(\dfrac{C_i - C_{ss}}{C_{i+1} - C_{ss}}\right)}}$ $\omega_n = \dfrac{2\pi}{t_{i+1} - t_i}\dfrac{1}{\sqrt{1 - \xi^2}}$

Table B.1 (Continued)

Step response type	Continuous process model	Discrete time process model
  Note that there is little delay initially for overdamped system Highly damped second-order process	$\dfrac{1}{\omega_n{}^2}\dfrac{d^2c(t)}{dt^2}$ $+\dfrac{2\xi}{\omega_n}\dfrac{dc(t)}{dt}+c(t)$ $=Km(t)$ to be rewritten as: $\dfrac{1}{\tau_1\tau_2}\dfrac{d^2c(t)}{dt^2}$ $+(\tau_1+\tau_2)\dfrac{dc(t)}{dt}$ $+c(t)=Km(t)$ $D=t_{\tau+D}-\tau$	Here, $\xi>1$ $c(n)=a_1c(n-1)+a_2c(n-2)$ $+Kb_1m(n-1)+Kb_2m(n-2)$ with: $a_1=e^{-T/\tau_1}+e^{-T/\tau_2}$ $a_2=-e^{-(T/\tau_1+T/\tau_2)}$ $b_1=1+\dfrac{\left(\dfrac{1}{\tau_2}e^{-T/\tau_1}-\dfrac{1}{\tau_1}e^{-T/\tau_2}\right)}{\left(\dfrac{1}{\tau_1}-\dfrac{1}{\tau_2}\right)}$ $b_2=-e^{-(T/\tau_1+T/\tau_2)}$ $+\dfrac{\left(\dfrac{1}{\tau_2}e^{-T/\tau_1}-\dfrac{1}{\tau_1}e^{-T/\tau_2}\right)}{\left(\dfrac{1}{\tau_1}-\dfrac{1}{\tau_2}\right)}$ However, this can be approximated as a first-order process with dead time. Between the two time constants τ_1 and τ_2 with $\tau_1\gg\tau_2$, the longer τ_1 dominates the response, while the effects of the shorter τ_2 can be $d=\dfrac{\tau_2}{T}$ Hence, this results in a first order plus dead time: $c(n)=e^{-T/\tau_1}c(n-1)+$ $K(1-e^{-T/\tau_1})m(n-d-1)$ with $\tau_2=2(t_{\tau+D}-t_{\tau/2+D})$
  Critically damped second-order process	$\dfrac{1}{\omega_n{}^2}\dfrac{d^2c(t)}{dt^2}$ $+\dfrac{2\xi}{\omega_n}\dfrac{dc(t)}{dt}+c(t)$ $=Km(t)$ which can be rewritten as: $\tau^2\dfrac{d^2c(t)}{dt^2}$ $+2\tau\xi\dfrac{dc(t)}{dt}$ $+c(t)=Km(t)$	For $\xi=1$ $c(n)=a_1c(n-1)+a_2c(n-2)$ $+Kb_1m(n-1)+Kb_1m(n-2)$ with: $a_1=2e^{-\omega_n T}$ $a_2=-e^{-\omega_n T}$ $b_1=1-e^{-\omega_n T}-\omega_n Te^{-\omega_n T}$ $b_2=e^{-\omega_n T}(e^{-\omega_n T}+\omega_n T-1)$ which can be approximated as: $c(n)=e^{-T/\tau_1}c(n-1)+$ $K(1-e^{-T/\tau_1})m(n-d-1)$

(Continued)

Table B.1 (Continued)

Step response type	Continuous process model	Discrete time process model
		The response reaches 59.4% of its steady-state value at $t = 2\tau$, where $\tau = 1/\omega_n$. If the time at which this level is reached is $t_{0.594}$, the time constant can be found from: $d = \dfrac{1}{2}t_{0.594}$ $\tau = 1/\omega_n = t_{0.594}/2$

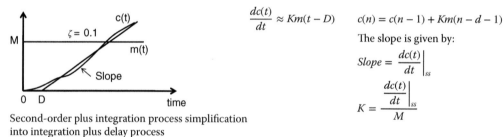

	$\dfrac{dc(t)}{dt} \approx Km(t - D)$	$c(n) = c(n - 1) + Km(n - d - 1)$
First-order plus integration process simplification into integration plus delay process		$Slope = \left.\dfrac{dc(t)}{dt}\right|_{ss}$ $K = \dfrac{\left.\dfrac{dc(t)}{dt}\right|_{ss}}{M}$

	$\dfrac{dc(t)}{dt} \approx Km(t - D)$	$c(n) = c(n - 1) + Km(n - d - 1)$
Second-order plus integration process simplification into integration plus delay process		The slope is given by: $Slope = \left.\dfrac{dc(t)}{dt}\right|_{ss}$ $K = \dfrac{\left.\dfrac{dc(t)}{dt}\right|_{ss}}{M}$

C

Z-Transform Tables

Table C.1 Z-transform table.

Laplace transform	Impulse response	Z-transform (impulse invariant)
$H(s) = L\{h(t)\}$	$h(t) = L^{-1}\{H(s)\}$	$H(z) = Z\{H(s)\}$
$\dfrac{1}{s}$	$u(t)$	$\dfrac{z}{z-1}$
$\dfrac{1}{s^2}$	t	$\dfrac{Tz}{(z-1)^2}$
$\dfrac{1}{s^3}$	$\dfrac{t^2}{2}$	$\dfrac{T^2 z(z+1)}{2(z-1)^3}$
$\dfrac{6}{s^4}$	t^3	$\dfrac{T^3 z^{-1}(1+4z^{-1}+z^{-2})}{2(1-z^{-1})^4}$
$\dfrac{1}{s+a}$	e^{-at}	$\dfrac{z}{z-e^{-aT}}$
$\dfrac{1}{(s+a)^2}$	te^{-at}	$\dfrac{Tze^{-aT}}{(z-e^{-aT})^2}$
$\dfrac{s}{(s+a)^2}$	$(1-at)e^{-at}$	$\dfrac{1-(1+aT)e^{-aT}z^{-1}}{(1-e^{-aT}z^{-1})^2}$
$\dfrac{a}{s(s+a)}$	$(1-e^{-at})$	$\dfrac{z[(1-e^{-aT})]}{(z-1)(z-e^{-aT})}$
$\dfrac{a}{s^2(s+a)}$	$t-\dfrac{1-e^{-at}}{s(s+a)}$	$\dfrac{z[(aT-1+e^{-aT})z+(1-e^{-aT}-aTe^{-aT})]}{a(z-1)^2(z-e^{-aT})}$

(continued)

Control Of Mechatronic Systems: Model-Driven Design And Implementation Guidelines,
First Edition. Patrick O.J. Kaltjob.
© 2021 Patrick O.J. Kaltjob. Published 2021 by John Wiley & Sons Ltd.

Table C.1 (Continued)

Laplace transform	Impulse response	Z-transform (impulse invariant)
$\dfrac{a^2}{s(s+a)^2}$	$1-(1+at)e^{-at}$	$\dfrac{z}{z-1}-\dfrac{z}{z-e^{-aT}}-\dfrac{aTe^{-aT}z}{(z-e^{-aT})^2}$
$\dfrac{b-a}{(s+a)(s+b)}$	$e^{-at}-e^{-bt}$	$\dfrac{z(e^{-aT}-e^{-bT})}{(z-e^{-bT})(z-e^{-aT})}$
$\dfrac{a}{(s^2+a^2)}$	$\sin(at)$	$\dfrac{z\sin(aT)}{z^2-2z\cos(aT)+1}$
$\dfrac{s}{(s^2+a^2)}$	$\cos(at)$	$\dfrac{z(z-z\cos(aT))}{z^2-2z\cos(aT)+1}$
$\dfrac{1}{(s+a)^2+b^2}$	$\dfrac{1}{b}e^{-at}\sin(bt)$	$\dfrac{1}{b}\left[\dfrac{z(e^{-aT}\sin(bT))}{z^2-2ze^{-aT}\cos(bT)+e^{-2aT}}\right]$
$\dfrac{s+a}{(s+a)^2+b^2}$	$\dfrac{1}{b}e^{-at}\cos(bt)$	$\dfrac{1}{b}\left[\dfrac{z^2+ze^{-aT}\cos(bT)}{z^2-2ze^{-aT}\cos(bT)+e^{-2aT}}\right]$
$\dfrac{a^2+b^2}{s[(s+a)^2+b^2]}$	$1-e^{-at}\left(\cos(bt)+\dfrac{a}{b}\sin(bt)\right)$	$\dfrac{z(Az+B)}{(z-1)(z^2-2ze^{-aT}\cos(bT))+e^{-2aT}}$ $A=1-e^{-aT}(\cos(bT))+\dfrac{a}{b}\sin(bT)$ $B=e^{-a2T}+e^{-a2T}\left(\dfrac{a}{b}\sin(bT)\right)-\cos(bT)$
$\dfrac{1}{s(s+a)(s+b)}$	$\dfrac{1}{ab}+\dfrac{e^{-at}}{a(a-b)}+\dfrac{e^{-bt}}{b(a-b)}$	$\dfrac{z(Az+B)}{(z-1)(z-e^{-aT})(z^2-e^{-bT})}$ $A=\dfrac{b(1-e^{-aT})-a(1-e^{-bT})}{ab(a-b)}$ $B=\dfrac{ae^{-aT}(1-e^{-aT})-be^{-aT}(1-e^{-bT})}{ab(a-b)}$
$\dfrac{2}{(s+a)^3}$	t^2e^{-at}	$\dfrac{T^2e^{-aT}(1+e^{-aT}z^{-1})z^{-1}}{(1-e^{-aT}z^{-1})^3}$
$\dfrac{\omega}{(s+a)^2+\omega^2}$	$e^{-at}\sin\omega t$	$\dfrac{z^{-1}e^{-aT}\sin\omega T}{(z^2-2ze^{-aT}\cos(\omega T))+e^{-2aT}z^2}$
$\dfrac{s+a}{(s+a)^2+\omega^2}$	$e^{-at}\cos\omega t$	$\dfrac{1-z^{-1}e^{-aT}\cos\omega T}{1-2z^{-1}e^{-aT}\cos(\omega T)+e^{-aT}z^2}$

Table C.2 Delay-included z-transform table.

Laplace transform	Impulse response	Delay-included z-transform
$H(s) = L\{h(t)\}$	$h(t) = L^{-1}\{H(s)\}$	$H_m(z) = Z\{H(s)\}$ where m = 1.0 for no delay
$\dfrac{1}{s}$	$u(t)$	$\dfrac{1}{z-1}$
$\dfrac{1}{s^2}$	t	$\dfrac{mT}{z-1} + \dfrac{T}{(z-1)^2}$
$\dfrac{1}{s^3}$	$\dfrac{t^2}{2}$	$\dfrac{T^2}{2}\left[\dfrac{m^2}{z-1} + \dfrac{2m+1}{(z-1)^2} + \dfrac{2}{(z-1)^3}\right]$
$\dfrac{1}{s+a}$	e^{-at}	$\dfrac{e^{-amT}}{z-e^{-aT}}$
$\dfrac{1}{(s+a)^2}$	te^{-at}	$\dfrac{Te^{-amT}[e^{-aT} + m(z - e^{-aT})]}{(z-e^{-aT})^2}$
$\dfrac{a}{s(s+a)}$	$1 - e^{-at}$	$\dfrac{1}{z-1} - \left[\dfrac{e^{-amT}}{z-e^{-aT}}\right]$
$\dfrac{a}{s^2(s+a)}$	$t - \dfrac{1-e^{-at}}{a}$	$\dfrac{T}{(z-1)^2} + \dfrac{amT-1}{a(z-1)} + \dfrac{e^{-amT}}{a(z-e^{-aT})}$
$\dfrac{a^2}{s(s+a)^2}$	$1 - (1+at)e^{-at}$	$\dfrac{1}{z-1} - \left[\dfrac{1+amT}{z-e^{-aT}} + \dfrac{aTe^{-aT}}{(z-e^{-aT})^2}\right]e^{-amT}$
$\dfrac{b-a}{(s+a)(s+b)}$	$e^{-at} - e^{-bt}$	$\dfrac{e^{-amT}}{z-e^{-aT}} - \dfrac{e^{-bmT}}{z-e^{-bT}}$
$\dfrac{a}{(s^2+a^2)}$	$\sin(at)$	$\dfrac{z\sin(amT) + \sin(1-m)aT}{z^2 - 2z\cos(aT) + 1}$
$\dfrac{s}{(s^2+a^2)}$	$\cos(at)$	$\dfrac{z\cos(amT) + \cos(1-m)aT}{z^2 - 2z\cos(aT) + 1}$
$\dfrac{1}{(s+a)^2+b^2}$	$\dfrac{1}{b}e^{-at}\sin(bt)$	$\dfrac{1}{b}\left[\dfrac{e^{-amT}[z\sin bmT + e^{-aT}\sin(1-m)bT]}{z^2 - 2ze^{-aT}\cos(bT) + e^{-2aT}}\right]$
$\dfrac{s+a}{(s+a)^2+b^2}$	$\dfrac{1}{b}e^{-at}\cos(bt)$	$\dfrac{e^{-amT}[z\cos bmT + e^{-aT}\sin(1-m)bT]}{z^2 - 2ze^{-aT}\cos(bT) + e^{-2aT}}$
$\dfrac{a^2+b^2}{s[(s+a)^2+b^2]}$	$1 - e^{-at}\left(\cos(bt) + \dfrac{a}{b}\sin(bt)\right)$	$\dfrac{1}{z-1} - \dfrac{e^{-amT}[z\cos bmT + e^{-aT}\sin(1-m)bT]}{z^2 - 2ze^{-aT}\cos(bT) + e^{-2aT}}$ $+ \dfrac{\dfrac{a}{b}(e^{-amT}[z\sin bmT + e^{-aT}\sin(1-m)bT]}{z^2 - 2ze^{-aT}\cos(bT) + e^{-2aT}}$
$\dfrac{1}{s(s+a)(s+b)}$	$\dfrac{1}{ab} + \dfrac{e^{-at}}{a(a-b)} + \dfrac{e^{-bt}}{b(a-b)}$	

D

Boolean Algebra, Bus Drivers, and Logic Gates

Compiled Boolean theorems.

$A + B = B + A$	$\overline{\overline{A}} = A$
$A \bullet B = B \bullet A$	$(A + B) + C = A + (B + C)$
$A + 0 = A$	$(A \bullet B)C = A(B \bullet C)$
$A \bullet 0 = 0$	$(A \bullet B) + (A \bullet C) = A(B + C)$
$A + 1 = 1$	$(A + B) \bullet (A + C) = A + BC$
$A + A = A$	$A + AB = A$
$A \bullet A = A$	$A + \overline{A}B = A + B$
$A + \overline{A} = 1$	$A(A + B) = A$
$A \bullet 1 = A$	$\overline{(A \bullet B)} = \overline{A} + \overline{B}$
$A\overline{A} = 0$	$\overline{(A + B)} = \overline{A} \bullet \overline{B}$

D.1   Some Logic Gates, Flip-Flops, and Drivers

Inputs A, A **Output** Y

Two-input AND gate

AND logic truth table

INPUTS		OUTPUTS
A	B	Y
0	0	0
1	0	0
0	1	0
1	1	1

Control Of Mechatronic Systems: Model-Driven Design And Implementation Guidelines,
First Edition. Patrick O.J. Kaltjob.
© 2021 Patrick O.J. Kaltjob. Published 2021 by John Wiley & Sons Ltd.

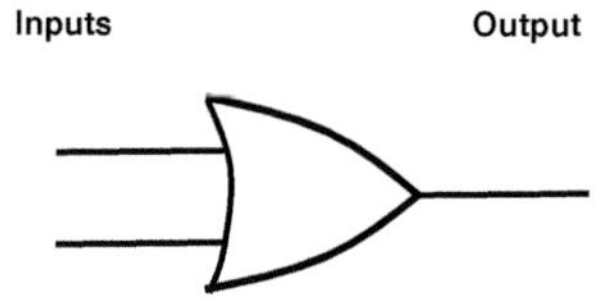

Inputs Output	OR logic truth table

Two-input OR gate

INPUTS		OUTPUTS
A	B	Y
0	0	0
1	0	1
0	1	1
1	1	1

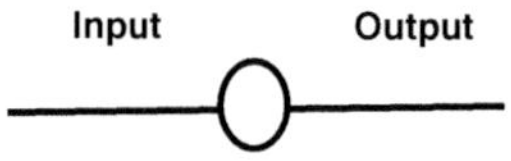

Input Output	NOT logic truth table

Two-input NOT gate

NOT truth table	
INPUT (A)	OUTPUT (NOT A)
0	1
1	0

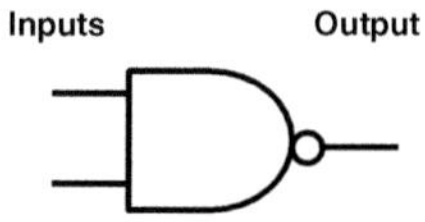

Inputs Output	NAND truth table

Two-input NAND gate

INPUTS		OUTPUTS
A	B	Y
0	0	1
1	0	1
0	1	1
1	1	0

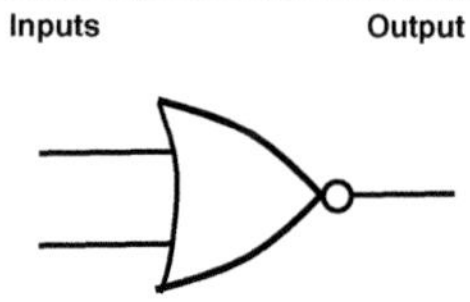

Inputs Output	NOR truth table

Two-input NOR gate

INPUTS		OUTPUTS
A	B	Y
0	0	1
1	0	0
0	1	0
1	1	0

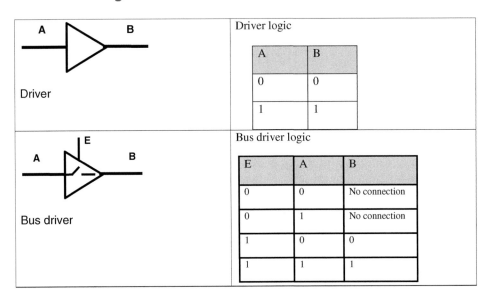

Inputs Output	EOR truth table
Two-input EOR gate	

EOR truth table

INPUTS		OUTPUTS
A	B	Y
0	0	0
1	0	1
0	1	1
1	1	1

D.2   Other Logic Devices: Drivers and Bus Drivers

Driver

Driver logic

A	B
0	0
1	1

Bus driver

Bus driver logic

E	A	B
0	0	No connection
0	1	No connection
1	0	0
1	1	1

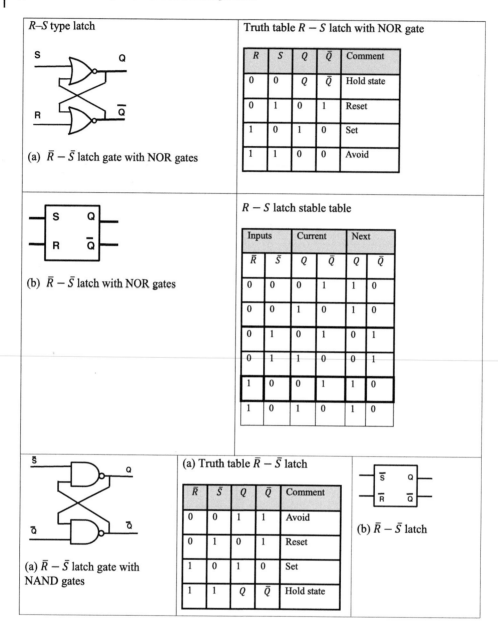

R–S type latch

(a) $\bar{R} - \bar{S}$ latch gate with NOR gates

Truth table $R - S$ latch with NOR gate

R	S	Q	$\bar{Q}$	Comment
0	0	Q	$\bar{Q}$	Hold state
0	1	0	1	Reset
1	0	1	0	Set
1	1	0	0	Avoid

(b) $\bar{R} - \bar{S}$ latch with NOR gates

$R - S$ latch stable table

Inputs		Current		Next	
$\bar{R}$	$\bar{S}$	Q	$\bar{Q}$	Q	$\bar{Q}$
0	0	0	1	1	0
0	0	1	0	1	0
0	1	0	1	0	1
0	1	1	0	0	1
1	0	0	1	1	0
1	0	1	0	1	0

(a) $\bar{R} - \bar{S}$ latch gate with NAND gates

(a) Truth table $\bar{R} - \bar{S}$ latch

$\bar{R}$	$\bar{S}$	Q	$\bar{Q}$	Comment
0	0	1	1	Avoid
0	1	0	1	Reset
1	0	1	0	Set
1	1	Q	$\bar{Q}$	Hold state

(b) $\bar{R} - \bar{S}$ latch

D.3   Gated $\bar{R} - \bar{S}$ Latch

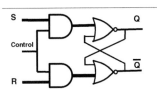

(a) $\bar{R} - \bar{S}$ flip flop gate with NAND gates

(b) Truth table $\bar{R} - \bar{S}$ flip flop

Control	S	R	Q(t+δ)
0	0	0	Q(t)
0	0	1	Q(t)
0	1	0	Q(t)
0	1	1	Q(t)
1	0	0	Q(t)
1	0	1	0
1	1	0	1
1	1	1	0

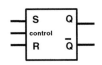

(c) $\bar{R} - \bar{S}$ flip flop gate with NAND gates

(d) State transition table of SR latch

	S(t)...R(t)			
$Q(t)$	00	01	11	10
0	0	0	d	1
1	1	0	d	1
		Q(t+1)		

D.4   D-Type (Delay-Flip-Flop)

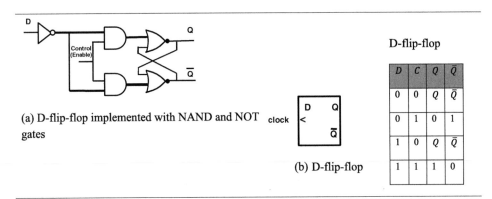

(a) D-flip-flop implemented with NAND and NOT gates

(b) D-flip-flop

D-flip-flop

D	C	Q	$\bar{Q}$
0	0	Q	$\bar{Q}$
0	1	0	1
1	0	Q	$\bar{Q}$
1	1	1	0

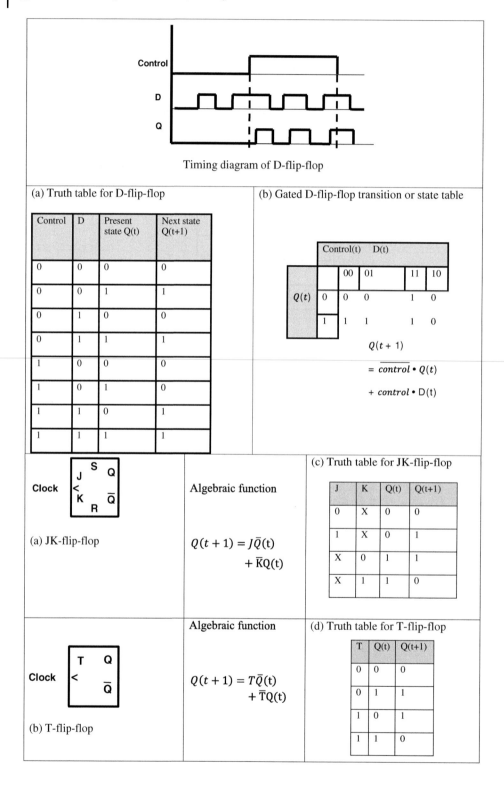

Timing diagram of D-flip-flop

(a) Truth table for D-flip-flop

Control	D	Present state Q(t)	Next state Q(t+1)
0	0	0	0
0	0	1	1
0	1	0	0
0	1	1	1
1	0	0	0
1	0	1	0
1	1	0	1
1	1	1	1

(b) Gated D-flip-flop transition or state table

		Control(t) D(t)			
		00	01	11	10
$Q(t)$	0	0	0	1	0
	1	1	1	1	0

$Q(t + 1)$

$$= \overline{control} \cdot Q(t)$$
$$+ control \cdot D(t)$$

(a) JK-flip-flop

Algebraic function

$$Q(t + 1) = J\bar{Q}(t) + \bar{K}Q(t)$$

(c) Truth table for JK-flip-flop

J	K	Q(t)	Q(t+1)
0	X	0	0
1	X	0	1
X	0	1	1
X	1	1	0

(b) T-flip-flop

Algebraic function

$$Q(t + 1) = T\bar{Q}(t) + \bar{T}Q(t)$$

(d) Truth table for T-flip-flop

T	Q(t)	Q(t+1)
0	0	0
0	1	1
1	0	1
1	1	0

D.5   Register or Buffer

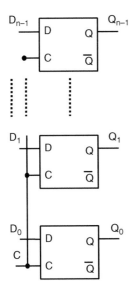

(c) *n*-bit register or buffer constructed with D-flip-flops

D.6   Adder

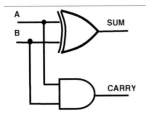

(a) Half-adder circuit

(a) Half-adder circuit logic

A	B	SUM	CARRY
0	0	0	0
1	0	1	0
0	1	1	0
1	1	0	1

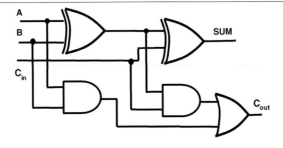

(b) Full-adder circuit

E

Solid-State Devices and Power Electronics

Electronics circuitry (e.g. power converters) is used to drive electromechanical motion actuating devices such as motors and solenoid valves. These devices are usually made of silicon (Si), germanium (Ge), and cadmium sulfide, and their ON/OFF switching defines the operation of any converter.

E.1 Power Diodes

With a *power diode*, when the supply voltage, V_{diode}, exceeds the forward biased voltage V_{fb} (around 0.7 V for silicon and 0.3 V for germanium), the diode is called "forward biased," similar to a closed switch. Conversely, when $V_{rb} < V_{diode} < V_{fb}$ (V_{rb} being the reverse breakdown voltage of the diode), the diode behaves as an open switch and is called "reverse biased." These diodes are suitable for signal rectification and peak.

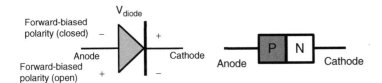

Figure E.1 Diode schematics and symbol.

A *Zener diode* is used to build AND and OR logic gates.

Figure E.2 *n*-input OR logic gate using a diode resistor.

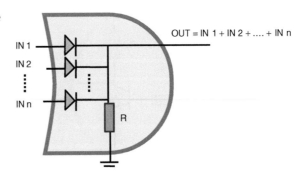

Control Of Mechatronic Systems: Model-Driven Design And Implementation Guidelines,
First Edition. Patrick O.J. Kaltjob.
© 2021 Patrick O.J. Kaltjob. Published 2021 by John Wiley & Sons Ltd.

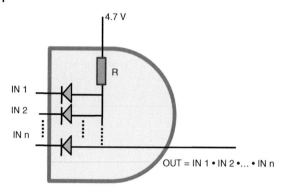

Figure E.3 *n*-input AND logic gate using a diode resistor.

E.2 Diode–Transistor Logic (DTL)

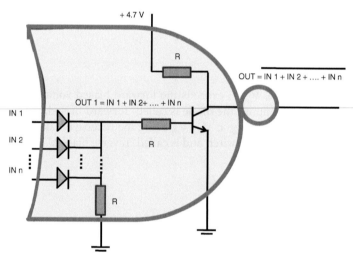

Figure E.4 *n*-input NOR logic gate using a diode transistor.

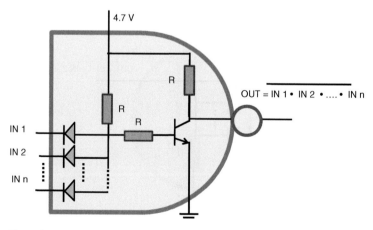

Figure E.5 *n*-input NAND logic gate using a diode transistor.

E.3 Power Transistors

Figure E.6 Power transistor schematics and symbol.

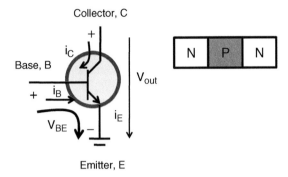

E.4 Resistor–Transistor Logic (RTL)

Figure E.7 NOT logic gate using a resistor transistor.

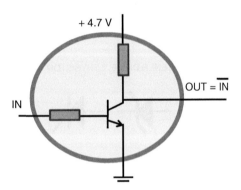

Transistor-based NOT logic gate

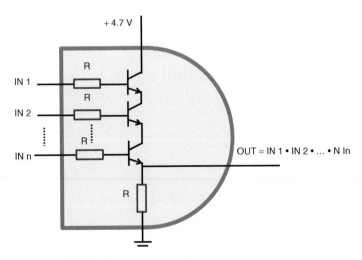

Figure E.8 AND logic gate using a resistor transistor.

E.5   Transistor–Transistor Logic (TTL)

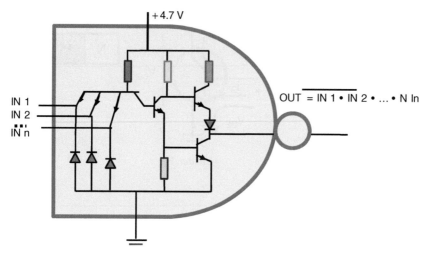

Figure E.9 Transistor–transistor logic implementation of a NAND gate.

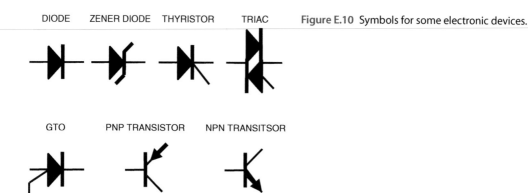

DIODE   ZENER DIODE   THYRISTOR   TRIAC   Figure E.10 Symbols for some electronic devices.

GTO   PNP TRANSISTOR   NPN TRANSITSOR

E.6   Metal Oxide Semiconductor FET (MOSFET)

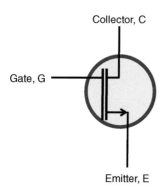

Figure E.11 MOSFET schematic.

E.7 Thyristors

The silicon-controlled rectifier (SCR), also called a reverse blocking triode thyristor, is a controllable time conduction diode.

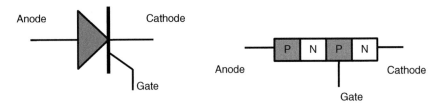

Figure E.12 Thyristor schematic and symbol.

The thyristor is ON when the voltage across the anode is higher than the voltage across the cathode and a current is allowed to flow into the gate for a few microseconds. It switches from a blocking phase to a conduction phase through a suitable gate pulse. Thyristors modulate the average output DC voltage by varying the firing timing of the voltage/current gate, which is given by:

$$V_D = 1.35 \times V_{RMS} \ (RMS - Phase\ Voltage) \tag{E.1}$$

Figure E.13 Thyristor switching due to voltage and current variation.

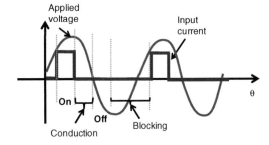

Index

Control Of Mechatronic Systems: Model-Driven Design And Implementation Guidelines,
First Edition. Patrick O.J. Kaltjob.
© 2021 Patrick O.J. Kaltjob. Published 2021 by John Wiley & Sons Ltd.

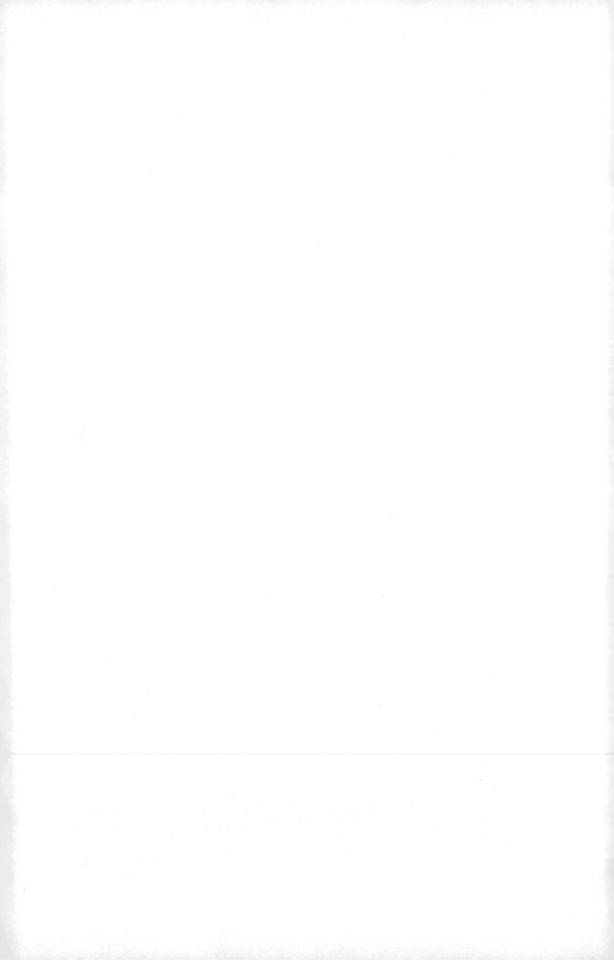

Printed and bound by CPI Group (UK) Ltd, Croydon, CR0 4YY

16/04/2025

14658553-0006